38 Springer Series in Chemical Physics
Edited by Fritz Peter Schäfer

Springer Series in Chemical Physics

Editors: V. I. Goldanskii   R. Gomer   F. P. Schäfer   J. P. Toennies

1 **Atomic Spectra and Radiative Transitions**
By I. I. Sobelman
2 **Surface Crystallography by LEED**
Theory, Computation and Structural
Results. By M. A. Van Hove, S. Y. Tong
3 **Advances in Laser Chemistry**
Editor: A. H. Zewail
4 **Picosecond Phenomena**
Editors: C. V. Shank, E. P. Ippen,
S. L. Shapiro
5 **Laser Spectroscopy**
Basic Concepts and Instrumentation
By W. Demtröder   2nd Printing
6 **Laser-Induced Processes in
Molecules**   Physics and Chemistry
Editors: K. L. Kompa, S. D. Smith
7 **Excitation of Atoms and Broadening
of Spectral Lines**   By I. I. Sobelman,
L. A. Vainshtein, E. A. Yukov
8 **Spin Exchange**
Principles and Applications in
Chemistry and Biology
By Yu. N. Molin, K. M. Salikhov,
K. I. Zamaraev
9 **Secondary Ions Mass Spectrometry
SIMS II**   Editors: A. Benninghoven,
C. A. Evans, Jr., R. A. Powell,
R. Shimizu, H. A. Storms
10 **Lasers and Chemical Change**
By A. Ben-Shaul, Y. Haas,
K. L. Kompa, R. D. Levine
11 **Liquid Crystals of One- and
Two-Dimensional Order**
Editors: W. Helfrich, G. Heppke
12 **Gasdynamic Laser**   By S. A. Losev
13 **Atomic Many-Body Theory**
By I. Lindgren, J. Morrison
14 **Picosecond Phenomena II**
Editors: R. M. Hochstrasser,
W. Kaiser, C. V. Shank
15 **Vibrational Spectroscopy of
Adsorbates**   Editor: R. F. Willis
16 **Spectroscopy of Molecular Excitons**
By V. L. Broude, E. I. Rashba,
E. F. Sheka
17 **Inelastic Particle-Surface Collisions**
Editors: E. Taglauer, W. Heiland
18 **Modelling of Chemical Reaction
Systems**   Editors: K. H. Ebert,
P. Deuflhard, W. Jäger

19 **Secondary Ion Mass Spectrometry
SIMS III**
Editors: A. Benninghoven, J. Giber,
J. László, M. Riedel, H. W. Werner
20 **Chemistry and Physics of Solid
Surfaces IV**   Editors: R. Vanselow, R. Howe
21 **Dynamics of Gas-Surface Interaction**
Editors: G. Benedek, U. Valbusa
22 **Nonlinear Laser Chemistry**
Multiple-Photon Excitation
By V. S. Letokhov
23 **Picosecond Phenomena III**
Editors: K. B. Eisenthal, R. M. Hochstrasser
W. Kaiser, A. Laubereau
24 **Desorption Induced by Electronic
Transitions DIET I**   Editors: N. H. Tolk,
M. M. Traum, J. C. Tully, T. E. Madey
25 **Ion Formation from Organic Solids**
Editor: A. Benninghoven
26 **Semiclassical Theories of Molecular
Scattering**   By B. C. Eu
27 **EXAFS and Near Edge Structures**
Editors: A. Bianconi, L. Incoccia, S. Stipcich
28 **Atoms in Strong Light Fields**
By N. B. Delone, V. P. Krainov
29 **Gas Flow in Nozzles**
By U. Pirumov, G. Roslyakov
30 **Theory of Slow Atomic Collisions**
By E. E. Nikitin, S. Ya. Umanskii
31 **Reference Data on Atoms, Molecules,
and Ions**   By A. A. Radzig, B. M. Smirnov
32 **Adsorption Processes on Semiconductor
and Dielectric Surfaces I**
By V. F. Kiselev, O. V. Krylov
33 **Surface Studies with Lasers**
Editors: F. R. Aussenegg, A. Leitner,
M. E. Lippitsch
34 **Inert Gases**
Potentials, Dynamics, and Energy Transfer
in Doped Crystals.   Editor: M. L. Klein
35 **Chemistry and Physics of Solid
Surfaces V**   Editors: R. Vanselow, R. Howe
36 **Secondary Ion Mass Spectrometry,
SIMS IV**   Editors: A. Benninghoven,
J. Okano, R. Shimizu, H. W. Werner
37 **X-Ray Spectra and Chemical Binding**
By A. Meisel, G. Leonhardt, R. Szargan
38 **Ultrafast Phenomena IV**
By D. H. Auston, K. B. Eisenthal
39 **Laser Processing and Diagnostics**
Editor: D. Bäuerle

Ultrafast Phenomena IV

Proceedings of the Fourth
International Conference
Monterey, California, June 11–15, 1984

Editors: D. H. Auston and K. B. Eisenthal

With 370 Figures

Springer-Verlag
Berlin Heidelberg New York Tokyo 1984

Dr. David H. Auston

AT & T Bell Laboratories, 600 Mountain Avenue
Murray Hill, NJ 07974, USA

Professor Dr. Kenneth B. Eisenthal

Department of Chemistry, Columbia University,
New York, NY 10027, USA

Series Editors

Professor Vitalii I. Goldanskii

Institute of Chemical Physics
Academy of Sciences
Kosygin Street 4
Moscow V-334, USSR

Professor Robert Gomer

The James Franck Institute
The University of Chicago
5640 Ellis Avenue
Chicago, IL 60637, USA

Professor Dr. Fritz Peter Schäfer

Max-Planck-Institut für
Biophysikalische Chemie
D-3400 Göttingen-Nikolausberg
Fed. Rep. of Germany

Professor Dr. J. Peter Toennies

Max-Planck-Institut für Strömungsforschung
Böttingerstraße 6–8
D-3400 Göttingen
Fed. Rep. of Germany

ISBN 3-540-13834-X Springer-Verlag Berlin Heidelberg New York Tokyo
ISBN 0-387-13834-X Springer-Verlag New York Heidelberg Berlin Tokyo

Offset printing: Beltz Offsetdruck, 6944 Hemsbach/Bergstr. Bookbinding: J. Schäffer OHG, 6718 Grünstadt
2153/3130-543210

Preface

The motivating idea of the first Topical Meeting on Picosecond Phenomena, which took place at Hilton Head Island in 1978, was to bring together scientists and engineers in a congenial setting who were developing picosecond lasers with those who were applying them to problems in chemistry, physics, electronics, and biology. The field has advanced remarkably in the following six years. This is reflected in the size of the conference which has more than doubled in the past six years and now includes scientists from many countries around the world. As evidenced by the papers in this volume, the application of ultrafast light pulses continues to grow in new and diverse directions encompassing an increasingly wide range of subject areas. This progress has gone hand-in-hand with the development of new and more precise methods of generating and measuring ultrafast light pulses, which now extend well into the femtosecond time domain. It was this latter advance which was responsible for changing the name of the conference to Ultrafast Phenomena.

The 1984 meeting was held at the Monterey Conference Center in Monterey, California from June 11 to 15 under the sponsorship of the Optical Society of America. A total of 320 registered participants, including 65 students, attended the three and one-half day conference. The overall enthusiasm of the participants, the high quality of the research presented, and ambiance of the setting combined to produce a successful and enjoyable conference.

Many people worked hard to make the conference a success. Special thanks are due to Joan Carlisle and her colleagues at the Optical Society of America for their vital assistance in smoothly implementing the meetings arrangements, and to the program committee for their advice and support. We gratefully acknowledge a grant from the United States Air Force Office of Scientific Research which provided the major source of funding for the conference. Additional grants from the following industrial sources are also appreciated: Coherent, E. I. du Pont de Nemours & Company, Hammamatsu Corporation, Klinger Scientific Corporation, IBM Corporation, Marco Scientific, Newport Corporation, Quantronix Corporation, Spectra-Physics, and Standard Oil Company (Ohio).

Murray Hill, NJ,
and New York, NY
July 1984

D.H. Auston
K.B. Eisenthal

Contents

Part I   Generation and Measurement Techniques

The Soliton Laser. By L.F. Mollenauer and R.H. Stolen (With 5 Figures)   2

Soliton Shaping Mechanisms in Passively Mode-Locked Lasers and Negative
   Group Velocity Dispersion Using Refraction
   By O.E. Martinez, J.P. Gordon, and R.L. Fork (With 4 Figures)   7

Compression and Shaping of Femtosecond Pulses
   By A.M. Weiner, J.G. Fujimoto, and E.P. Ippen (With 5 Figures)   11

Generation of 0.41-Picosecond Pulses by the Single-Stage Compression of
   Frequency Doubled Nd:YAG Laser Pulses
   By A.M. Johnson, R.H. Stolen, and W.M. Simpson (With 4 Figures) ...   16

Compression of Mode-Locked Nd:YAG Pulses to 1.8 Picoseconds
   By B.H. Kolner, J.D. Kafka, D.M. Bloom, and T.M. Baer
   (With 3 Figures)   19

Effects of Cavity Dispersion on Femtosecond Mode-Locked Dye Lasers
   By S. De Silvestri, P. Laporta, and O. Svelto (With 2 Figures)   23

3 MHz Amplifier for Femtosecond Optical Pulses
   By M.C. Downer and R.L. Fork (With 1 Figure)   27

Colliding Pulse Femtosecond Lasers and Applications to the Measurement
   of Optical Parameters. By J.-C. Diels, W. Dietel, E. Döpel,
   J. Fontaine, I.C. McMichael, W. Rudolph, F. Simoni, R. Torti,
   H. Vanherzeele, and B. Wilhelmi (With 5 Figures)   30

High Average Power Mode-Locked Co:MgF$_2$ Laser. By B.C. Johnson,
   M. Rosenbluh, P.F. Moulton, and A. Mooradian (With 3 Figures)   35

High Power Picosecond Pulses in the Infrared
   By P.B. Corkum (With 4 Figures)   38

Stimulated VUV Radiation from HD Excited by a Picosecond ArF* Laser
   By T.S. Luk, H. Egger, W. Müller, H. Pummer, and C.K. Rhodes
   (With 1 Figure)   42

Procedure for Calculating Optical Pulse Compression from Fiber-Grating
   Combinations
   By R.H. Stolen, C.V. Shank, and T.W. Tomlinson (With 1 Figure)   46

Generation of Infrared Picosecond Pulses Between 1.2μm and 1.8μm Using
a Traveling Wave Dye Laser. By H.-J. Polland, T. Elsaesser,
A. Seilmeier, and W. Kaiser (With 3 Figures) 49

A New Picosecond Source in the Vibrational Infrared
By A.L. Harris, M. Berg, J.K. Brown, and C.B. Harris(With 2 Figures)   52

Generation of Intense, Tunable Ultrahort Pulses in the Ultraviolet Using
a Single Escimer Pump Laser. By S. Szatmári and F.P. Schäfer
(With 1 Figure) 56

Travelling Wave Pumped Ultrashort Pulse Distributed Feedback Dye Laser
By B. Szabó, B. Rácz, Zs. Bor, B. Nikolaus, and A. Müller
(With 2 Figures) 60

Picosecond Pulses from Future Synchrotron-Radiation Sources
By R.C. Sah, D.T. Attwood, and A.P. Sabersky (With 2 Figures) 63

High Repetition Rate Production of Picosecond Pulses at Wavelength
<250 nm . By D.B. McDonald (With 1 Figure) 66

Ultrafast Self-Phase Modulation in a Colliding Pulse Mode-Locked Ring
Dye Laser. By Y. Ishida, K. Naganuma, T. Yajima, and L.H. Lin
(With 3 Figures) 69

Electro-Optic Phase-Sensitive Detection of Optical Emission and
Scattering. By A.Z. Genack (With 2 Figures) 72

Theoretical Studies of Active, Synchronous, and Hybrid Mode-Locking
By J.M. Catherall and G.H.C. New (With 3 Figures) 75

Technique for Highly Stable Active Mode-Locking
By D. Cotter (With 3 Figures) 78

Continuous Wave Mode-Locked Nd:Phosphate Glass Laser. By S.A. Strobel,
P.-T. Ho, C.H. Lee, and G.L. Burdge (With 3 Figures) 81

Active Mode-Locking Using Fast Electro-Optic Deflector. By A. Morimoto,
S. Fujimoto, T. Kobayashi, and T. Sueta (With 5 Figures) 84

Stable Active-Passive Mode Locking of an Nd:Phosphate Glass Laser Using
Eastman #5 Saturable Dye
By L.S. Goldberg and P.E. Schoen (With 3 Figures) 87

Limits to Pulse Advance and Delay in Actively Modelocked Lasers
By R.S. Putnam (With 2 Figures) 90

Novel Method of Waveform Evaluation of Ultrashort Optical Pulses
By T. Kobayashi, F.-C. Guo, A. Morimoto, T. Sueta, and Y. Cho
(With 6 Figures) 93

Noise in Picosecond Laser Systems: Actively Mode Locked CW Nd^{3+}:YAG and
Ar^+ Lasers Synchronously Pumping Dye Lasers. By T.M. Baer and
D.D. Smith (With 4 Figures) 96

High Power, Picosecond Phase Coherent Pulse Sequences by Injection
Locking. By F. Spano, F. Loaiza-Lemos, M. Haner, and W.S. Warren
(With 2 Figures) 99

Passive Mode-Locking with Reverse Saturable Absorption
By D.J. Harter and Y.B. Band (With 3 Figures) 102

Part II   Solid State Physics and Nonlinear Optics

Imaging with Femtosecond Optical Pulses
By M.C. Downer, R.L. Fork, and C.V. Shank (With 1 Figure) 106

Femtosecond Multiphoton Photoelectron Emission from Metals
By J.G. Fujimoto, J.M. Liu, E.P. Ippen, and N. Bloembergen
(With 3 Figures) 111

Time-Resolved Laser-Induced Phase Transformation in Aluminum
By S. Williamson, G. Mourou, and J.C.M. Li (With 3 Figures) 114

Picosecond Photoemission Studies of Laser-Induced Phase Transitions in
Silicon. By A.M. Malvezzi, H. Kurz, and N. Bloembergen
(With 3 Figures) 118

Dynamics of Dense Electron-Hole Plasma and Heating of Silicon Lattice
Under Picosecond Laser Irradiation. By L.A. Lompré, J.M. Liu,
H. Kurz, and N. Bloembergen (With 3 Figures) 122

Dynamics of the Mott Transition in CuCl with Subpicosecond Time
Resolution. By D. Hulin, A. Antonetti, L.L. Chase, J. Etchepare,
G. Grillon, A. Migus, and A. Mysyrowicz (With 3 Figures) 126

Picosecond Dynamics of Hot Dense Electron-Hole Plasmas in Crystalline
and Amorphized Si and GaAs
By P.M. Fauchet, and A.E. Siegman (With 3 Figures) 129

Picosecond Optical Excitation of Phonons in Amorphous As_2Te_3
By C. Thomson, J. Strait, Z. Vardeny, H.J. Maris, J. Tauc, and
J.J. Hauser (With 3 Figures) 133

Femtosecond Studies of Intraband Relaxation of Semiconductors and
Molecules. By A.J. Taylor, D.J. Erskine, and C.L. Tang
(With 5 Figures) 137

Picosecond Laser Studies of Noneqilibrium Electron Heating in Copper
By G.L. Eesley (With 3 Figures) 143

Picosecond Measurement of Hot Carrier Luminescence in $In_{0.53}Ga_{0.47}As$
By K. Kash and J. Shah (With 3 Figures) 147

Picosecond Carrier Dynamics in Semiconductors
By E.O. Göbel, J. Kuhl, and R. Höger (With 3 Figures) 150

Picosecond Dephasing and Energy Relaxation of Excitons in Semiconductors. By Y. Masumoto (With 6 Figures) 156

Femtosecond Dynamics of Nonequilibrium Correlated Electron-Hole Pair
Distributions in Room-Temperature GaAs Multiple Quantum Well
Structures. By W.H. Knox, R.L. Fork, M.C. Downer, D.A.B. Miller,
D.S. Chemla, C.V. Shank, A.C. Gossard, and W. Wiegmann
(With 1 Figure) 162

Femtosecond Transient Anisotropy in the Absorption Saturation of GaAs
By J.L. Oudar, A. Migus, D. Hulin, G. Grillon, J. Etchepare, and
A. Antonetti (With 2 Figures) 166

Holographic Interferometry Using Twenty-Picosecond UV Pulses to Obtain
Time Resolved Hydro Measurements of Selenium and Gold Plasmas
By G.E. Busch, R.R. Johnson, and C.L. Shepard (With 5 Figures) 170

Temporal Development of Absorption Spectra in Alkali Halide Crystals
Subsequent to Band-Gap Excitation. By W.L. Faust, R.T. Williams,
and B.B. Craig (With 1 Figure) 173

Kinetics of Free and Bound Excitons in Semiconductors
By X.-C. Zhang, Y. Hefetz, and A.V. Nurmikko (With 3 Figures) 176

Determination of Surface Recombination Velocities for CdS Crystals
Immersed in Electrolyte Solutions by a Picosecond Photoluminescence
Technique. By D. Huppert, S. Gottesfeld, Z. Harzion, M. Evenor,
and S. Feldberg (With 1 Figure) 181

Pulsewidth Dependence of Various Bulk Phase Transitions and
Morphological Changes of Crystalline Silicon Irradiated by 1 Micron
Picosecond Pulses. By S.C. Moss, I.W. Boyd, T.F. Boggess, and
A.L. Smirl (With 2 Figures) 184

Subthreshold Picosecond Laser Damage in Silicon Associated with Charge
Emission. By Y.K. Jhee, M.F. Becker, and R.M. Walser
(With 1 Figure) 187

Third-Order Nonlinear Susceptibilities of Dye Solutions Determined by
Non-Phasematched Third Harmonic Generation. By A. Penzkofer and
W. Leupacher (With 1 Figure) 190

Excitation Transport and Trapping in a Two-Dimensional Disordered
System: Cresyl Violet on Quartz. By P. Anfinrud, R.L. Crackel,
and W.S. Struve (With 1 Figure) 193

Temporal Dependence of Third-Order Non-Linear Optical Susceptibilities
of Fused Quartz and Liquid CCl_4. By J. Etchepare, G. Grillon,
I.Thomazeau, J.P. Chambaret, and A. Orszag (With 2 Figures) 196

High Excitation Electron Dynamics in GaInAsP. By A. Miller,
R.J. Manning, A.M. Fox, and J.H. Marsh (With 2 Figures) 199

Picosecond Nonlinear-Optical Limiting in Silicon
By T.F. Boggess, S.C. Moss, I.W. Boyd, and A.L. Smirl
(With 3 Figures) 202

Nonlinear Absorption and Nonlinear Refraction Studies in MEBBA
By M.J. Soileau, W.E. Williams, E.W. Van Stryland, S. Guha,
H. Vanherzeele, J.L.W. Pohlmann, E.J. Sharp, and G.L. Wood
(With 3 Figures) 205

High Density Carrier Generation in Indium Antimonide
By M. Sheikbahae, P. Mukherjee, M. Hasselbeck, and H.S. Kwok 208

Measurement of Two-Photon Cross-Section in DABCO with the Use of Picosecond Pulses. By G. Arjavalingam, J.H. Glownia, and P.P. Sorokin . 211

Part III   Coherent Pulse Propagation

Coherence Effects in Pump-Probe Measurements with Collinear,
   Copropagating Beams. By S.L. Palfrey, T.F. Heinz, and
   K.B. Eisenthal (With 2 Figures) 216

Observation of the 0π Pulse. By J.E. Rothenberg, D. Grischkowsky, and
   A.C. Balant (With 2 Figures) 220

Picosecond Two-Color Photon Echoes in Doped Molecular Solids
   By D.A. Wiersma, D.P. Weitekamp, and K. Duppen (With 1 Figure) 224

Subpicosecond Accumulated Photon Echoes with Incoherent Light in
   Nd^{3+}-Doped Silicate Glass. By H. Nakatsuka, S. Asaka, M. Fujiwara,
   and M. Matsuoka (With 3 Figures) 226

Femtosecond Dephasing Measurements Using Transient Induced Gratings
   By A.M. Weiner, S. De Silvestri, and E.P. Ippen (With 4 Figures) .. 230

Picosecond Pulse Multiphoton Coherent Propagation in Vapors
   By H. Vanherzeele, and J.-C. Diels (With 2 Figures) 233

Stimulated Photon Echo for Elastic and Depolarizing Collison Studies
   By J.-C. Keller and J.-L. Le Gouet (With 3 Figures) 236

Coherent Transient Spectroscopy with Ultra-High Time Resolution Using
   Incoherent Light. By N. Morita, T. Yajima, and Y. Ishida
   (With 3 Figures) 239

Part IV   Stimulated Scattering

Interaction-Induced, Subpicosecond Phenomena in Liquids
   By P.A. Madden (With 4 Figures) 244

Transient Infrared Spectroscopy on the Picosecond and Sub-Picosecond
   Time Scale. By H.-J. Hartmann and A. Laubereau (With 5 Figures) ... 252

Quantum Fluctuations in Picosecond Transient Stimulated Raman Scattering
   By N. Fabricius, K. Nattermann, and D. von der Linde (With 2 Figures) 258

Transient Coherent Raman Spectroscopy: Two Novel Ways of Line Narrowing
   By W. Zinth, M.C. Nuss, and W. Kaiser (With 2 Figures) 263

Picosecond Transient Raman Spectroscopy: The Excited State Structure of
   Diphenylpolyenes. By T.L. Gustafson, D.A. Chernoff, J.F. Palmer, and
   D.M. Roberts (With 3 Figures) 266

Time-Resolved Nonlinear Spectroscopy of Vibrational Overtones and Two-
   Phonon States. By G.M. Gale, M.L. Geirnaert, P. Guyot-Sionnest, and
   C. Flytzanis (With 4 Figures) 270

Direct Picosecond Determination of the Character of Virbrational Line
   Broadening in Liquids. By G.M. Gale, P. Guyot-Sionnest,
   and W.Q. Zheng (With 2 Figures) 274

Picosecond Time-Domain Coherent Active Raman Spectroscopy of Free
  Nitrogen Jet. By S.A. Akhmanov, N.I. Koroteev, S.A. Magnitskii,
  V.B. Morozov, A.P. Tarassevich, V.G. Tunkin, and I.L. Shumay
  (With 2 Figures) 278

Part V   Transient Laser Photochemistry

Picosecond Chemistry of Collisionless Molecules in Supersonic Beams
  By A.H. Zewail (With 4 Figures) 284

Energy Transfer in Picosecond Laser Generated Compressional Shock
  Waves. By A.J. Campillo, L.S. Goldberg, and P.E. Schoen
  (With 2 Figures) 289

The Role of A and A' States in the Geminate Recombination of Molecular
  Iodine. By D.F. Kelley and N.A. A.-Haj (With 4 Figures) 292

Molecular Dynamics of I_2 Photodissociation in Cyclohexane: Experimental
  Picosecond Transient Electronic Absorption. By P. Bado, C.G. Dupuy,
  J.P. Bergsma, and K.R. Wilson (With 3 Figures) 296

Iodine Photodissociation in Solution: New Transient Absorptions
  By M. Berg, A.L. Harris, J.K. Brown, and C.B. Harris
  (With 3 Figures) 300

Photodissociation of Triarylmethanes
  By L. Manring and K. Peters (With 4 Figures) 304

Femtosecond Time Resolved Multiphoton Ionization: Techniques and
  Applications. By B.I. Greene (With 8 Figures) 308

Threshold Ionization in Liquids
  By G.W. Robinson, J. Lee, and R.A. Moore (With 1 Figure) 313

Picosecond Multiphoton Laser Photolysis and Spectroscopy of Liquid
  Benzenes. By H. Miyasaka, H. Masuhara, and N. Mataga
  (With 3 Figures) 317

Chemical Reactions in Condensed Media
  By D. Statman, W.A. Jalenak, and G.W. Robinson (With 3 Figures) ... 320

Subpicosecond and Picosecond Time Resolved Laser Photoionization of
  Phenothiazine in Micellar Models. By Y. Gauduel, A. Migus,
  J.L. Martin, J.M. Lemaître, and A. Antonetti (With 2 Figures) 323

Isomerization Intermediates in the Photochemistry of Stilbenes
  By F.E. Doany and R.M. Hochstrasser (With 2 Figures) 326

Part VI   Molecular Energy Redistribution, Transfer, and Relaxation

Picosecond Laser Studies on the Effect of Structure and Environment on
  Intersystem Crossing in Aromatic Carbenes. By E.V. Sitzmann,
  J.G. Langan, Z.Z. Ho, and K.B. Eisenthal (With 4 Figures) 330

Energy Flow from Highly Excited CH Overtones in Benzene and Alkanes
By E.L. Sibert III, J.S. Hutchinson, J.T. Hynes, and W.P. Reinhard
(With 7 Figures) 336

Pump-Pump Picosecond Laser Techniques and the Energy Redistribution
Dynamics in Mass Spectrometry. By M.A. El-Sayed, D. Gobeli, and
J. Simon (With 4 Figures) 341

Intramolecular Electronic and Vibrational Redistribution and Chemical
Transformation in Isolated Large Molecules - S_1 Benzene
By K. Yoshihara, M. Sumitani, D.V. O'Connor, Y. Takagi, and
N. Nakashima (With 5 Figures) ....

Ultrafast Intramolecular Redistribution and Energy Dissipation in
Solutions. The Application of a Molecular Thermometer
By P.O.J. Scherer, A. Seilmeier, F. Wondrazek, and W. Kaiser
(With 4 Figures) 351

Picosecond Time-Resolved Fluorescence Spectra of Liquid Crystal:
Cyanooctyloxybiphenyl. By N.Tamai, I. Yamazaki, H. Masuhara,
and N. Mataga (With 3 Figures) 355

Excited-State Solvation Dynamics in 4-Aminophthalimide
By S.W. Yeh, L.A. Philips, S.P. Webb, L.F. Buhse, and J.H. Clark
(With 3 Figures) 359

The Pyrazine Mystery: A Resolution
By A. Lorincz, F.A. Novak, D.D. Smith, and S.A. Rice
(With 1 Figure) 362

Picosecond Laser Studies of Photoinduced Electron Transfer in
Porphyrin-Quinone and Related Model Systems. By N. Mataga,
A. Karen, T. Okada, Y. Sakata, and S. Misumi (With 2 Figures) 365

The Excited-State Proton Transfer Reactions of Flavonols in Alcoholic
Solvents. By K.-J. Choi, B.P. Boczer, and M.R. Topp 368

Excited-State Proton-Transfer Reactions in 1-Naphthol
By S.P. Webb, S.W. Yeh, L.A. Philips, M.A. Tolbert, and J.H. Clark
(With 3 Figures) 371

Structural and Solvent Effects on the Excited State Dynamics of
3-Hydroxyflavones. By P.F. Barbara and A.J.G. Strandjord
(With 1 Figure) 374

Electron Transfer Reactions from the First Excited Singlet State of a
Polymethine Cyanine Dye. By D. Doizi and J.C. Mialocq
(With 3 Figures) 377

Photoinduced Electron Transfer in Polymethylene Linked Donor-Acceptor
Compounds: $A-(CH_2)_n-D$. By H. Staerk, W. Kühnle, R. Mitzkus,
R. Treichel, and A. Weller (With 2 Figures) 380

Four Wave Mixing Studies and Molecular Dynamics Simulations
By M. Golombok and G.A. Kenney-Wallace (With 1 Figure) 383

Relaxation of Large Molecules Following Ultrafast Excitation
By A. Lorincz, F.A. Novak, and S.A. Rice 387

Picosecond Time-Resolved Spectroscopy of Electronically Excited
tris(2,2'-Bipyridine) Ruthenium(II) Dichloride. By L.A. Philips,
W.T. Brown, S.P. Webb, S.W. Yeh, and J.H. Clark (With 3 Figures) .. 390

Part VII Electronics and Opto-Electronics

Modelocking at Ti:LiNbO$_3$-InGaAsP/InP Composite Cavity Laser Using a
High-Speed Directional Coupler Switch. By R.C. Alferness,
G. Eisenstein, S.K. Korotky, R.S. Tucker, L.L. Buhl, I.P. Kaminow,
and J.J. Veselka (With 5 Figures) 394

Picosecond Optical Measurements of Circuit Effects on Carrier Sweepout
in GaAs Schottky Diodes. By A. Von Lehmen and J.M. Ballantyne
(With 4 Figures) 398

Color Center Formation and Recombination in KBr and LiF by Picosecond
Pulsed Electrons. By K. Fujii, R. Kikuchi, S. Katagiri, K. Tsumori,
and M. Kawanishi (With 4 Figures) 402

Subpicosecond Electro-Optic Sampling Using Coplanar Strip Transmission
Lines. By K.E. Meyer and G.A. Mourou (With 2 Figures) 406

Čerenkov Radiation from Femtosecond Optical Pulses in Electro-Optic
Media. By K.P. Cheung, D.H. Auston, J.A. Valdmanis, and
D.A. Kleinman (With 5 Figures) 409

Ultraviolet Photoemission Studies of Surfaces Using Picosecond Pulses
of Coherent XUV Radiation. By R. Haight, J. Bokor, R.H. Storz,
J. Stark, R.R. Freeman, and P.H. Bucksbaum (With 3 Figures) 413

Synchronous Mode-Locking of a GaAs/GaAlAs Laser Diode by a Picosecond
Optoelectronic Switch. By J. Kuhl and E.O. Göbel (With 3 Figures) . 417

Photochron Streak Camera with GaAs Photocathode
By C.C. Phillips, A.E. Hughes, and W. Sibbett (With 4 Figures) 420

Sequential Waveform Generation by Picosecond Optoelectronic Switching
By C.S. Chang, M.C. Jeng, M.J. Rhee, C.H. Lee, A. Rosen, and
H. Davis (With 2 Figures) 423

Picosecond Gain Measurements of a GaAlAs Diode Laser
By W. Lenth (With 3 Figures) 425

Transient Response Measurements with Ion-Beam-Damaged Si-on-Sapphire,
GaAs, and InP Photoconductors. By R.B. Hammond, N.G. Paulter, and
R.S. Wagner (With 2 Figures) 428

Picosecond Optoelectronic Studies of Microstrip Dispersion
By D.E. Cooper (With 2 Figures) 430

Measurement of the Soft X-Ray Temporal and Spectral Response of InP:Fe
Photoconductors. By D.R. Kania, R.J. Bartlett, P. Walsh, R.S. Wagner,
R.B. Hammond, and P. Pianetta (With 1 Figure) 433

Dynamic Response of Millimeter Waves in a Semiconductor Waveguide to
Picosecond Illumination. By A.M. Yurek, M.-G. Li, C.D. Striffler,
and C.H. Lee (With 2 Figures) 436

Part VIII Photochemistry and Photophysics of Proteins, Chlorophyll,
Visual Pigments, and Other Biological Systems

Time Resolution of Tryptophans in Myoglobin
By D.K. Negus and R.M. Hochstrasser (With 3 Figures) 440

Resolution of the Femtosecond Lifetime Species Involved in the Photo-
dissociation Process of Hemeproteins and Protoheme. By J.L. Martin,
A. Migus, C. Poyart, Y. Lecarpentier, A. Astier, and A. Antonetti
(With 3 Figures) 447

Picosecond Vibrational Dynamics of Peptides and Proteins
By T.J. Kosic, E.L. Chronister, R.E. Cline, Jr., J.H. Hill, and
D.D. Dlott (With 5 Figures) 452

New Investigations of the Primary Processes of Bacteriorhodopsin and of
Halorhodopsin. By H.-J. Polland, W. Zinth, and W. Kaiser
(With 2 Figures) 456

Primary Events in Vision Probed by Ultrafast Laser Spectroscopy
By A.G. Doukas, and R.R. Alfano (With 2 Figures) 459

Picosecond Time-Resolved Polarized Emission Spectroscopy of
Biliproteins (Influence of Temperature and Aggregation)
By S. Schneider, P. Hefferle, P. Geiselhart, T. Mindl, F. Dörr,
W. John, and H. Scheer (With 1 Figure) 462

Dynamics of Energy Transfer in Chloroplasts and the Internal Dynamics
of an Enzyme. By R.J. Gulotty, L. Mets, R.S. Alberte, A.J. Cross,
and G.R. Fleming (With 4 Figures) 466

Picosecond Single Photon Fluorescence Spectroscopy of Nucleic Acids
By R. Rigler, F. Claesens, and G. Lomakka (With 4 Figures) 472

Excited-State Dynamics of NADH and 1-N-Propyl-1,4-Dihydronicotinamide
By D.W. Boldridge, T.H. Morton, G.W. Scott, J.H. Clark, L.A. Philips
S.P. Webb, S.M. Yeh, and P. van Eikeren (With 2 Figures) 477

Primary Process in the Photocycles of the Low pH Bacteriorhodopsin
By T. Kobayashi, H. Ohtani, J. Iwai, and A. Ikegami
(With 2 Figures) 481

Picosecond Spectroscopy on the Primary Process in the Photoconversion
of Protochlorophyllide to Chlorophyllide a. By T. Kobayashi,
J. Iwai, M. Ikeuchi, and Y. Inoue (With 2 Figures) 484

Energy Transfer in Photosynthesis: The Heterogeneous Bipartite Model
By S.J. Berens, J. Scheele, W.L. Butler, and D. Magde 487

Excitation Energy Transfer in Phycobilin-Chlorophyll. A System of Algal
Intact Cells. By I. Yamazaki, N. Tamai, T. Yamazaki, M. Mimuro,
and Y. Fujita (With 2 Figures) 490

Analysis of Fluorescence Kinetics and Energy Transfer in Isolated
α Subunits of Phycoerythrin from *Nostoc* Sp. By A.J. Dagen,
R.R. Alfano, B.A. Zilinskas, and C.E. Swenberg
(With 3 Figures) 493

XV

Picosecond Time-Resolved Fluorescence Spectra of Hematoporphyrin Derivative and Its Related Porphyrins. By M. Yamashita, M. Nomura, S. Kobayashi, T. Sato, and K. Aizawa (With 1 Figure) 497

Femtosecond Spectrosopy of Bacteriorhodopsin Excited State Dynamics By M.C. Downer, M. Islam, C.V. Shank, A. Harootunian, A. Lewis (With 4 Figures) 500

Time-Resolved Picosecond Fluorescence Spectra of the Antenna Chlorophylls in the Green Alga Chlorella Vulgaris By J. Wendler, W. Haehnel, and A.R. Holzwarth (With 1 Figure) 503

Index of Contributors 507

Part I

Generation and Measurement Techniques

The Soliton Laser

L.F. Mollenauer and R.H. Stolen

AT & T Bell Laboratories, Crawford Corners Road, Holmdel, NJ 07733, USA

The soliton laser, a novel concept in ultrashort-pulse lasers, is a mode-locked laser using pulse compression and solitons in a single-mode fiber to force the laser itself to produce pulses of a well-defined shape and width. Thus the fiber is in one way or another involved in the laser's feedback loop. Although the basic concept is a general one, we report here primarily on the first successful version[1], based on a sync-pumped, mode-locked color-center laser operating in the 1.5 μm region. To date this color-center soliton laser has directly produced pulses as short as 130 fsec, and has allowed for the production of pulses of as little as 50 fsec FWHM, by compression in a second, external fiber. Other advantages include wide tunability (limited only by power requirements for soliton production in the fiber), output pulses that are always transform limited, easy adjustment for production of ~sech^2 pulse shape, and a relative simplicity of construction.

Pulse compression and solitons[2] in single mode fibers result from the interaction of nonlinearity with "negative" group velocity dispersion (derivative with respect to wavelength <0). Simple pulse compression can be understood as follows: Self phase modulation (due to the nonlinearity) leads to a "chirping" of the pulse, such that frequencies in the leading half of the pulse are lowered, while those in the trailing half are raised. The negative dispersion will then tend to retard the leading (low frequency) half of the pulse, and advance the trailing (high frequency) half, thereby causing a sufficiently intense pulse to collapse upon itself.

There exists a critical, lower intensity for which the pulse-narrowing effect described above is just great enough to balance the usual pulse-broadening effect of dispersion. At that intensity, one obtains the fundamental soliton- a pulse that never (in the limit of zero loss) changes its sech^2 shape. For an input pulse of the same shape and width, but of twice the peak amplitude (4 times the peak intensity), one obtains the first periodic behavior, an $N = 2$ soliton. (See Fig. 1). However, it is important to note[3] that there is a continuum of $N = 2$ solitons, i.e., there exist other input pulse shapes that result in periodic behavior qualitatively similar to that shown in Fig. 1. The soliton laser is based on such $N = 2$ solitons, and on the fact that the soliton period (z_0) scales with (the square of) the input pulse width (t).

The laser itself is shown in Fig. 2. The mode-locked color-center laser is coupled through beam splitter S and micro-

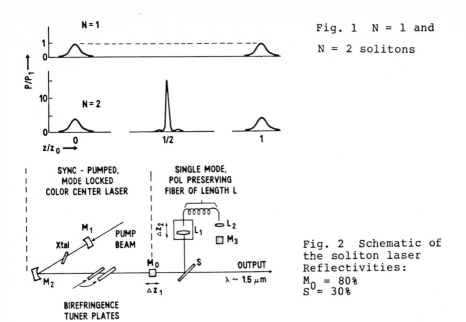

Fig. 1 N = 1 and N = 2 solitons

Fig. 2 Schematic of the soliton laser Reflectivities:
$M_0 = 80\%$
$S = 30\%$

scope objective L_1 to a length L of single-mode, polarization-preserving fiber; L_2 and M_3 form an efficient and stable cat's eye retroreflector at the other end of the fiber. It is important to note that the space between each end of the fiber and the corresponding lens surface is filled with a special index matching oil. (The oil is a completely halogenated paraffin that has no detectable absorption in the 1.5 μm region.) Without this precaution, reflections from the fiber ends produce feedback that interferes with the soliton laser action. The input end of the fiber and L_1 are mounted on a common translation stage to facilitate final adjustment (Δz_2) of the optical path length in the fiber arm to be an integral multiple of the main cavity length. (Pulses returned from the fiber must be made coincident with those already present in the main cavity.)

The $Tl^0(1)$ color-center laser and its mode-locking behavior have been described elsewhere[4]. For present purposes, its significant features are as follows: When pumped with ~5 W at 1.064 μm, the laser is tunable from ~1.4 to ~1.6 μm and produces a stable, non-fading output, up to ~1-W time-average power at band center, and finally, by itself, the sync-pumped laser produces pulses of >8 psec FWHM.

The low-loss, high-birefringence (hence polarization-preserving) fibers used in our laser were made in-house by the technique of preform deformation. The drawn fiber (core diameter, ~8.6μm) has the external form of a flat ribbon, making its polarization axes easy to locate. The fiber's polarization-preserving ability is vital to successful operation, for otherwise feedback into the the polarization-sensitive laser would tend to fluctuate wildly with fiber length, wavelength, and other factors.

With addition of the fiber arm, the device operates as follows: As the laser action builds up from noise, the initially broad pulses are considerably narrowed by passage through the fiber. The narrowed pulses, reinjected back into the main cavity, force the laser itself to produce narrower pulses. This process builds upon itself until the pulses in the fiber become $N = 2$ solitons whose period (z_0) matches the double fiber length (2L); at this point the pulses have substantially the same shape following their double passage through the fiber as they had upon entry.

Operation of the laser on $N = 2$ solitons is clearly indicated by the empirically determined dependence of the produced pulse width (t) on the square root of fiber length (see Fig. 3) as required for those solitons and the condition $z_0 = 2L$. Also, the values of peak power ($\hat{P}$) implied by measurement of time-average powers in the fiber correspond, to within experimental error, to values required for $N = 2$ solitons. Note that although for the shortest pulses, the peak powers in the fiber are rather large (nearly 10 kW), the corresponding time average powers remain modest (<100 mW), due to the long time (10 nsec) between pulses.

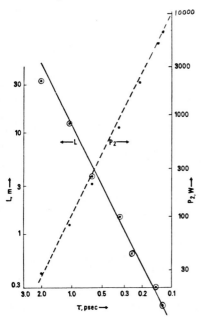

Fig. 3 Dots: fiber length and peak pwr versus pulse width; solid line: $z_0/2$, dashed line: $\hat{P}_2$, both calculated for sech $N = 2$ solitons

Although the fiber length is the primary determinant of pulse width, we have also found it necessary (as one would expect) to reduce the "gain dispersion" (tuning element selectivity) as the pulses become shorter. Of the two birefringence plates used here (sapphire, 4 and 1 mm thick), the thicker was removed for pulse widths <0.5 psec, and for the 150 and 130 fsec pulse widths, the 1 mm plate had to be rotated to a "shallow" angle such that its selectivity was considerably less than maximum.

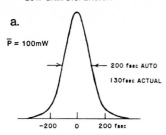

PULSE SHAPES WITH 20cm CONTROL FIBER

LOW GAIN DISPERSION

a.

$\overline{P}$ = 100mW

200 fsec AUTO

130fsec ACTUAL

−200   0   200 fsec

b.

$\overline{P}$ = 135mW

−400 −200   0   200   400 fsec

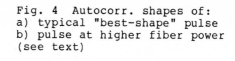

Fig. 4 Autocorr. shapes of:
a) typical "best-shape" pulse
b) pulse at higher fiber power
(see text)

It should be noted that the data points of Fig. 3 represent adjustment (through focusing or defocusing of L_1) of power in the fiber to obtain the narrowest and "best-shape" pulses. However, corresponding to each fiber length represented in Fig. 3, there is in fact a finite span of powers for which stable soliton-laser action is obtained, and a corresponding range of pulse shapes and widths. For example, Fig. 4a shows a typical "best-shape" pulse, while the triple-peaked autocorrelation trace of Fig. 4b corresponds to a double-peaked pulse that is produced at higher power levels in the fiber. This variation of pulse shape corresponds to the continuum of possible N = 2 solitons, as has recently been explained by the Haus-Islam theory[3] of the soliton laser.

It should also be noted that pulses returned from the fiber are required not only to be coincident with, but also to be in phase with, those circulating in the main cavity. To assist in maintainence of the precise relative cavity lengths (fiber arm and main cavity), recently we have mounted M_3 on a piezoelectric transducer driven by a HV op-amp. The error signal to drive the op-amp is derived from second-harmonic light (generated in a non-linear crystal) of the laser output pulses. A slight dithering motion of M_3 at ~400 kHz and phase-sensitive detection result in an error signal proportional to the slope of the curve of second-harmonic intensity versus mirror displacement. Thus, the resultant servo system constantly adjusts the position of M_3 such that the second-harmonic light is always at a maximum. This system has proven quite successful in maintaining stable soliton laser action, even in the face of considerable mirror vibration and drift.

The effects of compression in an external fiber on 150 fsec FWHM soliton laser output pulses are shown in Fig. 5. The resultant ~50 fsec FWHM pulses have very little pedestal, and it should be possible to remove even that by using the fiber itself

5

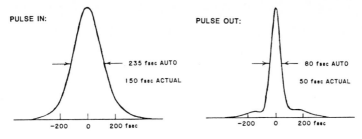

PULSE IN:

235 fsec AUTO

150 fsec ACTUAL

-200   0   200 fsec

PULSE OUT:

80 fsec AUTO

50 fsec ACTUAL

-200   0   200 fsec

Fig. 5   Pulse compression in 35 cm fiber

as a pulse-height discriminator[5]. It should be noted that to obtain this result, we have found it mandatory to use index-matching oil at both ends of the compressor fiber, just as described above for the control fiber. Without that precaution, reflection from either or both of the fiber ends produces feedback that is destructive of the soliton laser action. However, the 35 cm compressor- fiber length is longer than optimal. With optimal fiber length[6] and with perhaps even shorter pulses from the soliton laser, it should be possible to achieve pulse widths in the 20 to 30 fsec range in the very near future. The bandwidths of such pulses (approaching 500 cm^{-1}) will match (or nearly match) the greatest known homogeneous line widths of various semiconductor, color center, or dye transitions. Thus they should allow for the measurement of relaxation phenomena in those systems to limits of resolution set only by the uncertainty principle itself.

Finally, let it be noted that it should be possible to make a very simple but effective version of the soliton laser from an semiconductor diode laser, sync-pumped by electrical pulses and tightly coupled to a length of fiber. Such a device could be a very convenient and inexpensive source of transform-limited pulses of various widths down to perhaps 1 or 2 psec.

1. L. F. Mollenauer and R. H. Stolen, "The Soliton Laser," Opt. Lett. 9, 13 (1984)
2. L. F. Mollenauer, R. H. Stolen, and J. P. Gordon, "Experimental observation of picosecond pulse narrowing and solitons in optical fibers," Phys. Rev. Lett. 45, 1095 (1980)
3. H. A. Haus and M. N. Islam, "Theory of the soliton laser," unpub. Also see M. N. Islam and H. A. Haus, postdeadline paper PD2, 1984 Conf. on Ultrafast Phenomena, Monterey, Ca.
4. L. F. Mollenauer, N. D. Vieira, and L. Szeto, "Mode locking by synchronous pumping using a gain medium with microsecond decay times," Opt. Lett. 7, 414 (1982)
5. R. H. Stolen, J. Botineau, and A. Ashkin, "Intensity discrimination of optical pulses with birefringent fibers," Opt. Lett. 7, 512 (1981)
6. L. F. Mollenauer, R. H. Stolen, J. P. Gordon, and W. J. Tomlinson, "Extreme picosecond pulse narrowing by means of soliton effect in single-mode optical fibers," Opt. Lett. 8, 289 (1983)

Soliton Shaping Mechanisms in Passively Mode-Locked Lasers and Negative Group Velocity Dispersion Using Refraction

O.E. Martinez

CEILAP, CITEFA-CONICET, Zufriategui y Varela, 1603 V. Martelli, Argentina

J.P. Gordon and R.L. Fork

AT & T Bell Laboratories, Crawford Corners Road, Holmdel, NJ 07733, USA

We have developed analyses of soliton shaping mechanisms in a colliding pulse passively mode-locked laser [1] and of the adjustable group velocity dispersion obtainable from pairs of prisms [2,3,4]. We find soliton mechanisms aid in generating pulses which are shorter by the order of a factor of two than otherwise obtainable and also provide improved stability of the laser against changes in net gain. The negative group velocity dispersion, which we show necessarily accompanies any angular dispersion, provides an experimental means of introducing adjustable low loss group velocity dispersion in the laser resonators. This is useful in examining these soliton shaping mechanisms.

Our calculations of self phase modulation and group velocity dispersion in a colliding pulse laser [1] suggest that soliton mechanisms help shape the short stable pulses observed from these lasers. Recent experimental evidence of downchirping in such lasers [5,6], and novel methods for introducing adjustable group velocity dispersion in laser resonators [2,3,4] provide further evidence that the same mechanisms which shape N=1 solitons in optical fibers [7] can occur and cause pulse shaping in passively mode-locked lasers. We describe closed form analytical solutions for passive mode-locking in the presence of group velocity dispersion and self phase modulation. We include in particular self phase modulation due both to time-dependent saturation of the absorber and to the fast optical Kerr effect in the dye solvent.

Our analysis follows closely that of Haus [8]. The principal departure is that the frequency transfer functions of the components of the cavity are not assumed centered with respect to the emission frequency. New terms result in the expansion to second order in the detuning.

The results obtained for a laser using Rhodamine 6G for gain and DODCI as a saturable absorber [1] are shown in Figs. 1 and 2. To one side of zero group velocity dispersion the pulse broadens rapidly (see Fig. 1) because the broadening effects of chirp and group velocity dispersion ($-a_2$) are additive. The chirp in that region also grows (see Fig. 2) correspondingly. For the opposite sign of a_2 the pulsewidth is less sensitive to the group velocity dispersion. A less pronounced broadening is observed after passing the optimum a_2. The chirp is also small in this region. This occurs because the self phase modulation and group velocity dispersion compress the pulse in a manner similar to that encountered in soliton formation in optical fibers. Note that this shaping mechanism not only provides a narrower pulse but also introduces significantly reduced sensitivity of the pulsewidth to fluctuations of the gain. The minimum pulsewidth is also shifted towards positive group velocity dispersion (negative a_2).

In summary, closed form analytical solutions have been found for passively mode-locked lasers, where self phase modulation and group velocity dispersion

7

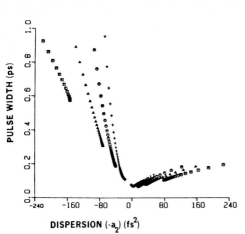

Fig. 1. Pulse width as a function of group velocity dispersion $(-a_2)$. Curves are for gains of 6.0%, 6.2%, 6.4%, and 6.6% with the highest lying curve corresponding to the largest gain.

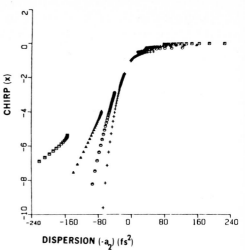

Fig. 2. Chirp as a function of intracavity group velocity dispersion for the same laser parameters as above.

are included. The solutions found are chirped secant hyperbolic pulses. If these two additional mechanisms (self phase modulation and group velocity dispersion), are properly adjusted a reduced minimum pulsewidth can be achieved with improved stability against fluctuations in the gain and loss. The inadequate balance of these phase modulation terms gives rise to broad pulses sensitive to fluctuation in gain and loss parameters.

We also describe negative group velocity dispersion obtainable from prisms. In the visible region of the spectrum most glasses have positive group velocity dispersion. The availability of optical elements having adjustable negative dispersion is therefore important to many applications involving extremely short pulses. Such elements, essentially free from loss, can be achieved using prisms [3]. Consider for example a pair of identical prisms, arranged antiparallel so that the second one undoes the ray bending of the first. Loss can be minimized by operating so that the rays of interest pass through all of the prism faces approximately at Brewster's angle. This requires that tan(a)=1/n, where 2a is the apex angle of the prisms, and that tan(e)=n, where e is the external angle of incidence of the beam on the prism faces. Here n is the refractive index of the prism material. Collimated rays of different frequencies incident on the prism pair emerge dispersed laterally, but all parallel to one another. Group velocity dispersion is proportional to the second derivative of the optical path through the prism pair near the prism apexes (the prisms must be placed properly).

Two components of group velocity dispersion are revealed. The first, which is positive, is mainly a result of the length of prism material traversed by the beams. The second, which is negative, is proportional to $(dn/d\lambda)^2$ and to the distance between the apexes of the prisms. Using values

8

appropriate to quartz at a wavelength near .620 μm, we find that a separation of the prisms by 25 cm produces a negative component sufficient to balance the positive contribution from 7.4 mm of quartz. Since the beam need pass through only a few millimeters of quartz in traversing the prisms, we see that there can be considerable negative dispersion left over. The positive contribution to the net dispersion can be adjusted by moving the prisms so that the beams traverse more material, without noticeably changing the beam paths outside of the prisms.

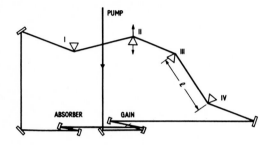

Fig. 3. Colliding pulse laser including a four prism sequence which provides adjustable group velocity dispersion.

We have included adjustable group velocity dispersion in a colliding pulse laser in the form of a four prism sequence as shown in Fig. 3. The four prism sequence is advantageous in that the transmitted beam is collinear with the incident beam and the mode exhibits no transverse displacement of the different frequency components [3]. We also continuously adjust the intracavity group velocity dispersion by translating one prism, e.g., prism II, normal to its base. Tuning the intracavity dispersion in this manner yields plots of pulsewidth vs. group velocity dispersion with an asymmetry and magnitude closely approximating those shown in Fig. 1. We obtain a minimum pulsewidth, ~65 fsec, similar to that previously reported, suggesting that earlier laser configurations closely approached the optimum relationship between group velocity dispersion and self phase modulation simply by selection of mirrors and other laser parameters, Fig. 4.

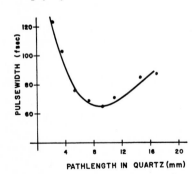

Fig. 4 Experimental plot of pulsewidth vs. intracavity pathlength in quartz.

In conclusion we have: (1) shown that a negative contribution to group velocity dispersion always accompanies refraction, (2) shown that soliton shaping mechanisms similar to those which cause N=1 soliton formation in optical fibers are important in obtaining stable trains of short pulses in passively mode-locked lasers, and (3) constructed a laser which uses these two findings to generate stable trains of short pulses.

References

1. R. L. Fork, B. I. Greene, and C. V. Shank, Appl. Phys. Lett. 38, 671 (1981).

2. J. P. Gordon and R. L. Fork, Optics Letters 9, 153 (1984).

3. R. L. Fork, O. E. Martinez, and J. P. Gordon, Optics Letters 9, 150 (1984).

4. O. E. Martinez, R. L. Fork, and J. P. Gordon, Optics Letters 9, 156 (1984).

5. W. Dietel, E. Dopel, D. Kuhlke, and B. Wilhelmi, Optics Comm. 43, 433 (1982).

6. W. Dietel, J. J. Fontaine, and J. C. Diels, Optics Lett. 8, 4 (1983).

7. L. F. Mollenauer, R. H. Stolen, and J. P. Gordon, Phys. Rev. Lett. 45, (13), 1095 (1980).

8. H. A. Haus, IEEE J. Quantum Electron, QE-11, No. 9, p. 736 (1975).

Compression and Shaping of Femtosecond Pulses

A.M. Weiner, J.G. Fujimoto, and E.P. Ippen

Department of Electrical Engineering and Computer Science, Research Laboratory of Electronics, Massachusetts Institute of Technology
Cambridge, MA 02139, USA

During the past several years ultrashort pulse technology has advanced dramatically from the picosecond to the femtosecond time domain. The colliding-pulse modelocked (CPM) cw dye laser [1] is now a reliable source of pulses shorter than 100 fsec; and pulses from other laser systems are being compressed to similar durations by nonlinear optical fiber methods [2]. Fiber compression of amplified pulses from a CPM-based system has already been used to generate 30 fsec pulses [3]. In the first part of this paper we describe experiments in which this same combination of techniques has been extended to produce pulses as short as 16 fsec [4]. These pulses, with a center wavelength of about 625 nm, are comprised of only 8 optical cycles. Then, as an application for such ultrashort pulses, we describe time-resolved reflectometry studies of multilayer dielectric mirror coatings. Dramatic pulse distortions can be observed after a single reflection from a broadband mirror. Finally, we discuss the use of transient four-wave mixing for pulse shortening and shaping at much higher power levels.

The pulses for all of the experiments reported here were derived from a colliding-pulse-modelocked (CPM) ring oscillator [1] and amplified in a high power, femtosecond dye amplifier chain [5]. With the first two stages of the chain, we achieve pulsewidths of about 65 fsec (following grating-pair compensation) with energies of about 2 μJ, more than enough for our fiber compression experiments. For high power experiments we can utilize all four stages to obtain 75 fsec pulses with energies of 250 μJ.

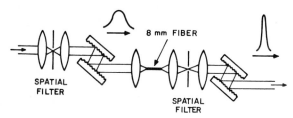

Fig. 1 Experimental arrangement for fiber compression

Our experimental arrangement for fiber compression is shown in Fig. 1. The amplified pulses are filtered spatially before the fiber to minimize excitation of unwanted, cladding modes and to avoid physical damage to the fiber due to beam hot spots. A pair of gratings compensates for dispersion in the amplifier, the spatial filter components, and the input coupling lens. A second spatial filter at the output of the 8 mm long fiber discriminates against unguided modes; a second grating-pair provides the actual pulse compression and is adjusted to precompensate for dispersion in the autocorrelator. Both the parallelicity and relative rotational align-

11

ment of these gratings were found to be especially critical for successful pulse compression.

Spectral broadening of the pulse by self-phase modulation in the 8 mm fiber is observed by monitoring the spectrum at the output with an optical multichannel analyzer. Optimum compression was achieved when the bandwidth had increased a factor of four to about 35 nm. The power coupled into the fiber is adjusted with neutral density filters and by moving the input objective in and out of focus. Best results were obtained with approximately 5 nJ per pulse coupled into the fiber. The experimental parameters and the compression factor are close to those calculated for optimum compression in the presence of group velocity dispersion [6]. Further spectral broadening was easily observed by increasing the input power but did not result in shorter pulses.

Measurement is performed by autocorrelation using the conventional noncollinear SHG method [7]. The relative delay was computer controlled with a 0.1 μm/step translation stage in one arm of the autocorrelator. In order to eliminate any pulse distortion from the mirrors, only front-surface aluminum mirrors are used after the fiber. Because the mandatory beam splitter introduces a 1 mm path length of glass in one arm of the interferometer, a similar beamsplitter was inserted into the other arm to restore the dispersive balance. To minimize geometrical broadening effects due to the noncollinear geometry, the angle between the beams is kept less than 2° and care is taken to insure diffraction-limited focusing. KDP crystals of several different thicknesses were used for the measurements. When oriented for maximum SHG, crystals with thicknesses of 0.1 and 0.3 mm yield essentially identical pulse measurements although small differences can be observed in the wings of the autocorrelation measurement.

A typical autocorrelation trace is shown in Fig. 2. Pulse durations in the range 15-18 fsec are obtained regularly and consistently. The pulsewidth is 16 fsec full-width half-maximum, assuming a secant hyperbolic pulse shape, and corresponds to a factor of four compression. Pulse coherence is corroborated by measured time-bandwidth products $\Delta\nu\cdot\Delta t$ of about 0.43.

As an application of our compressed pulses, we have performed time-resolved reflectometry measurements of broadband dielectric mirror coatings. These experiments were motivated by the sensitivity of the CPM ring dye laser pulsewidth to the selection of cavity mirrors and by calculations that

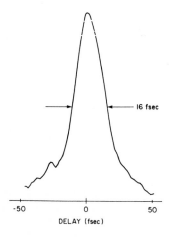

16 fsec

-50   0   50
DELAY (fsec)

Fig. 2   Autocorrelation trace of a 16 fsec pulse

dispersive effects due to the mirrors become important for pulsewidths of 100 fsec or less [8]. Pulse distortion by a broadband mirror can be expected to depend on the phase spectrum of the mirror, an attribute which is generally unspecified for commerical mirror coatings.

A dramatic example of what can happen· upon reflection from a broadband dielectric mirror is illustrated in Fig. 3. These traces, which represent the effect of a single reflection at 45° incidence from a type BD.1 coating obtained from Newport Research Corp., are representative of what can happen with any broadband coating. The curves are autocorrelation traces taken before and after refelction; the severe distortion of the multiply peaked post-reflection data suggests that not only $d^2\phi/d\omega^2$ but also higher order derivatives of the phase must be considered. This fact was confirmed by varying the separation of the grating-pair in the compressor; successful compensation of the mirror distortion could not be achieved.

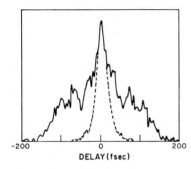

DELAY(fsec)

Fig. 3 Autocorrelation traces taken before and after reflection from a broad-band dielectric mirror. dashed curve: before reflection; solid curve: after reflection

Our data may be understood qualitatively as follows. Broadband dielectric mirror coatings are typically constructed by cascading two or more quarter-wave stacks, each designed to reflect a different wavelength. Because the reflectivity of a quarter-wave stack may have sidelobes outside its high reflectivity region, a short pulse which is reflected primarily by a particular stack may also experience partial reflection by other off-resonant stacks. Partial reflections originating at different stacks will tend to broaden the pulse and may lead to satellite pulses.

To put our results in perspective, we should remark that not all mirrors tested caused noticeable pulse distortion. Neither aluminum reflectors nor single-stack dielectric coatings used near their center wavelength affected the reflected pulse shape. Broadband dielectric reflectors, as illustrated above, can lead to a variety of distortion effects which vary with the angle of incidence and which can differ markedly for two mirrors nominally of the same manufacture and design. The fact that pulse distortion upon reflection is strongly and directly manifest with 16 fsec pulses suggests that this technique may be applied to characterize mirrors for femtosecond laser cavities.

For many other applications it is necessary to produce ultrashort pulses of much higher intensity than is possible with fiber compression. To this end we have also investigated transient four-wave mixing as an approach to pulse shortening and optical gating at high powers ⌊9⌋. The experimental arrangement is illustrated in Fig. 4. Two high power pulses, incident on a thin nonlinear sample, interfere and produce a spatially periodic excitation, which scatters part of the incident energy into new directions. Previous authors [10] have proposed measurement of the scattering efficiency

13

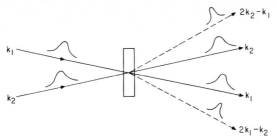

Fig. 4 Experimental arrangement for transient four-wave mixing

versus the delay between the two incident pulses as a means of observing polarization dephasing in nonlinear media. Measurements of the scattered pulse shapes may provide an even more sensitive technique for such observations [9]. The scattered pulse durations depend upon the material time constants in addition to the initial pulse shapes and the relative incident delay.

From the point of view of ultrashort pulse generation, a direct application of this two pulse technique is its use for shaping and shortening pulses. Using a thin dye cell as the nonlinear medium, relatively efficient scattering can be achieved. Theoretical plots of the scattered pulsewidths are shown in Fig. 5 as a function of relative incident delay, assuming that polarization dephasing (T_2) is very rapid for these room-temperature dye solutions [11,12] but that absorption recovery (T_1) is slow compared to the pulses. Note that very different behavior is expected for Gaussian and secant hyperbolic pulse shapes. The experimental data obtained with our high intensity amplified pulses show a variation with delay that is more rapid than that expected for either shape. This may be evidence of pulse asymmetry or a sharp leading or trailing edge on our amplified pulses. For 68 fsec input pulses, we have been able to demonstrate shortening down to 37 fsec. An interesting extension of this technique might be the appropriate tailoring of pulse shapes for further compression with an optical fiber.

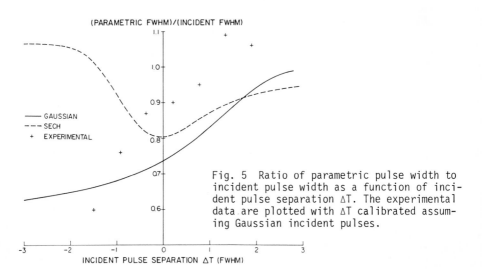

Fig. 5 Ratio of parametric pulse width to incident pulse width as a function of incident pulse separation ΔT. The experimental data are plotted with ΔT calibrated assuming Gaussian incident pulses.

14

This work was supported by the Joint Services Electronics Program under DAAG 29-83-K-003. A.M. Weiner was a Fannie and John Hertz Foundation Fellow.

References

1. R.L. Fork, B.I. Greene and C.V. Shank, Appl. Phys. Lett. 38, 671 (1981).
2. B. Nikolaus and D. Grischkowsky, Appl. Phys. Lett. 43, 228 (1983).
3. C.V. Shank, R.L. Fork, R. Yen, R.H. Stolen, and W.J. Tomlinson, Appl. Phys. Lett. 40, 761, (1982).
4. J.G. Fujimoto, A.M. Weiner and E.P. Ippen, Appl. Phys. Lett. 44, 832 (1984).
5. R.L. Fork, C.V. Shank and R.T. Yen, Appl. Phys. Lett. 41, 223 (1982).
6. W.J. Tomlinson, R.H. Stolen and C.V. Shank, J. Opt. Soc. Am. B1, 139 (1984).
7. E.P. Ippen and C.V. Shank, in Ultrashort Light Pulses: Picosecond Techniques and Applications, S.L. Shapiro, ed. (Berlin: Springer Verlag, 1977).
8. S. De Silvestri, P. Laporta and O. Svelto, IEEE J. Quantum Electron. QE-20, 533 (1984).
9. J.G. Fujimoto and E.P. Ippen, Opt. Lett. 8, 446 (1983).
10. T. Yajima, Y. Ishida, and Y. Taira, Picosecond Phenomena II, R.M. Hochstrasser et al., eds. (Berlin: Springer Verlag, 1980), 190.
11. A.M. Weiner and E.P. Ippen, Opt. Lett. 9, 53 (1984).
12. A.M. Weiner, S. De Silvestri and E.P. Ippen, "Femtosecond Dephasing Measurements Using Transient Induced Gratings", also in this volume.

Generation of 0.41-Picosecond Pulses by the Single-Stage Compression of Frequency Doubled Nd:YAG Laser Pulses

A.M. Johnson, R.H. Stolen, and W.M. Simpson

AT & T Bell Laboratories, Crawford Corners Road, Holmdel, NJ 07733, USA

The technique of optical pulse compression utilizing self-phase modulation (SPM) to chirp the pulse in a single-mode fiber followed by a grating-pair dispersive delay line has been very successful in compressing dye laser pulses. A 3X single-stage compression of the colliding pulse modelocked laser has resulted in optical pulses as short as 0.03 psec [1]. A 65X two-stage compression of a synchronously modelocked dye laser has resulted in optical pulses as short as 0.09 psec [2]. We report an 80X single-stage compression of 240W, 33 psec duration optical pulses from a cw modelocked frequency doubled Nd:YAG laser to 3.6 kW, 0.41 psec duration optical pulses. The compression of these 0.532 μm optical pulses has resulted in the largest single-stage compression factor reported to date. These results demonstrate the production of subpicosecond optical pulses directly from relatively long optical pulses without the use of a modelocked dye laser.

In setting up the compressor it is important to have an initial estimate of the proper fiber length and grating separation. Recently, TOMLINSON et al. [3] have presented numerical calculations of the optimum fiber length, grating separation, and achievable pulse compression in normalized form, permitting translation of the results to any given input pulse parameters. Calculations show that it is possible to achieve subpicosecond pulses of high quality starting with pulses as long as 100 psec duration, but that the required fiber length and grating separation increase rapidly with pulsewidth, and that the optimum pulse shape becomes a sensitive function of peak power, fiber length, and grating separation.

In standard grating compressor configurations [1,2] large grating separations result in a large transverse displacement of the different spectral components. The compression factor will depend critically on the ability to transform the resultant cylindrical beam into a usable circular beam without losing the high frequency components. A grating pair configuration that was used by DESBOIS et al. [4] to expand the optical pulses of a modelocked Nd:glass laser has been modified for use in these compression experiments (see Fig. 1). The modified grating compressor

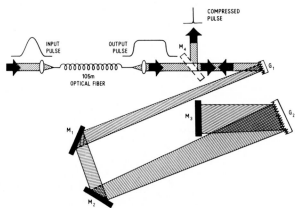

Fig. 1: Schematic drawings of the optical pulse compressor. The dispersive delay line consists of gratings G_1 and G_2, and mirrors M_1, M_2, and M_3. The compressed pulse is deflected out by mirror M_4.

cancels the transverse displacement of the different spectral components and provides a high quality circular beam with all of its frequency components intact. The cost of using this configuration is reduced transmission due to the fact that the beam is diffracted four-times instead of twice. On the other hand, use of the standard configuration [1,2] for long input pulses could result in poor compression.

The 0.532 μm input optical pulses [5] were obtained from a Quantronix 116 cw Nd:YAG laser that was harmonically modelocked at a modulation frequency of 100 MHz and then frequency doubled in a crystal of $KTiOPO_4$ (KTP). This system is capable of delivering short (33 psec) 0.532 μm optical pulses of average power approaching 1.5W (455W peak) at a repetition rate of 100 MHz.

The input pulses had an average power of 780 mW, a peak power of 240W, and a bandwidth of ~0.3Å. The optical pulses were coupled into a 105 m long single-mode polarization-preserving optical fiber [6] with a core diameter of 3.8 μm. The frequency-doubling crystal (KTP) provides excellent isolation between the Nd:YAG laser and scattered light from the optical fiber. The spectrally broadened output pulse had a bandwidth of ~20A (see Fig. 2) and an average power of 350 mW, for an overall fiber transmission, including coupling loss, of 45%. This output power included a small stimulated Raman scattering contribution of 5%. The output pulse was injected into the dispersive delay line shown schematically in Fig. 1. We used a pair of holographic gratings with 1800 grooves/mm and a diffraction efficiency of 84%. The total transmission of the fiber optic compressor was 19%.

The autocorrelation functions of the input and compressed pulses are displayed on the same scale in Fig. 3. The input pulse had an autocorrelation half-width of 47 psec or a pulsewidth of 33 psec assuming a Gaussian pulse shape. The autocorrelation half-width of the compressed pulse was 0.64 psec or a pulsewidth of 0.41 psec assuming a $sech^2$ pulse shape. The structure observed far out on the wings of the compressed pulse had an amplitude of ~1% of the peak height. The compressed pulses had an average power of 150 mW and a peak power of 3.6 kW. The peak power of the compressed pulses was increased by a factor of 15 over the input pulses.

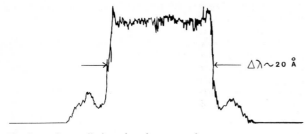

Fig. 2: Spectrally broadened output pulse.

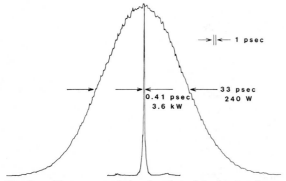

Fig. 3: The autocorrelation function of the input and compressed pulses displayed on the same scale. This represents a compression of 80X in a single stage.

17

This high-repetition rate source of high peak power optical pulses is directly useful for applications involving subpicosecond pulses in the visible and is a potential pumping source for a large class of visible, tunable subpicosecond dye lasers and a source for generating harmonics in the uv. In preliminary experiments, 0.5 psec compressed 0.532 μm pulses were used to synchronously pump a Rhodamine 6G dye laser. Pumping with an average power of 110 mW, the dye laser output had an autocorrelation half-width of 0.46 psec (see Fig. 4) and an average power of 25 mW (η = 23%) operating at 592 nm. Assuming a sech^2 pulse shape, we calculate a pulsewidth of 0.30 psec, the shortest yet obtained *directly* from a synchronously modelocked dye laser (containing no saturable absorber), and can be directly attributed to the ultrashort pumping pulses.

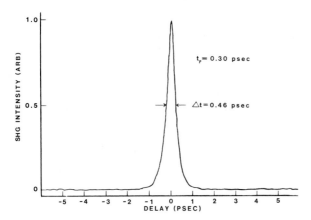

Fig. 4: Autocorrelation function of a synchronously modelocked Rh 6G dye laser pumped with 0.5 psec compressed pulses.

References

1. C. V. Shank, R. L. Fork, R. Yen, R. H. Stolen, and W. J. Tomlinson, Appl. Phys. Lett. *40*, 761 (1982).
2. B. Nikolaus and D. Grischkowsky, Appl. Phys. Lett. *43*,, 228 (1983).
3. W. J. Tomlinson, R. H. Stolen, and C. V. Shank, J. Opt. Soc. Am. B *1*, 139 (1984).
4. J. Desbois, F.Gires, and P. Tournois, 1EEE J. Quantum Electron. *QE-9*, 213 (1973).
5. A. M. Johnson and W. M. Simpson, Opt. Lett. *8*, 554 (1983).
6. R. H. Stolen, V. Ramaswamy, P. Kaiser, and W. Pleibel, Appl. Phys. Lett. *33*, 699 (1978).

Compression of Mode-Locked Nd:YAG Pulses to 1.8 Picoseconds

B.H. Kolner and D.M. Bloom

Edward L. Ginzton Laboratory, Stanford University, Stanford, CA 94305, USA

J.D. Kafka and T.M. Baer

Spectra-Physics Corporation, Laser Products Division, 1250 W. Middlefield Road, Mountain View, CA 94042, USA

In this paper we report the compression of 80 picosecond long mode-locked Nd:YAG pulses to 1.8 picoseconds by using self-phase modulation in single-mode optical fibers and a grating pair dispersive delay line. Although we generated pulses as short as 1.0 picoseconds, pulsewidths of 1.8 picoseconds were more routinely obtained. Our compressor incorporated a novel grazing incidence delay line which greatly reduced the grating separation. We achieved pulse compression ratios of 45:1 and generated visible picosecond pulses by second harmonic generation. For fixed wavelength applications the compressed cw mode-locked Nd:YAG laser provides a completely solid state source of picosecond pulses. In addition, the compressed pulses should improve the performance of synchronously pumped tunable lasers.

Pulse compressors based on self-phase modulation in single-mode optical fibers have been demonstrated in the visible spectrum [1-4]. Previous workers typically used short duration pulses and grating configurations exhibiting low dispersion. The results of these experiments suggested that compression of longer pulses would require inconveniently large grating separations [5]. However, analysis of grating pair dispersive delay lines shows that the dispersion is a widely varying function of input angle, groove density and wavelength [6-8]. We discovered that grazing incidence and high diffraction angles increase the dispersion by more than two orders of magnitude. Thus, only tens of centimeters of grating separation are required to compress 100 picosecond Nd:YAG pulses. To realize high efficiency the gratings are used close to the Littrow condition. At any wavelength it is possible to choose a grating ruling that satisfies the Littrow condition at grazing incidence (80 degrees).

In our experiment a Spectra-Physics Model 3000 cw mode-locked Nd:YAG laser generated 80 picosecond pulses at 82 MHz with 8 watts average power. The beam was attenuated to 2 watts to prevent stimulated Raman scattering and focused into 300 meters of single-mode fiber. The coupling efficiency was typically 50%. At the fiber output the pulsewidth was unchanged but the bandwidth broadened from 0.3 Å to 35 Å due to self-phase modulation. The broadened frequency spectrum is shown in Figure 1a.

The output pulses passed through a pair of 1800 lines/mm gratings with a separation of 33 cm. We achieved compressed pulses with an average power of 500 mW and duration of 1.8 picoseconds, measured by second harmonic autocorrelation (Figure 2). Using a 5 mm long KTP crystal, an average power of 40 mW was generated at 532 nm.

We predicted the amount of spectral broadening and subsequent pulse compression by assuming that the fiber causes self-phase modulation only

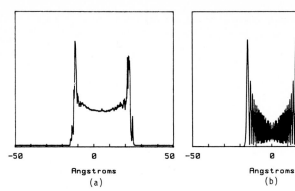

-50   0   50   -50   0   50

Angstroms   Angstroms

(a)   (b)

Fig. 1   a) Measured frequency spectrum of the self-phase modulated pulse. b) Calculated frequency spectrum for A=150.

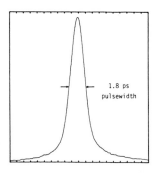

1.8 ps
pulsewidth

Fig. 2 Measured autocorrelation waveform of compressed Nd:YAG pulse (1 ps/div). The indicated FWHM pulsewidth assumes a Gaussian envelope.

and that group velocity dispersion is negligible. These assumptions are justified in our experiments because a 100 picosecond pulse at 1.06 microns requires a 400 km fiber to double in width due to group velocity dispersion. The effect of the self-phase modulation is to introduce into the phase of the electric field a contribution $\phi(t)$ that is proportional to the intensity envelope $I(t)$.

$$E(L,t) = E_0(L,t) \bullet \exp(k_0 L + \phi(t) - \omega t)$$

$$\phi(t) = A \bullet I(t)/I_0 .$$

The dimensionless parameter A is defined as:

$$A = \frac{2\pi L n_2 I_0}{\lambda_0 n_0 c \epsilon_0}$$

where: L = fiber length
                  n_0 = linear index of refraction
                  n_2 = nonlinear index of refraction
                  I_0 = peak intensity
                  λ_0 = free space wavelength
                  ϵ_0 = free space permittivity.

We obtain the compressed time domain waveform by Fourier transforming the electric field, multiplying by the grating pair transfer function [6], and inverse transforming the product. The spectral broadening, optimum grating spacing, and minimum pulsewidth are characterized by the parameter A. For the power densities in our experiments, A is approximately 150. We used this value to predict a spectral broadening ratio of 155 and a final pulsewidth of 1.6 picoseconds assuming a Gaussian pulse shape. This is in good agreement with our experimental results. The theoretical spectra of the chirped pulse is shown in Figure 2b.

Self-phase modulation in single-mode fibers in the absence of group velocity dispersion (GVD) produces a frequency sweep (chirp) across the pulse that is linear in the center of the pulse only [1,2]. The grating pair is a matched filter for a linear frequency sweep and thus parts of the pulse with a nonlinear chirp do not contribute to an optimally compressed pulse. Furthermore, when the parameter A increases, more of the pulse is chirped nonlinearly and the expected increase in compression ratio suffers. This explains why we obtained a pulse compression ratio of 45:1 while the frequency spectrum broadened by 117:1. Figure 3 shows the theoretical spectral broadening and pulse compression ratios as a function of A. This situation is not as severe in fibers with group velocity dispersion because the GVD tends to linearize the chirp over more of the pulse.

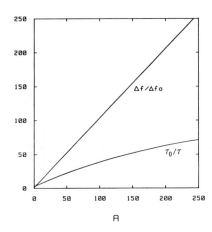

Fig. 3 Theoretical spectral broadening ratio $\Delta f / \Delta f_0$ and pulse compression ratio τ_0 / τ as a function of the parameter A.

Δf = final FWHM bandwidth
Δf_0 = initial FWHM bandwidth
τ = final FWHM pulsewidth
τ_0 = initial FWHM pulsewidth

In conclusion, we have built an efficient and compact pulse compressor using a single-mode optical fiber and a grazing incidence grating delay line. With this device we compressed 80 picosecond pulses from a cw mode-locked Nd:YAG laser to 1.8 picoseconds. We also generated synchronized pulses at 532 nm with 8% conversion efficiency. The compressor offers a simple, compact alternative to synchronously pumped dye or color center lasers for the generation of picosecond pulses in the near infrared.

1. B. Nikolaus and D. Grischkowsky, App. Phys. Lett. 42,1 (1983)
2. B. Nikolaus and D. Grischkowsky, App. Phys. Lett. 43,228 (1983)
3. C.V. Shank, R.L. Fork, R Yen, R.H. Stolen, App. Phys. Lett. 40, 761 (1982)
4. A.M. Johnson, R.H. Stolen, W.M. Simpson, App. Phys. Lett., 44, 729 (1984)

5. W.J. Tomlinson, R.H. Stolen and C.V. Shank, J. Opt. Soc. Am. B, $\underline{1}$,139 (1984)
6. E.B. Treacy, IEEE J. Quant. Elect., $\underline{QE-5}$, 454 (1969)
7. E.B. Treacy, Physics Letters, $\underline{28A}$, 34 (1968)
8. J.D. McMullen, Applied Optics $\underline{18}$, 737 (1979)

22

Effects of Cavity Dispersion on Femtosecond Mode-Locked Dye Lasers

S. De Silvestri, P. Laporta, and O. Svelto

Centro di Elettronica Quantistica e Strumentazione Elettronica del C.N.R.,
Istituto di Fisica del Politecnio, P.zza Leonardo da Vinci 32
I-20133 Milano, Italy

1. Introduction

Significant progress has recently been made in the generation of femtosecond dye laser pulses, in particular by means of the colliding pulse technique [1]. Some observed features of the mode-locked (ML) laser still remain largely unexplained, however. In a few cases, for instance, a strong negative frequency chirp has been experimentally observed [2] and attributed to a phenomenon of self-phase-modulation (SPM) arising from the saturation of the saturable absorber [3]. In other cases, the choice of the cavity mirrors [4] and their layout in the laser cavity [5] have been shown to influence the pulse duration.

In a previous work [6] we have suggested that the above experimental features may arise from the dispersive contribution of the cavity elements and, notably, of the mirrors. In this work we present a comparison between the dispersive contribution of a $\lambda/4$ multilayer mirror (whose bandwidth is usually 120-200 nm) and that of a broadband mirror (whose bandwidth is usually more than 400 nm). The effect of these contributions on the behavior of a ML laser and a comparison with the effect arising from SPM in the saturable absorber will also be discussed.

2. Approach

The response of an unsaturated dispersive element to an input pulse of frequency ω can be analyzed either in the time or in the frequency domain. For our purpose, especially in the case of multilayer dielectric mirrors, we have found more convenient to operate in the frequency domain. The element has accordingly been described by a transfer function with a constant amplitude and a phase shift $\phi(\omega)$. Assuming that $\phi(\omega)$ varies slowly with ω, it can be expanded up to second order around the central oscillating frequency ω_L, viz: $\phi(\omega) = \phi(\omega_L) + (\omega-\omega_L)\phi' + (\omega-\omega_L)^2\phi''/2$. The output pulse after the dispersive element is obtained by transforming the input pulse into the frequency domain, multiplying it by the transfer function of the element and transforming back into the time domain. For an unchirped Gaussian input pulse of duration τ_{in} (FWHM), the output is a broadened Gaussian pulse of duration

$$\tau_o = \tau_{in} \{1 + [(16\ell n^2 2)\phi''^2/\tau_{in}^4]\}^{\frac{1}{2}} \tag{1}$$

with a liner frequency chirp $\delta\omega$ given by

$$\delta\omega = - (16\ell n^2 2)\phi'' t/[(16\ell n^2 2)\phi''^2 + \tau_{in}^4] . \tag{2}$$

Equations (1) and (2) show that, for a given value of τ_{in}, the amplitude broadening and the frequency sweep depend only on ϕ''. This is therefore the only quantity which needs to be specified in order to characterize the dispersion property of the element. One may further notice that, according to

23

Eq. (2), a positive value of ϕ'' results in a negative frequency chirp in the output pulse.

3. Results

According to the previous discussion, the cavity mirrors have been characterized by the corresponding values of ϕ''. As a particularly relevant example, Fig. 1 shows the case of a $\lambda/4$ narrowband multilayer dielectric mirror. We see that ϕ'' is zero at resonance ($\omega=\omega_m$) and it rapidly increases away from resonance with a sign depending on whether one is working toward the red-shifted or blue-shifted sides of the reflectivity curve. Note that in some experiments one of the cavity mirrors is used near to its long wavelength cut-off to limit the oscillation bandwidth toward the red side. Assuming in this case a 4% transmission for the center wavelength, a relatively high value of $\phi''(\simeq170\times10^{-30}\,s^2)$ is obtained from Fig. 1b. Note also that the calculated values of ϕ'' depend on the refraction indices of the layers and increase if the layer thickness has some errors from the theoretical $\lambda/4$ values or if the mirror used is tilted. For instance, the value of ϕ'' can increase by about a factor 2 for an error of $\simeq8\%$ in the first three layers [7].

For a broadband multilayer mirror the values of ϕ'' will obviously depend upon the way by which the mirror is made. As an example, Fig. 2 shows the case of two superposed and nearly contiguous $\lambda/4$ stacks [8] with the films of one stack 1.25 times thicker than those of the other stack. Figure 2a indeed shows that the high reflectivity bandwidth has now been increased by approximately a factor 2 compared to that of Fig. 1a. Figure 2b then shows that, within the high reflectivity band, ϕ'' acquires an oscillatory behavior with very large peak values. This behavior occurs predominantly on the high frequency side in Fig. 2b while it occurs on the opposite side by reversing the order of the two stacks. In general it occurs when the light beam traverses, out of resonance, the first stack before being reflected from the deeper stack.

As a summary of the results obtained in the case cavity mirrors, we can say that values of ϕ'' up to $300\times10^{-30}\,s^2$ and up to at least $2\,000\times10^{-30}\,s^2$ can be expected for a narrowband and for a broadband mirror, respectively. Table 1 shows, for comparison, the values calculated for all cavity elements which may be present in a femtosecond ML laser cavity. The effect of the SPM in the saturable absorber has also been evaluated in term of an equivalent ϕ'' [6]. In all cases listed in Table 1 the calculation has been performed at $\lambda=610$ nm.

4. Discussion

From Table 1 and from the results of Figs. 1 and 2 it is apparent that the dispersion due to the cavity mirrors, that arising from pieces of quartz or

Table 1 Dispersive contribution of cavity components

Cavity component	$\phi''[10^{-30}\,s^2]$
Ethylene glycol jet stream (100 μm)	− 8.4
Quartz (1 mm)	− 54
Flint F2 glass (1 mm)	− 160
Anomalous dispersion in DODCI	7.5
Anomalous dispersion in DODCI photoisomer	− 32
Self-phase modulation in DODCI	up to 300
Narrowband multilayer dielectric mirror	up to 300
Broadband multilayer dielectric mirror	up to 2000

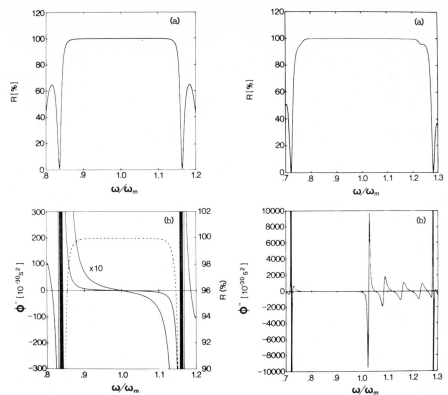

Fig. 1 Dispersive contribution of a narrowband multilayer dielectric mirror. The figure shows a plot both of the reflectance R (Fig. 1a) and of the second derivative ϕ'' of the phase shift (Fig. 1b) vs the normalized frequency ω/ω_m, where ω_m is the mirror resonance frequency. The calculation has been performed for the following sequence of the layers: $(\text{Air})(\text{HL})^{11}$ (H)(Substrate), where the corresponding refractive indices are $n_H = 2.28$ and $n_L = 1.45$.

Fig. 2 Dispersive contribution of a broadband multilayer dielectric mirror. The calculation has been performed for two superposed $\lambda/4$ stacks with the following sequence of the layers: $(\text{Air})(\text{H'L'})^9(\text{HL})^9$ (H)(Substrate), where the film thickness ratio is $L'/L = H'/H = 1.25$ and where the corresponding refractive indices are $n_H = n_{H'} = 2.28$ and $n_L = n_{L'} = 1.45$. The normalizing frequency ω_m has now been taken to be the central frequency of the high reflectivity band.

glass of a length exceeding a few millimeters, and the self-phase modulation in DODCI play the most important role. To assess the influence of the corresponding values of ϕ'' on ML behavior, we can observe that they will begin to be important when the corresponding broadening per pass, as obtained from Eq. (1), becomes comparable (say 30%) to the pulse shortening per pass arising from the combined effect of saturation of the absorber and amplifier. For a pulse shortening of 1% [9] and for $\phi'' = 200 \times 10^{-30}$ s^2 and $\phi'' = 2,000 \times 10^{-30}$ s^2 we get from Eq. (1) that these dispersive contributions become important for $\tau_{in} \leq 85$ fs and $\tau_{in} \leq 270$ fs, respectively. It should be noted that for a given value of ϕ'', the contribution becomes increasingly more important as the pulse duration is decreased. As an example, for $\phi'' = 2,000 \times 10^{-30}$ s^2 and

τ_{in} = 30 fs we obtain that the pulsewidth increases by about a factor 6 for a single reflection from the mirror. Actually, for such a short pulsewidth, the corresponding bandwidth becomes so wide to be comparable with the oscillation period for ϕ'' in Fig. 2b. This means that an expansion of $\phi(\omega)$ up to its second derivative is not adequate any more. In this case, more detailed calculations show that, on reflection, the pulse becomes generally asymmetric and develops a substructure [10], similar to what has already been shown, a few years ago, to occur for ML solid state lasers [11].

In conclusion we can say that these considerations on cavity dispersion should be helpful in designing a femtosecond ML laser.

References

1   R.L.Fork, B.I.Greene, and C.V.Shank, Appl. Phys. Lett. 38, 671 (1981)
2   W.Dietel, J.J.Fontaine, and J.-C.Diels, Optics Lett. 8, 4 (1983)
3   D.Kühlke, W.Rudolph, and B.Wilhelmi, IEEE J. Quantum Electr. QE-19, 526 (1983)
4   C.V.Shank, R.L.Fork, and R.T.Yen, in "Picosecond Phenomena III" edited by K.B.Eisenthal, R.M.Hochstrasser, W.Kaiser, and A.Laubereau (Springer Verlag, Heidelberg, 1982) pp. 2-5
5   E.P.Ippen, personal communication
6   S.De Silvestri, P.Laporta, and O.Svelto, IEEE J. Quantum Electr., May issue (1984)
7   S.De Silvestri, P.Laporta, and O.Svelto, Optics Letters, August issue (1984)
8   A.F.Turner and P.W.Baumeister, Appl. Optics 5, 69 (1966)
9   M.S.Stix and E.P.Ippen, IEEE J. Quantum Electr. QE-19, 520 (1983)
10 S.De Silvestri, P.Laporta, and O.Svelto, unpublished results
11 O.Svelto, Appl. Phys. Letters 17, 83 (1970)

3 MHz Amplifier for Femtosecond Optical Pulses

M.C. Downer and R.L. Fork

AT & T Bell Laboratories, Crawford Corner Road, Holmdel, NJ 07733, USA

M. Islam

Massachusetts Institute of Technology, Cambridge, MA 02139, USA

We describe a system designed specifically for amplification of femtosecond optical pulses [1,2] at high (3 MHz) repetition rate. A key problem in high repetition rate amplification is that the energy of the available pump pulses is only a few microjoules as compared to the hundreds of millijoules available for low repetition rate amplifiers [2]. We resolve this problem, in part, by using multiple collinear passes to extract efficiently the pump pulse energy while also maintaining the transverse coherence and femtosecond duration of the amplified pulse. In addition we have for the first time cavity dumped the source laser, a colliding pulse mode-locked ring laser [3] while maintaining the femtosecond pulse duration. The pulse energy incident on the 3 MHz amplifier has thereby been increased by an order of magnitude, thus reducing the amplification requirements. Our amplifier, like previous high repetition rate amplifiers [4,5], employs a cavity dumped argon ion laser as the pump. The unique features of our system are the amplification of femtosecond, rather than picosecond, pulses and the use of an exactly collinear pumping geometry.

Previous attempts to cavity dump the colliding-pulse laser have broadened the pulses to several hundred femtoseconds because of the positive group velocity dispersion introduced by the intracavity acousto-optic Bragg cell. We have now compensated this additional positive dispersion using an intracavity configuration of four prisms [6,7], which has been shown [6,7] to provide an adjustable negative dispersion with negligible loss. We thereby achieve a cavity dumped pulse train at 3 MHz with 1 nJ/pulse and 100 fsec. pulse duration.

Synchronization of the cavity dumpers in the pump and source lasers with each other and with the 100 MHz source laser pulse train is achieved by means of an electronic AND gate. This gate detects a coincidence between the pulse train from the source laser and an independent 3 MHz electronic pulse train from a signal generator. A coincidence signal is then sent simultaneously to the RF drivers for the two cavity dumpers. Independent variable electronic delays permit centering of the RF pulse on a source laser pulse for most efficient dumping as well as compensation for differences in optical path length of the pump and source laser pulses in reaching the amplifier.

The amplifier itself consists of two stages, each providing two exactly collinear passes, as shown in Figure 1. The linearly polarized femtosecond pulse enters the first stage

through a polarizing prism, and is then focused by a 10 cm
radius mirror to a spot diameter of 20 μm in a jet containing
Rhodamine 640. A second 10 cm mirror recollimates the beam
and a dichroic mirror directs it to a quarter-wave plate and
mirror combination, which retroreflects the beam and rotates
the polarization by 90°. The pulse returns along the same
path for a second pass, then exits the first stage through the
polarizing prism. Second stage amplification proceeds in an
analogous fashion.

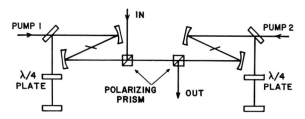

Figure 1   Amplifier Schematic

The pump pulse (1 μJ, 15 nsec. duration) enters collinear with
the source beam through the dichroic mirror source beam and is
focused in the jet to a similar spot size. The optical delay
between the first and second passes of the femtosecond pulse
is less than the pump pulse duration but greater than the dye
lifetime (~3 ns), allowing for repumping the gain. The jet is
produced by sapphire nozzles made by Lexel or Kyburz [8]
designed to yield an optically flat jet of 300 μm thickness.
Dye concentration (~10^{18} molecules/cm^3) is adjusted to absorb
95 percent of the pump energy.

Further amplification is available in principle by
retroreflecting the two-pass first stage output for two
additional passes, then separating the four-pass output from
the input with an acousto-optic switch before entering the
second stage. In practice we have had difficulty in tightly
focussing the third and fourth passes in the jet, possibly
because of gain shaping which has occurred in the first two
passes. We therefore prefer the simpler two pass arrangement
described above.

Our preliminary operation of this amplifier has produced pulse
energies of 10 to 12 nJ at 3 MHz, corresponding to 30 to 35 mW
average power. We observe gain saturation although full gain
saturation should not occur until 100 nJ pulse energies are
reached. Further amplification should therefore still be
possible. The ratio of amplified pulse power to amplified
spontaneous emission is about 50 to 1 without the use of
saturable absorbers. We attribute this favorable ratio to the
pencil shaped geometry of the gain region and to the low
single pass gain [9]. The amplified pulse is less than
15 percent broader than the input pulse both temporally and
spectrally. Good transverse mode quality is achieved by
careful mode matching of the pump and source laser beams. The
use of collinear pumping and optically flat jets also improve
the transverse coherence of the beam.

We have demonstrated a system for amplifying femtosecond pulses to energies greater than 10 nJ at a 3 MHz repetition rate. Even without further improvements, this pulse energy should be sufficient to achieve several-fold pulse compression [10], and possibly continuum generation, in an optical fiber. A variety of nonlinear optical experiments with femtosecond time resolution and powerful signal averaging capability should become possible despite comparatively low amplified pulse energies.

M. Islam is a Fannie and John Hertz fellow.

References

[1] R. L. Fork, C. V. Shank, R. Yen and C. A. Hirlimann, IEEE Jour. of Quant. Elect., QE-19, 500 (1983).

[2] R. L. Fork, C. V. Shank and R. Yen, Applied Physics Letters 41, 223, (1982).

[3] R. L. Fork, B. I. Greene and C. V. Shank, Appl. Phys. Lett. 38, 671 (1981).

[4] A. H. Waldeck, W. T. Lotshaw, D. B. McDonald and G. R. Fleming, Chem. Phys. Lett. 88, 257 (1982).

[5] T. L. Gustafson and D. M. Roberts, Opt. Commun., 43, 141 (1982).

[6] R. L. Fork, O. E. Martinez and J. P. Gordon, Opt. Lett. 9, 150 (1984).

[7] J. P. Gordon and R. L. Fork, Opt. Lett. 9, 153 (1984).

[8] H-P Harri, S. Leutwyler and E. Shumacher, Rev. Sci. Inst. 53, 1855 (1982).

[9] A. Migus, C. V. Shank, E. P. Ippen and R. L. Fork, IEEE Jour. of Quant. Elec., QE-18, 101 (1982).

[10] J. G. Fujimoto, A. M. Weiner and E. P. Ippen, Appl. Phys. Lett. 44, 832 (1984).

Colliding Pulse Femtosecond Lasers and Applications to the Measurement of Optical Parameters

J.-C. Diels[1], I.C. McMichael, F. Simoni[2], R. Torti, and H. Vanherzeele
Center for Applied Quantum Electronics, Department of Physics, North Texas
State University, P.O. Box 5368, Denton, TX 76203, USA

W. Dietel, E. Döpel, W. Rudolph, and B. Wilhelmi
Friedrich-Schiller-Universität, DDR-6900 Jena, German Democratic Republic

J. Fontaine
GREPA, Université Louis Pasteur, F-67000 Strasbourg, France

I. THE RING LASER APPLIED TO NEW DIAGNOSTIC TECHNIQUES

Propagation of short coherent pulses through media poses new constraints on
materials, as well as on the design of experimental set ups, in order to
maintain the best possible temporal resolution. Hence the need for accurate
and sensitive techniques to measure the change in pulse parameters transmitted through (or reflected off) optical samples. We show that the mode
locked ring laser is in itself an accurate tool to investigate linear and
nonlinear optical properties of its components. Measurements of transmission and reflection can be performed intra or extracavity. Intracavity measurements are more accurate because of the multiple passages through the
sample. As in cw intracavity spectroscopy the enhancement of sensitivity
when operating in the stationary mode-locking regime is of the order of the
mean number of cycles in the cavity. The absorber jet is in ideal configuration to study the phase relaxation time of the dye by Degenerate Four
Wave Mixing (DFWM). Because of the colliding pulse mode locking process, the
two "pump pulses" for the DFWM interaction are simultaneously present in the
jet, while the laser output itself can be used for the probe pulse. The
"thin sample" condition required to obtain high temporal resolution in DFWM
|1| is also met in such a laser.

I.1. DIRECT TRANSMISSION MEASUREMENTS

While conceptually obvious, the direct transmission method is the most difficult to implement, since it requires an accurate determination of the
incident and transmitted pulse shape and phase modulation. We have shown
|2| that the pulse envelope, in amplitude and phase, can be inferred from
simultaneous measurements of the interferometric autocorrelation, the
intensity autocorrelation, and the spectrum. The pulse shape is determined
by successively fitting calculated autocorrelation and spectra of a trial
function with the corresponding measurement. We developed numerical and
analytical methods to calculate the autocorrelations and spectra. Since the
pulse shape is generally asymmetric |2|, it is most convenient to use in-
cident pulses which have a symmetric spectrum (a necessary condition for

---

[1]Present Address: Lab. d'Optique Moléculaire, Université de Bordeaux I

[2]Present Address: Universita della Calabria, Cosenza

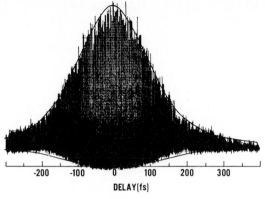

DELAY(fs)

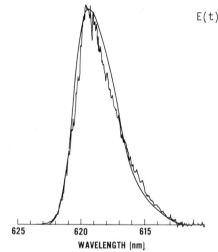

WAVELENGTH (nm)

Figure 1. Fitting of an asymmetric pulse shape with a linear chirp, to an interferometric autocorrelation (above), an asymmetric spectrum (below) and an intensity autocorrelation. The solid lines indicate the calculated envelope of the interferometric autocorrelation (above) and the calculated spectrum (below) for the pulse electric field envelope :

$$E(t) = \frac{\exp\{-0.2i(t/\tau)^2\}}{\exp(-0.4t/\tau)+\exp(1.6t/\tau)}$$

The parameter τ was 100 fs, corresponding to a pulse duration of 140 fs (**FWHM**). The amount of downchirp was undercompensated, with 1.3mm less glass (Fused silica) than the optimum value (9mm) in that particular configuration.

the pulses to be free of phase modulation). The design of our ring laser |3| enables us to generate pulses of symmetric spectrum through intracavity chirp compensation. After transmission and reflection through absorbing and dispersive media, reshaping and chirping can again be determined by successive fitting iterations. An example of how a pulse of arbitrary shape and chirp can be handled is shown in Fig.1.

This method of direct transmission measurements is rather tedious and should be limited to the study of nonlinear effects inducing large changes in amplitude or phase.

I.2. INTRACAVITY INSERTION

The extreme sensitivity of mode locked dye lasers to spectral properties of the intracavity elements makes them accurate tools to measure dispersive properties of components. A simple method is to insert the element to be analysed in the cavity, and to deduce the properties of the sample from the change in output pulse characteristics.

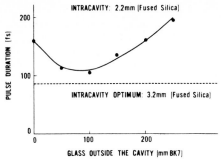

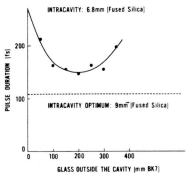

Figure 2. Pulse duration as a function of extracavity glass thickness, for two different intracavity glass lengths and focussing in the absorber.

To illustrate this technique, we measured the pulse duration as a function of extracavity glass path, for different lengths of intracavity glass and focussing parameters in the absorber jet(Fig.2).

The difference between the minimum achievable pulse duration by intracavity versus extracavity compensation is a measure of the nonlinearity of the chirp induced in the absorber.

I.3. INTRACAVITY COMPENSATION

The disadvantage of the previous technique is that the nonlinearity of the mode-locked operation makes it often very difficult to retrace the properties of the sample from the change in output pulse characteristics. The alternative that we demonstrate is a "zero method", by which a "calibrated" intracavity optical component is modified in such a way as to restore the properties of the laser prior to insertion of the sample. As an example, we have measured the dispersive properties of several coatings, by compensating their action with glass samples of variable length (Fig.3).

The "unperturbed" situation is reached with all broadband mirrors and a certain amount of glass compensating the downchirp induced by the saturable absorber. Next, one of the broadband mirrors is replaced by a "sample" mirror, and the intracavity glass compensation is readjusted, until the laser output becomes bandwidth limited again. The measurement is repeated at various wavelengths, making use of the laser tunability |4|.

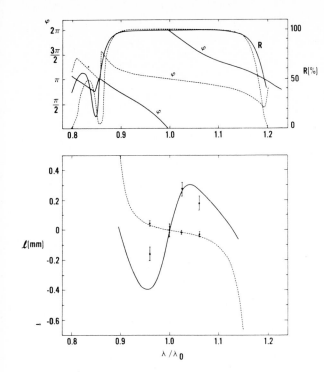

Figure 3.
Calculated reflection coefficient R, phase factor (above) and equivalent glass thickness (below) for an 18 $\lambda/4$ layer dielectric mirror (dashed) lines) and a 17 $\lambda/4$ + one $\lambda/2$ layer mirror (solid lines). Experimental data points obtained with three mirrors of different central frequency are shown on the lower figure.

The method is very sensitive partly because the effect of the sample is multiplied by the mean number of cycles in the cavity, and also because of the sensitivity of the diagnostic techniques (in determining the "zero chirp" point). In agreement with theoretical predictions, we find that mirrors operating near any transmission edge can create either a "down chirp" or an "up chirp" depending on the particular layer structure. It should be noted that the "compensation" is made on the second derivative of the phase versus frequency. A complete analysis of the output pulse shape and residual modulation (with sample) can lead to the determination of higher order terms.

II.   A LINEAR COLLIDING PULSE LASER

The basic advantage of the mode locked ring laser over previous linear cavity designs results from counterpropagating interaction in the absorber jet, of the two pulses that circulate in the cavity. It has been shown that the enhanced saturation and induced grating causes pulse shortening |5| (for pulses larger than the absorber jet thickness) and stabilisation |6| (for pulse lengths of the order of the jet thickness). We create a similar situation by terminating a linear cavity with an antiresonant ring |7|. The latter consists in a 50% beam splitter, followed by a simple optical system returning each of the two (reflected and transmitted) beams into each other(Fig.4). When perfectly aligned, this optical system acts as a 100% broadband reflector.

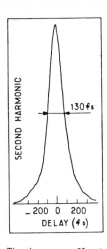

Figure 4. Antiresonant Ring Laser.

Figure 5. Autocorrelation of a 130 fs pulse (FWHM).

The beams reflected and transmitted by the beam splitter are focussed by two 3 cm curvature mirrors in a saturable absorber dye jet. The position of these curved mirrors is adjusted in order to position the common focal point at equal optical distance from the 50% reflecting interface. The laser output is taken through a curved mirror at the amplifier end of the cavity. Stable pulses of 130 fs duration were generated (Fig.5). The pulse spectra and autocorrelation exhibit the same characteristics as the "downchirped" pulses of the ring dye laser |4| with insufficient chirp compensation. Therefore we believe that as in the case of the ring laser, accurate intra-cavity chirp compensation can be made to generate much shorter pulses. A significant advantage of this linear laser over the ring laser is ease of alignment, because the optics around the amplifying jet and absorber jet can be adjusted independently. A unique feature of the antiresonant ring laser is the possibility to adjust continuously the pulse repetition rate (by translating the antiresonant ring assembly along the laser beam axis). The ability to adjust the cavity length continuously makes this laser an ideal candidate for hybrid mode-locking.

This work was supported by the National Science Foundation under Grant Nb ECS 8119568. We are indebted to Spectra Physics for partial support of the linear colliding pulse laser research, through the loan of a |7| argon ion laser.

1. J.-C. Diels, W.C. Wang and Winful, Appl. Phys. B26 105 (1981).
2. J.-C. Diels, J.J. Fontaine and F. Simoni, "Phase Sensitive Measurements of Femtosecond Laser Pulses from a Ring Cavity", Lasers 83, San Francisco (1983).
3. W. Dietel, J.J. Fontaine and J.-C. Diels, Optics Letters 8, 4 (1983).
4. J.J. Fontaine, W. Dietel and J.-C. Diels, IEEE J. of Quantum Electronics QE-19, 1467 (1983).
5. J. Hermann, F. Weider, B. Wilhelmi, Appl. Phys. B26, 105 (1982).
6. J.-C. Diels, I.C. Mc Michael, J.J. Fontaine and C.Y. Wang 3rd Topical meeting on Picosecond Phenomena, Garmisch-Partenkirchen (June, 1982).
7. A.E. Siegman, Opt. Lett. 6, 334 (1981).

High Average Power Mode-Locked Co:MgF$_2$ Laser[1]

B.C. Johnson, M. Rosenbluh[2], P.F. Moulton, and A. Mooradian

Lincoln Laboratory, Massachusetts Institute of Technology
Lexington, MA 02173-0073, USA

The use of Co:MgF$_2$ as a laser material was first demonstrated in the early years of laser development [1]. More recently, vibronic lasers such as Co:MgF$_2$ have proven to be versatile, broadly tunable, cw and pulsed sources of infrared radiation [2,3]. We report here the first tunable, cw mode-locked operation of this laser.

The most interesting property of Co:MgF$_2$ is the large gain bandwidth. Lasers based on Co:MgF$_2$ are tunable over most of the range covered by the fluorescence curve of Fig. 1. Continuously tunable cw laser operation between 1.56 and 2.15 µm has been demonstrated. The peak gain section of Co:MgF$_2$ is small ($\sim 10^{-21}$ cm^2), and the lifetime is correspondingly long (1.3 ms at 77 K). Therefore, practical cw mode locking requires active intracavity modulation. The structure in the fluorescence curve indicates that the local bandwidths that determine mode-locked pulse widths are variable and are much less than the overall bandwidth. This leads to wavelength-dependent pulse widths. However, even the sharpest peaks in Fig. 1 have bandwidths at least 10 times broader than that of Nd:YAG.

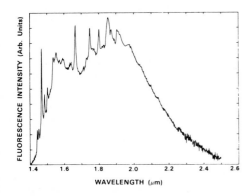

Fig. 1. Fluorescence spectrum of Co:MgF$_2$ at 77 K.

Initial pulse-width measurements were made using a laser system consisting of a three-mirror cavity containing a liquid-nitrogen-cooled, Brewster cut Co:MgF$_2$ crystal, a birefringent tuning element, and an acousto-optic loss modulator. The laser was longitudinally pumped with a cw 1.33 µm Nd:YAG laser. Average mode-locked output powers up to 100 mW were obtained with this configuration.

---

[1]This work was supported by the Department of the Air Force.
[2]Permanent Address: Bar Ilan University, Physics Department, Ramat Gan, Israel.

Autocorrelation pulse-width measurements for the three-mirror Co:MgF$_2$ system are shown in Fig. 2. The pulse-width data, while scattered, were reproducible. Each point corresponds to a different wavelength and tuning plate angle, although some points are closely spaced in wavelength around peaks in the gain curve. The structure in the fluorescence curve of Fig. 1, which is also present in the gain spectrum, influences the pulse widths. Longer pulses occur at the zero-phonon-line wavelengths.

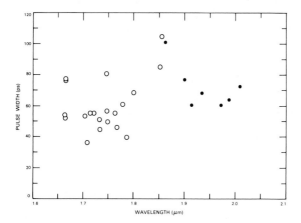

Fig. 2. Pulse width vs wavelength for Co:MgF$_2$ laser. The open and filled-in symbols represent data from different sets of cavity mirrors.

According to theory [4], the bandwidths of the tuning element and the gain medium at zero-phonon lines indicate that pulse widths under 20 ps are achievable. The difference between the results of Fig. 2 and theory are primarily due to the bandwidth-limiting effects of weak etalons formed by reflections from second surfaces of cavity mirrors and the tuning element. Reflections as small as 10^{-6} times the circulating power can produce these bandwidth-limiting effects.

The laser system shown in Fig. 3 was an improvement over the three-mirror cavity in several respects. First, higher output power was possible. The mode waist in the Co:MgF$_2$ crystal was large (270 μm) so a 20 W, multiple-transverse-mode Nd:YAG beam could be focused into the active region. Average mode-locked powers up to 2 W were measured. Second, many of the optical problems affecting the performance of the three-mirror laser were solved. Cavity mirrors with a 5 degree wedge and apertures on either side of the tuning element eliminated unwanted etalon effects, and increased the bandwidth of the laser cavity. When tuned to the 1.667 μm zero-phonon line wavelength, 18 ps pulses with a bandwidth of 23 GHz were observed. Pulse widths of 19 ps were expected given the ~ 1500 GHz gain-medium bandwidth. This result shows that the improved optical design reduced the pulse width to the theoretical limit. However, timing jitter

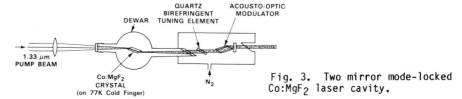

Fig. 3. Two mirror mode-locked Co:MgF$_2$ laser cavity.

of ~ 100 ps was observed for these pulses, indicating that the bandwidth was too large for very stable operation.

More stable operation was obtained at the expense of longer pulse widths by placing a thin, fused silica etalon in the cavity. Etalons ranging in thickness from 0.1 mm to 1.3 mm were tried. Although they were uncoated, their finesse was sufficient to significantly restrict the laser bandwidth, since the fixed cavity losses were less than 1%. The resulting pulse widths ranged from 50 to 250 ps in proportion to the square root of the etalon thickness. Although the pulses were longer, they were more stable, and not dependent on the operating wavelength.

The lasers described above have produced pulses shorter than most similar actively mode-locked lasers, such as the cw Nd:YAG laser. The high power capability of these lasers makes possible extended tunability by nonlinear techniques. The high energy storage possible with the long lifetime of the gain medium should allow higher peak powers with simultaneous Q-switching and mode-locking.

References

1. L. F. Johnson, R. E. Dietz and H. J. Guggenheim, Appl. Phys. Lett. 5, 21 (1964).
2. P. F. Moulton and A. Mooradian, Appl. Phys. Lett. 35, 838 (1979).
3. P. F. Moulton, IEEE J. Quantum Electron. QE-18, 1185 (1982).
4. D. J. Kuizenga and A. E. Seigman, IEEE J. Quantum Electron. QE-6, 691 (1970).

High Power Picosecond Pulses in the Infrared

P.B. Corkum

National Research Council of Canada, Division of Physics
Ottawa, Ontario, Canada, K1A OR6

Progress in short pulse generation in the visible and near infrared spectral regions has traditionally resulted from improvements in actively or passively mode-locked lasers. At other frequencies, however, this need not be the case. Although not usually explicitly stated, a high power picosecond (or subpicosecond) visible pulse can be used in conjunction with any of a number of nonlinear techniques to generate low power pulses of similiar duration over all the visible, near or mid-infrared spectral region.

The ability to generate low power picosecond pulses over a wide spectral range has two important consequences. (1) From the point of view of picosecond technology, other lasers (e.g. CO_2 and excimer lasers) will be more effective amplifiers than dye lasers and (2) from the point of view of laser physics, the highest power radiation available from laser sources will result from amplifying ultrashort pulses in efficient, high energy storage, large gain-bandwidth (e.g. CO_2, Nd:glass) lasers.

Of the various lasers potentially capable of amplifying picosecond very high powers, multi-atmosphere CO_2 lasers are among the most attractive. The large gain bandwidth of multi-atmosphere CO_2 lasers, combined with their high energy storage, the relatively low dispersion of most optical materials in the infrared, and the gaseous nature of the gain medium all suggest an ideal amplifying medium for picosecond pulses.

A high power 2 picosecond 0.616 µm pulse was generated by amplifying the output of a synchronously pump cw dye laser in an XeCl pumped dye amplifier chain. (An early version of the dye laser has been described previously.[1])

Given such a pulse, semiconductor switching[2] provides a process for producing picosecond pulses at any wavelength in the mid-infrared at which there is an existing source of coherent radiation. The experimental configuration shown in Fig. 1 was used to generate high contrast ($>10^6$:1 power contrast ratio) 10 µm pulses with duration that could be chosen anywhere between 2 and 40 ps. It basically consists of a two-element reflection switch which "turns on" the 10 µm pulse, followed by a transmission switch which acts to "turn off" the infrared pulse. A similar configuration has been used previously.[3]

For the short pulse measurements (<8 ps), this pulse was contrast-enhanced (by an additional factor of 10^2) and differentiated on a semiconductor etalon.[4]

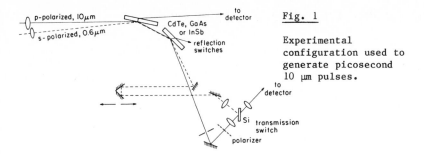

Fig. 1

Experimental configuration used to generate picosecond 10 μm pulses.

The resulting 10 μm pulses were regeneratively amplified in a 20 Hz, Lumonics 880 multi-atmosphere CO_2 laser. The three-element resonator consisted of a 1.5 m radius-of-curvature concave gold-coated mirror, an f = 1 m AR coated ZnSe lens and a plane 80% reflectivity ZnSe output coupler. The separation between the two curved elements was 240 cms, ensuring a beam waist in the gain medium of ~.01 cm^2. A slightly wedged beam splitter was placed between the lens and the output coupler, both for pulse injection and for monitoring the regenerative amplification process.

A single pulse (Fig. 2) containing up to 1.5 mJ was selected from the output train of the multi-atmosphere laser with a 1 cm^2 x 4.5 cm CdTe Pockels cell. Figure 3 is a schematic of the regenerative amplifier and pulse selection optics.

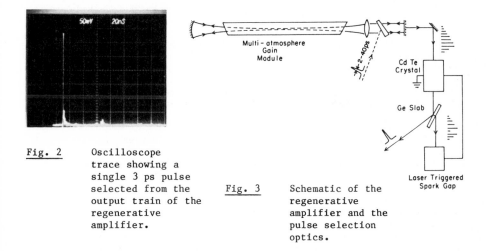

Fig. 2   Oscilloscope trace showing a single 3 ps pulse selected from the output train of the regenerative amplifier.

Fig. 3   Schematic of the regenerative amplifier and the pulse selection optics.

Either the output train or the output pulse from the regenerative amplifier could be autocorrelated with an infrared autocorrelation using collinear beams and tellurium as a nonlinear crystal. Type II phase matching insured a zero background autocorrelation trace.

Figure 4 shows the peak pulse in the mode-locked train obtained with a 3 ps injected pulse. The laser was operated on both the 10 μm P branch (right trace, 14 atmospheres laser pressure) and the 9 μm R branch (left trace, 10 atmospheres laser pressure).

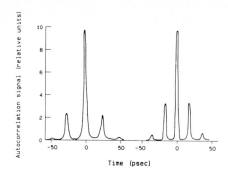

Fig. 4   Autocorrelation
traces of the
regeneratively
amplified pulse.
Left trace: 9 μm R
branch, Right trace:
10 μm P branch.

Both traces show evidence of satellite pulses. These are due to
the residual gain modulation resulting from insufficient overlap of
the pressure broadened rotation lines. The 18 ps pulse separation in
Fig. 4 (right trace) and the 25 ps pulse separation in Fig. 4 (left
trace) are characteristic of the rotational line separation on the P
and R branch respectively.

Two important conclusions are obtained from these picosecond
regenerative amplification studies.

(i) Multi-atmosphere CO_2 lasers are capable of amplifying pulses
as short as ~2.5 ps through a total gain of 10^{10} without significant
pulse broadening, and

(ii) Nearly all of the stored energy ($\geqslant 1$ J/cm^2 for our device) in
a multi-atmosphere CO_2 laser within the mode volume of our resonator
(0.01 cm^2 x 40 cm) can be concentrated in a single picosecond,
regeneratively amplified pulse.

Although single pass amplification studies of multi-atmosphere
CO^2 lasers are required to determine accurately the energy extraction
efficiency under scalable conditions, it is possible roughly to
estimate the saturation energy from the time dependence of the
regeneratively amplified output train. Within the limitations of the
procedure we obtain a saturation energy in the range 200 mJ/cm^2 $\leqslant$
E_{sat} $\leqslant$ 600 mJ/cm^2. Even the smallest value is well in excess of the
saturation energy of any other picosecond amplifying medium with the
exception of Nd:glass where the nonlinear index of refraction at
present imposes a ~8 x 10^9 W/cm^2 limit on the instantaneous output
power for a well designed laser.*

With the availability of mJ, picosecond 10 μm pulses and the
promise of 100 mJ pulses in the near future, it is important to
investigate the possibility of efficient harmonic generation to the
5 μm and shorter wave length region. We find that at an incident
power density of ~100 mJ/cm^2, second harmonic conversion efficiency of

---

*This nonlinear limit is not a relevant long-term consideration. In
Nd:glass lasers, it is possible to amplify chirped pulses and these
pulses (or their harmonics) can be compressed in a dispersing medium.
In this configuration, Nd:glass lasers will have a higher E_{sat} and a
much higher effect I_{sat} than even multi-atmosphere CO_2 lasers.

~25% is obtained with proustite. Since optical-quality proustite crystals can be grown with apertures in excess of 1 cm^2, the energy available in 5 μm pulses should grow in proportion to the 10 μm pulse energy, up to a limit of ~25 mJ. Extension of these results to higher harmonics has not yet been investigated.

Single picosecond pulses have also been used to investigate optical damage to infrared optical components. Although more detailed measurements must be performed, we find that optical damage occurs in NaCl at a damage threshold of ~500 mJ/cm^2 with a 2.5 ps pulse. Gold-coated mirrors, illuminated with s-polarized light at a large angle of incidence, have a damage threshold of 3 J/cm^2 (at 70^0).

References

1. Paul B. Corkum and Roderick S. Taylor: IEEE Journal of Quantum. Electron., 18, 1962 (1982).
2. A.J. Alcock and P.B. Corkum: Can. J. of Phys, 57, 1280 (1979).
3. S.A. Jamison and A.V. Nurmikko: Appl. Phys. Lett. 33, 598 (1978).
4. P.B. Corkum and D. Keith: Proceedings of the Conference on Infrared Physics, July 23-27, 1984 Zurich, Switzerland.

Stimulated VUV Radiation from HD Excited by a Picosecond ArF* Laser

T.S. Luk, H. Egger, W. Müller, H. Pummer, and C.K. Rhodes

Department of Physics, University of Illinois at Chicago, P.O. Box 4348
Chicago, IL 60680, USA

Stimulated emission on E,F → B, B → X, and C → X transitions of
HD is observed following two-photon excitation in the E,F state
with a picosecond ArF* laser. The efficiency of the strongest
B → X line is ∿ 1%. Electron collisions are found to play an
important role for C → X emissions.

1. Summary

Rare gas halogen lasers have been used extensively as a fundamental source for wavelength conversion by various nonlinear processes [1]. Multiphoton excitation of appropriate gain media has been
shown to be a promising method for the production of coherent short
wavelength radiation. Since these processes have a strongly nonlinear scaling, the efficiency of the wavelength conversion is
strongly influenced by the intensity. However, the presence of
photoionization of the excited state commonly limits the energy
fluence for efficient conversion of pump radiation to short wavelength radiation [2]. Therefore, short pulse excimer lasers are
ideal pump sources for multiphoton excited short wavelength lasers. In a previous experiment, a picosecond ArF* (193 nm) laser [3] has been used to two-photon excite the E,F state of molecular hydrogen [4]. The observation of subsequent stimulated
emission in the vacuum ultraviolet (VUV) has confirmed the importance of the aforementioned factors. In this paper, it is shown
that the same technique can be used to excite the E,F state of the
related heteronuclear molecule HD. Strong stimulated emission in
the VUV spectral range is observed following excitation of the
E,F (v = 6) J = 0,1 states of HD, with energy conversion efficiencies in one line approaching 1%.

The experimental apparatus consists of a 1-m vacuum monochromator (McPherson 225) with a PAR optical multichannel detector
and analyser (OMA) at the exit port. A 3-m tube was attached to
the entrance slit chamber. The tube has a fused silica entrance
window and a LiF exit window and was filled with HD at pressures
between a few Torr and 100 Torr. A 2 GW ArF* laser beam with a
10 ps pulse duration and 5 cm^{-1} bandwidth is focused with a 1.6
meter focal length lens to the middle of the tube and directed
into the monochromator through the entrance slit. Because of the
traveling wave character of the excitation, the amplified spontaneous VUV emission (ASE) is expected to be mainly in the forward
direction, and the use of a resonator does not relax the gain requirements. Therefore, no mirrors were used. For saturation of
the gain medium by ASE, a small signal gain-length product $g_o\ell$
of ∿ 20 is necessary.

Several attempts to produce stimulated emission from E,F levels excited via S transitions ($\Delta J = -2$) were unsuccessful. However, the tuning range of the ArF* laser also allows selective population of the E,F $v = 6$, $J = 0$ and $J = 1$ states via Q(0) and Q(1) two-photon transitions. In this case, strong, directional radiation was observed in the infrared and vacuum ultraviolet spectral ranges, originating from E,F $\rightarrow$ B, B $\rightarrow$ X and C $\rightarrow$ X transitions. For identification of observed emissions, Refs. 5-8 have been used. The shortest wavelength transition seen was at 84039 cm^{-1}, on the C $\rightarrow$ X (2 $\rightarrow$ 6) R(1) line. The most intense emission registered was at 73644 cm^{-1}, the B $\rightarrow$ X (0 $\rightarrow$ 5) P(3) transition. The excitation and emission pathway is similar to that which applies to H_2 [4]. Using Frank-Condon factors, which are available for E,F $\rightarrow$ B and B $\rightarrow$ X transitions [9], and Hönl-London factors [10], the relative strengths of different rovibrational transitions can be predicted qualitatively and are found to agree with observed line strengths. Naturally, a quantitative comparison can only be made if the exponential character of a laser process, the effects of saturation, and population depletion are taken into account.

As in previous measurements [4,11] in H_2, the optical Stark effect causes the optimum excitation frequency to be blue-shifted with respect to the unperturbed value, whereas no shift could be detected in the emission spectra, indicating that the stimulated emission occurs slightly later than the excitation. With the unperturbed transition frequencies for the Q(0) and Q(1) transition at 103219 cm^{-1} and 103171 cm^{-1}, respectively, the actual two-photon frequencies for optimum excitation are $\sim$ 103290 cm^{-1} and $\sim$ 103200 cm^{-1}, respectively.

Besides the E,F $\rightarrow$ B $\rightarrow$ X cascade emissions, radiation in the 120 nm range is observed, originating from C $\rightarrow$ X transitions. As reported previously [4,11] in H2, electron collisions are very efficient in transferring population from the E,F to the C state. Since the E($2s\sigma$) and C($2p\pi$) states resemble their atomic counterparts, the atomic mechanism discussed by PURCELL [12] has been used to estimate a collisional rate constant of $k_e \sim 10^{-6}$ cm^3 sec^{-1}, which depends only on electron temperature and energy difference between the E,F and C states. Since these two parameters and all the other experimental conditions do not change significantly for HD, the rate constant will be approximately the same as in the case of H_2, and electron collisions are considered to be the most probable transfer mechanism. As shown in Fig. 1, this collisional transfer process correctly predicts the C $\rightarrow$ X transitions observed. Other possible processes which have been considered are (a) stimulated emission in the far infrared, which is excluded because, in some cases, an energy increase is required, and (b) in the case of HD, direct two-photon excitation of the C state, which is allowed because of the broken symmetry of the heterogeneous molecule. However, it is expected that the two-photon transition rate is not large enough to result in strong stimulated emission because the C state wave function still has mainly a "u" character with only a rather small admixture of a "g" type wave function [7]. In addition, when krypton gas is mixed with HD in order to increase the electron density, improved performance is observed on the C $\rightarrow$ X transitions. This is a further indication that electron collisions play the dominant role in the E,F $\rightarrow$ C transfer.

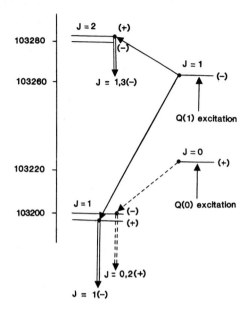

Fig. 1. The excited states participating in the electron collisional transfer from the E,F state to the C state. For Q(0) excitation one path 0(+) → 1(-) is observed, as shown by the dashed
arrow, while for Q(1) excitation two paths, 1(-) → 1(+) and 1(-)
→ 2(+), are seen, as shown by the solid arrows. All the electron
collisional amplitudes observed conform to dipole selection rules.

In conclusion, strong stimulated emission has been observed
in HD following two-photon excitation of the E,F state. In the
VUV spectral range, 28 lines were identified, which brings the
total number of hydrogen laser lines (H_2 and HD) in the VUV to
greater than 50. Peak efficiencies on individual lines approach
1%. The AC Stark effect has been observed in the excitation spectrum but not in emission, and electron collisions are found to
play a dominant role in the population of the C state. The high
power and short wavelengths available from the H_2 and HD sources
make them attractive for the performance of a wide range of studies
The H_2 source, for example, is currently being used in
photoemission studies of GaAs surfaces.

2. Acknowledgements

The authors wish to acknowledge the expert technical assistance of
M. J. Scaggs, J. R. Wright and D. M. Gustafson. This work was supported by the Air Force Office of Scientific Research under contract number F49630-83-K-0014, the Department of Energy under grant
number DeAS08-81DP40142, the Office of Naval Research, the National
Science Foundation under grant number PHY81-16626, the Defense
Advanced Research Projects Agency, and the Avionics Laboratory, Air

Force Wright Aeronautical Laboratories, Wright Patterson Air Force Base, Ohio.

References

1. T. Srinivasan, H. Egger, H. Pummer, C. K. Rhodes: Generation of extreme ultraviolet radiation at 79 nm by sum frequency mixing. J. Quantum Electron. QE-19, 1270 (1983)

2. T. Srinivasan, Generation of Coherent Vacuum Ultraviolet and Extreme Ultraviolet Radiation, Thesis (Univ. of Illinois at Chicago 1983) pp. 24-27

3. H. Egger, T. S. Luk, K. Boyer, D. F. Muller, H. Pummer, T. Srinivasan, C. K. Rhodes: Picosecond, tunable ArF* excimer laser source. Appl. Phys. Lett. 41, 1032 (1982)

4. H. Pummer, H. Egger, T. S. Luk, T. Srinivasan, C. K. Rhodes: Vacuum-ultraviolet stimulated emission from two-photon excited molecular hydrogen. Phys. Rev. A28, 795 (1983)

5. I. Dabrowski, G. Herzberg: The absorption and emission spectra of HD in the vacuum ultraviolet. Can. J. Phys. 54, 525 (1976)

6. L. Wolniewicz, K. Dressler: The EF and GK $^1\Sigma_g^+$ states of hydrogen: adiabatic calculation of vibronic states in H_2, HD, and D_2. J. Mol. Spectrosc. 67, 416 (1977)

7. J. D. Alemar-Rivera, A. Lewis Ford: Nonadiabatic effects in the B, C, and E,F states in HD. J. Mol. Spectrosc. 67, 336 (1977)

8. P. Quadrelli: Nichtadiabatische Kopplungen in Angeregten Zuständen des Wasserstoffmoleküls, Thesis (ETH Zurich, No. 7319, 1983)

9. Private Communication: Computer calculation by Louis DiMauro at AT&T Bell Laboratories using parameters from Constants of Diatomic Molecules by K. P. Huber and G. Herzberg

10. G. Herzberg: Spectra of Diatomic Molecules (Van Nostrand Reinhold, New York 1950)

11. T. Srinivasan, H. Egger, T. S. Luk, H. Pummer, C. K. Rhodes: The influence of the optical stark shift on the two-photon excitation of the molecular hydrogen E,F $^1\Sigma_g^+$ state. IEEE J. Quantum Electron. QE-19, 1874 (1983)

12. E. M. Purcell: Collisional and radiative properties of the H_2 E,F $^1\Sigma_g^+$ state. Astrophys. J. 116, 457 (1952)

Procedure for Calculating Optical Pulse Compression from Fiber-Grating Combinations

R.H. Stolen and C.V. Shank

AT & T Bell Laboratories, Crawford Corners Road, Holmdel, NJ 07733, USA

W.J. Tomlinson

Bell Communications Research, Holmdell, NJ 07733, USA

An optical pulse can be compressed by a dispersive grating pair after first inducing a frequency chirp in a single-mode optical fiber [1,2]. It is of interest to know how much compression is possible and what will be the necessary fiber length and grating separation, given some particular initial optical pulses. This is not immediately obvious, because the achievable pulse compression and the quality of the compressed pulse are sensitive functions of the input pulse parameters, and also depend critically on fiber length and grating separation.

In this paper we present results for fiber-grating compression in a normalized form, from which one can easily calculate the optimum fiber length, the achievable compression, and proper grating separation, for any given input pulse and fiber. These results come from a theoretical analysis, using numerical simulation, of self-phase modulation in a fiber with group-velocity dispersion (GVD) and the action of a grating compressor [3]. By extrapolation, the calculations can be extended to cover a wide range of input pulse parameters.

The frequency chirp is introduced by self-phase modulation in the fiber due to the intensity-dependent refractive index. An important feature of the process in a fiber is that GVD acts to linearize the chirp, which permits almost ideal compression by a grating pair [4]. For a particular input pulselength, peak power, wavelength, and fiber core area, this optimal chirp occurs, however, for only one fiber length. This optimal length varies as $\tau_o^2/\sqrt{P}$ where τ_o and P are the input pulsewidth and peak power. Input pulsewidths typically range between 100 fsec and 100 psec which, for a 600 nm wavelength, leads to optimal fiber lengths between about 1 cm and 10 km. Group-velocity dispersion decreases as the wavelength approaches the dispersion minimum near 1.3 μm, and optimal lengths can become even longer. It is clear that it will often be necessary to use a fiber length much less than the optimum.

We can identify two limiting regimes of practical interest. The first and most desirable is that of a fiber of the optimal length to provide the best linear chirp. The second regime is that for which the length is much less than optimal and the effects of GVD can be neglected. Some of the properties of the chirp and compression in these two limiting regimes are illustrated in Fig. 1 where the compression factor is about 12.5 in each case. We show the pulse shape exiting the fiber, the frequency spectrum, the chirp, and the compressed pulse for both optimum quadratic and ideal compressors. An ideal compressor adds all frequency components in phase; the quadratic compressor, which is a good approximation to an actual grating pair, adds all components in phase only if the chirp is linear. The pulse exiting the fiber has the same shape and intensity as the pulse entering the fiber if GVD is negligible, while GVD broadens the pulse by about a factor of three in a fiber of optimal length. The overall width of the frequency spectrum is about the same in the two cases, but GVD acts to fill in the spectrum. The spectrum flattens as the chirp becomes linear over most of the pulse. For each length, the grating separation was optimized to give the maximum peak intensity [3]. The optimum fiber length was chosen to maximize the energy in the compressed pulse [3]. It is interesting to note that when the fiber length and grating separation are optimized, the quality of the compressed pulse is better than that with an ideal compressor in the absence of GVD.

The procedure for calculating the compression, the optimal fiber length, and the grating separation is presented in Tables I and II. There are two important normalized parameters.

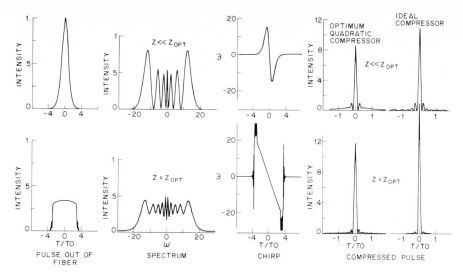

Fig. 1   Pulse shapes before and after compression, frequency spectra, and chirp for the limiting regimes of optimal fiber length and of negligible group-velocity dispersion. The lower curves are for $A=20$ and the corresponding optimum fiber length of $z/z_o = 0.075$. The upper curves are for the case of negligible GVD, and an intensity-length product $A^2 z/z_o = 12.5$, which was chosen to given approximately the same compression (with an optimum quadratic compressor).

TABLE I

$$z_o = \tau_o^2/C_1 \; ; \; A = \sqrt{P/P_1}$$

$$C_1 = \frac{D(\lambda)\lambda}{0.322\pi^2 c^2} = \begin{cases} 0.144 & (\lambda=514.5 \; nm) \\ 0.117 m^{-1}ps^2 (\lambda=600 \; nm) \\ 0.031 & (\lambda=1.06\mu m) \end{cases}$$

$$P_1 = \frac{nc\lambda A_{eff}}{16\pi z_o n_2} = 7.92 \left[\frac{\lambda(cm) A_{eff}(cm^2)}{z_o(cm)} \right] \times 10^{14} Watts$$

The first is the normalizing length z_o which depends on the input pulse length and the magnitude of the GVD. The second parameter is the normalized amplitude A. To determine A it is first necessary to find the normalizing power P_1 which depends on z_o, the nonlinear index n_2, and the fiber effective core area A_{eff}. Table 2 then gives approximate expressions for the compression factor τ_o/τ, the optimum length z_{opt} and the grating separation b in both limits of fiber length (for amplitudes $A \leqslant 3$, see Ref. 3). In the equation for grating separation we have chosen a groove spacing d of 5.56×10^{-5} cm (1800 lpm), and a wavelength of 600 nm as standard. The angle γ' is between the normal to the grating and the diffracted beam and is usually around $30°$. For example, for a 1 kW, 6 ps input pulse at 600 nm, and a fiber core area of 10^{-7}cm^2, the compressed pulse width is 117 fs and z_{opt} and b are 6 m and 28 cm respectively. (The normalizing length is 308 m, the normalized

TABLE II

$$z = z_{opt}$$

$$\tau_o/\tau \approx 0.63A$$

$$z_{opt}/z_o \approx 1.6/A$$

$$b = 84 \, C_2 \, \frac{cm}{ps^2} \left[\frac{\tau_o^2}{A} \right]$$

$$z \ll z_{opt}$$

$$\tau_o/\tau \approx 1+0.9[A^2z/z_o]$$

$$b = 13 \, C_2 \, \frac{cm}{ps^2} \left[\frac{\tau_o^2}{A^2z/z_o} \right]$$

$$C_2 = \left[\frac{d\,(cm)}{5.56\times10^{-5}} \right]^2 \cdot \left[\frac{600}{\lambda(nm)} \right]^3 \cos^2\gamma'$$

amplitude is 80.6, and P_1 is 154 mW.) It is interesting to note that provided one can use a fiber of optimum length, the relations in Tables I and II can be manipulated to show that the compressed pulse length is independent of the input pulse length.

Recent experimental measurements of 80X single-pass pulse compression agree remarkably well with the predictions of Table II [5]. An experimental verification of these relations is useful because Table II represents an extrapolation of our calculations. The experimental results also do not seem sensitive to the input pulse shape.

At present, there is little opportunity to vary the GVD of the fibers used in pulse compression experiments, but this situation may change as improvements are made in dispersion-modified fibers [6]. The relations in Tables I and II form a basis for dealing with the desirability of higher or lower dispersion. In general, for long pulses at around 1.0 μm increasing the dispersion would improve pulse quality at the expense of a slightly longer compressed pulse. For extremely short input pulses in the visible, a lower dispersion would lead to shorter compressed pulses for a given input power. There are other considerations involved with fiber-grating compression which are not discussed here in detail. It is necessary to insure good isolation between the mode-locked laser and the fiber to prevent degradation of the input pulse quality. Competition with stimulated Raman scattering is often present, although so far this has not been a major limitation. However, the effects of stimulated Raman scattering will become more important at longer wavelengths and larger initial pulse widths because the Raman threshold is governed by the walkoff distance between the pump and generated Stokes pulses [3]. The pulse quality becomes extremely sensitive to fluctuations in input power for large compression factors. There is also the possibility of optimization of various parameters in the limiting regime of negligible GVD. These questions are, at present, under active investigation and their resolution is expected to extend the general usefulness of pulse compressors using optical fibers.

[1]   C. V. Shank, R. L. Fork, R. Yen, R. H. Stolen, and W. J. Tomlinson, Appl. Phys. Lett. **40**, 761 (1982).

[2]   B. Nikolaus and D. Grischkowsky, Appl. Phys. Lett. **42**, 1 (1983); **43**, 228 (1983).

[3]   W. J. Tomlinson, R. H. Stolen, and C. V. Shank, J. Opt. Soc. Am. B *1*, 139 (1984).

[4]   D. Grischkowsky and A. C. Balant, Appl. Phys. Lett. **41**, 1 (1982).

[5]   A. M. Johnson, R. H. Stolen, and W. M. Simpson, Appl. Phys. Lett. **44**, 729 (1984).

[6]   L. G. Cohen, W. L. Mammel, and S. J. Jang, Electron. Lett. **18**, 1023 (1982).

Generation of Infrared Picosecond Pulses Between 1.2 μm and 1.8 μm Using a Traveling Wave Dye Laser

H.-J. Polland, T. Elsaesser, A. Seilmeier, and W. Kaiser
Physik Department der Technischen Universität München
D-8000 München, Fed. Rep. of Germany

Recently, the synthesis of infrared dyes of high photochemical stability became possible /1/. One of these dyes was used to operate a synchronously pumped cw dye laser system tunable around 1.3 μm /2/. Here we demonstrate a new method for effective pumping of dyes with small fluorescence quantum yield (10^{-4}). Laser emission up to 1.8 μm is now accessible by a traveling wave pumping system.

Fig. 1 shows the experimental system consisting of a mode-locked Nd:glass pump laser (pulse duration 4,5 ps, energy 1 mJ), a diffraction grating GR1, and a transversely pumped dye cell G. The grating GR1 introduces a time delay across the horizontal diameter of the laser beam which is adjusted to the traveling time of the stimulated emission along the dye cell. In the generator cell G a picosecond laser emission with a spectral width of several 100 wavenumbers is produced. Tunable pulses with narrow bandwidth are generated extending the system by a frequency selector, which works like a grating spectrometer (grating GR2, lenses L4,L5, and slits S1,S2), and by an amplifier cell A. Tuning is achieved by rotating grating GR2.

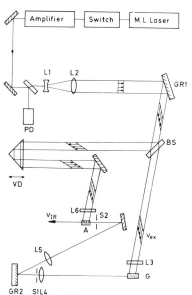

Fig. 1
Experimental system for the generation of narrow-band tunable dye laser pulses

49

First we discuss data obtained by our traveling wave system without frequency selector and amplifier /3/. In our investigations we used different new IR dyes with absorption maxima between 1.1 μm and 1.4 μm. As an example, the result for dye S501 is presented in Fig. 2. We find the absorption maximum at 1.42 μm. The amplified spontaneous emission of the dye is shown on the r.h.s.; it peaks at 1.8 μm. The bandwidth of the emission is approximately 320 cm^{-1}. For infrared dyes with a fluorescence lifetime of several picoseconds we measured energy conversion efficiencies of $\approx$ 2 % and pulse durations close to that of the pump pulse (4 ps).

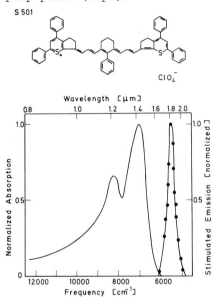

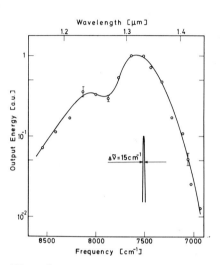

Fig. 2
Molecular structure, absorption and stimulated emission spectrum of dye S501 (without frequency selector)

Fig. 3
Tuning range and spectral width of IR pulses employing dye No. 5 (with frequency selector and amplifier)

To demonstrate the performance of the complete system including the wavelength selector (Fig. 1) we present results on dye No. 5 /4/, which has a fluorescence lifetime of 2.7 ps. In Fig. 3 the output energy is plotted as a function of the emission frequency. Picosecond pulses tunable from 7000 cm^{-1} and 8500 cm^{-1} are generated with a spectral width of 15 cm^{-1}. The parameters of the dye laser emission are: Energy conversion 10 % (at 1.32 μm), pulse duration 4.5 ps, divergence 3 mrad. Operating the system for many days with the same dye solution, no deterioration of emission efficiency was observed.

Conclusions: The traveling wave arrangement proves to be a very reliable system for effective pumping of dyes of low quantum efficiency. Its easy alignment makes the system a practical device. The high intensity of the dye laser emission allows additional frequency conversion via nonlinear optical processes.

Acknowledgement:
The authors acknowledge valuable contributions by Professor
K.H. Drexhage.

References

1   B. Kopainsky, P. Qiu, W. Kaiser, B. Sens, and K.H. Drexhage,
    Appl. Phys. B29 (1982) 15
2   A. Seilmeier, W. Kaiser, B. Sens, and K.H. Drexhage, Optics
    Lett. 8 (1983) 205
3   H.-J. Polland, T. Elsaesser, A. Seilmeier, W. Kaiser, M. Kussler,
    N.J. Marx, B. Sens, and K.H. Drexhage, Appl. Phys. B32 (1983) 53
4   T. Elsaesser, H.-J. Polland, A. Seilmeier, and W. Kaiser, IEEE
    J. Quant. Electron. QE-20 (1984) 191

A New Picosecond Source in the Vibrational Infrared

A.L. Harris, M. Berg, J.K. Brown, and C.B. Harris

Department of Chemistry, University of California at Berkeley, and
Materials and Molecular Research Division of Lawrence Berkeley Laboratory
Berkeley, CA 94720, USA

The successful generation of picosecond pulses in the vibrational
infrared by stimulated electronic Raman scattering (SERS) of visible
pulses from an amplified synchronously pumped dye laser operating with
rhodamine dyes is reported[1]. The simplicity and efficiency of SERS as a
frequency shifting technique have been amply demonstrated with nanosecond
pulses[2-4]. SERS has the advantages of using a non-damageable medium and
having no phase-matching requirements. Cotter and Wyatt have shown that
picosecond SERS from the cesium 6s-7s transition (see Fig. 1a) can
generate infrared even more efficiently than the corresponding nanosecond
process[5]. However, the practical utility of their system was limited by
the difficulty in generating high power picosecond pump pulses in the
blue-green. Because of this difficulty, a SERS technique based on a
reliable synchronously pumped dye laser operating with rhodamine dyes is
desirable.

The 6s-5d transition in atomic cesium (see Fig. 1a) was chosen for
this work because its 14597 cm^{-1} transition frequency shifts red pulses

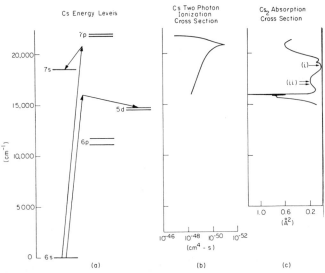

Fig. 1. Cs atom and Cs$_2$ dimer spectroscopy important for picosecond
infrared generation. a) Partial energy level diagram of the Cs atom
showing the 6s-7s Raman transition used in previous work[5], and the 6s-5d
Raman transition discussed here. b) Theoretical two photon ionization
cross-section of the Cs atom as a function of visible frequency[8]. c)
Absorption cross-section of the Cs$_2$ dimer in the visible.

into the vibrational infrared, and because it is predicted to have an exceptionally wide tuning range[6]. The major impediment to using the 6s-5d transition has been the competing absorption of cesium dimers in the red region of the spectrum[6] as shown in Fig. 1c. This difficulty was overcome in two ways: (i) by maximizing the pump pulse energy and (ii) by using a specially constructed superheated heat pipe to minimize the cesium dimer concentration.

The picosecond amplifier consists of three amplifier stages longitudinally pumped with a total of 250 mJ from a frequency-doubled Nd:YAG laser. Isolation between stages is provided by jets of Crystal Violet in ethylene glycol. This amplifier can produce pulses of 1-2 mJ with a stability of ±10% (standard deviation) over most of the tuning range of a synchronously pumped dye laser using R6G (see Fig. 2). The amplified pulse has an autocorrelation FWHM of 1.5 ps (1.0 ps sech^2 pulse). The average spectrum is 18 cm^{-1} wide, but part of the width is due to pulse-to-pulse shifts in the wavelength. The average single pulse bandwidth is 11 cm^{-1}, giving a bandwidth product of 0.33 compared to 0.32 for a bandwidth limited sech^2 pulse shape. A small amount of phase modulation in the third amplifier is indicated by a low intensity shoulder on the red side of the pulse's spectrum.

A one-meter column of cesium vapor was generated in a split-wick heat pipe[6] constructed of Inconel 601. The cesium dimer concentration was reduced by thermal dissociation in a cesium column superheated to 1200°C. Under these conditions the majority of the remaining cesium dimers are in the equilibrium regions at the ends of the vapor column. With the current design, 20% of the visible pump light is transmitted through a 50 torr column of cesium.

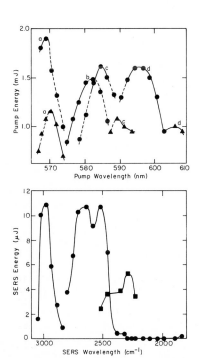

Fig. 2. Top: Visible pump-pulse energy versus wavelength: a, R590; b, R610; c, KR620 ; d, R640; ● ,methanol solutions; ▲ ,water solutions (no surfactants). Points used for best SERS generation are outlined with the solid portions of the curves. Bottom: ● , SERS energy generated with short pulses; ■ ,SERS energy generated with broad pulses.

Tunable infrared pulses are generated by simply focussing the visible beam into the superheated cesium vapor. Fixed frequency emissions from excited cesium atoms can generally be blocked by filters. A peak quantum efficiency of 4.6%, giving 11 μJ of infrared, was obtained[1] (see Fig. 2). The bandwidth of the infrared pulse varied from 6-10 cm^{-1}. An infrared pulse length of 1-2 ps is implied if a Gaussian bandwidth-limited pulse is assumed. The theory of stimulated Raman scattering and previous experimental results indicate that the generated infrared-pulse length should be similar to or shorter than the input-pulse length, so that this is a reasonable value.

The conversion efficiency is relatively constant above 2350 cm^{-1} (see Fig. 2). The dip near 2850 cm^{-1} is only due to the lack of a good amplifier dye. Below 2350 cm^{-1}, however, the efficiency drops suddenly. This is unexpected based on the very wide tuning range (3500-900 cm^{-1}) recently demonstrated for nanosecond pulses on the cesium 6s-5d transition[2].

The cause of the efficiency drop is revealed by the dependence of the infrared energy on pulse width. For frequencies above 2350 cm^{-1} the conversion is insensitive to the pulse width. By contrast, below 2350 cm^{-1} the infrared energy increases dramatically from <0.5 μJ to 4 μJ as the input pulse to the amplifier is broadened from 2 to 20 ps with the energy of the visible pulse kept constant. Since SERS in the transient regime[7] should not be affected by changes in pulse width at constant energy[7], we conclude that competing non-linear losses cause the drop in infrared energy below 2350 cm^{-1}.

The likely sources of non-linear loss are two-photon ionization of cesium atoms or cesium dimers by the visible pulse. Although experimental cross-sections are not available at these wavelengths, the calculated cross-section[8] for two-photon ionization of cesium atoms is of the correct magnitude to cause significant losses under our conditions. Furthermore the cross-section is calculated to increase with decreasing wavelength, as shown in Fig. 1.

Fig. 2 shows a tuning curve for one dye with broad (~10 ps) pulses which illustrates the increased tuning range available. It may be possible to tune these broader pulses to even longer wavelengths using other amplifier dyes.

It should be noted that a small amount of SERS was observed at the far end of the tuning range (1950 cm^{-1}) with short pulses. The increase in infrared generation may result from approaching a minimum in the dimer absorption at 615 nm. This suggests that even short pulses may be tuned much further if the dimer losses can be further reduced.

In conclusion, a source of tunable infrared picosecond pulses based on a reliable rhodamine dye laser has been demonstrated. Further reductions in cesium dimer concentration should increase the tuning range by reducing both linear and non-linear losses. Only when the extent of atomic non-linear losses is determined will the full capabilities of the technique be known.

This research was supported by the National Science Foundation.

References

1. M. Berg, A. L. Harris, J. K. Brown, and C. B. Harris, Optics Lett. 9, 50 (1984).
2. A. L. Harris, J. K. Brown, M. Berg, and C. B. Harris, Optics Lett. 9, 47 (1984).
3. D. S. Bethune, A. J. Schell-Sorokin, J. R. Lankard, M. M. T. Loy, and P. P. Sorokin, in Advances in Laser Spectroscopy, B. A. Garetz and J. R. Lombardi, eds. (Wiley, New York, 1983) Vol. II, p. 1.
4. D. Cotter, D. C. Hanna, and R. Wyatt, Opt. Commun. 16, 256 (1976).
5. R. Wyatt and D. Cotter, Opt. Commun. 37, 421 (1981).
6. R. Wyatt and D. Cotter, Appl. Phys. 21, 199 (1980).
7. A. Laubereau and W. Kaiser, Rev. Mod. Phys. 50, 607 (1978).
8. H. Bebb, Phys. Rev. 149, 25 (1966).

Generation of Intense, Tunable Ultrashort Pulses in the Ultraviolet Using a Single Excimer Pump Laser

S.Szatmári, F.P.Schäfer

Max-Planck-Institut für Biophysikalische Chemie, Abteilung Laserphysik
D-3400 Göttingen, Fed. Rep. of Germany

The cost and complexity of experimental arrangements for the generation of ultrashort, high-power, tunable excimer laser pulses is considerable (1,2). In this paper we describe a simple method and experimental arrangement for the generation of ultrashort excimer laser pulses of GW peak power with <5 % amplified spontaneous emission (ASE) pedestal, using only a single commercial excimer laser for the generation of a single ultrashort dye laser pulse at twice the excimer laser wavelength and amplification of the frequency-doubled dye laser pulse.

Figure 1 shows the experimental arrangement. A slightly modified commercial excimer laser (Model EMG 150, Lambda Physik, Göttingen) is used to pump the various dye cells with the output from its oscillator stage, while the amplifier stage of this laser (with the mirrors for regenerative amplification removed) serves to amplify the frequency-doubled output of the dye laser amplifier chain. The oscillator stage is usually filled with the standard XeCl-gas mix for operation at 308 nm, whereas the amplifier stage is filled with the excimer laser gas mix deliv-

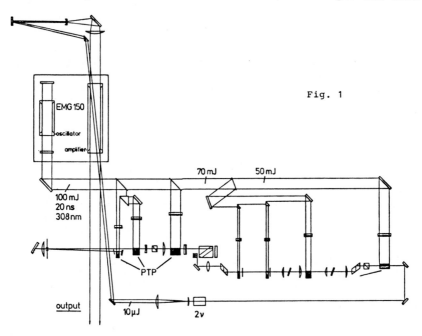

Fig. 1

ering the wanted wavelength. Up to now we have worked with XeCl at 308 nm and KrF at 248.5 nm.

The available pump energy from the oscillator is distributed among the various dye cells to be pumped by beamsplitters, mirrors, and prisms, as indicated in Fig.1. The first few mJ of pump energy are focussed into a 5 mm wide dye cell containing a $3 \cdot 10^{-3}$ M solution of p-terphenyl (PTP) in cyclohexane. This dye cell is surrounded by two slightly tilted mirrors, resulting in a variant of the quenched dye laser described recently (3). This laser yields a single pulse of <300 ps duration when pumped by a 20 ns long pump pulse. The spectral bandwidth is several nanometers centered around 340 nm. Since this bandwidth seems too large for pumping a distributed-feedback dye laser (DFDL), it is reduced by the slit-lens-grating combination shown in the lower left of Fig.1. The spectrally narrowed beam is then amplified in two cells with PTP solution. The output is used to pump a DFDL of the type described earlier (4) consisting of a cylindrical lens, a quartzglass block, a holographic grating, and a dye cell. This version is of outstanding simplicity and particularly well suited for fixed-wavelength operation. Fine tuning of the laser wavelength is possible by a suitable choice of the refractive index of the dye solution (mixing ratio of two or more solvents) and/or temperature control, the latter giving a tuning range of several Ångstroms. If pumped high above threshold a DFDL will emit several pulses. As recently observed (5) these pulses show a directional sweep of several 10 mrad from one pulse to the next so that the first pulse, which always exhibits the shortest pulse duration, can be singled out by a suitably placed aperture. The cross-section of the pumped region of the first amplifier stage will already serve as such an aperture.

The amplifer chain consists of four stages with two saturable absorbers as shown in Fig.1 and is of more or less standard design (6). The only important change is the method that was applied here for the suppression of residual satellite pulses, as described in detail elsewhere (7). It consists of putting a glassplate behind the amplifier cell in such distance that the partially reflected first pulse will deplete the inversion in the dye cell and thus eliminate the gain for the eventual following pulses. This device is applied in the first two amplifier stages, which is entirely sufficient for suppression of all eventual trailing pulses.

The output pulse is frequency-doubled in a suitable crystal and the beam magnified by a telescope to about 12 mm diameter. This beam with a pulse energy of about 10 μJ is then passed through the amplifier section of the excimer laser in a direction just avoiding the electrodes as shown in Fig.1. The amplified beam is then magnified to fill the discharge cross-section of the amplifier in a second pass. Between the first and the second pass a delay path is adjusted to give maximum amplification in the second pass.

For XeCl the output is a pulse of 5 ps duration (as measured by a streak camera) with a pulse energy of 10 mJ. The ASE background, which is several ns long, has less than 5 % of the pulse energy. For KrF the output obtained is 30 ± 10 mJ in a pulse of <10 ps duration and with <6 % ASE energy.

Conclusions

We have described a method of generating in a very simple way high-power, tunable picosecond pulses in the ultraviolet, using a single commercial excimer laser oscillator-amplifier combination. The XeCl-excimer oscillator pumps a quenched dye laser to give a short pulse of 300 ps FWHM at 340 nm, which in turn pumps a DFDL giving a <10 ps pulse at 616 nm. After amplification, frequency-doubling and a double-pass amplification in the XeCl-amplifier the output pulse has an energy of 10 mJ, a FWHM of <5 ps, and an ASE content of <5 % of the pulse energy, in the KrF amplifier the output is 30 mJ in <10 ps and <6 % ASE energy.

While in the present set-up most of the dye cells were simple, static, unstirred spectrophotometer cells, thus limiting the pulse repetition rate to at most 2 Hz, replacement of the static dye cells by flow cells should essentially result in a repetition rate limited only by the excimer laser, i.e. in our set-up 25 Hz. An optimized version of the quenched dye laser should reduce its pulse width and concomitantly that of the DFDL. A much greater reduction in pulse width should be possible by pulse compression of the dye laser fundamental using the method recently described by Nikolaus and Grischkowsky (8). This could lead to a pulse width of the 308 nm pulse of 150 fs, which could be amplified in the excimer laser amplifier with almost the same efficiency as pulses of a few ps duration (1).

The extension of the method described here for a XeCl- and KrF-excimer laser to other laser wavelengths is straightforward, necessitating only a change of dye solutions, frequency-doubling crystal, and grating in the DFDL. For wavelengths shorter than the transmission limit of presently available frequency-doubling crystals (<190 nm (9)), frequency-shifting of the dye laser fundamental to obtain input pulses for the excimer laser amplifier can be done by using the stimulated anti-Stokes Raman emission in a gas cell filled with hydrogen or other suitable gas (10).

We would like to emphasize as an important aspect of the ready availability of high-energy, picosecond UV pulses that excite-and-probe experiments should now be much easier to perform and also in a much extended wavelength region of 308.5 nm to 1.85 μm by using part of the UV-pulse energy to pump one or several travelling-wave dye lasers (11,12) as probe lasers at the wavelengths of interest.

References

1. P.B. Corkum, R.S. Taylor: IEEE J. Quant. Electron. QE-18, 1962 (1982).
2. H. Egger, T.S. Luk, K. Boyer, D.F. Muller, H. Pummer, T. Srinivasan, C.K. Rhodes: Appl. Phys. Lett. 41, 1032 (1982).
3. F.P. Schäfer, Lee Wenchong, S. Szatmári: Appl. Phys. B 32, 123 (1983).
4. Zs. Bor: Optics Comm. 39, 383 (1981).
5. S. Szatmári, Zs. Bor: Appl. Phys. B 34, 29 (1984).
6. S. Szatmári, F.P. Schäfer: Appl. Phys. B 33, 95 (1984).

7. S. Szatmári, F.P. Schäfer: Opt. Quant. Electron. $\underline{16}$, 277 (1984).
8. B. Nikolaus, D. Grischkowsky: Appl. Phys. Lett. $\underline{43}$, 228 (1983).
9. Chen Chuangtian, Wu Bochang, Jiang Aidong, You Guiming: contribution to the Intern. Conf. on Lasers, 1983, Guangzhou, Digest p. 358.
10. N. Morita, K.H. Lin, T. Yajima: Appl. Phys. B $\underline{31}$, 63 (1983).
11. Zs. Bor, S. Szatmári, A. Müller: Appl. Phys. B $\underline{32}$, 101 (1983).
12. H.J. Polland, T. Elsaesser, A. Seilmeier, W. Kaiser, M. Kussler, N.J. Marx, B. Sens, K.H. Drexhage: Appl. Phys. B $\underline{32}$, 53 (1983).

Travelling-Wave Pumped Ultrashort Pulse Distributed Feedback Dye Laser

G.Szabó*, B.Rácz*, Zs.Bor*, B.Nikolaus, and A.Müller

Max-Planck-Institut für Biophysikalische Chemie, Abteilung Laserphysik
D-3400 Göttingen, Fed. Rep. of Germany

Distributed feedback dye lasers (DFDLs) are convenient sources of wavelength tunable transform limited picosecond laser pulses [1]. Single pulses are obtained by proper choice of the pumping power [2,3]. As we have shown earlier, the duration of the pulses from a standing wave DFDL can be as short as 2 ps when pumped by 16 ps long pulses [4]. The shortest pulse in this case is limited by the transit time of light through the DFDL. It should, however, be possible to overcome this limitation by travelling wave pumping. We have recently introduced this concept experimentally and demonstrated that considerable shortening of a generated ASE pulse with respect to the pump pulse can be achieved in this way [5].

The experimental arrangement of the travelling wave pumped DFDL is shown in Fig.1. A pump pulse of about 5 ps duration

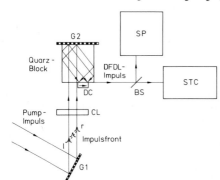

with an energy of 5 mJ at 308 nm was generated by the excimer-dye-laser-oscillator-amplifier combination reported recently by SZATMARI and SCHÄFER [6]. A holographic grating (G1) in first order of diffraction is used to create a continuous spatial delay across the diffracted beam as discussed in [5]. On the surface of the dye cell a diffraction pattern is formed such that the spatial position of

Fig.1 Experimental arrangement of the travelling wave pumped DFDL.
G1 = holographic diffraction grating for producing the delayed pulse front, G2 = holographic diffraction grating (2442 mm⁻¹) serving as beamsplitter in the DFDL, CL = cylindrical lens, DC = dye cell, BS = beam splitter, SP = spectrographic recording system, STC = streak camera system

*Permanent address: JATE University, Dept. of Experim. Physics,
            Dóm tér 9, H-6720 Szeged, Hungary

fringes is unchanged but their envelope moves from left to right (Fig.1) with a speed

$$v = c/\tan\gamma \tag{1}$$

where c is the speed of light in vacuum and γ is the angle between the pulsefront and the normal to the wavevector. Travelling wave excitation is achieved when

$$\tan\gamma = n \tag{2}$$

where n is the refractive index of the dye solution. The effect of mismatch between the sweep velocity of the excitation and the DFDL pulse was studied by varying the refractive index of the dye solution. The latter was a $5 \cdot 10^{-3}$ mol/l solution of rhodamine 6G in a mixture of DMSO and methanol. Due to the change of refractive index from 1.4059 to 1.4809 the DFDL was tuned from 576 to 606 nm [1]. This tuning, however, should exert only a negligible effect on the pulse formation mechanism, since it occurs in the range of the maximum of the gain curve. The temperature of the laser medium was constant within 1 K during the experiments. The DFDL output was monitored simultaneously by a monochromator (Jarrell-Ash, 1200 lines/mm, f = 500 mm) equipped with an optical multichannel analyzer (B & M Spectronic OSA 500) and a high resolution streak camera system (Hamamatsu C1370-01) yielding, resp., spectral resolution of 1.2 A and temporal resolution of 1.9 ps.

Figure 2 shows the streak camera traces of travelling wave DFDL pulses. The pump energy was kept constant during the whole experiment and corresponded to twenty times the threshold pump energy of the DFDL. The shortest pulse (Fig.2c) which was shorter than the resolution of the streak camera was obtained when condition (2) was satisfied. Small deviations from this condition resulted in broadening of the pulses to 5 - 6 ps. It is interesting to note that for $\tan\gamma > n$ multiple pulses were generated (Fig.2f), while for $\tan\gamma < n$ this was not observed. For comparison we studied a standing wave DFDL using the same pump

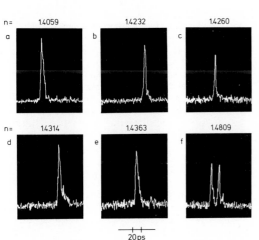

Fig.2 Variation of the duration of the travelling wave pumped DFDL output pulses as measured with a streak camera.
Time base 0.7 ps/channel. $5 \cdot 10^{-3}$ M rhodamine 6G in a mixture of DMSO and methanol. The refractive index of the solution indicated above the traces was adjusted by varying the mixing ratio

pulse. In this case multiple pulsing began at a pump energy corresponding to only four times the DFDL threshold. Thus, travelling wave excitation has the additional desirable feature of favoring generation of a single DFDL pulse.

The time resolution of the streak camera system is 1.9 ps, corresponding to 5.4 channels of the optical multichannel analyzer at the highest streak speed. With properly matched refractive index of the dye solution pulses of 6 channels width were recorded, indicating that the pulse duration is definitely shorter than the time resolution of the streak camera, perhaps 1 ps or even less.

While travelling wave excited ASE exhibits a broad, structured spectrum [5], that of the travelling wave excited DFDL consists of a narrow line with a halfwidth of about 4 A. In the absence of autocorrelation measurements, which are presently in preparation, the time-bandwidth product provides additional information supporting our conclusion reached above, if one assumes that the pulses of the travelling wave excited DFDL are transform-limited, as is the case for the standing wave DFDL [4].

In conclusion we have demonstrated that travelling wave excited distributed feedback dye lasers favour generation of single ultrashort pulses with a duration of 1 ps or less.

Acknowledgements

This work has been supported by a joint project of the "Deutsche Forschungsgemeinschaft" and the Hungarian Academy of Sciences. We thank Prof. F.P. Schäfer for his interest and we are greatly indebted to Dr. K. Hohla and Mr. R. Schwarzwald of Lambda Physik Göttingen for making the streak camera system available to us.

References

1. Zs. Bor, B. Rácz, G. Szabó, A. Müller, H.-P. Dorn: Helv. Phys. Acta 56, 383 (1983)
2. Zs. Bor: IEEE J. Quant. Electron. QE-16, 517 (1980)
3. Zs. Bor, A. Müller, B. Rácz, F.P. Schäfer:
   a) Appl. Phys. B 27, 9 (1982)
   b) Appl. Phys. B 27, 77 (1982)
4. G. Szabó, Zs. Bor, A. Müller: Appl. Phys. B 31, 1 (1983)
5. Zs. Bor, S. Szatmári, A. Müller: Appl. Phys. B 32, 101 (1983)
6. S. Szatmári, F.P. Schäfer: Opt. Commun. 48, 279 (1983)

Picosecond Pulses from Future Synchrotron-Radiation Sources*

R.C. Sah and D.T. Attwood

Lawrence Berkeley Laboratory, University of California, Berkeley, CA 94720, USA

A.P. Sabersky

P.O. Box 2483, Santa Cruz, CA 95063, USA

1 Summary

Intense, quasi-coherent photon pulses with pulse lengths on the order of
10 picoseconds will be available from future high-brilliance synchrotron-
radiation sources. Photon energies will span the range from the VUV to
soft x-rays.

2 Synchrotron-Radiation Sources

Existing synchrotron facilities, such as those at the Stanford Synchrotron
Radiation Laboratory (SSRL) and at the National Synchrotron Light Source
(NSLS), provide intense and tunable sources of electromagnetic radiation.
In these facilities, short bunches of electrons circulate in electron
storage rings and emit synchrotron radiation when they pass through the
bending magnets of the storage rings. In a few instances, special magnetic insertion devices called "wigglers" and "undulators" [1] are used to
provide even more intense photon beams. Wigglers and undulators are characterized by periodic spatial alternations in magnetic-field direction,
and they produce no net deflections of the electron beam. Photon energies
over a wide range, up to the x-ray region, are available from these
machines. Typical photon pulse lengths are 100 to 400 picoseconds, and
repetition rates can be as high as 10 MHz or more.

A new generation of high-brightness synchrotron-radiation facilities
is now in the planning stages. In these new designs, most photon beams
will originate from wiggler and undulator magnets rather than from
storage-ring bending magnets. As an example of one of these new storage-
ring designs, we shall consider the Advanced Light Source (ALS), which has
been proposed by Lawrence Berkeley Laboratory [2]. The ALS has been designed to provide pulse lengths of 20-90 picoseconds at a repetition rate
of up to 500 MHz. Extremely intense photon beams from the VUV to soft
x-rays will be available from wiggler and undulator magnets. Interference
effects in the undulators will produce quasi-monochromatic radiation (in
the sense that the emitted radiation will be largely confined to a few
sharp spectral lines), and undulators will provide extraordinarily high
spectral brilliance, up to 10^{18} photons/(sec)(mm^2)(mr^2)(0.1 percent
bandwidth). Because high-brilliance photon beams originate from a small
area and are emitted into a small solid angle, they have a substantial
coherent power, as shown in Fig. 1. Figure 2 shows the spatial pattern
of 500-eV radiation from a 5.7 kG (K = 1.87) ALS undulator, as seen on a
screen located 10 meters from the source. The high-brilliance central
spot provides 10^{16} photons per second in a 1-percent bandwidth.

*This work was supported by the Office of Energy Research, Office of Basic
Energy Sciences, Department of Energy under Contract No. DE-AC03-76SF00098.

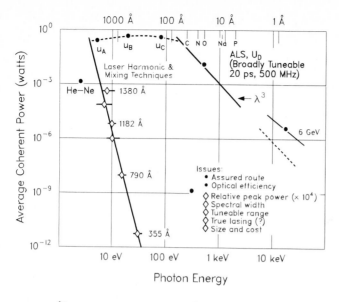

Fig. 1. Broadly tunable coherent power* will be available in an interesting spectral region

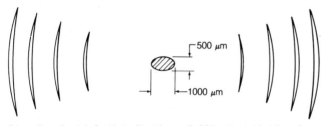

Fig. 2. Spatial distribution of 500 eV radiation from an ALS undulator

Special measures may produce photon pulses as short as 3 picoseconds at the ALS. First, the storage-ring magnets can be re-tuned for a low "momentum-compaction factor," in order to reduce the natural bunch length. Unfortunately, this re-tuning would have adverse consequences on the electron beam: larger emittance, non-zero momentum dispersion at the insertion devices, and reduced quantum lifetime. Furthermore, the ALS must be operated at greatly reduced beam currents to avoid the bunch-lengthening effects of the microwave instability. A second approach would be to use a high-frequency, high-power RF system to provide high-current, ultra-short-bunch operation. However, the higher impedance of this RF system may cause electron-beam instabilities.

3 Short-Pulse Diagnostic Instrumentation

Electron bunches in high-current storage rings are subject to many temporal instabilities. Pulse jitter (an instability in which an entire electron bunch oscillates as a whole) is best measured by spectral analysis of the signal from a beam pick-up electrode in the storage-ring vacuum chamber. Pulse-length and pulse-shape measurements are best studied using a streak camera [3], which can provide 3-picosecond time resolution and 10-picosecond stability. Visible-light intensities from synchrotron-radiation facilities are adequate for single-pulse measurements. The streak camera can be triggered with signals from beam pick-up electrodes. To make full use of the short pulse structure of synchrotron radiation, one must work below the instability threshold in beam current or use measurement methods [4] which are insensitive to instabilities.

4 New Applications

The short pulse lengths, high brilliance, and quasi-coherent nature of photon beams from modern x-ray synchrotron sources will permit new applications across the scientific spectrum. These include time-resolved absorption and fluorescence studies of chemical and material properties with small samples and improved spectroscopic resolution, studies of highly ionized plasmas whose lifetimes are measured in tens of picoseconds, as well as picosecond imaging and diffraction studies. Unique applications will certainly appear for coherent VUV and soft x-ray probing on these time scales, including off-axis x-ray holography of microscopic samples. Depending on system and pulse shape stability it may be possible to extend some of the fluorimetry techniques below one picosecond using frequency domain techniques.

5 References

1. H. Winick et al., "Wiggler and Undulator Magnets," Physics Today, May 1981, pp. 50-63.
2. R.C. Sah, "The Advanced Light Source," Proc. of the 1983 Particle Accelerator Conference, IEEE Trans., NS-30, No. 4, pp. 3100-3102.
3. A.P. Sabersky and M.H.R. Donald, "Modes on a Short SPEAR Bunch as Observed on a Streak Camera," proc. of the 1981 Particle Accelerator Conference, IEEE Trans., NS-28, No. 3, pp. 2449-2451.
4. E. Gratton, D.M. Jameson, R.D. Hall, "Multifrequency Phase and Modulation Fluorimetry," Univ. of Illinois (Urbana-Champaign), Dept. of Physics, P-83-8-108 (1983).

High Repetition Rate Production of Picosecond Pulses at Wavelength < 250 nm[1]

D.B. McDonald

Chemistry Division, Argonne National Laboratory, Argonne, IL 60439, USA

1 Introduction

Two of the major challenges to ultrafast laser spectroscopists are the generation of the needed wavelengths and the production of a sufficient transient population for observation. As laser pulses become shorter, this second requirement becomes evermore difficult to meet; to create the same excited population in picoseconds rather than nanoseconds may require 1000 times the peak intensity and a corresponding increase in unwanted nonlinear processes in the sample. In this paper we describe a laser system for generating picosecond pulses with wavelength <250nm at a 5kHz repetition rate. Short pulses from a synchronously pumped dye laser are amplified to a peak power >5MW in a dye medium pumped by a copper vapor laser. The high peak powers allow nonlinear optical techniques to shift the wavelength of the pulses well into the UV as in many of the high-power low repetition-rate laser systems [1-5]. The high repetition rate of the system described here, however, allows for sophisticated signal averaging needed in detecting extremely small populations of transients using pump-probe picosecond spectroscopy.

2 The Laser System

A schematic is given in Fig. 1. Low intensity picosecond pulses are created by a conventional mode-locked argon-ion laser synchronously pumping a cavity-dumped dye laser (Coherent Radiation) [6,7]. Pulses are typically 20-30ps in duration with 10-50mJ of energy per pulse. The copper vapor laser (Plasma Kinetics Model 451) produces 30ns pulses with about 8mJ per pulse divided between 510nm and 578nm with a ratio of about 2:1. Several laser dyes absorb both of these wavelengths nearly equally.

Since a typical lifetime for a laser dye is less than 5ns, most of the long CVL light pulse would be wasted in a single pass amplifying stage. The system described here makes use of a multiple pass amplifier to take full advantage of the pulse energy of the CVL. Multiple passes through an amplifying medium necessitate a uniform gain region of a suitable geometry. For this reason, a prism amplifier cell similar to that described by Bethune [8] is used. The 12.5mm fused silica prism has a 2mm hole in such a position that light incident perpendicular to the hypotenuse converges on the cylinder of dye solution from four sides because of the total internal reflection. The dye concentration can be chosen so that the gain region is almost uniformly illuminated. The CVL beam, initially about 6cm in diameter, is focused and recollimated to about 1cm. Flowing the dye solution of ethanol and water (1:1) at about one liter per minute is sufficient to prevent appreciable thermal disruption.

---

[1]Work performed under US-DOE.

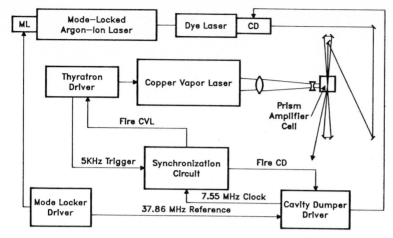

Fig. 1 Schematic of the laser system. Only three of the five passes of amplification are shown.

An optical delay has been inserted between the cavity dumper and the amplifier cell. Without this delay spontaneous emission from the gain medium could reflect back from the cavity dumper and be amplified, and in the five pass configuration several hundred milliwatts of laser output were observed from such a one mirror laser. The roughly 25ns of round trip delay causes the reflected light to be too late for appreciable amplification. The 3ns of delay between each pass through the amplifier cell allows the gain medium to recover and provides space for mirrors (2 between each pass). On every second pass the beam encounters a 2m radius of curvature mirror to keep it from diverging severely. The total time from the first stage of amplification to the fifth is 12ns, well within the pulse width of the copper vapor laser. Synchronization of the CVL and dye laser pulses can be accomplished using 74LS series integrated circuitry described in more detail elsewhere [9].

Third harmonic generation is accomplished by mixing the fundamental and second-harmonic wavelengths within phase matched nonlinear optical crystals. The fundamental and second harmonic polarizations are made nearly parallel by using a single quartz rotation plate as used by Kato [10]. KD*P and KB5 were used to generate pulses tunable around 210nm. Wavelengths were separated using a high power laser grating optimized for 300nm.

3 Results

In the amplifier, single pass gains of 20 have been observed in the best cases, for small pulses and the best synchronization. The largest five-pass gain observed was 10^4, when the dye laser pulses were very long due to a large (about 1cm) cavity mismatch. With good pulses of about 25ps duration and 20nJ energy the best overall gain observed was 5×10^3, corresponding to about 100μJ per pulse. Under these circumstances, the gains for passes 1 through 5 were about 18, 12, 5, 2.8 and 1.5. The apparent saturation was augmented by beam divergence, the energy added on the last pass (about 35μJ) is only about 5% of the total energy available in the gain medium at any time. The gain is rather strongly peaked at 620nm with Kiton Red, indicating that saturation may not be the main limi-

67

tation to the presently obtained output power. Considerable improvement
of the optics may be possible, but the beam quality after amplification
seems to be excellent.

At approximately 50cm after the last stage of amplification the dye
laser beam comes to a mild beam waist. At this point it is possible to
generate stimulated Raman scatter in 2cm of liquid benzene. This ability
is strongly dependent on the cavity-length matching between the dye laser
and the argon-ion laser. Since the total energy is quite insensitive to
cavity length, the change in stimulated Raman intensity must reflect the
length of the dye-laser pulses. The sharply peaked dependence indicates
that the dye-laser pulses have not been appreciably lengthened by the five
stages of amplification. The brightness of the disruptive scatter in the
benzene cell serves as an extremely easy way to adjust the dye laser cavity length for minimum pulse duration. Autocorrelation measurements have
verified this conclusion. If the pulses were assumed to be bandwidth-
limited 25ps pulses (characteristic of an optimally aligned cavity pumped-
dye laser), the peak powers would be about 5MW. The time-bandwidth
product is, however, likely to be nearly an order to magnitude higher than
for a bandwidth limit pulse [11], so the peak power of the noise-burst
pulse is probably about 10MW.

For third harmonic generation we have also used a 0.25mm etalon to decrease the laser bandwidth so that the pulses are nearer the bandwidth
limit. A net conversion efficiency to the third harmonic is in excess of
1%. We feel that such pulses at 5kHz repetition rate will be useful for
pump-probe experiments in picosecond spectroscopy.

4 References

1 W. R. Green, J. Lukasik, J. R. Willison, M. D. Wright, J. F. Young
       and S. E. Harris, Phys. Rev. Lett. 42, 970 (1979).
2 D. E. Cooper, R. W. Olson, R. D. Wieting, and M. D. Fayer, Chem. Phys.
       Lett. 67, 41 (1979).
3 Y. Taira and T. Yajima, Opt. Commun. 29, 115 (1979).
4 A. Wokaun, P. F. Liao, R. R. Freeman and R. H. Storz, Opt. Lett. 7, 13
       (1982).
5 P. B. Corkum and R. S. Taylor, IEEE J. Quantum Electron. QE-18, 1962
       (1982).
6 D. B. McDonald, D. Waldeck and G. R. Fleming, Opt. Commun. 34, 127
       (1980).
7 D. B. McDonald, J. L. Rossel and G. R. Fleming, IEEE J. Quantum
       Electron. QE-17, 1134 (1981).
8 D. S. Bethune, Appl. Opt. 20, 1897 (1981).
9 D. B. McDonald and C. D. Jonah, to be published in Rev. Sci. Instrum.
10 K. Kato, IEEE J. Quantum Electron. QE-13, 544 (1977).
11 V. Sundstrom and T. Gilbro, Appl. Phys. 24, 233 (1981).

Ultrafast Self-Phase Modulation in a Colliding Pulse Mode-Locked Ring Dye Laser

Yuzo Ishida, Kazunori Naganuma, T. Yajima, and L.H. Lin*

Institute for Solid State Physics, University of Tokyo
Roppongi, Minato-ku, Tokyo 106, Japan

In recent years, generation and application of ultrashort pulses from colliding pulse mode-locked (CPM) ring dye lasers [1-3] have attracted much attention for the study of ultrafast phenomena in the femtosecond region. We found that both spectral and temporal characteristics of these pulses are not simple but are largely affected by self-phase modulation (SPM), strongly depending on the concentration of the absorber dye.

In a hybridly mode-locked dye laser of Fabry-Perot type, we have already revealed experimentally and theoretically that the output pulses have large spectral asymmetry and associated temporal features and that their origin is due to slow SPM (the relaxation time of refractive index τ_r being longer than the pulse width t_p) in the dye jet stream [4]. This paper presents new experimental results on the peculiar characteristics of the output pulses from a CPM ring dye laser and its interpretation based on the process of fast SPM ($\tau_r \ll t_p$).

The CPM ring dye laser used has a rhodamine 6G dye jet stream of 180 μm thickness and a saturable absorber dye (DODCI or DQOCI) jet stream of 30 μm thickness, and was pumped by a cw argon ion laser with a power of 1-2 W. The average power of the dye laser output was 5-10 mW around 630 nm. The temporal characteristics of the pulses of 0.13-0.35 ps duration were examined by using the conventional SHG intensity correlator and a new method of spectrum-resolved (SR) SHG correlator [4]. In the latter, the second harmonic (SH) output is spectrally resolved, and the correlation function $G_s(\omega_s, \tau)$ is measured with respect to two variables ω_s (SH frequency) and τ (delay time). The function is strongly dependent on both the shape and phase characteristics of the pulse and therefore provides some information on them.

Figs.1a, 1b and 1c show the dependence of the spectrum on the absorber dye concentration. In every measurement, the CPM laser was operated with a pump power at about 20 % above the threshold to get stable mode-locking. As can clearly be seen, three very different types of spectra were observed, depending on the dye concentration. In the lower concentration regime below 5×10^{-5} M (Fig.1a), the asymmetric character is very similar to the case of the hybrid mode-locked dye laser. The SR-SHG correlation width $\Delta\tau$ as a function of SH wavelength also showed the same type of asymmetry, i.e., $\Delta\tau$ at longer SH wavelength is narrower than that at shorter SH wavelength. These properties are supposed to arise mainly from the asymmetry of phase in time caused by slow SPM ($\tau_r \gg t_p$) as before.

---

*Permanent address : Shanghai Institute of Optics and Fine Mechanics, Shanghai, Peoples Republic of China

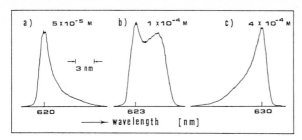

Fig.1 Asymmetric spectra of CPM pulses for different concentrations of absorber dye

On the other hand, at higher concentration over 1×10^{-4} M, the shape of the spectrum drastically changed and finally the asymmetry reversed, i.e., a broad wing appeared on the anti-Stokes side only. The SR-SHG correlation width as a function of SH wavelength for the case of Fig.1c is shown in Fig.2 together with the SH spectrum. The width decreases monotonically with decreasing SH wavelength, in contrast with the result at low concentrations. These spectral and temporal behaviors of the pulse are of new types not reported previously, and are peculiar to the CPM laser, allowing the operation at high concentrations and the generation of extremely short pulses.

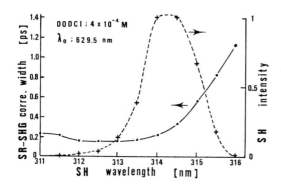

Fig.2 Spectrum-resolved SHG correlation width and SH spectrum for the case of high dye concentration

These results could also be explained by including the effect of SPM in the dye solution as in the previous study [4] but with very different regime of parameters. Figs.3a and 3b show the calculated power spectrum $I(\omega)$ of the fundamental pulse and the calculated SR-SHG correlation function $G_s(\omega_s, \tau)$, respectively, where various parameters have been adjusted to be consistent with the observed results such as the degree of spectral asymmetry and the characteristics of the SR-SHG correlation traces. The pulse is assumed to have an asymmetric double-sided exponential shape with the ratio of trailing and leading widths being $\Delta t^+/\Delta t^- = 5$. For the Kerr material, used parameters are $\tau_r/t_p = 0.01$ and $S_{pm}(= \omega_0 n_2 z_{eff} A_0^2/2c) = 2.5$, where ω_0, A_0, n_2 and z_{eff} are the peak frequency and the peak field of the input pulse, the nonlinear refractive index and the effective propagation length of the Kerr material, respectively.

It can be seen that calculated results reproduce well the essential parts of the experimental results. It is to be noted that under the condition $\tau_r/t_p > 1$ or symmetric pulse shape with $\tau_r/t_p < 1$, we can hardly explain

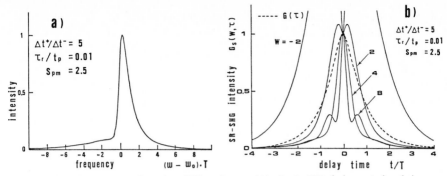

Fig.3 Calculated results for CPM pulses with fast SPM ($\tau_r \ll t_p$) :(a) power spectrum and (b) spectrum-resolved SHG correlation traces for different SH frequency deviations $W = (\omega_s - 2\omega_0)T$. ($T = t_p/\ln2$). Dashed curve is the conventional SHG correlation trace

the results in Fig.1c and Fig.2. This means that the characteristic behavior of the CPM pulses at high absorber concentrations would be the results of a combined effect of very fast relaxation process ($\tau_r < 10$ fs) such as of electronic origin and asymmetric pulse shape. In addition, it was found that the calculated spectral profile with double peaks corresponding to Fig.1b occurs only under the condition $\tau_r/t_p \simeq 1$.

Further, some special features of SR-SHG correlation traces shown in Fig.3b, such as the appearance of a dip at zero delay-time for the low SH frequency components and the structure on the wings for the high SH frequency components, have already been observed experimentally. The feature of the calculated conventional SHG intensity correlation trace $G(\tau)$ (dashed curve in Fig.3b) with sharp peak and broad wings was also often observed. These results further support our model of explanation.

Because the CPM lasers tend to operate at higher dye concentrations to achieve shorter pulses, the SPM effect as presented here must be considered in their design and applications.

References

1. R.L. Fork, B.I. Greene and C.V. Shank : Appl. Phys. Lett. 38, 671 (1981)
2. J.M. Halbout and C.L. Tang : Appl. Phys. Lett. 40, 765 (1982)
3. W. Dietel, J.J. Fontaine and J.-C. Diels : Opt. Lett. 8, 4 (1983)
4. Y. Ishida, K. Naganuma and T. Yajima : Technical Digest of CLEO, 1983, p.202; to be published

Electro-Optic Phase-Sensitive Detection of Optical Emission and Scattering

A.Z. Genack

Exxon Research and Engineering Company, Clinton Township, Route 22 East
Annnandale, NJ 08801, USA

The limits of time resolution are being reduced dramatically with the development of ever shorter laser pulses. In this communication I discuss the potential for approaching these limits by experiments in the frequency domain using electro-optic (EO) phase-sensitive detection. The method is used to separate Raman scattering (RS) from spontaneous emission and to measure the rate of fluorescence decay.

Phase resolution is achieved using the experimental arrangement shown schematically in Fig. 1. The intensity of a cw laser is modulated by a Pockels cell. The emission excited by the laser is passed through a second EO intensity modulator with adjustable phase relative to the laser modulation. The light is then dispersed in a Spex Triplemate spectrometer and detected by an EG&G PAR 1420 intensified photodiode array. The use of the emission EO modulator and integrating detector serves to demodulate the light and the detected signal depends upon the phase difference between the rf driving the modulators. In our initial experiments with 30 MHz modulation a phase shift is introduced by delaying the pulse generator which triggers the rf applied to the laser modulator. In subsequent experiments, we will attempt to expand the frequency range and correspondingly compress the time scale by using microwave modulation.

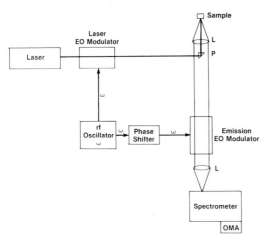

Fig. 1 Schematic illustration of apparatus for EO phase-sensitive detection. The EO modulator may be a Pockels cell and polarization analyzer or a Sears Debye acousto-optic modulator followed by a slit. The demodulated light is dispersed in a spectrometer and detected in an OMA. The phase shift between the modulators is introduced electronically.

The phase shift of the modulation will be varied using an optical delay line between the first modulator and the sample.

The phase lag of fluorescence, Φ, induced by light modulated at a radial frequency, ω, for a species with fluorescence decay time, τ, is [1]

$$\tan \Phi = \omega\tau \tag{1}$$

and the depth of modulation, m, is

$$m = (1 + \omega^2\tau^2)^{-1/2}. \tag{2}$$

By measuring the modulation parameters, τ can be determined. Short life-times can be inferred from the measurement of a small phase shift at high modulation frequencies. Since the transient fluorescence signal is the Fourier transform of the modulation observed in the frequency domain, a more complex decay in the time domain can be reconstructed from the frequency dependence of m and Φ.

In the case of a single fluorescing species, for which τ is well defined, the fluorescence background is nulled by subtracting spectra detected $\pm$ 90° out of phase with the fluorescence. This corresponds approximately to being in and out of phase with RS. Since the fluores-cence is detected with equal sensitivity in the two sequences, it is eliminated in the difference spectrum. For multicomponent fluorescent decay, the phase lag may be forced to approach 90° by modulating at high frequencies, $\omega \gg 1/\tau_i$, for decay times τ_i and the fluorescence may again be nulled. This makes possible measurements of RS in cases where it is ordinarily swamped by fluorescence. [2]

Phase-resolved RS with modulation at 30 MHz is used to observe RS from toluene solvent which is masked by fluorescence from 5 X 10^{-5} molar fluorol 555. [3] The phase shift of the emission modulator is adjusted until the fluorescence levels detected in the two sequences are equalized. This condition is obtained with a delay, T_d = 3 nsec, in the pulse generator, which triggers the rf applied to the laser modulator, relative to the in-phase delay for peak transmission of demodulated scattered light. The phase shift of fluorescence is thus $\phi = \pi/2 - \omega T_d$, and corresponds to a lifetime of 8.3 nsec. This is in agreement with the fluorescent lifetime of 8.5 nsec measured with an SLM phase fluorometer.

The spectrum collected in 100 sec with 35 mW of 4880Å light and 22 cm^{-1} resolution with the laser modulator delayed 3 nsec from the in phase configuration is shown in Fig. 2a. The difference betwen this spectrum and one with the emission modulator shifted 180° is shown in Fig. 2b and reveals the CH stretching region of the Raman spectrum. The spectrum of neat toluene with the two modulators set in phase is shown in Fig. 2c. The larger noise level in the fluorescence suppressed RS spectrum of toluene reflects statistical noise in the fluorescence which is retained in the difference spectrum.

The use of EO phase-sensitive detection expands the utility of fluorometric measurements since   a) increased modulation frequencies, related to the frequency of EO modulation rather then the frequency response of the photodetector can be used; b) the frequency range is further extended since harmonics of the applied voltage appear in the modulated light; c) sensitive optical multichannel detection is used and d) purely electronic noise is absent. For a modulation frequency of 8 GHz

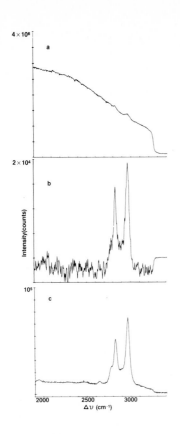

Fig. 2a. Fluorescence from 5 X 10^{-5} M fluorol 555 in toluene with modulators set nearly in phase. Sample irradiated with 35 mW of 4880 Å laser light.

Fig. 2b. Difference spectrum of spectra taken ± 90° out of phase with fluorescence. RS from CH modes observed.

Fig. 2c In-phase spectrum of neat toluene.

the measurement of a 1° phase shift would correspond to a lifetime of .35 psec, using eq. 1. Determining the phase of emission induced by higher harmonics introduced by EO modulation would give even better definition to the transient phenomena. In similar fashion the dephasing time of the transition excited by the laser could be obtained from measurements of phase shifts of resonant RS. The high sensitivity and multichannel detection used in these experiments should make it possible to resolve small phase shifts at a number of modulation frequencies and, thereby, measure ultrafast optical phenomena.

1. W. R. Ware; Creation and Detection of the Excited State, A. A. Lamola, Ed. (Marcel Dekker, New York, 1971)
2. A. Z. Genack, Anal. Chem., to be published
3. M. L. Lesiecki and J. M. Drake, Appl. Opt. 21, 557 (1982)

Theoretical Studies of Active, Synchronous, and Hybrid Mode-Locking

J.M. Catherall and G.H.C. New

Department of Physics, Imperial College of Science and Technology
Prince Consort Road, London, SW7 2BZ, United Kingdom

In previous work, we have shown how self-consistent solutions for both active mode-locking (AML) and mode-locking by synchronous pumping (MLSP) may be derived from simple difference equations [1-2]. Using a rate-equation model for the gain and a unidirectional ring cavity with the bandwidth controlled by a Fabry-Perot etalon, we demonstrated in the case of MLSP that for positive values of the cavity mismatch Δ (= pump period - cavity period), steady-state profiles can be generated from a first-order difference equation (the "stepping" algorithm). The simplicity of this solution arises from the fact that for $\Delta > 0$, the mismatch (like the filter) introduces delay, and both processes therefore transfer information across the pulse profile from front to back. For $\Delta < 0$ however, the information flows are opposed; the profile is then governed by a second-order difference equation and recourse to numerical methods is unavoidable. The set of steady-state solutions presented in Fig. 1 indicates that as Δ is decreased, the profiles are forced into the region ahead of threshold, until a point is reached where they broaden abruptly; this effect has frequently been observed experimentally (e.g. [3]).

We have investigated the effect of systematic perturbation of the control parameters by tracing the pulse evolution, numerically showing, for example, resonant response to pump power modulation and the masking of satellite pulse autocorrelations. Attention has also been paid to the effect of random perturbations induced by spontaneous emission. We recall that to initiate the stepping algorithm, a value for the field far out on the leading edge of the pulse profile has to be supplied, the magnitude of which is determined by the

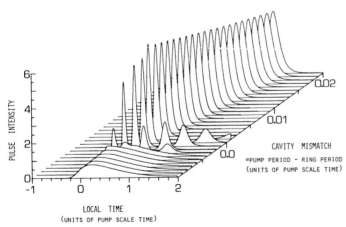

Fig. 1: Typical set of steady-state pulses as a function of cavity mismatch.
(Filter memory time = 0.005 x pump scale time)

75

level of spontaneous emission. The need within the stepping model to obtain a truly steady-state solution requires us to ignore the stochastic nature of this process, and to assume that a profile based on the time-averaged boundary value represents the average of actual, continuously-fluctuating pulses. Numerical simulations, in which spontaneous emission is modelled by a weak stochastic source term [4], suggest that optimum mismatch occurs when Δ is slightly negative, and that the pulses become noisy outside this region.

The behaviour of a hybrid system in which a saturable absorber is included in a synchronously mode-locked laser cavity has been studied. A weak absorber merely modifies the steady-state profiles which, for $\Delta > 0$, can again be predicted by a stepping algorithm. A strong absorber, however, can lead to self-pulsing solutions under certain conditions (see Fig. 2). The studies also indicate the possible importance of the absorber recovery time in achieving good mode-locking.

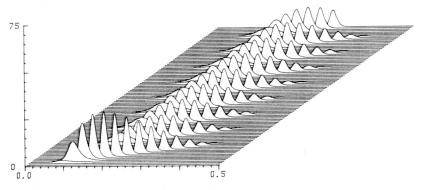

Fig. 2:   Self-pulsing solution in a synchronously-pumped laser containing a strong saturable absorber. (1000 transits at 10-transit intervals, filter memory time = 0.002, Δ = .001)

Similar difference equation techniques to those described above have also been applied to an actively mode-locked laser with sinusoidal loss modulation and, for a laser medium with a long relaxation time (Nd:YAG is an example), periodic solutions have been obtained for negative as well as positive values of [2]. In a dye laser or a semiconductor laser however, the relaxation time is comparable to the cavity transit time; in this case, it is no longer possible to represent gain saturation by a constant factor, and the complete depletion and recovery cycle of the gain within the cavity transit time must be followed. As shown in the left frames of Fig. 3 (profiles a and b), the effect of gain modulation in a typical case is to intensify, narrow, and shift the pulse.

Since these solutions fill the entire cavity, no boundary value has to be supplied; but in the absence of spontaneous emission, the field in the dark region of the cavity may fall to an unrealistically low value. While profile (c) in Fig. 3 shows the effect on (b) when a weak coherent signal is injected, a full-scale numerical simulation is necessary to see the effect of a stochastic source. The jitter that can result is evident in the right frames of Fig. 3 and occurs despite the fact that the "spontaneous emission" is many orders of magnitude weaker than the intensity near the peak of the mode-locked pulse. The jitter can be eliminated by injecting a coherent signal that is sufficiently strong to swamp the stochastic sources, a feature that has been demonstrated experimentally by VAN GOOR [5]. Recent results also suggest that the jitter decreases as optimum cavity mismatch is approached.

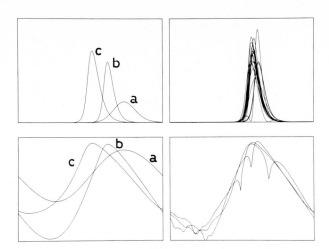

Fig. 3: An actively mode-locked laser at zero mismatch. Left upper: Steady-
         state solution for (a) T(relax) >> T(cav) and (b) T(relax) = T(cav).
         Profile (c) shows the effect on (b) of a coherent injected signal.
         Left lower: Profiles (a)-(c) on 8-decade log scale
         Right upper: Jitter induced in (b) by a weak stochastic source
                       (20 pulses at 10-transit intervals)
         Right lower: Log plot of 3 pulses from the upper frame

References

[1] J.M. Catherall, G.H.C. New and P.M. Radmore, Opt. Lett., $\underline{7}$, 319 (1982).
[2] G.H.C. New, L.A. Zenteno and P.M. Radmore, Opt. Commun., $\underline{48}$, 149 (1983).
[3] C.P. Ausschnitt, R.K. Jain and J.P. Heritage, IEEE J. Quantum Electron.,
     $\underline{QE-15}$, 912 (1979).
[4] J.A. Fleck, Phys. Rev. B, $\underline{1}$, 84 (1970).
[5] F.A. van Goor, Opt. Commun., $\underline{45}$, 404 (1983).

Technique for Highly Stable Active Mode-Locking

D. Cotter

British Telecom Research Laboratories, Martlesham Heath, Ipswich
Suffolk IP5 7 RE, United Kingdom

We describe a method for producing highly stable active mode-locking of lasers, in which the generated optical pulse train is phase-locked to a RF generator of high spectral purity. This technique results in a dramatic reduction in pulse timing jitter (to as little as ±0.3 ps), which is essential for certain measurement and signal processing applications. We also describe a technique for precise measurement of pulse timing jitter using a monomode optical fibre as a dispersion-free delay line.

Our initial step was to investigate the phase coherence of the pulse train generated by acousto-optic mode-locking in an argon ion laser (an unmodified Coherent CR-18 operating at 514.5 nm). Figure 1 shows the arrangement used. The mode-locked laser pulse train was detected and the RF signal at 76 MHz corresponding to the longitudinal mode spacing was extracted. This signal was compared in phase with the frequency-doubled output of a highly coherent RF source (Programmed Test Sources PTS200). Frequency doubling is necessary because in acousto-optic mode-locking the applied RF frequency is equal to half the pulse repetition frequency. Figure 2 shows a typical output signal from the phase detector when the argon laser was carefully aligned to produce mode-locked pulses of 80 ps FWHM. These measurements show that the output pulse train is subject to a timing jitter of as much as ±20 ps, with slew rate as great as 0.2 ps/μs (the measurement bandwidth was 10 MHz).

We identified that small amplitude fluctuations of the RF power dissipated in the acousto-optic modulator result in phase fluctuations of

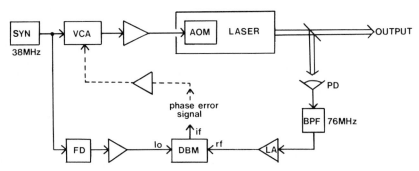

Fig 1: Measurement of phase error between the optical pulse train and a precision low-noise RF generator. The addition of the feedback loop shown dotted provides active stabilisation. SYN:frequency synthesiser; VCA:voltage-controlled attenuator; AOM:acousto-optic modulator; PD:photodetector; BPF:band-pass filter; LA:limiting amplifier; FD:frequency doubler; DBM:double-balanced mixer

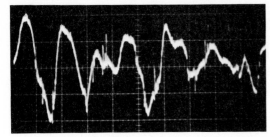

Fig 2:

Phase error signal for an
unstabilised mode-locked laser
(vertical: pulse jitter 10 ps/div,
horizontal: time 5 ms/div)

the optical pulse train, and we believe this AM to PM conversion is one
critical factor which determines the quality of the mode-locked pulse train.
For this investigation the RF power was deliberately modulated in amplitude
using a thin film voltage-controlled attenuator. Figure 3 shows the phase
error signal measured for 1.9 dB peak-peak RF power modulation, as a
function of modulation frequency. The low frequency roll-off below 0.15 Hz
is characteristic of the ∿1 s thermal time constant of the modulator prism.
The position and intensity of the resonant peak at ∿6 kHz depend on the RF
power dissipation.

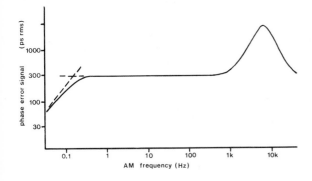

Fig 3:

Measured phase error
signal as a function of
the frequency of amplitude
modulation of RF drive to
acousto-optic modulator

It is already known that temperature fluctuations in the acousto-optic
modulator lead to unstable mode-locking. This can now be understood on the
basis of Fig.3. Temperature fluctuations cause a shift in the acoustic
resonance frequency, leading to changes in the RF power dissipated in the
modulator and hence phase fluctuations. These fluctuations can be
compensated by using a feedback loop (shown dotted in Fig.1) to phase-lock
the optical ouput to the RF source. This is more direct than previous
methods based on negative-feedback power stabilisation [1] or frequency-
locking the modulator acoustic resonance [2]. Moreover the optical phase-
lock technique has the ability to compensate phase perturbations caused by
mechanical vibration of the laser cavity, instabilities in the gain medium
(e.g. plasma instabilities or thermal lensing), and gain modulation (e.g.
due to power supply ripple). Active phase-stabilisation improves the
temporal coherence of the laser output giving somewhat cleaner pulses with
much reduced timing jitter. Using the arrangement shown in Fig.1, better than
30 dB low-frequency phase noise suppression was obtained.

To increase the locking bandwidth and also to permit phase control down
to zero frequency, an improved arrangement was devised in which an
electronic phase modulator was used as the control element in place of the

VCA shown in Fig.1. Using this system the optical pulse train has been locked to the RF source for periods of several hours, with phase fluctuations maintained at less than ±0.3 ps. Precise tailoring of the amplitude-phase/frequency response and operating point of the control loop was essential to obtain these results. These control techniques can be implemented without any modification of the optical laser cavity itself.

The mode-locked ion laser was used to pump synchronously a standing-wave R6G dye laser. Phase stabilisation was achieved by the same method as above, or alternatively by repositioning the photodetector to receive the dye laser pulses. This latter technique allows compensation of any additional phase fluctuations due to acoustic vibration and flow instabilities in the dye jet. Again, reduction of pulse timing jitter to less than ±0.3 ps has been obtained. The autocorrelation traces of the dye laser pulses were somewhat cleaner and also indicated ⌄40% increase in peak power.

In addition to the electronic method of phase noise measurement described above, we have devised a technique for precision measurement of the timing jitter in a train of pulses from a synchronously pumped dye laser. This technique provides sub-picosecond resolution on a time scale up to tens of microseconds, using a monomode silica fibre as a dispersion-free optical delay line. The ultrashort pulses obtained from a synchronously-pumped dye laser are first down-converted to the 1.3-1.5 μm wavelength range by nonlinear mixing [3], and these IR pulses are then transmitted through a long length of monomode fibre (up to several km). The IR wavelength is tuned to match precisely the wavelength of zero group-velocity dispersion in the particular fibre used. The IR peak pulse power in the fibre (1-10 mW) is low enough that pulse broadening due to the influence of third-order dispersion and optical Kerr effect is negligible [4]. The IR pulses which emerge from the fibre are then cross-correlated with the visible pulse train direct from the laser, using sum-frequency mixing in a lithium iodate crystal [3]. Thus the cross-correlation width provides a measure of the pulse timing jitter. In preliminary experiments pulse timing jitter of less than 0.2 ps after 0.5 μs delay (cross-correlation between pulses separated by 38 pulse train periods), and 3.1 ps after 12 μs delay (910 pulse periods) has been measured. These preliminary results were obtained without using active phase-stabilisation; measurements using active stabilisation are in progress.

The optical phase-lock loop stabilisation techniques which have been described here can also be applied to mode-locked cw Nd:YAG lasers, and this is being investigated. Active phase-stabilisation may also improve the properties of synchronous gain-modulated mode-locked semiconductor lasers.

I acknowledge permission of the Director of Research, British Telecom,to publish this work.

References
1. S Kishida, K Inoue and K Washio: Optics Letters 5, 191 (1980)
2. H Klann, J Kuhl and D Von der Linde: Optics Comm 38, 390 (1981)
3. D Cotter and K I White: Optics Comm 49, 205 (1984)
4. K J Blow, N J Doran and E Cummins: Optics Comm 48, 181 (1983)

Continuous Wave Mode-Locked Nd:Phosphate Glass Laser

S.A. Strobel, Ping-Tong Ho, and Chi H. Lee

Department of Electrical Engineering, University of Maryland
College Park, MD 20742, USA

G.L. Burdge

Laboratory for Physical Sciences, College Park, MD 20740, USA

We report the first successful generation of picosecond pulses from a cw neodymium phosphate glass laser by active modelocking. Neodymium phosphate
glass combines the advantages of Nd:YAG (cw operation) and Nd:glass (broader
bandwidth). Our experiment opens the door to cw subpicosecond pulse trains
from neodymium.

CW lasing has been reported in a linear cavity [1]. We use a ring cavity
for modelocking as shown in figure 1. The pump source is an Argon laser
(Spectra Physics 170) operating single line at 514 nm. The Nd:phosphate
glass rod is anti-reflection coated at the ends. Many improvements in laser
performance are obtained with the end pump geometry. Rod temperature,
thermally induced birefringence and tangential stress are all substantially
decreased with argon laser pumping as compared with lamp excitation. Athermal
phosphate glass is chosen so that thermal lensing effects can be avoided.
With up to two watts of pump power, no changes of output beam quality were
noticed. In addition, the polarization of the ring laser always followed the
polarization of the pump laser.

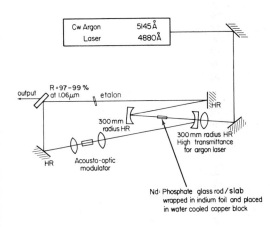

Figure 1:
cw modelocked Nd:phosphate glass
ring laser

Modelocking is achieved by a travelling wave acousto-optic modulator
(Intra-Action model AOM-70R) placed between two AR coated 85 mm focal length
lenses. The modulator is driven at a frequency equal to the cavity round
trip time. For a travelling wave crystal the modulation risetime is proportional to the beam diameter; therefore focussing the beam is necessary. The
modulator driver must have very good frequency stability and the cavity
length must be controlled to within ten microns. It is best to sample the
beam with a high speed (∿100 ps risetime) detector and a sampling oscilloscope
to adjust the cavity length properly.

81

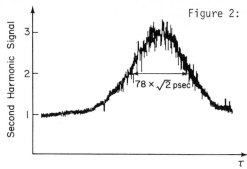

Figure 2: Autocorrelation trace with 0.2 mm intracavity etalon

$78 \times \sqrt{2}$ psec

In modelocked operation up to 15 mw average power from one arm has been obtained with 2 watts of pump power. Pulsewidths were measured with the standard Michelson-type autocorrelator with a type I $LiIO_3$ crystal. In earlier experiments we generated 33 picosecond pulses with a Schott LG 760 rod. Recently, we have used a new neodymium phosphate glass slab [2], and have added an intracavity etalon. The etalon has a free spectral range of 500 GHz and a finesse of 1. Using this etalon, we generated very stable and reproducible pulses (figure 2). We can fit these pulses extremely well by a Gaussian with a FWHM of 78 picoseconds. The spectral width, $\lesssim 0.3\text{Å}$, was measured with a Morris-McIlrath spectrometer [3], yielding $\Delta\nu\Delta t \lesssim .6$. The pulse repetition frequency is 85 MHz, with an oscillogram of the output pulse train shown in figure 3.

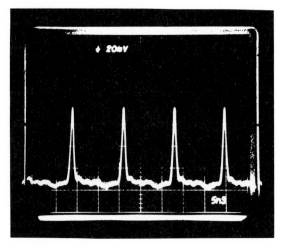

Figure 3: Oscillogram of pulse train taken with a high speed detector

In conclusion, we have demonstrated the first cw modelocking in a neodymium phosphate glass laser, and picosecond pulses have been obtained.

Acknowledgements

We would like to thank Dr. Robert Gammon and Dr. Thomas D. Wilkerson of the University of Maryland for their help. This paper is from a dissertation to be submitted to the Graduate School, University of Maryland, by Scott A. Strobel in partial fulfillment of the requirements for the Ph.D. degree in Physics. We acknowledge the support of the Graduate School in presenting this paper.

References

1) S. Kishida, K. Washio, S. Yoshikawa, and Y. Kato, "CW oscillation in a Nd: phosphate glass laser", Appl. Phys. Lett., Vol. 34, pp. 273-275, 1979.
2) We are grateful to the Shanghai Institute of Optics and Fine Mechanics for the slab, anti-reflection coated on both ends.
3) M. B. Morris and T. J. McIlrath, "Portable high resolution laser monochromator-interferometer with multichannel electronic readout", Appl. Optics, Vol. 18, pp. 4145-4151, 1979.

Active Mode-Locking Using Fast Electro-Optic Deflector

A. Morimoto, S. Fujimoto, T. Kobayashi, and T. Sueta
Faculty of Engineering Science, Osaka University, Toyonaka, Osaka 560, Japan

Recently, an ultrafast electro-optic deflector (EOD) has developed into a picosecond pulse generator [1], an optical streak camera [2], a picosecond optical Fourier transformer [3], and so on. Here, we propose to use the EOD as an active mode-locking modulator. Advantage of the EOD is in its desirable loss modulation property. Stable 6ps pulses were experimentally generated from a purely active mode-locked Nd:glass laser using the EOD.

An example of the active mode-locking of a laser using an EOD is shown schematically in Fig.1. The EOD is placed inside the laser cavity close to a concave mirror. It is driven with sinusoidal-wave voltage of which the period is just double the cavity round-trip time. Light deflected upward by the EOD at a time will be deflected downward after one cavity round trip. In the stationary state, the light travels in the cavity along two broken lines shown in Fig.1, and the angle between two lines oscillates periodically at the cavity round trip frequency. By inserting an aperture in the cavity close to the concave mirror, the diffraction loss increases except for the light which passes through the EOD at a time of no deflection.

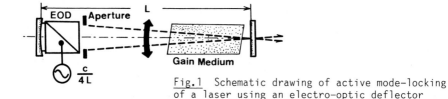

Fig.1 Schematic drawing of active mode-locking of a laser using an electro-optic deflector

The EOD for the mode-locking can be constructed of $LiTaO_3$ or $LiNbO_3$ electro-optic crystals which show larger Pockels effect than KDP or ADP used in ordinary electro-optic mode-locking loss modulators. The EOD requires only one polarization for its operation, and it is not sensitive to the variation of the crystal temperature. The more the modulating power, the wider the laser beam spreads at the concave mirror. The shutter opening time of the active modulator decreases as the modulating power increases. When using ordinary optical loss modulators, such as electro-optic modulators or acousto-optic modulators, there are many excess transmission peaks within one cavity round trip under a large modulation depth.

An experimental arrangement of the actively mode-locked Nd:Glass laser using the EOD is shown in Fig.2. The Nd:glass rod is phosphate (Hoya LHG-8). The total energy discharged through a flashlamp at threshold was about 60J. A double prism EOD, placed in the cavity close to the concave mirror, is shown schematically in Fig.3. Two trapezoidal $LiTaO_3$ crystals are in contact optically, and their optical axes are inverse with each other. Both end surfaces

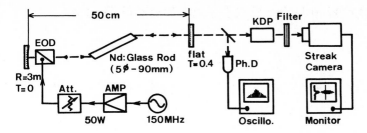

Fig.2 Experimental arrangement for the active mode-locking of a Nd:glass laser

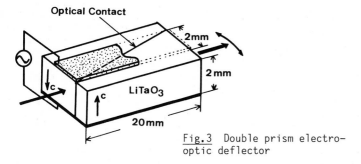

Fig.3 Double prism electro-optic deflector

are AR-coated for 1.06μm light. Deflection angle of the EOD was 1.5mrad/kV. The EOD is driven with a 500μs duration 150MHz sinusoidal-wave voltage using an LC matching circuit up to 50W modulating power corresponding to 1kV peak to peak of the EOD voltage. The aperture of the EOD (2x2mm), itself, provides the loss modulation through the deflection. Temporal measurements were made by a 9ps resolution streak camera (HTV C979) using second-harmonic pulses from a KDP crystal, because its photocathode is not sensitive to the 1.06μm light.

Stable mode-locked pulse trains were obtained when the cavity length was matched to the modulation frequency with an accuracy of better than ±50μm. Figure 4 shows a streak photograph of a second-harmonic pulse in the train with the 50W modulating power. The pulse duration is close to the temporal resolution 9ps of the streak camera. Taking into account the pulse shortening effect of the second-harmonic generation and the temporal resolution of the streak camera, the duration of the fundamental pulse is estimated as about 6ps. It is likely, however, that the mode-locking is still in a transient region. With the smaller modulating power, a pulse splits into a bunch of several separate pulses. Figure 5 shows the pulse duration as a function of the modulating power.

Though the principle of the mode-locking using the EOD predicts that the output light is deflected, there was no evidence of the output beam distortion on the experiment. The pulses evolve short enough and the angle of deflection within the pulse duration is negligible compared to the output beam diffraction width. The observed pulse duration decreases very rapidly as the modulating power increases. The simple theory of transient buildup of mode-locked pulses [4] predicts that a pulse duration is inversely proportional to a modulation depth. It is indicated that the buildup of these pulses is more complicated than described by the simple model.

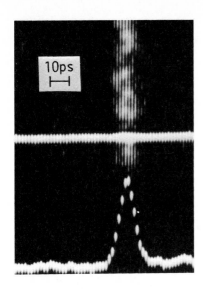

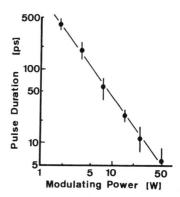

Fig.4 A streak photograph of a second harmonic pulse

Fig.5 Dependence of the pulse duration on the modulating power

Stable active mode-locking of a Nd:glass laser was achieved using an EOD as an active modulator. The shortest pulse duration obtained experimentally was about 6ps. It is comparable to the value of a passively mode-locked Nd:glass laser. The use of the EOD with the large modulation depth would be effective for many lasers which do not allow a sufficient buildup time for the active mode-locking. The use of the EOD is promising to generate subpicosecond pulses from a purely active mode-locked Nd:glass laser.

References

1. T. Kobayashi, H. Ideno, and T. Sueta, IEEE J. Quant. Electron. QE-16, 132 (1980).
2. C.L.M. Ireland, Optics Commun. 30, 99 (1979).
3. T. Kobayashi, H. Ibe, S. Fujimoto, A. Morimoto, and T. Sueta, IEEE J. Quant. Electron. QE-19, 674 (1983).
4. D.J. Kuizenga, D.W. Phillion, T. Lund, and A.E. Siegman, Optics Commun. 9, 221 (1973).

Stable Active-Passive Mode Locking of an Nd:Phosphate Glass Laser Using Eastman # 5 Saturable Dye

L.S. Goldberg and P.E. Schoen

Naval Research Laboratory, Washington, DC 20375, USA

While the stability of passively mode locked Nd:YAG lasers has been greatly improved in recent years through addition of intracavity acoustooptic loss modulators, relatively little attention has been given to the Nd:glass laser, especially in the very short pulse regime [1-6]. In the hybrid mode locking approach, the active loss modulation reduces the inherent statistical nature of the early-stage pulse build up, while retaining the effect of the saturable absorber as nonlinear loss element and Q switch for final-stage pulse shortening. We have investigated use of the new fast-relaxing saturable absorbing dye Eastman #5 in active-passive mode locking of an Nd:phosphate glass laser. Its use has led to stable mode-locking performance and generation of ~ 6ps duration pulses, largely free of occurrences of satellite pulse structure.

The heptamethine pyrylium dye #5 [7,8] is chemically more stable than the conventional saturable absorbers #9860 and #9740. It has a considerably shorter relaxation time (2.7 ps $\underline{vs}$ 7 ps) and, consequently, saturates at a significantly higher laser flux. Although ALFANO et al. [9] have reported passive mode locking and short pulse generation with dye #5 in a multi-transverse mode glass laser, KOLMEDER and ZINTH [10] found that because of its higher saturation flux, they could not obtain passive mode locking in a standard TEM_{00} mode cavity, requiring instead a folded-cavity geometry to provide a focused, higher intensity region for the dye. We have confirmed the results of Kolmeder and Zinth in our experiments by attempting purely passive modelocking with our acoustooptic modulator turned off. Even at fairly high pumping levels and high dye concentrations only conventional lasing was observed with dye #5, whereas #9860 would readily saturate and mode lock. However, when used together with the acoustooptic modulator, dye #5 gave excellent mode locking results.

The lasing medium was Owens-Illinois EV-4 phosphate glass, whose athermal properties have enabled pulse repetition rates above 1 Hz [11]. An IntraAction acoustooptic modulator operating at 57 MHz RF drive frequency (114 MHz optical) in pulsed mode was placed adjacent to the output mirror. The rear cavity mirror had a 6-m concave radius and was in contact (400 μm spacer) with the flowing dye (1,2 dichloroethane solvent). The flat output coupling mirror (R=55%) was mounted on a translation stage for precise cavity length adjustment. The cavity also contained 2 Brewster-angle polarizing plates and a mode selecting iris.

Stable well-formed TEM_{00} mode pulse trains (Fig. 1) of about 4 mJ energy were obtained under active-passive mode locking with dye #5 at relatively low concentrations (13% single-pass absorption). Below about 9% absorption mode locking became markedly less stable and lasing frequently was missed. Higher concentrations, while leading to shorter pulse durations, also led to occasional damage to the rear mirror from the more intense TEM_{00} mode beam. A sharp lasing (mode locking) threshold was observed, characteristic

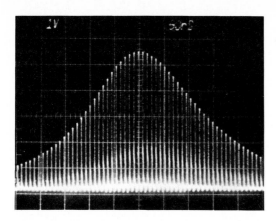

Fig. 1
Laser fundamental pulse train

of active-passive mode locking. The laser second harmonic (Fig.2) was found to be most stable and of maximum amplitude within a few volts above threshold. At higher pumping levels the second-harmonic signal became erratic from shot-to-shot and decreased in amplitude, indicating a lengthening or breakup in the pulse. Under the above optimum conditions, the amplitude stability of the laser fundamental pulse trains was ±10%, with stability of single pulses switched-out low in the train (1/5 peak) of ±5%. At this level the single pulses showed no significant spectral broadening due to self-phase modulation.

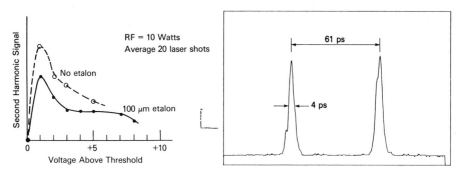

Fig. 2   Second harmonic signal intensity vs flashlamp voltage above lasing threshold.   The intracavity etalon (solid curve) was used to lengthen the laser pulse

Fig. 3   Streak camera trace of single pulse (replica pair) after frequency doubling to 527 nm

Streak camera data (Fig.3) obtained with an Imacon 500 showed pulse widths typically in the range of 6-8 ps (at 1054 nm). In addition, the consistency of obtaining single, well-formed pulses with dye #5 was very high, with only occasional occurrences of satellite pulses or multiple pulse structure as observed by SEKA and BUNKENBURG [3] in their earlier work.   In contrast, for dye #9860, while stable mode locked pulse trains could readily be obtained over a wide range of dye concentrations, when examined with the streak camera the individual pulses were of markedly longer duration, typically from 10-14 ps. Most importantly, the streak data showed a considerably greater incidence and degree of satellite pulse structure in the output.

References

1. S. Kishida and T. Yamane: Optics Commun. 18, 19 (1976).
2. I.V. Tomov, R. Fedosejevs, and M.C. Richardson: Appl. Phys. Lett. 30, 164 (1977).
3. W. Seka and J. Bunkenburg: J. Appl. Phys. 49, 2277 (1978).
4. B.B. Craig, W.L. Faust, L.S. Goldberg, P.E. Schoen, and R.G. Weiss: in Chemical Physics 14: Picosecond Phenomena II, R.M. Hochstrasser, W. Kaiser and C.V. Shank Eds., Berlin, Springer-Verlag (1980), pp. 253-258.
5. M.A. Lewis and J.T. Knudtson: Appl. Opt. 21, 2897 (1982).
6. H.P. Kortz: IEEE J.Quantum Electron. QE-19, 578 (1983).
7. G.A. Reynolds and K.H. Drexhage: J. Organic Chem. 42, 885 (1977).
8. B. Kopainsky, W. Kaiser, and K.H. Drexhage: Opt. Commun. 32, 451 (1980).
9. R.R. Alfano, N.H. Schiller, and G.A. Reynolds: IEEE J. Quantum Electron. QE-17, 290 (1981).
10. C. Kolmeder and W. Zinth: Appl. Phys. 24, 341 (1981).
11. L.S. Goldberg, P.E. Schoen, and M.J. Marrone: Appl. Opt. 21, 1474 (1982).

Limits to Pulse Advance and Delay in Actively Modelocked Lasers

R.S. Putnam

Department of Electrical Engineering and Computer Science, and

Research Laboratory of Electronics, Massachusetts Institute of Technology Cambridge, MA 02139, USA

1. Introduction

The time domain approach to modeling modelocked lasers involves following a pulse around the cavity, recording the gain, loss, pulse broadening, pulse shortening and pulse delay, and finding a pulse shape that will reproduce itself after one complete roundtrip [1-3]. Such a self-consistent solution does not automatically guarantee stability against noise nor against second order effects which are ignored but may accumulate anomalously.

Experimental verification of the modelocking models has been based primarily on autocorrelation measurements [2,4], and to a lesser degree on streak camera measurements [5] which have detected multiple pulsing and provide very useful interpretations of typical autocorrelation traces.

2. Cavity Length Mismatch in Actively Modelocked Lasers

A mismatch between a modelocked laser's cavity length and the modulation rate is corrected by preferential amplification of the front or rear of a laser pulse, producing a new pulse that is advanced or delayed relative to the old pulse. At this point an assumption is usually made that the intracavity bandwidth limitation smooths over any irregularity in shape that the pulse may have acquired. This assumption fails with large cavity length mismatches that require a considerable corrective advance or delay of the pulse. Consider a longer cavity that requires a compensating pulse advance. The front wing of the laser pulse is preferentially amplified by an off-center gain peak which produces a new pulse that is slightly advanced in the laser cavity. The front wing of the laser pulse is acting as the seed energy for the next pulse, and the important question is what was the source of this seed energy. The two choices are noise and optical energy diffused away from the peak of the pulse by repeatedly passing through any bandwidth limitation in the laser cavity. However the intracavity filter can only extend the wings of the laser pulse by a limited distance in one pass. If the pulse advance required due to the mismatched cavity length is larger than the distance that the front wing of the laser pulse can be extended in one pass through the bandwidth limited cavity, then the front wing will dry up. In this case the front wing of the pulse is replaced by amplified noise. Proper modelocking involves producing the new pulse primarily by a transformation of the old pulse with a minimum of additional spontaneous emission noise.

The diffusion distance produced by the intracavity filter is limited to the width of one impulse response, per roundtrip. Figure 1a shows the low-pass equivalent impulse response of an arbitrary intracavity filter. The filter extends the front wing of the laser pulse relative to the peak of the pulse by a "distance" equal to the intrinsic group delay of the filter. The rear wing of the laser pulse is extended by the remaining width of the impulse response. Figure 1b shows the real impulse response for an intracavity etalon. The intracavity bandwidth limitation produces an advance

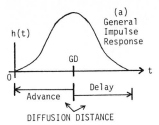
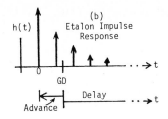

Figure 1a The width of the impulse response gives the pulse spreading of a filter. The diffusion distance is divided into advance and delay relative to the group delay (GD) of the filter. Figure 1b The etalon impulse response gives a sharply limited pulse advance, and an extended and attenuated pulse delay.

and delay of the laser pulse, whether or not it is developed by an off center gain peak.

The roundtrip cavity length sensitivity of a sync-pumped dye laser with $\omega_c\tau = 100$ [2] and 1 ps pulses is about $1/\omega_c = \tau/100$ or 3 microns. Micron control of a cavity length would be required. Therefore, even micron fluctuations of the dye jet thickness would be a problem.

3. Short Noise Pulses
Increasing the cavity length by more than the inverse bandwidth forces the front wing of the laser pulse to be reconstituted by amplified spontaneous emission. The repetitive amplification plus the regular time shift produces the rising edge of the pulse from noise, while the gain, swollen due to the relative delay of the laser pulse, saturates and cuts off the end of the laser pulse. Thus short noise pulses, filtered by a few dozen roundtrips through the intracavity filter, can be produced with an excessively long cavity in a sync-pumped dye laser. These tend to produce a coherent spike in the autocorrelation trace [5].

4. Recovery Time
The pulse compression $\delta\tau/\tau$ per roundtrip is balanced by the intracavity bandwidth limitation $1/(\omega_c\tau)^2 = \delta\tau/\tau$. This leads to the intrinsic time constant for the pulsewidth to return to its steady state as $(\omega_c\tau/2)^2$ roundtrips [6]. Thus the recovery time for a passively modelocked laser that has a large $\delta\tau/\tau = .02$ [3] can be 12 roundtrips. A sync-pumped dye laser [2] with $\omega_c\tau = 100$ would require 2500 roundtrips.

5. Pulse Compression Due to the Intracavity Bandwidth Limitation
The function of bandwidth limitation in the operation of a sync-pumped modelocked laser can be better understood by considering an archetypical ring dye laser having a fully infinite bandwidth and a perfect cavity length match [6]. Figure 2 shows the pump pulse, the integrated gain shape and the cavity loss line. One solution for the resulting laser pulse, as shown, is a very broad pulse shaped like the end of the pump pulse, which removes just enough energy to keep the gain saturated at the loss line. In effect this laser is operating quasi-cw, that is, each point in time along the laser pulse returns to the same point in the gain profile, and saturates the gain as if it were acting alone, since there is no temporal mixing with an infinite bandwidth.

Producing short pulses involves arranging for the gain regularly to exceed the loss line, such that the gain can then be saturated below the loss

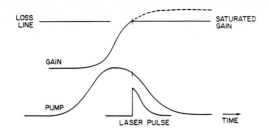

LOSS LINE ――――――― SATURATED GAIN

GAIN

PUMP

LASER PULSE TIME

Figure 2 A broad pulse replica of the tail of the pump pulse is produced with a perfect cavity match and an infinite intracavity bandwidth. The pulse acts quasi-cw.

line and provide a concave gain shape. Providing a longer cavity to delay the returning pulse will provide excess gain and produce shortening at the end of the pulse. However the pulse walks off, and the front of the pulse will be continually replaced by amplified spontaneous emission.

Alternatively a bandwidth limitation can be introduced which will affect the sharp front edge of the laser pulse of Fig. 2 and produce a separation between the rising edge of the pulse and the peak of the pulse. This allows the gain to exceed the loss temporarily, until the peak of the laser pulse hits the gain medium. In the steady state, the rising edge will sit about where the gain first crosses the loss line. Thus the intracavity bandwidth limitation softens the front edge of the laser pulse, which allows the gain temporarily to exceed the loss in the steady state.

Supported by the Joint Services Electronics Program, DAAG29-83-K-0003.

References

1. H. A. Haus, IEEE J. Quant. Electron. QE-11, 736 (1975).
2. C. P. Ausschnitt, R. K. Jain and J. P. Heritage, IEEE J. Quant. Electron. QE-15, 912 (1979).
3. M. S. Stix and E. P. Ippen, IEEE J. Quant. Electron. QE-19, 520 (1983).
4. E. W. VanStryland, Opt. Commun. 31, 93 (1979).
5. S. L. Shapiro, R. R. Cavanagh, and J. C. Stephenson, Opt. Lett. 6, 470, (1981).
6. R. S. Putnam, PhD Dissertation, MIT, (1983).

Novel Method of Waveform Evaluation of Ultrashort Optical Pulses

Tetsuro Kobayashi, Feng-Chen Guo*, Akihiro Morimoto, and Tadasi Sueta

Engineering Science, Osaka University, Machikaneyama-Cho
Toyonaka, Osaka 560, Japan

Yoshio Cho

Scientific and Industrial Research, Osaka University
Yamadagaoka, Suita, Osaka 565, Japan

1. Introduction

For the measurement of ultrashort optical pulses SHG auto-correlation method has been most popularly used [1,2]. For this case, an adequate pulse shape such as $sech^2$, Gauss, etc., is presumed and then the correlation width is converted into the pulsewidth. Since one SHG correlation curve does not correspond identically to one pulse waveform, this method includes inevitable ambiguity and is, particularly, not applicable to the pulse with unfamiliar or unanticipa waveform.

In this paper we propose a novel method which is applicable to any kind of pulse and makes possible to estimate the pulse shape, the amount of included frequency chirping, and the pulsewidth, using SHG auto-correlation curve together with the frequency spectra of the pulses without any presumption for the shape.

2. Principle of this Method.

For an isolated single optical pulse or a single pulse picked out from a periodic modelocked pulse train, the instantaneous optical field can be written by Fourier integral as

$$E(t) = \int A(\omega)e^{j\{\omega t + \phi(\omega)\}} d\omega \qquad (1)$$

where ω is the optical angular frequency, $A(\omega)(\geq 0)$ and $\phi(\omega)$ are the amplitude and phase of ω-frequency component of the pulse optical field, respectively, and are the continuous functions of ω. The amplitude $A(\omega)$ can be experimentally determined by the measurement of the frequency spectral profile. Since for the periodic pulse train only line spectra are measured, then $A(\omega)$ should be determined from the envelope of line spectra or from broadened spectra having a resolution wider than the interval of the line spectra. In order to specify the optical fields of Eq. (1), other than $A(\omega)$ obtained above, we must determine $\phi(\omega)$. The continuous function $\phi(\omega)$ can be expanded by Taylor expansion around $\omega = \omega_0$ as

$$\phi(\omega) = \phi(\omega_0) + (\omega - \omega_0)p + (\omega - \omega_0)^2 q + (\omega - \omega_0)^3 r + \dots \qquad (2)$$

where $p(=[\partial\phi/\partial\omega]_{\omega_0})$ corresponds to the time delay of the pulse envelope and $q(= 1/2[\partial^2\phi/\partial\omega^2])$ and $r(=1/6[\partial^3\phi/\partial\omega^3])$ are the factors corresponding to linear and nonlinear frequency chirping, respectively.

Using Eqs. (1) and (2), the intensity pulse shape $I(t)$ and SHG auto-correlation function $G_2(\tau)$ can be written as follows:

---

*Present Address: Yongchuan Optoelectronics Research Institute
P.O. Box 1102, Yongchuan, Sichuan, China.

$$I(t) \propto |E(t)|^2 = |\int A(\omega)e^{j\{(\omega-\omega_0)(t-p)+(\omega-\omega_0^2)q+(\omega-\omega_0)^3r + \ldots\}}d\omega|^2, \quad (3)$$

$$G_2(\tau) = \int I(t)I(t-\tau)dt/ \int I^2(t)dt . \quad (4)$$

As can be seen from these equations, the terms of $\phi(\omega_0)$, and p are not significant for I(t) and $G_2(\tau)$. If the values of q, r, etc., are given, pulse waveform can be determined except for the pulse position and the constant phase. If $\phi(\omega)$ is a slowly varying function of ω within the spectral width, $\phi(\omega)$ can be well approximated by taking account of only two terms, i.e., q and r in Eq. (2) (third-order approximation). The values of q and r can be determined by best-fitting of calculated (by substituting Eq. (3) into Eq. (4)) correlation curve to experimental one. Pulse waveform obtained through this procedure satisfies both the frequency spectra and the SHG auto-correlation experimentally obtained. Therefore, the pulse waveform thus obtained can be regarded to be most probable among ones estimated from these experimental results.

3. Application of Actively Modelocked Diode Laser Pulse

The experimental setup we used is shown in Fig. 1 where the optical pulses are obtained by active modelocking of GaAlAs/GaAs DH laser [3,4]. The spectral profile of the output pulse, which was observed with a scanning Fabry-Perot interferometer, is shown in Fig. 2. The SHG auto-correlation traces ($\propto 1+2G_2(\tau)$) were also measured simultaneously. The simultaneous SHG auto-correlation to Fig. 2 is shown in Fig. 3. The 3:1 contrast ratio of the correlation peak to background shows the laser was well modelocked. Figure 4 shows the example of the SHG auto-correlation curve $G_2(\tau)$ and the corresponding pulse shape I(t) which have been calculated numerically using the observed spectra (Fig.2) with the best-fitting parameters: $q=2.08\times10^{-22}$, r=0 (determined by the least square method). As a result, a slightly unsym-

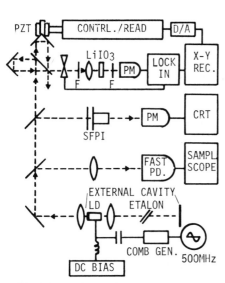

Fig.1 Experimental arrangement for the generation and measurement of ultrashort pulses from an actively modelocked semiconductor laser.

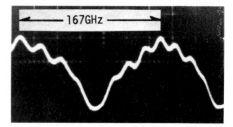

Fig.2 Typical spectral profile of output pulses observed with a scanning F-P Interferometer.

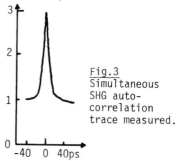

Fig.3 Simultaneous SHG auto-correlation trace measured.

94

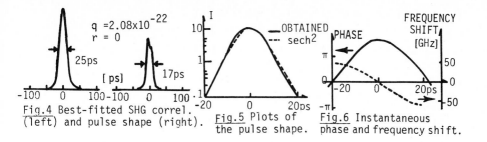

Fig.4 Best-fitted SHG correl. (left) and pulse shape (right).

Fig.5 Plots of the pulse shape.

Fig.6 Instantaneous phase and frequency shift.

metrical 17ps (FWHM) pulse is composed. Figure 5 shows the semilogarithmic plot of the obtained (calculated) pulse shape. It is seen from the figure that the pulse waveform was fairly close to sech^2 shape.

Our method can give the phase information of the pulse field. Figures 6(a) and 6(b) show the instantaneous phase and frequency (shift) of the pulse field calculated using the above parameters. In the figure a down-frequency chirping of 60 GHz during the pulsewidth (FWHM) is seen. The value of the time-bandwidth product ($\Delta\nu\Delta t$) is evaluated to be 1.3 using the 17 ps pulse width and 80 GHz spectral width (FWHM). This value is about four times wider than one for transform-limited sech^2 shaped pulse.

On the other hand we also measured the full SHG auto-correlation including the interference terms[6], which gives some information on the amount of the chirping. The results of this measurement obviously support the validity of our method.

4. Conclusions

A novel method of waveform evaluation of ultrashort pulses using the SHG auto-correlation trace together with the frequency spectra has been proposed and applied to the evaluation of the pulses experimentally obtained by actively modelocked GaAlAs/GaAs DH laser. By this method, besides the pulsewidth, the pulse shape, the phase and frequency characteristics of the pulses can be evaluated. The usefulness of the method has been confirmed experimentally.

The method is also useful for the evaluation of ultrashort pulses from the other kind of lasers.

References

1.   D.J. Bradley and G.H.C. News, Proc. IEEE 62, 313 (1974).
2.   E.P. Ippen and C.V. Shank, Ultrashort Light Pulses, ed. by S.L. Shapiro (Springer-Verlag, Berlin 1977) Chap. 3.
3.   P.-T. Ho, L.A. Glasser, E.P. Ippen and H.A. Haus, Appl. Phys. Lett. 33, 241 (1978).
4.   P.L. Liu, Chinlon Lin, T.C. Damen, and D.J. Eilenberger, Picosecond Phenomena II, ed. by R.M. Hochstrasser, W. Kaiser, and C.V. Shank (Springer-Verlag, Berlin, 1980).
5.   T. Kobayashi and T. Sueta, Oyo Butsuri, 50, 951 (1981).
6.   J.C. Diels, E. VanStryland, D. Gold, Picosecond Phenomena, ed. by C.V. Shank, S.L. Shapiro and E.P. Ippen (Springer, Berlin, 1978).

Noise in Picosecond Laser Systems: Actively Mode Locked CW Nd^{3+}:YAG and Ar^+ Lasers Synchronously Pumping Dye Lasers

T.M. Baer

Spectra-Physics, Inc., 1250 West Middlefield Road, P.O. Box 7013
Mountain View, CA 94039-7013, USA

D.D. Smith

Purdue University, Department of Chemistry, West Lafayette, IN 47907, USA

1. Introduction

In the interest of increasing the sensitivity of laser spectroscopy, investigators are utilizing high frequency modulation [1,2] and polarization [3] techniques. In the case of high-resolution-high-frequency modulation laser spectroscopy, beating between the modulation side bands allows formidable sensitivity [1]. In high frequency modulation pump-probe picosecond spectroscopy [2], the side bands produced are negligibly displaced relative to the laser spectral bandwidth; thus the carrier frequency (or sums and differences of carrier frequencies on different laser beams [4,5]) are detected directly.

To improve our understanding of the laser physics and aid in the application of high frequency modulation techniques, we have measured laser intensity/phase fluctuations on different types of synchronously pumped dye laser systems. We avoid the term "noise" to describe fluctuations, as the power spectrum can exhibit structure.

2. Experimental Apparatus

The measurements were conducted using a Spectra Physics series 3000 mode locked CW Nd^{3+}:YAG laser and a model 165 Ar^+ laser. Similar stabilized mode locking electronics and acousto-optics (Spectra Physics model 451, 452A, 453) were used on both lasers. Each pump laser was used to drive the *same* 375B dye laser with a 376B circulator. Intensity fluctuations to a 100 Hz bandwidth were measured by a Hewlett Packard 8553B, 8552B and 141T RF spectrum analyzer and a photodiode. The laser illuminated the full area of the photodiode to systematize frequency, intensity and position-dependent sensitivity changes. The photodiodes were an amplified EG&G Ortec DT-25 and a custom-modified Ortec HFD-1100.

3. Results

Figure 1 shows the fluctuation spectrum of an Ar^+ laser at 3 KHz detection bandwidth as a function of the acousto-optic mode locking frequency at a fixed cavity length. The top trace is a typical power spectrum for the optimized cavity and mode locking frequency. "Optimized" implies maximal average power with minimal pulse width and amplitude fluctuations without multiple pulsing (to extent that a precision pyroelectric power meter and a 35 ps rise time photodiode with a sampling oscilloscope permits). The peak at zero frequency in Fig. 1 does not represent average laser power as it is out of the range of the spectrum analyzer. The lowest trace is a spectrally filtered incandescent lamp with the same optical power as the laser to approximate a shot noise limited source (adequate given other limitations; not shown is the electronic noise floor, a few dB below the lamp trace). The second trace from the bottom is a 1 KHz *decrease* in the synthesizer frequency (cavity too short) and the third trace from the bottom is a 1 KHz *increase* in the synthesizer frequency (cavity too long). With the cavity short fluctuations are reduced and a "comb" split by 375 KHz emerges extending to the detection bandwidth limit. Interpretation of the fluctuation spectrum and results of new sub-Hertz bandwidth measurements are the subject of a coming paper [6].

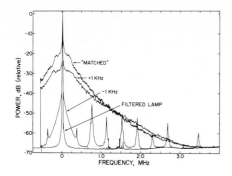

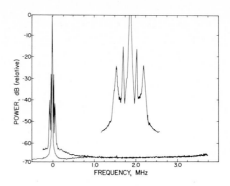

Fig. 1 Fluctuation spectrum of
a mode locked Ar⁺ laser

Fig. 2 Fluctuation spectrum of a
modelocked Nd³⁺:YAG laser

The upper trace of Fig. 2 shows the spectrum of the 1.06 μm beam of an optimally mode locked CW Nd³⁺:YAG laser at 1 KHz detection bandwidth and the lower trace is the power-matched filtered lamp. The structure at 10's of KHz (inset, Fig. 2) may be due to relaxation oscillations [6] and can be eliminated for certain operating conditions. The point is a well-adjusted Nd³⁺:YAG system   only a few dB above shot noise limit at 500 KHz vs. the Ar⁺ laser which is approximately 30 dB above.

Figure 3 illustrates the fluctuation spectrum of a dye laser synchronously pumped by an ion laser as a function of dye laser cavity length mismatch for optimized pump laser conditions with 350 mW pump power, 45 mW output power, Rh590 dye and a 15%T output coupler. Independent of the type of pump source, the dye laser develops regular fluctuations for length mismatches. A dye laser 40 μm too short produces structure split by 200 KHz and 40 μm too long has structure split by 100 KHz. The splitting differences for too short and too long is   sensitive to many operating parameters and is different for cavity dumped operation; the fluctuation "comb" structure is pump power-dependent and tuning etalons result in quieter performance than 3 plate birefringent filters at high pump powers [6].

Figure 4 shows the "modulation transfer function" for the dye laser. An acousto-optic modulator was placed between the mode locked pump source and the dye laser. Small modulation of -60 dBc (decibel below the optical power of the carrier) is imposed by the acousto-optic modulator and the fluctuation at the same frequency relative to dye laser

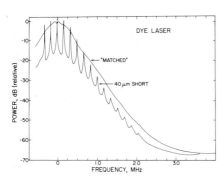

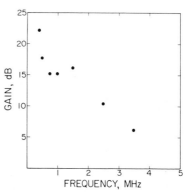

Fig. 3 Sync-pumped dye laser
fluctuation vs. cavity length

Fig. 4 Synchronously pumped dye
laser "modulation transfer function"

power is measured. The dye laser amplifies modulation up to about 4 MHz. The dye laser is a *nonlinear* amplifier and for large modulation depths gain is reduced and the gain curve shape changes. Since pump laser fluctuations will be *amplified or passed* under typical operating conditions, the dye laser is quieter pumped by Nd^{3+}:YAG than Ar^+. At 500 KHz the YAG pumped dye laser is typically (depending on operating parameters) 20 dB quieter.

4. Conclusions

The mode locked CW Nd^{3+}:YAG laser has lower amplitude noise than its Ar^+ counterpart at frequencies over approximately 100 KHz. When the Nd^{3+}:YAG laser synchronously pumps a dye laser, dye laser noise is reduced but technical noise prohibits reaching shot noise limit until 3 MHz. The dye laser modulation transfer function peaks at low frequencies, decreasing to 4 MHz and its fluctuation spectrum is sensitive to cavity length mismatch and resonator misalignment serving as a real-time performance diagnostic of a synchronously pumped systems to approach shot noise-limited performance. One of us (DDS) would like to acknowledge helpful discussions with Dr. A. Lorincz.

5. References

1. G. C. Bjorklund, Opt. Lett., 5 (1980) 15.

2. B. F. Levine, C. V. Shank and J. P. Heritage, J. Quant. Elec., 15 (1979) 1418, B. F. Levine and C. G. Bethea, J. Quant. Elec., 16 (1980) 85.

3. R. E. Teets, F.V. Kowalski, W. T. Hill, N.Carlson and T. W. Hansch, "Proceedings of the Society on Photo-Optical Instruments and Engineering," San Diego (1977), pp. 88ff.

4. P. Bado, S. B. Wilson and K. R. Wilson, Rev. Sci. Inst., 53 (1982) 706.

5. L. Andor, A. Lorincz, J. Siemion, D. D. Smith and S. A. Rice, Rev. Sci. Inst. 55 (1984) 64.

6. D. D. Smith and T. M. Baer and A. Lorincz, to be published.

High Power, Picosecond Phase Coherent Pulse Sequences by Injection Locking

F. Spano, F. Loaiza-Lemos, M. Haner, and W.S. Warren
Department of Chemistry, Princeton University, Princeton, NJ 08544, USA

Complex multiple pulse sequences have proven extremely valuable in NMR as probes of intermolecular and intramolecular interactions. Optical analogs of many of these sequences can be useful in elucidating the effects of collisional perturbation in gases, guest-guest and guest-host interactions in mixed molecular crystals, or any general perturbations which make the usual two-level approximations invalid. But all these sequences require phase coherence (the phase of each pulse must be specified relative to the first) which was more difficult to achieve in optical spectroscopy than at radio-frequencies [1]. We present here a new technique which extends the applicability of phase coherent pulse sequences to high powers and $\sim$ 200 ps resolution, and should be extendable to even shorter pulses.

The experimental arrangement is shown in Fig. 1. Pulses of approximately 250 kW peak power and 300 ps duration are produced by several independently triggerable, low jitter (± 2 ns) atmospheric nitrogen lasers (PRA LN-103). For short pulse sequences this jitter can be virtually eliminated by using only one laser and splitting its output. The UV pulses transversely pump a dye cell laser consisting of a 1 cm^2 cross section cuvette filled with Rhodamine 6G and placed between a 99% reflective mirror and a weakly reflective (5%) window. The cavity length is approximately 2 cm. A cylindrical lens focuses the UV radiation into the dye where it penetrates about 1 mm into the lasing medium. External adjusting knobs allow one to change the orientation of the front window, the dye cell angle and the distance between the cylindrical lens and the cuvette. The resulting dye cell pulse has an energy of ca. 10 μJ as measured with a Molectron J3 and is 200 ps long. However its frequency spectrum is of course very wide, corresponding basically to the dye fluorescence curve. In order to narrow the bandwidth to within the "Fourier transform limit" we use an injection locking technique [2] whereby light pulses generated by chopping the cw beam of a ring dye laser are focused into the cavity of the tunable dye cell laser. When the spatial and temporal overlap of the UV pulse and the injected radiation pass a certain threshold value the stimulated emission in the mode dictated by the injected pulse dominates the other possible modes of the cavity, and all the energy deposited by the UV laser pulse goes into that particular frequency. However previous injection experiments have generally used long ($\sim$10 nsec) pumping pulses which would compromise the pulse sequence resolution.

The chopping of the cw beam coming from the ring dye laser is accomplished with an acousto-optic modulator (AOM), which receives a radio-frequency pulse (RF) synchronously with the nitrogen laser triggers. The piezoelectric material of the AOM vibrates at the frequency of the RF pulse and generates a travelling diffraction grating while the RF pulse is on. Because of the $\vec{k}$ vector matching conditions the chopped beam suffers a frequency shift but retains the phase relationship of the RF sequence. We have used this approach in previous experiments [1,3] to generate pulse sequences with $\sim$1 W peak power and 4 ns risetimes. The alignment of the AOM chopped

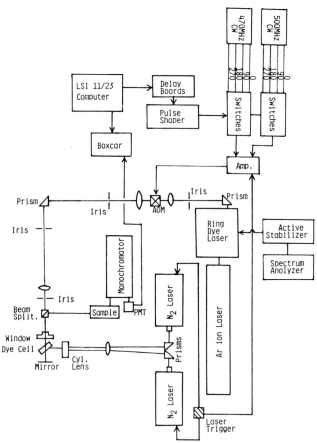

Fig. 1. Schematic diagram of the injection locking experiment. The pulses of the two Nitromite N_2 lasers are directed into the dye cell using quartz prisms while the pulse from the ring dye laser is diffracted by the AOM. The RF pulses are generated with the indicated phase shifts and can be gated into the AOM using homemade electronics. The whole set-up is computer controlled.

pulses into the dye cell cavity is critical for the occurrence of injection locking. A rough alignment using two diaphragms guarantees that the inclination of the pulses is correct. For the final adjustment a 12 inch focal length convex lens on a fine adjustable X-Y-Z mount brings the diameter of the pulses as well as their depth into the dye cell to maximum overlap.

The dependence of output power at the ring laser frequency on the strength of the injected pulses is shown in Fig. 2. The onset of injection was detected by dispersing the pulses through a 0.4 m monochromator with 10 μm slits (0.6 Å resolution). The peak width in this figure is resolution limited with more than 95% of the 10 μJ pulse energy within the monochromator bandwidth. Tests with a 1 cm^{-1} FSR Fabry-Perot etalon show that the pulse bandwidth is in fact less than 5 GHz. The pulse width is roughly 200 psec (FWHM). Sweeping the frequency of the ring dye laser over a range of 30 GHz gives a steady decrease in the intensity of the injected pulse signal, which eventually dies away when the frequency mismatch is too large. Re-

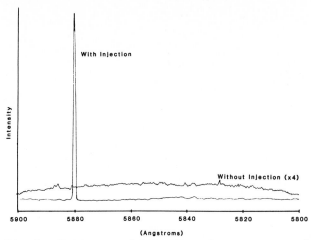

With Injection

Without Injection (x4)

Intensity

| 5900 | 5880 | 5860 | 5840 | 5820 | 5800 |

(Angstroms)

Fig. 2. Percentage of injection pulse energy as a function of the injected power. The threshold value for the experiment discussed here lies around 25 mW.

adjustment of the dye cell laser mirrors is then required to recover the injected power. In addition, removal of the weakly reflecting front window in the dye cell cavity totally destroys the injection locking. We conclude from these points that our laser is not simply acting as a double pass amplifier, but is instead actively promoting the selected mode.

The output pulses reflect the phase of the injected radiation, as demonstrated by simple phase sensitive detection experiments on I_2 in the bulb. The amount of polarization produced on resonance by a single pulse is proportional to sin θ, where θ is just the product of the pulse area and the electric dipole moment. But the total polarization measured by a photodiode does not show this simple oscillatory behavior, since off resonance effects dominate. When we apply a single pulse from our injected laser and detect the on resonance polarization in quadrature with the pulse the sinusoidal pattern is recovered.

The most important feature of this experimental setup is that the combination of high power and phase coherence permits application of these pulse sequences to systems with subnanosecond relaxation times. We expect that phase coherent techniques will prove useful in a number of complex systems, including low temperature pure and mixed crystals. We wish to thank Professor M. Littman for his assistance in laser design, and the Petroleum Research Fund and Research Corporation for their financial support.

References

1. W. S. Warren and A. H. Zewail, J. Chem. Phys. 75, 2278 (1981); W. S. Warren and A. H. Zewail, J. Chem. Phys. 78, 2297 (1983).

2. L. E. Erickson and A. Szabo, Appl. Phys. Lett. 18, 433 (1971).

3. W. S. Warren and M. A. Banash, Proc. 5th Rochester Conf. on Coherence and Quantum Optics (1983); M. A. Banash and W. S. Warren, J. Chem. Phys. (submitted).

Passive Mode-Locking with Reverse Saturable Absorption

D.J. Harter and Y.B. Band*

Allied Corporation, Research and Development, 7 Powderhorn Drive
Mt. Bethel, NJ 07060, USA

In the passive mode-locking of a laser, the leading edge of the mode-locked
pulse is shaped by the rapid reduction of loss due to absorption saturation of
the absorbing material. There are two mechanisms in general use for shaping
the trailing edge of a mode-locked pulse. For gain media which saturate
easily, such as organic dyes, the saturation of the gain which follows the
peak of the pulse shapes the lagging edge of the pulse [1]. For gain media
which cannot be saturated by an individual mode-locked pulse, because the
saturation energy is too high, the trailing edge of the pulse continues to see
gain and is not suppressed. For solid-state lasers (e.g., alexandrite, ruby,
Nd:YAG, and Nd:Glass, etc.) which have high saturation energy, a saturable
absorber with a fast relaxation time is used since a pulse with energy greater
than the absorber's saturation energy will experience the least loss in propagating through the cavity when the pulse length is shorter than the absorber's
relaxation time [2]. Therefore, by this mechanism the pulsewidths of broadly
tunable, solid-state mode-locked lasers are limited by the relaxation times of
saturable absorbers rather than the lasing bandwidth of the media.

Here, we present a different technique for shortening the lagging edge of a
passively mode-locked pulse. An additional passive element, whose absorption
increases with increasing pulse energy, is added to the cavity. In particular, we consider a reverse saturable absorber, which is a material with a
larger excited state cross section than ground state cross section in certain
wavelength regions. Figure 1 shows input pulses and normalized output pulses
after traversing through a reverse saturable absorber with the properties of
rhodamine 6G at 435 nm, which are given in Table 1, except a higher concentration of dye is used for this figure. The t exp-t pulse in Fig. 1a is
shortened considerably while the sech pulse in Fig. 1b is not shortened but is
translated forward by the reverse saturable absorber in a manner similar to a
pulse propagating through a saturating gain medium. The theoretical details
for pulse shortening and the criteria for a material to be a reverse saturable
absorber are given in reference [3].

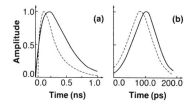

Fig. 1 Dashed temporal pulse shape
is the calculated output (normalized)
after traversing through a reverse
saturable absorber. The solid curve
is the input pulse, a t exp-t pulse in
(a) and sech (t) pulse in (b).

---

* Present Address: Ben Gurion University, Department of Chemistry
Beer-Sheva, Israel

Table 1 - Properties and Parameter Values for Numerical Example

Alexandrite	Cryptocyanine	Reverse Saturable Absorber
g_o - 0.0497	σ_s - 2.0 x 10^{16} cm^2	σ_r - 5.0 x 10^{-18} cm^2
E_L - 18.1 J/cm^2 x a_L (cm^2)	N_s - 0.51 x 10^{16} cm^{-3}	σ_{ex} - 2.0 x 10^{-16} cm^2
T_L - 250.0 x 10^{-6} sec.	ℓ_s, ℓ_r - 0.12 cm	N_r - 15.8 x 10^{16} cm^{-3}
	T_s - 22.0 x 10^{-12} sec	T_r - 3.9 x 10^{-9} sec

L = .01, T_R = 3.25x10^{-9} sec, E_O = 8.5x10^{-5} J/cm^{2}xa_L, 10 a_L = a_s = a_r (cm^2)

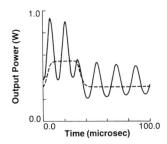

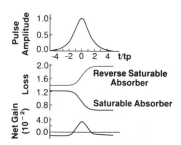

Fig. 2 Temporal response of a cw alexandrite laser to pump rate perturbation without a reverse saturable absorber (solid) and with a reverse saturable absorber (dotted).

Fig. 3 The pulse amplitude, net gain, and the loss from the saturable absorber and the reverse saturable absorber as a function of time for the numerical example.

An additional problem in passive mode-locking a laser with a gain medium which does not saturate easily is that such lasers tend to be unstable against the onset of relaxation oscillations. Such instabilities deleteriously affect the passive mode-locking of these lasers. Haus [4] has theoretically shown that this is a problem for the cw Nd:YAG laser. His arguments apply to the cw alexandrite laser as well. The addition of a reverse saturable absorber into a laser cavity suppresses the onset of relaxation oscillations as is shown in Fig. 2. A hypothetical reverse saturable absorber with properties at 750 nm similar to rhodamine 6G at 435 nm is placed in the cavity of a cw alexandrite laser. The solid line shows the output of the laser when the pumping rate is changed 20% for 30 microseconds without an intracavity reverse saturable absorber. The dashed line shows the laser output with an intracavity reverse saturable absorber illustrating that the population relaxation oscillations are damped while the initial output power is reduced only 15%.

For a laser to be passively mode locked a number of criteria must be met. The most stringent criteria are that the net gain be positive for the duration of the pulse and negative at all other times. To show that it is feasible to meet these criteria on the net gain for mode-locking with a reverse saturable absorber, a numerical example has been studied for a cw alexandrite laser with cryptocyanine as the saturable absorber and a hypothetical reverse saturable absorber which has properties at 750 nm that rhodamine 6G has at 435 nm (see Table 1). Figure 3 shows the temporal behavior of the pulse amplitude, the net gain, and the loss from the saturable and reverse saturable absorber for this example. It is clear from the plot of the net gain that it is positive for the duration of the pulse and is negative at all other times. This example also meets the other criteria for passive mode-locking.

The equation for the temporal pulsewidth of the mode-locked pulse is the same as derived by Haus [1] for the case of mode-locking with a slow saturable absorber (eqn.(33)). In applying eqn. (33) to the numerical example of a cw alexandrite laser, it is found that the temporal pulsewidth will be 0.5 pico-seconds when a 34 Å bandwidth of the 1000 Å lasing bandwidth of alexandrite (~ 7000 Å - 8000 Å) is mode locked. The details of this numerical example and the derivation of the equations describing passive mode locking with a reverse saturable absorber will be given elsewhere [5].

In spite of the fact that we presently do not have a reverse saturable absorber at 750 nm, there are many dyes that have larger excited state cross sections than ground state cross sections in certain wavelength regions.

Hammond [6] showed that the excited state cross section was larger than the ground state cross section in rhodamine 6G from 465 nm to at least 400 nm. We have confirmed experimentally that rhodamine 6G behaves as a reverse saturable absorber at 435 nm and that many rhodamines behave as reverse saturable absorbers throughout the blue and the blue-green. Decker [7] reported that the excited state cross section was 200 times larger than the ground state cross section for IR-140 at 532 nm and in reference [8] sudanschwarz B and sulfonated indanthrone were shown to behave as reverse saturable absorbers at 694.3 nm.

In conclusion, we have described a new technique to mode lock passively using a reverse saturable absorber (in addition to a saturable absorber). This method is particularly suitable for lasers in which a single mode-locked pulse cannot saturate the gain medium. We have also shown that the reverse saturable absorber will stabilize a laser from population relaxation oscilla-tions and that dyes which behave as reverse saturable absorbers exist.

References

1. H.A. Haus, IEEE J. Quantum Electron. QE-11, 736 (1975).
2. H.A. Haus, J. Appl. Phys. 46, 3649 (1975).
3. D.J. Harter, M.L. Shand, and Y.B. Band, J. Appl. Phys. 55, xxx (1984).
4. H.A. Haus, IEEE J. Quantum Electron., QE-12, 169 (1976).
5. D.J. Harter, Y.B. Band, E. Ippen, to be published.
6. P.R. Hammond, IEEE J. Quantum Electron., QE-11, 1157 (1980).
7. C.D. Decker, Appl. Phys. Lett., 27, 607 (1975).
8. C.R. Giuliano and L.D. Hess, IEEE J. Quantum Electron. QE-3, 358 (1967).

Part II

Solid State Physics and Nonlinear Optics

Imaging with Femtosecond Optical Pulses

M.C. Downer, R.L. Fork, and C.V. Shank

AT & T Bell Laboratories, Crawford Corners Road, Holmdel, NJ 07733, USA

The interaction of short laser pulses with semiconductors has been studied by a variety of techniques including time-resolved reflectivity, [1-5] transmission,]1-5] photoluminescence, [1-5] surface ellipsometry, [6] and surface second harmonic generation. [7] In the present work, we report an imaging technique used to obtain the first time-resolved photographs of a silicon surface at fixed time delays ranging from 100 fsec. to 600 psec following excitation with an intense ultrashort optical pulse. When the fluence E of the excitation pulse exceeds a threshold value E_{TH} (approximately 0.1 J/cm^2, under our experimental conditions) a rapid increase in surface reflectivity occurs which has been widely interpreted [8] as thermal melting. [1-5,9] The photographs depict the evolution of the surface reflectivity during and following melting with a time resolution of 100 fsec. and a spatial resolution of 5 μm. Using a movie camera and elementary synchronization electronics, we have also made a motion picture which shows the continuous sequence of melting, boiling, and material ejection over a 600 psec period slowed in time by as much as a factor of 10^{13}. The still photographs presented here depict the major events in this sequence.

Our photographs provide a detailed study of the physics of a highly excited semiconductor surface as a function of time, position on the silicon surface, and reflected light frequency. The optical pump pulse initially excites an electron-hole plasma which alters the real part of the dielectric constant, and therefore the reflectivity and transmission. Within a few hundred femtoseconds following excitation, much of the energy of the excited carriers is transferred to the lattice via LO phonon emission. This results in melting for $E > E_{TH}$. This process is evident in the present photographs as a rapid rise in surface reflectivity up to the known value (R ~ 70% at optical frequencies) for molten silicon. Material ejected from the hottest parts of the molten surface further alters the observed reflectivity at time delays of ten to several hundred picoseconds by scattering and absorbing the illuminating light. We have obtained strong evidence that this material is ejected in the form of liquid droplets several hundred angstroms in diameter which atomize in less than a nanosecond by using selective imaging and spectral analysis of the light reflected from these hot regions.

The time-resolved images were obtained using a variation of the pump and probe technique. A 10 Hz train of 80 fsec., 0.2 mJ pulses at 620 nm. was produced by a colliding-pulse mode-locked (CPM) dye laser [10] followed by a four-stage optical amplifier. [11] A 50% beam splitter divided this output into a pump beam, which was focussed at normal incidence onto a silicon [111] surface (spot diameter ~ 150 μm.), and a second beam, which was focussed into a 1-cm. path length cell containing water to produce a white light continuum pulse, [12] which then served as a probe. The continuum beam was focussed at near normal incidence to a spot of approximately 400 μm diameter overlapping and illuminating the area excited by the pump pulse. The silicon wafer was translated 300 μm after each laser shot by an electronically controlled stage which moved the wafer in a raster pattern. Consequently, each laser shot interrogated a fresh region of the sample.

A magnified image of the excited surface region was obtained at a given pump-probe time delay by collecting the specularly reflected continuum light with an objective lens at f/5 and then imaging the collected light with a second lens onto a screen or photographic film. Typical magnifications were 100x. The time delay between pump and probe was then varied during filming by a standard stepper motor controlled optical delay. The stepper motor increments were 1 μm, corresponding to an optical delay of $6\frac{2}{3}$ fsec.

The images of the excited silicon surface at time delays ranging from $\Delta t = -0.5$ to $+600$ psec. are shown in Figures 1a to 1h. Here the pump fluence was approximately $5E_{TH}$. Before arrival of the excitation pulse, only a uniformly illuminated region of the surface is seen (see Figure 1a). After arrival of the excitation pulse, the reflectivity of the excited region is selectively increased because significant melting has occurred in this region (see Figures 1b through 1d). At $\Delta t = +0.1$ psec, this region is faint (see Figure 1b) for two reasons. First, melting is not complete by this time [7,9] and second, there is a frequency sweep on the probe pulse because of dispersive optics in the path of the probe beam. Thus, only the trailing blue frequencies see the molten silicon at the earliest pump-probe delays. In Figures 1c and 1d, where $\Delta t = 0.5$ and 1.0 psec., respectively, the excited region becomes brighter and whiter as melting is completed and all frequencies in the continuum pulse are reflected by the molten layer.

The appearance of the highly reflective molten spot at later times ($\Delta t > 1.0$ psec.) depends dramatically on the pump fluence. For fluences between E_{TH} and about 2.5 E_{TH}, its appearance remains essentially unchanged from $\Delta t = 1.0$ psec. out to $\Delta t = 600$ psec. This is consistent with earlier results [9] showing that for these fluence levels the reflectivity levels off after about 1.0 psec [9] and remains unchanged until later than a nanosecond, [13] when re-solidification begins.

At higher fluence levels, a dark region begins to appear in the center of the molten spot at $\Delta t = 5$ to 10 psec, as shown in Figure 1e. This central region continues to darken, becoming darkest between $\Delta t = 50$ and 100 psec., although the edge remains bright (see Figure 1f). At still later times, the center of the dark spot begins to become transparent again (Figure 1g) and by $\Delta t = 600$ psec., it has substantially dissipated (Figure 1h), except for a narrow dark ring at the outer edge of the original dark spot.

We believe that this dark region originates from a gradually thickening cloud of material ejected from the hot molten silicon surface in the form of liquid droplets. Maximum particle emission occurs in the hot, central portion of the molten silicon, while the edge remains cool enough that a substantial absorptive cloud never develops, thus explaining the bright outer ring of unobscured molten silicon which persists throughout the time of observation. Earlier studies have demonstrated substantial emission above melting threshold of both charged particles [14] and neutral silicon atoms. [15,16]

Using simple thermodynamic considerations, we calculate that the maximum

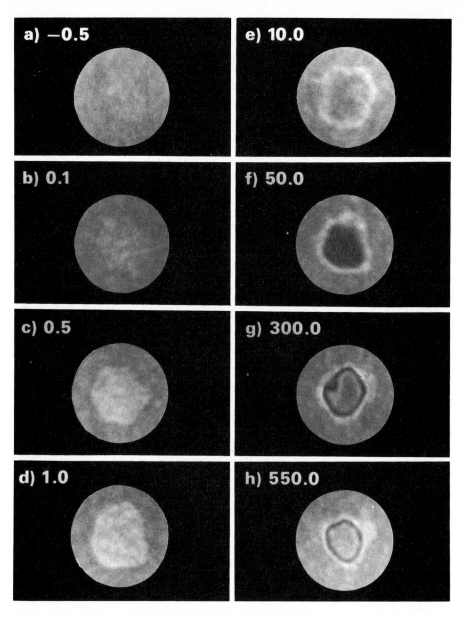

Figure 1 - Time-resolved photographs of a silicon [111] surface
following photo-excitation by an 80 fsec. optical pulse of 0.5
J/cm². Numbers indicate pump-probe optical delay in psec.
Note the rapid appearance of the highly reflective molten
silicon (b through d), followed by the ejection and
dissipation of evaporated material (dark central spot in e
through h).

depth of silicon that can be melted and vaporized by a pulse of 0.5 J/cm^2 is approximately 3600 A. We believe, however, that the actual depth of material removed from the surface is substantially less than 3500 A. Since a highly reflective molten layer is still evident after dissipation of the cloud, it is proven that some of the pulse energy has been deposited in melted, but unvaporized, silicon.

The decrease in optical density at later times, ($\Delta t > 50$ psec.) can be interpreted as the atomization of the liquid droplets into single atoms or small aggregates of atoms, which have negligible absorption and scattering cross-sections at optical wavelengths. This change is equivalent to the removal of effectively absorbing particles from the cloud.

The spatial variations in the optical density at the later time delays are also explained in a natural way by this model. Because of the approximately Gaussian intensity profile of the excitation pulse, the absorbing cloud is hottest in the center. Consequently, the liquid droplets atomize most rapidly in the center, while near the cooler edges of the cloud the droplets remain large for a longer time. This explains why the cloud becomes transparent first in the center, while a dark ring remains at the edge (inside the bright ring of unobscured molten silicon) even at the latest time delays studied. A detailed analysis of the particle emission process will appear in a forthcoming publication. [17]

We have demonstrated a simple technique for filming the progress of an ultrafast melting and material ejection process, with a time resolution of 100 fsec. The results suggest that for excitation fluences several times the melting threshold, silicon is ejected from the melted surface in the form of droplets with radii of several hundred angstroms, which atomize in several hundred picoseconds. Applications of this femtosecond imaging technique to the study of other ultrafast processes, such as carrier transport in semiconductors, should be possible. Infrared illumination frequencies would be particularly sensitive to the presence of photo-excited carriers. With optically thin samples, transmitted as well as reflected light could be imaged.

We are grateful to A. Ashkin and P. L. Liu for helpful discussions, to D. W. Taylor for the computer interface and to F. Beisser for technical support.

REFERENCES

1. Laser and Electron Beam Processing of Materials, ed. by C. W. White and P. S. Peercy (Academic Press, New York, 1980).
2. Laser and Electron Beam Solid Interactions and Material-Processing, eds. J. F. Gibbons, L. D. Hess, and T. W. Sigmon (North-Holland, Amsterdam, 1981).
3. Laser and Electron Beam Interactions with Solids, eds. B. R. Appleton and G. K. Celler (North-Holland, Amsterdam, 1982).
4. Laser-Solid Interactions and Transient Thermal Processing of Materials, eds. J. Narayan, W. L. Brown, and R. A. Lemons (North-Holland, Amsterdam, 1983).
5. Picosecond Phenomena III, eds. K. B. Eisenthal, R. M. Hochstrasser, W. Kaiser, and A. Lauberau (Springer-Verlag, Berlin, 1982).
6. D. H. Auston and C. V. Shank, Phys. Rev. Lett., 32, 1120 (1974).
7. C. V. Shank, R. T. Yen and C. Hirlimann, Phys. Rev. Lett., 51, 900 (1983).

8. For alternative interpretations, see J. A. Van Vechten, R. Tsui, and F. W. Sans, Phys. Lett., 74A, 422 (1979); see also J. A. Van Vechten, in Ref. 3, pp. 49–60.

9. C. V. Shank, R. Yen, and C. Hirlimann, Phys. Rev. Lett., 50, 454 (1983).

10. R. L. Fork, B. I. Greene, and C. V. Shank, Appl. Phys. Lett., 38, 671 (1981).

11. R. L. Fork, C. V. Shank, and R. Yen, Appl. Phys. Lett., 41, 223 (1982).

12. R. L. Fork, C. V. Shank, C. Hirlimann, R. Yen, and W. J. Tomlinson, Opt. Lett., 8, 1 (1983).

13. R. Yen, J. M. Liu, H. Kurz, and N. Bloembergen, in Ref. 3, pp. 37–42.

14. J. M. Liu, R. Yen, H. Kurz, and N. Bloembergen, Appl. Phys. Lett., 39, 755 (1981).

15. M. Hanabusa, M. Suzuki, and S. Nishigaki, Appl. Phys. Lett., 38, 385 (1981).

16. B. Stritzker, A. Pospieszczyk, and J. A. Tagle, Phys. Rev. Lett., 47, 356 (1981).

17. M. C. Downer, R. L. Fork, and C. V. Shank, J. Opt. Soc. Am. B, to be published.

Femtosecond Multiphoton Photoelectron Emission from Metals

J.G. Fujimoto and E.P. Ippen
Department of Electrical Engineering and Computer Science and Research
Laboratory of Electronics, Massachusetts Institute of Technology
Cambridge, MA 02139, USA
J.M. Liu
GTE Laboratories, Incorporated, Waltham, MA 02254, USA
N. Bloembergen
Gordon McKay Laboratory, Division of Applied Sciences
Harvard University, Cambridge, MA 02139, USA

Multiphoton photoelectron emission from metals has been the subject of experimental and theoretical investigation for several years [1]. The development of high intensity ultrashort pulse laser sources has made possible the extension of these studies into the picosecond time regime [2,3]. For intense pulses of short enough duration, it has been postulated that a transient thermal nonequilibrium between the electrons and phonons may be generated [4]. This phenomenon has been termed anomalous heating and is predicted when the laser pulse durations are comparable to or shorter than the electron-phonon energy relaxation time. Because of the smaller heat capacity of the electron gas, heating of the electrons to temperatures in excess of the lattice melting temperature would then be possible. Previous experimental investigations of photoelectron emission have been performed with high intensity picosecond pulses in attempts to observe anomalous heating [2,3]. To date, none have achieved the temporal resolution necessary for such an observation. However, recently, indirect evidence of transient heating has been reported using picosecond reflectivity measurements in copper [5].

In this paper we report on the observation of multiphoton and thermally assisted photoemission from a tungsten metal surface using high intensity 75 fsec optical pulses at 620 nm. Measurements of the power-law dependence of the photoemitted charge vs. incident pulse intensity provide evidence for a thermal nonequilibrium between the electrons and lattice. Pump-probe measurements indicate an electron-phonon energy relaxation time of several hundred femtoseconds.

The processes of multiphoton and thermally assisted photoemission in tungsten for pulses at our wavelength are illustrated in Fig. 1. For a work function of 4.3 eV and a photon energy of 2 eV, photoemission occurs via a three-photon process at low temperatures. Thermally assisted lower order processes should result in increased photoemission with electron heating.

In our experiments, 75 fsec pulses produced by a CPM ring dye laser and 4-stage dye amplifier [6,7] were focussed at normal incidence onto the surface of a polished polycrystalline tungsten sample mounted in a vacuum chamber maintained at 10^{-6} Torr. A 1 mm diameter wire held 2 mm above the metal surface and biased at 5 kV functioned as an anode and collected the photoemitted charges. Incident pulse energies and total photoemitted charge were measured and recorded by computer on a shot-to-shot basis. For pump-probe measurements a Michelson-like delay line was used to produce two equal intensity pulses which were incident nearly collinearly and focussed to the same spot on the sample surface. Relative delay between the pulses was controlled by computer which also discriminated against pulse energy fluctuations.

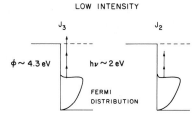

LOW INTENSITY

THERMALLY ASSISTED

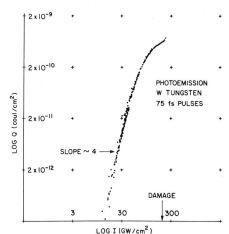

Fig. 1 Multiphoton photoemission processes in tungsten

PHOTOEMISSION
W TUNGSTEN
75 fs PULSES

SLOPE ~ 4

DAMAGE

LOG I (GW/cm^2)

Fig. 2 Log-log plot of photo-emitted charge vs. laser pulse intensity. Each point is a single laser shot

Experimental measurements of total photoemitted charge vs. laser pulse intensity are displayed in the log-log plot of Fig. 2. Note that even at low incident intensities the photoemission behavior is still more nonlinear than the intrinsic three-photon process. A dependence of approximately slope four is observed at intensities of $\sim$30 GW/cm^2. If this high slope is the result of thermally assisted photoemission, then energy arguments imply that there must be a thermal nonequilibrium between the electrons and lattice. At higher intensities a decrease in the nonlinearity of the photoemission is observed up to the damage threshold for the metal. This decrease in slope may be explained by a space-charge effect which screens the sample from the extracting anode field and thus suppresses the photo-emission. The magnitude of the saturation effect is in fact dependent on the extraction field. In addition, pump-probe studies show that the space-charge saturation recovers on a time scale of $\sim$100 picoseconds.

Evidence for a thermal nonequilibrium between the electrons and lattice is further provided by femtosecond pump-probe measurements as shown in Fig. 3. The total photoemitted charge is plotted as a function of the delay between two equal intensity pulses. The two traces are obtained for different pairs of pulse energies: 13 GW/cm^2 and 27 GW/cm^2. When the two pulses are temporally overlapped, separated by delays of the order of the pulsewidth, the photoemission is determined by the instantaneous intensity of the overlapping pulses and a high order autocorrelation function of the pulse intensity is obtained. The space-charge saturation effect causes the decrease in peak to background contrast ratio in the higher intensity pump-probe trace. When the delay between the two pulses is several hundred femtoseconds and the pulses are not temporally overlapping, the enhanced photoemission in the wings of the pump-probe correlation data indicate the presence of thermally assisted photoemission with temperature increases lasting for several hundred femtoseconds. Thus the high in-tensity 75 fsec laser pulses produce a transient nonequilibrium heating

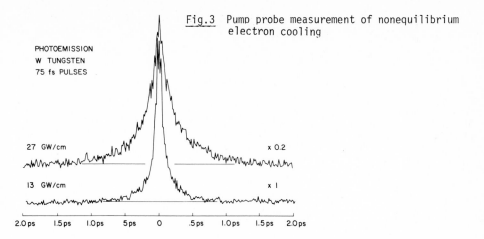

Fig.3 Pump probe measurement of nonequilibrium electron cooling

PHOTOEMISSION
W TUNGSTEN
75 fs PULSES

27 GW/cm   x 0.2

13 GW/cm   x 1

2.0 ps   1.5 ps   1.0 ps   5 ps   0   .5 ps   1.0 ps   1.5 ps   2.0 ps

of the electron gas which then cools by energy transfer to the lattice with an electron-phonon energy transfer time of several hundred femtoseconds.

References

1. S.I. Anisimov, V.A. Benderskii, and G. Farkas, Sov. Phys. Usp. 20, 467 (1977).
2. R. Yen, J. Liu, and N. Bloembergen, Opt. Commun. 35, 277 (1980).
3. R. Yen, J.M. Liu, N. Bloembergen, T.K. Yee, J.G. Fujimoto, and M.M. Salour, Appl. Phys. Lett. 40, 185 (1982).
4. S.I. Anisimov, B.L. Kapeliovich, and T.L. Perel'man, Sov. Phys. -JETP, 39, 375 (1975).
5. G.L. Eesley, Phys. Rev. Lett. 51, 2140 (1983).
6. R.L. Fork, B.I. Greene, and C.V. Shank, Appl. Phys. Lett. 38, 671 (1981).
7. R.L. Fork, C.V. Shank, and R.T. Yen, Appl. Phys. Lett. 41, 223 (1982).

Time-Resolved Laser-Induced Phase Transformation in Aluminum

S. Williamson and G. Mourou

Laboratory for Laser Energetics, University of Rochester, 250 East River Road
Rochester, NY 14623, USA

J.C.M. Li

Department of Mechanical Engineering, University of Rochester
250 East River Road, Rochester, NY 14623, USA

Phase transformation in condensed matter is an important area of study in solid state physics since it relates to the genesis and evolution of new microstructure. The mechanisms responsible in such critical phenomena are still not fully understood. Previous experimental information has left unmeasured such important parameters as the minimum number of nuclei and their common critical radius for a transformation to occur as well as the interphase velocity. Several probe techniques have now been developed to time-resolve phase transformations in semiconductors during laser annealing. However most of these probes (eg. electrical conductivity,[1,2] optical reflection,[3,4] optical transmission,[5] Raman scattering,[6] and time of flight mass spectrometry[7]) supply no direct information about the atomic structure of the material. Probing the structure can reveal when and to what degree a system melts as it is defined by degradation in the long range order of the lattice. True structural probes based on x-ray[8] and low energy electron[9] diffraction and EXAFS[10] with nanosecond time resolution have been developed offering fresh insight into both the bulk and surface dynamics of material structure. Also a subpicosecond probe based on structural dependent second harmonic generation[11] has been demonstrated. But at present, only the technique of picosecond electron diffraction[12] can produce an unambiguous picture of the structure on the picosecond timescale. In this letter we report on the results of using this probe to directly observe the laser induced melting of aluminum.

The burst of electrons is generated from a modified streak camera that, via the photoelectric effect, converts an optical pulse to an electron pulse of equal duration.[13] Equal in importance is the fact that the electron pulse can be synchronized with picosecond resolution to the laser pulse.[14] The experimental arrangement is illustrated in Fig. 1. A single pulse from an active-passive modelocked Nd+3:YAG laser is spatially filtered and amplified to yield energies up to 10 mJ. The portion of the laser irradiating the photocathode is first up-converted to the fourth harmonic of the fundamental wavelength in order to produce the electrons efficiently. The duration of the UV pulse and thus the electron pulse is ~ 20 ps. Once the electron pulse is generated it accelerates through the tube past the anode and then remains at a constant velocity. The specimen is

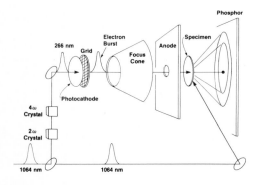

Fig. 1 Schematic of Picosecond electron diffraction apparatus. A streak camera tube (deflection plates removed) is used to produce the electron pulse. The 25 keV electron pulse passes through the Al specimen and produces a diffraction pattern of the structure with a 20 ps exposure.

located in this drift region. A gated microchannel plate image intensifier in contact with the phosphor screen amplifies the electron signal ~ 10^4 times.

A metal was chosen over a semiconductor as the specimen because of the ease with which metals can be fabricated in ultra-thin polycrystalline films. Free standing films 250 ± 20 Å in thickness were required so that the electrons sustain, on the average, one elastic collision while passing through the specimen. This thickness of Al corresponds to twice the 1/e penetration depth at 1060 nm. It is worth noting that the penetration depth for a metal does not vary significantly from solid to liquid as is the case with a semiconductor where a change in absorption of one to two orders of magnitude is possible. The films were fabricated by first depositing Al onto formvar substrates and then vapor-dissolving away the formvar. Since the diffusion length $(D\tau)^{\frac{1}{2}}$, where D is the thermal diffusivity coefficient (0.86 cm^2/s) and τ is time, is limited to 250 Å, the temperature in the Al is uniformly established in less than 10 ps. The absorption of the laser by aluminum is 13 ± 1 percent. Because the diffracted electrons lie in concentric rings at discrete radii from the zero order we can circular average to recover the signal. This is accomplished by rapidly spinning the photograph of the signal about its center. The process acts to accentuate the real signal occurring at fixed radii while smoothing out the randomly generated background noise. The diameter of the laser stimulus is ~ 4 mm ($1/e^2$) and is centered over the 2 mm specimen. Synchronization between the electron pulse and the laser stimulus is achieved by means of a laser-activated deflection plate assembly.[15]

The experimental procedure is then to stimulate the aluminum sample with the laser while monitoring the lattice structure at a given delay. The films are used only once even though for low fluence levels (< 8 mJ/cm^2) the films could survive repeated shots. Figure 2 shows the laser-induced time-resolved phase transformation of aluminum at a constant fluence of ~ 13 mJ/cm^2. The abrupt disappearance of rings in the diffraction pattern occurs with a delay of 20 ps. As is evident, the breakdown of lattice order can be induced in a time shorter than the resolution of our probe. However, the fluence required for this rapid transition exceeds F_{melt}, the calculated fluence required to completely melt the Al specimen under equilibrium conditions (~ 5 mJ/cm^2). At a constant fluence of 11 mJ/cm^2 the phase transition was again observed but only after a probe delay of 60 ps. Figure 3 shows the melt metamorphosis of Al where the points represent the delay time before the complete phase transition is observed for various fluence levels. We see that the elapsed time increases exponentially with decreased fluence and at 7 mJ/cm^2 the delay is ~ 1 ns. Because the fluence level that is applied is always in excess of F_{melt}, the observed delay time suggests that the Al is first driven to a superheated solid state before melting.

Let us consider the melt to originate from a spherical cluster of atoms that begin to vibrate incoherently to the point where structural order is lost. The nucleation theory then tells us that beyond a critical radius, r* (on the order of tens of Angstroms) the

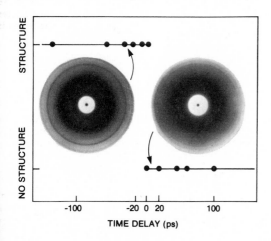

TIME DELAY (ps)

Fig. 2 Time-resolved laser-induced phase transition in aluminum. The pattern on the left is the diffraction pattern for Al and represents the points along the top line -where the electron pulse arrives before the laser stimulus. The pattern on the right shows the loss of structure in the Al 20 ps (or more) after applying the laser stimulus at a fluence of 13 mJ/cm^2. The fine line background structure occuring in both pictures is an artifact of the circular averaging technique.

115

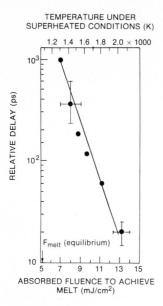

TEMPERATURE UNDER
SUPERHEATED CONDITIONS (K)

1.2 1.4 1.6 1.8 2.0 × 1000

RELATIVE DELAY (ps)

F_{melt} (equilibrium)

ABSORBED FLUENCE TO ACHIEVE
MELT (mJ/cm²)

Fig. 3 Laser-induced melt metamorphosis for aluminum. The points mark the elapse time for the diffraction rings to completely disappear. It must be pointed out that as the fluence was decreased the abruptness with which the rings disappeared became less dramatic. Consequently, determination of the precise delay increases in difficulty with decreasing fluence. The vertical error bar represents the degree of uncertainty in defining the moment of complete melt. The region beneath the curve represents the conditions under which the Al is left in a superheated solid state.

sphere will rapidly expand throughout the volume until the entire Al specimen is transformed into the liquid phase. According to the theory, the observed transition time (τ_{ob}) between the driving pulse and the complete disappearance of the diffraction rings should correspond to the time necessary for the expanding liquid spheres to fill the volume between nuclei. The average distance between nuclei is $2v\,\tau_{ob}$, where v is the radial rate of expansion. Taking $v = 5\times10^5$ cm/s and $\tau_{ob} = 20$ ps, this distance is 2000 Å, a value much larger than the film thickness. Hence, the expansion is primarily cylindrical rather than spherical. The nuclei density is thus given by

$$N = \frac{1}{\pi(2v\tau_{ob})^2} .$$

(1)

Eq. (1) is valid regardless of whether the nuclei originate on the surface or in the bulk. If we assume v to be constant with lattice temperature, we find that the density of nuclei increases from 10^5 cm^{-2} to 10^9 cm^{-2} when the fluence is increased from 7 mJ/cm² to 13 mJ/cm².

In conclusion, we have demonstrated that the picosecond electron diffraction technique can be used to time-resolve the laser-induced melt metamorphosis in aluminum. It was found possible to completely melt the aluminum in a time shorter than 20 ps if sufficient laser fluence is applied ($\gtrsim 2.6$x F_{melt}). The time required to melt the aluminum increases exponentially with decreasing fluence and at 1.4xF_{melt} the phase transition time increases to ~1 ns. During this time, the two phases coexist as a heterogeneous melt while the superheated solid is being continuously transformed into liquid. We believe that these results show for the first time in an unambiguous way, the relationship between superheating and delayed melting and demonstrate the important role that the technique of picosecond electron diffraction can play in the study of the genesis of melting.

Acknowledgment

This work was partially supported by the Laser Fusion Feasibility Project at the Laboratory for Laser Energetics which has the following sponsors: Empire State Electric Energy Research Corporation, General Electric Company, New York State Energy

Research and Development Authority, Northeast Utilities Service Company, Southern California Edison Company, The Standard Oil Company (Ohio), the University of Rochester. Such support does not imply endorsement of the content by any of the above parties.

We would like to acknowledge the support of Jerry Drumheller who assisted in the fabrication of the Al films as well as to Hsiu-Cheng Chen for her help during the experiment.

References

1. M. Yamada, H. Kotani, K. Yamazaki, K. Yamamoto, and K. Abe, **J. Phys. Soc. Japan,** 49, 1299 (1980).
2. G.J. Galvin, M.O. Thompson, J.W. Mayer, R.B. Hamond, N. Paulter and P.S. Peercy, **Phys. Rev. Lett.,** 48, 33 (1982).
3. D.H. Auston, C.M. Surko, T.N.C. Venkatesan, R.E. Slusher, and J.A. Golovchenko, **Appl. Phys. Lett.,** 33, 437 (1978).
4. C.V. Shank, R. Yen, and C. Hirlimann, **Phys. Rev. Lett.,** 50, 454 (1983).
5. J.M. Liu, H. Kurz, and N. Bloembergen, **Appl. Phys. Lett.,** 41, 643 (1982).
6. H.W. Lo and A. Compaan, **Phys. Rev. Lett.,** 44, 1604 (1980).
7. A. Pospieszczyk, M.A. Harith, and B. Stritzker, **J. Appl. Phys.,** 54, 3176 (1983).
8. B.C. Larson, C.W. White, T.S. Noggle, and D. Mills, **Phys. Rev. Lett.,** 48, 337 (1982).
9. R.S. Becker, G.S. Higashi, and J.A. Golovchenko, **Phys. Rev. Lett.,** 52, 307 (1984).
10. H.M. Epstein, R.E. Schwerzel, P.J. Mallozzi, and B.E. Campbell, **J. Am. Chem. Soc.,** 105, 1466 (1983).
11. C.V. Shank, R. Yen, and C. Hirliman, **Phys. Rev. Lett.,** 51, 900 (1983).
12. G. Mourou and S. Williamson, **Appl. Phys. Lett.,** 41, 44 (1982).
13. D.J. Bradley and W. Sibbett, **Appl. Phys. Lett.,** 27, 382 (1975).
14. G. Mourou and W. Knox, **Appl. Phys. Lett.,** 36, 623 (1980).
15. S. Williamson, G. Mourou, and S. Letzring, **Proc. 15th Int. Cong. on High Speed Photography,** Vol. 348 (I), 197 (1983).

Picosecond Photoemission Studies of Laser-Induced Phase Transitions in Silicon

A.M. Malvezzi, H. Kurz, and N. Bloembergen

Division of Applied Sciences, Harvard University, Cambridge, MA 02138, USA

Quantitative evaluation of the energy transfer processes from optically excited electron-hole pairs to lattice phonons has become an important issue in the study of laser-induced phenomena at the surface of semiconductors. This goal has been mostly pursued through time-resolved optical diagnostics. Electron-hole plasma kinetics and evolution of the lattice temperature have been explored in a wide range of time scales [1]. However, lack of information on the energy dependence of several critical parameters in the dielectric response of the medium prevent an evaluation of the energy content of the carriers with these methods. The second question of major importance for the understanding of the laser-induced phase transitions is whether the solid state lattice structure becomes unstable by electronic or vibronic excitation. Even with ps laser excitation, however, strong evidence for an ultrafast energy transfer from the plasma to the lattice phonons has been found [2] and the validity of the thermal model confirmed.

Photoemission provides an alternative approach to address the problem of energy transfer and to study the onset of the phase transition. In contrast to optical measurements, whose interpretation is complicated by strong spatial gradients of plasma density and temperature, here only the outermost layer of the surface is probed. According to the well-known Fowler-Dubridge theory, high excitation photoelectric processes are governed by power laws corresponding to the number of photons required to bridge the gap ϕ between valence band and vacuum level via virtual or intermediate states. The strict power law is violated if the threshold for photoemission is broadened by hot carrier distributions. At very high carrier temperatures thermionic emission is expected, described by the Richardson-Dushmann equation. The high energy tail of the hot carrier distribution reaches the vacuum level.

In previous picosecond photoemission experiments a striking coincidence between the appearance of a high reflectivity phase and the emission of positive particles has been found [3]. Close to the threshold for the transition to a highly reflecting structure, the photoemission increases strongly. Recently it has been shown that this drastic increase of electron emission is due to the neutralization of space charge fields by injected positive charged ions [4].

To use picosecond photoemission techniques for quantitative investigations of the energy transfer processes as well as of the ultrafast phase transition to the metallic state, the photoelectric response has to be studied in a wide range of excitation levels. We report here the results of extensive measurements on picosecond photoemission from crystalline silicon with photon energies corresponding to the second harmonic ($h\nu$ = 2.33 eV) and fourth harmonic ($h\nu$ = 4.66 eV) of a 30 ps Nd:YAG laser pulse. The photoelectric response is studied as a function of the incident laser fluence from $\sim 10^{-4}$ F_{th} to $2F_{th}$, where F_{th} is the threshold value for the formation of amorphous material on the surface. Using a two-beam temporal correlation technique, information about the kinetics of the nonlinear effects is obtained. In particular, evidence of plasma-assisted photoemission is observed. Both Si (111) and Si (100) samples have been used with comparable results. A single vacuum diode configuration, as described in [4], has been used with a biasing voltage up to ∓ 4 KV.

The emitted charge density versus UV laser fluence is illustrated in Fig. 1 where three distinct photoelectron emission regimes can be observed. At lower fluences up to 0.4 mJ/cm² (regime I), we observe a progressive departure from linearity. The collected charge density is independent of the electric field applied to the collector wire. In this regime a quadratic contribution is superimposed on the linear single-photon ionization contribution [4]. The total collected charge density q(C/cm²) is described by

$$q = \kappa_1 \frac{e}{h\nu} F + \kappa_2 \frac{e^2}{2\sqrt{2}\,(h\nu)^2} \frac{1}{\tau} F^2$$

where τ is the laser pulse duration, F the laser fluence (J/cm²), e the electron charge, hν the photon energy and κ_1, κ_2 the coefficient for linear and quadratic photoemission, respectively. The dashed line in Fig. 1 shows the calculated charge density fitted with $\kappa_1 = 2.7 \times 10^{-8}$ C/J and $\kappa_2 = 9 \times 10^{-15}$ C cm² s/J. The experimental linear quantum yield of 8 ∓ 2 × 10⁻⁷ electrons per absorbed photon is in agreement with data found in the literature for silicon samples exposed to air [5].

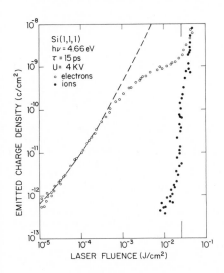

Fig. 1. Emitted charge density versus laser fluence at 266 nm for Si (111) samples. The dashed line refers to the calculated linear and quadratic photoelectron effect with the data given in the text

A second photoelectron emission regime (region II) extends up to the fluence $F_{th}^{4\omega}$. The photoemission is here progressively blocked by space charge effects. The collected charge density is now dependent on the voltage applied to the collector wire. The rapid accumulation of charges on the surface screens the applied field, and the electron emission is limited to the leading edge of the picosecond laser pulse.

When the laser fluence reaches regime III, the critical value for structural changes of the surface, $F_{th}^{4\omega} \simeq 25$ mJ/cm², the electron emission exhibits an extremely nonlinear increase. This behavior occurs with the simultaneous appearance of a positive ion signal, observable when the bias voltage is reversed. We attribute the sharp increase of photoelectrons and the positive ion emission to a structural charge of the surface. The threshold for photoemission is abruptly changed to lower values and ions are generated during this highly disordered phase. This picture is consistent with melting of the surface within the laser pulse. The presence of positive ions certainly contributes to lower drastically the effect of the space charge, so that an increasing number of electrons is detected. At $\sim 2F_{th}^{4\omega}$ equal amounts of electrons and ions are collected. Similar behavior at fluences above the critical value $F_{th}^{2\omega} = 0.2$ J/cm^2 are observed using 532 nm (2.33 eV) ps pulses.

To explore the nature of the quadratic contribution in regime I, we performed a photoemission correlations experiment. Two laser pulses separated by a variable time delay τ strike the surface at the same spot and the resulting electron emission is measured. In Fig. 2 the ratio of the total collected charge Q and the laser energy (Q/E_n) is plotted versus the sum of laser fluences (J/cm^2). The $\tau = \infty$ condition is simulated by separating the laser spot sufficiently on the target surface. Below $F_{th}^{2\omega} = 3$ mJ/cm^3 the quantum efficiency increases linearly as expected from the quadratic photoelectric dependence displayed in Fig. 1. The slope depends on the delay time between the two UV picosecond pulses. The first pulse obviously sensitizes the sample for the following pulse. The noticeable correlation between two pulses separated by 90 ps indicates the existence of intermediate states mediating the two-step process.

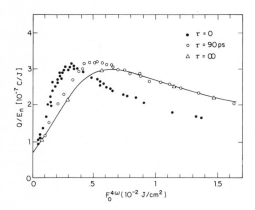

Fig. 2. Ratio Q/E_n of the emitted negative charge over total laser energy at 266 nm for two overlapping pulses on Si (111) samples temporally separated by the delays τ shown

In the space charge regime above 5 mJ/cm^2, the situation is reversed. The first pulse reduces the quantum yield of the second pulse by forming a space charge layer in front of the surface. The largest reduction occurs if the two pulses strike the surface simultaneously ($\tau = 0$). The blockage of the emission from the second pulse is totally lifted after time delays larger than 90 ps. Obviously the density of the space charge cloud generated by the first pulse is significantly reduced after 90 ps. This leads to the conclusion that the electron emission process is terminated immediately after the first picosecond pulse and the charge build-up during the pulse is removed from the cathode within 90 ps.

To obtain more information about the nonlinear process in regime I, the cross correlation of the collected charges between the first and the second pulse has been measured. Despite the large experimental errors in Fig. 3, the temporal behavior of the correlated photoelectric signal can be safely

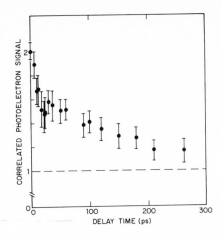

Fig. 3. Correlated photoelectron signal versus delay time τ for 266 nm illumination of Si (111) samples in region I. The data are normalized to the τ = ∞ results

attributed to intermediate excited electron states, whose lifetime is consistent with Auger recombination and ambipolar diffusion of an electron-hole plasma in silicon [1]. This plasma-assisted enhancement emission accounts for a major part of the quadratic photoelectric response in UV-picosecond irradiation experiments. Vice versa, this new result may open the way to further studies of the plasma kinetics in the outermost layer of the excited surface.

The lower photoelectron yield at 2.33 eV provides favorable conditions for detection of possible thermionic contributions. The steep dependence of the thermionic effect versus carrier temperature should give rise to an additional highly nonlinear contribution to the total collected charge. The photoelectric emission data at 2.33 eV versus laser fluence shows principally a quadratic dependence mixed with higher order contributions as soon as the threshold value of $F_{th}^{2\omega} = 200$ mJ/cm^2 for amorphous spot formation is reached. In contrast to the 4.66 eV irradiation, the photoelectric yield depends strongly on the surface conditions, as expected from a near-threshold excitation. Below $F_{th}^{2\omega}$ the collected charge density is considerably lower than in the case of UV excitation. Thus space charge limitations by excessive photoelectric emission are less severe. Transferring the UV results, they should be restricted to the region ~30% below $F_{th}^{2\omega}$. Application of the Richardson-Dushmann equation to the data taken at 0.1 J/cm^2, where space charge effects can be totally ruled out, reveals an upper limit of $T_c \sim 2200$ K for the carrier temperature averaged over the laser pulse duration. This upper limit of the averaged carrier temperature requires an energy relaxation time of $\tau_e = 1$ ps [6,7].

References

1. L.A. Lompré, J.M. Liu, H. Kurz, N. Bloembergen: Mat. Res. Soc. Symp. Proc. Boston 1983
2. L.A. Lompré, J.M. Liu, H. Kurz, N. Bloembergen: Appl. Phys. Lett. 44, 3 (1984)
3. J.M. Liu, R. Yen, H. Kurz, N. Bloembergen: Mat. Res. Soc. Symp. Proc. 4, 23 (1982)
4. A.M. Malvezzi, J.M. Liu, N. Bloembergen: Mat. Res. Soc. Symp. Proc. Boston 1983
5. R.M. Broudy, Phys. Rev. B 1, 3430 (1970)
6. A. Lietola, J.F. Gibbons, Mat. Res. Soc. Symp. Proc. 4, 163 (1982)
7. Work supported by the U.S. Office of Naval Research under contract N00014-83K-0030 and by the Joint Services Electronics Program of the U.S. Department of Defense under contract N00014-75-C-0648.

Dynamics of Dense Electron-Hole Plasma and Heating of Silicon Lattice Under Picosecond Laser Irradiation

L.A. Lompré*, J.M. Liu**, H. Kurz, and N. Bloembergen

Division of Applied Sciences, Harvard University, Cambridge, MA 02138, USA

Numerous investigations of the mechanisms for pulsed laser-induced phase transition provide ample evidence that the surface of metals and semiconductors
undergoes a solid-liquid phase change as soon as a critical laser fluence is
exceeded. There is no doubt about the thermal nature of the phase transition
as long as nanosecond pulses are being used [1]. It is the main purpose of
this contribution to clarify whether under picosecond irradiation the simple
thermal melting approach is still valid. This question is focused on the time
required to establish thermal equilibrium between carriers and phonons [2].

The conventional picosecond pump and probe technique is applied [3,4]. The
highly focused probe beam monitors the transmission and reflectivity of bulk
surfaces on thin silicon films on sapphire (SOS), induced by an exciting
picosecond pulse at 0.532 μm. The wavelength of the probe pulse is varied
between 0.532 and 2.8 μm. By comparing the changes in optical properties at
different wavelengths, the contributions due to a variation in lattice temperature T_L and those due to changes in carrier density N are separately
determined [5].

At the doubled frequency of a Nd:YAG laser ($\hbar\omega$ = 2.33 eV), the optical
properties of picosecond-excited silicon are mainly determined by the indirect
band gap transition. The phonon-assisted indirect absorption α depends
strongly on the lattice temperature T_L [6]. In SOS samples multiple reflections from air-silicon and silicon-sapphire interfaces enhance the optical
detectivity of changes $\Delta\varepsilon$ in the real part of the dielectric functions considerably. Time-resolved analysis of the reflectivity signatures of SOS sam-
ples reveal the interplay between thermal ($\Delta\varepsilon$ > 0) and free carrier ($\Delta\varepsilon$ < 0)
contributions to ε [9]. Due to Auger recombination the plasma density is
known to drop below 1.2×10^{20} cm^{-3} after 200 ps. At this time the plasma contributions become negligible at 0.532 μm and the amount of lattice heating
can be determined experimentally by comparing samples with different film
thicknesses [5].

In Fig. 1 the surface temperature developed 200 ps after the exciting pulse
is shown. Picosecond optical probing of lattice temperature reveals significantly higher values than those derived up to now from Raman scattering data
using ns pulses. Compared to bulk surfaces (F_{th} = 200 mJ/cm^2), SOS samples
exhibit a lower threshold value for surface melting (160 mJ/cm^2). Because of
multiple interferences in SOS samples, a larger amount of energy density is
absorbed in the silicon film. The absorbed energy density in 0.5 μm thick
silicon films is determined by reflectivity and transmission measurements of
the pump beam (see top of Fig. 1). The surface temperature increases non-

---

*Permanent address: C.E.N./Saclay, DPh.G/S.P.A., 91191 Gif-sur-Yvette Cedex,
 France
**Present address: GTE Laboratories, 40 Sylvan Road, Waltham, MA 02154, USA

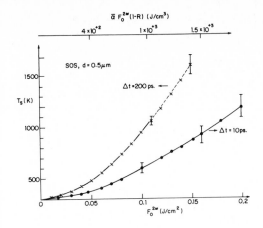

Fig. 1. Surface temperatures T_s versus incident laser fluence at 10 and 200 ps after the heating pulse at 0.532 μm. The data are evaluated from time-resolved reflectivity and transmission measurements on SOS samples using the thermo-optic data of ref. [6]

linearly with the laser fluence $F(2\omega)$ and reaches values close to the melting point of silicon (T_m = 1680 K) at a fluence level (160 mJ/cm^2) where the reflectivity increases rapidly to the liquid state value. Obviously lattice heating and melting occur during the pump pulse. The band gap shrinks nearly instantaneously, resulting in an increase of the pump beam absorbance. This picture is confirmed by measuring the transmission at time delays shorter than the pulse duration. At a time delay of Δt = 10 ps, significant lattice heating already occurs. The result shown in Fig. 1 is obtained by temporal deconvolution of the heating and probing pulse. Plasma contribution has been properly taken into account. Clearly, phonons participating in the indirect absorption process received a significant amount of energy deposited primarily in the electron-hole plasma. The observed transmission changes are in excellent agreement with the optical heating model, where an instantaneous energy transfer to the phonons is assumed.

At a probing frequency below the indirect band gap, free carrier contribution becomes dominant. Drude-like plasma resonances have been observed at 2.8 μm [7]. They have been used for a preliminary estimate of plasma densities far below the threshold for melting. At this probing wavelength the plasma edge is reached at a fluence level of 40 mJ/cm^2 already. Strong free carrier absorption causes transmission drops to zero, preventing a reliable determination of the maximum plasma density at higher fluence levels. For this purpose thin film and bulk crystals are carefully investigated at 1.064 and 1.9 μm. At these probe wavelengths the plasma contributions are less pronounced, and measurements up to the melting point can be performed. This study is completed by a novel three-pulse technique, where the first pulse at 0.532 μm creates carriers, a second pump pulse at 1.064 μm couples energy to the plasma by free carrier absorption without creating new carriers, and a third pulse monitors the induced optical changes [8]. The plasma heating of the second pulse does not affect the susceptibility associated with intraband (Drude) and interband transitions. Neither the reduced mass m^* nor the density of electron-hole pairs is changed by additional heating of the plasma. The probability of carrier multiplication by impact ionization is very low [8,9]. The energy loss rate of the carriers to the phonons is faster than the rate for impact ionization, which is expected to be comparable to the Auger recombination rate.

As a selected example, the transmission of bulk silicon at 1.064 μm is plotted versus incident laser fluence for two different time delays Δt in Fig. 2. At Δt = 200 ps the plasma density levels off to a constant value above $F(2\omega)$ = 50 mJ/cm^2 due to the nonlinearity of the Auger recombination

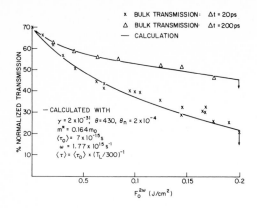

Fig. 2. Transmission of bulk
silicon at 1.064 μm versus incident laser fluence at 0.532 μm
for two different time delays
between pump and probe pulse

process. The lattice temperature, however, increases strongly with F(2ω).
Under these conditions the free carrier absorption at 1.064 μm associated
with intraband transitions is proportional to $[m^* <\tau> (2\gamma t)^{\frac{1}{2}}]^{-1}$, where m* means
the reduced optical mass for electron-hole pairs, $<\tau>$ the energy averaged
momentum relaxation time of the carriers, and γ the Auger recombination coefficient.

The solid line in Fig. 2 represents calculated transmission values of bulk
silicon using $\gamma = 2 \times 10^{-31}$, $m^* = 0.164 \, m_0$ and averaged momentum relaxation
time $<\tau> = 7 \times 10^{-15} \times 300/T_L$, as found in ns experiments [10]. The agreement
between measured and calculated data is excellent, indicating a strong phonon
participation in the carrier momentum relaxation, even at plasma density
levels at which electron-hole scattering dominates the relaxation process.
The same calculation provides satisfactory results for the transmission at
Δt = 10 ps, where the plasma density depends on the laser fluence F(2ω). In
this case the plasma density has been calculated with the standard equations,
including ambipolar diffusion and neglecting impact ionization. The calculated evolution of plasma density and temperature at the surface of silicon
irradiated with increasing laser fluence F(2ω) are finally summarized in
Fig. 3. These calculated data are cross checked with experimental data at
different probing wavelengths, excitation levels and time delays. They have
been used to analyze the dependence of the optical reduced mass on the carrier density in reflectivity measurements. Excellent agreement with the data
at 1.064, 1.9 and 2.8 μm has been found up to a fluence level of 150 mJ/cm²,
for a constant optical reduced mass of $(0.16 + 0.1)m_0$. The stability of the
Drude term indicates a "cold plasma" which is in thermal equilibrium with the

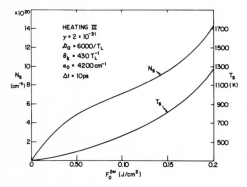

Fig. 3. Calculated plasma densities and temperatures at the surface of bulk silicon versus incident laser fluence at 0.532 μm
using the optical heating model
with instantaneous energy transfer during the picosecond pulse.

lattice. Close to the threshold value for melting, the maximum plasma density is limited to 1.4×10^{21} cm^{-3}, mainly due to Auger recombination. The measured values of lattice temperature and maximum plasma density are consistent with optical heating and thermal melting even on a time scale of picoseconds.

This research was supported by the U.S. Office of Naval Research under contract N00014-83K-0030 and by the Alexander von Humboldt Foundation, F. R. Germany.

References

1. D.H. Auston, C.M. Surko, T.N.C. Venkatesan, R.E. Slusher, J.A. Golovchenko: Appl. Phys. Lett. 33, 437 (1978)
2. J.M. Liu, R. Yen, H. Kurz, N. Bloembergen: Appl. Phys. Lett. 39, 755 (1981); R. Yen, J.M. Liu, H. Kurz, N. Bloembergen: Appl. Phys. Lett. A 27, 153 (1982)
3. J.M. Liu, H. Kurz, N. Bloembergen: Appl. Phys. Lett. 41, 643 (1982)
4. D. von der Linde, N. Fabricius: Appl. Phys. Lett. 41, 991 (1982)
5. L.A. Lompré, J.M. Liu, H. Kurz, N. Bloembergen: Appl. Phys. Lett. 43, 168 (1983)
6. G.E. Jellison, F.A. Modine: Appl. Phys. Lett. 41, 180 (1982)
7. H. van Driel, L.A. Lompré, N. Bloembergen: Appl. Phys. Lett. 44, 285 (1984)
8. L.A. Lompré, J.M. Liu, H. Kurz, N. Bloembergen: Appl. Phys. Lett. 44, 3 (1984)
9. L.A. Lompré, J.M. Liu, H. Kurz: Proc. Mat. Res. Soc. Symposium, Boston 1983
10. K.G. Svantesson: J. Phys. D, Appl. Phys. 12, 425 (1979)

Dynamics of the Mott Transition in CuCl with Subpicosecond Time Resolution

D. Hulin and A. Mysyrowicz
Groupe de Physique des Solides, Ecole Normale Supérieure, F-75005 Paris, France

L.L. Chase
Physics Department, Indiana University, Bloomington, IN, USA

A. Antonetti, J. Etchepare, G. Grillon, and A. Migus
Laboratoire d'Optique Appliquée, Ecole Polytechnique - ENSTA
F-91120 Palaiseau, France

It is well known that a strongly excited semiconductor crystal may become partly metallic, due to the formation in the medium of a dense plasma of free electron-hole pairs. This plasma phase represents the lowest electronically excited state of the crystal, as long as the free carriers density n exceeds a critical value n_c. If n falls below n_c, the excited system reverts to an insulating (excitonic) phase consisting of strongly correlated electron-hole pairs. The complete understanding of this Mott-like transition, particularly its dynamical aspect, is of great interest, both from a fundamental point of view as well as for potential applications, e.g. in fast switching devices.

A convenient method of investigation of the Mott transition in highly excited semiconductors is via time-resolved luminescence spectroscopy : by monitoring the amount of emission characteristic of each phase in function of time, following a brief excitation pulse, one gets direct access to the kinetics of the metal to insulator conversion, under the proper conditions of a free relaxing plasma.

We have applied this method to the case of CuCl, using an experimental apparatus giving a time resolution better than 10^{-12}sec. (see fig. 1). The crystal is initially excited with an intense UV pulse, of duration 150 femtoseconds, obtained by second harmonic generation from the amplified output of a passively mode-locked ring dye laser. Absorption of the incident UV light by the crystal gives rise to a population of hot electron-hole pairs, having a maximum density $n \sim 10^{20}$ cm^{-3}. The use of an ultrafast optical Kerr shutter [1] in the detection channel (time resolution ≤ 1 ps) (see fig. 1) allows one to record simultaneously the luminescence emitted by the sample over a wide spectral range, covering the entire region of interest (3850 - 4200 Å), with a time resolution better than 10^{-12}sec. Successive spectra, separated by intervals of 1 ps, have been recorded at T = 15 K.

Consider first the time integrated luminescence spectrum, shown in figure 2. It consists of two distinct emission lines. The narrow emission peaked at 392 results from the radiative recombination of excitonic molecu-

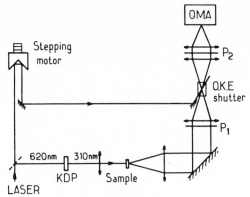

Fig. 1 – *Experimental set-up* ; P_1, P_2 *crossed polarizers* ; *OKE shutter :*
2 mm thick benzene cell ; OMA : *spectrograph and optical*
multichannel analyzer.

les, i.e., it represents a dielectric, insulating phase of the excited system [2]. The broader emission at lower energies is due to the recombination
of electron-hole pairs in the plasma (conducting phase) [3].

The time-resolved spectra indicate that the plasma occurs first, and that
it lasts for approximatively 5 ps. The plasma then converts abruptly into a
gas of biexcitons. It is found that the switching time between both states
of the system is 3 ps, and can be even less if the effects due to spatial
inhomogeneities in the excitation are taken into account. On the other hand,
if the input excitation intensity is reduced below threshold for plasma
appearance, keeping all other experimental parameters unchanged, the biexciton luminescence appears with no measurable delay, and the build-up time of
this luminescence is smaller than the experimental resolution. This indicates that the formation time of excitonic particles, from a pool of hot free
carriers of density $n < n_c$, is less than 0.5 ps.

The short (3 ps), Mott switching time found in CuCl is in sharp contrast
with similar results obtained in GaAs [4], where it was shown that the Mott
transition is smooth, with both phases coexisting over a period of several
hundred picoseconds. This difference may be accounted for by the difference in exciton binding energies E_x in both materials. In GaAs = $E_x \sim 4$ meV,
so that the ratio $kT/E_x \sim 1$ (T is the plasma effective temperature at the
time of conversion and is of the order 50 K). In such a case, there is
really no well-defined transition from a plasma to an insulating phase as
n reaches n_c from above. In CuCl, $E_x = 0.2$ eV, fulfilling the conditions
$kT/E_x \ll 1$.

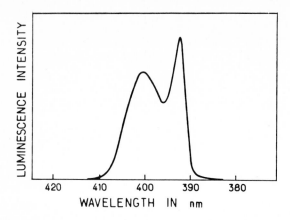

Fig. 2 – *Time integrated luminescence of CuCl held at T = 15 K. Emission at 410 nm is from the plasma; line at 392 nm is from excitonic molecules.*

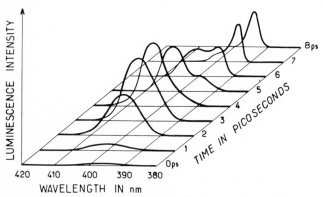

Fig. 3 – *Time-resolved luminescence spectra of CuCl at 15 K. The appearance of the excitonic molecules is delayed, due to the presence of the plasma at the initial stage of evolution.*

In summary, we have studied the dynamics of the Mott transition from the plasma to the excitonic phase in highly excited CuCl. It was observed that this transition is very rapid, requiring less than 3 ps for complete conversion.

References

1. J. Etchepare, G. Grillon, R. Astier, J.L. Martin, C. Bruneau and A. Antonetti. Picosecond Phenomena III, Springer-Verlag (Berlin,Heidelberg,New York), 217, (1982).
2. For a recent review of biexciton luminescence in CuCl, see for instance J.B. Grun, B. Hönerlage, R. Levy in Excitons, E.I. Rashba and M.D. Sturge, editors North Holland (1982).
3. D. Hulin, A. Antonetti, L.L. Chase, J.L. Martin, A. Migus, A. Mysyrowicz. Optics Comm. 42, 260 (1982).
4. E.O. Göbel, P.H. Liang, D. von der Linde. Solid State Comm. 37, 609 (1981).

Picosecond Dynamics of Hot Dense Electron-Hole Plasmas in Crystalline and Amorphized Si and GaAs

P.M. Fauchet[1] and A.E. Siegman

Edward L. Ginzton Laboratory, Stanford University, Stanford, CA 94305, USA

1. Introduction

The physics of highly excited semiconductors has recently been the subject of many studies, especially in connection with the problem of pulsed laser annealing. Careful experiments [1-3] have now been performed with the temporal resolution required to demonstrate that one single picosecond or femtosecond pulse can produce melting at the surface of silicon. Since the carrier relaxation time is very short ($\lesssim$ 1 ps), melting occurs during illumination with ~30 ps pulses, such as those available from doubled Nd:YAG lasers. Unfortunately, because even the very dense electron-hole plasma (EHP) produced by a strong visible pulse will cause only a small change in the reflectivity of a probe beam (at 1.06 μm), up to densities ~10^{21} cm^{-3}, it is difficult to use time-resolved reflection measurements to obtain detailed information on the dynamics of the very dense and hot EHP just before melting occurs. In this paper, we present results obtained using a novel technique that allows us to study in more detail the plasma close to but below the melting phase transition in Si and GaAs.

2. The Two-Color Excitation Technique

The major advantage of our novel two-step excitation technique is that it decouples the creation of the EHP from the heating of the lattice. A visible picosecond pulse at 532 nm with intensity below melting threshold is used to "prepare" the sample, i.e., to create a dense electron-hole plasma at the surface, while still only slightly increasing the lattice temperature. The intensity of an infrared picosecond pulse at 1.06 μm, which is delayed with respect to the visible pulse by variable increments up to plus or minus several nanoseconds, is then adjusted to just melt the surface, as revealed by post-mortem examination using a microscope.

The absorption of 1.06 μm radiation in silicon occurs predominantly via free-carrier absorption, and therefore measurements of melting threshold using 1.06 μm pulses alone are very sensitive to surface preparation and doping level. In our experiment, however, the initial free-carrier concentration is set by the 532 nm pulse, so that we can control the free-carrier absorption for the IR pulse. As the IR pulse delay is increased, the carrier density decreases under the combined action of recombination and diffusion, and the IR intensity required to melt thus increases. On the other hand, when the IR pulse precedes the visible pulse, little or no effect is expected, since no plasma is present. We thus have a method for probing the dynamics of electron-hole plasmas in semiconductors at densities below those obtained at melting threshold, a region where only small reflectivity changes are observed. (Our method must be distinguished from the earlier two-pulse

---

[1]Permanent address: Department of Electrical Engineering and Computer Science, Princeton University, Princeton, NJ 08544

work of AUSTON et al.[4] in which an initial 532 nm pulse melted the Si surface, and the addition of a 1.06 μm pulse prolonged the melt duration).

Results of detailed theoretical calculations quantifying this analysis have been published elsewhere [5]. They show the sensitivity of the two-color excitation technique to the free-carrier absorption cross-section, EHP relaxation and recombination times.

3. Experimental Results

We have performed such a series of two-pulse experiments on various types of Si and GaAs samples. An anti-resonant Q-switched and mode-locked Nd:YAG laser system provided the ~30 ps pulses at 1.06 μm and 532 nm that were used to melt the surface of the samples. Figure 1 shows typical results obtained on ion-implanted <100> Si [6]. For each delay, we plot the joint energy density at the two wavelengths required to just melt the surface, as determined by post-mortem examination. When the IR pulse precedes the visible pulse (top left), melting occurs only provided one of the beams is close to the melting threshold at that wavelength: the two pulses act more or less independently. For $\tau_d \gtrsim 0$ ps (top right), the absorption of the IR photons is greatly enhanced by the almost simultaneous illumination at 532 nm. For larger delays (bottom curves), the curves are modified towards higher thresholds, under the influence of recombination, carrier diffusion, and eventually heat diffusion. Figure 2 shows the sharp decrease in IR fluence needed to melt when an EHP is present. (The nanosecond time scale recovery is not produced by carrier diffusion or recombination but by heat diffusion, a rather slow process).

In Fig. 3, we show the results of a similar experiment performed on ion-implanted GaAs. For negative delays, the results are qualitatively similar to those obtained in Si, but for positive delays, the experimental data lie

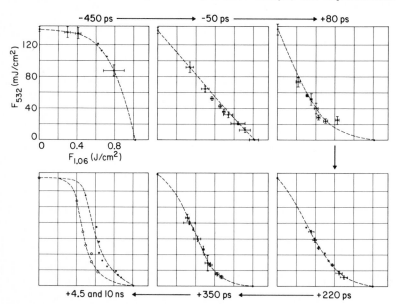

Fig. 1: Results of a two-color experiment performed on ion-implanted Si for several time delays

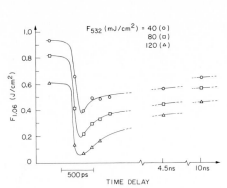

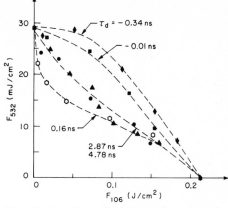

Fig. 2: IR fluence needed to melt
the sample versus time delay for
three visible fluences

Fig. 3: Results of a two-color experiment performed on ion-implanted GaAs
for several time delays

on a curve whose curvature seems to change for large IR intensity. Although
there is a rather strong induced IR absorption for $\tau_d > 0$, the shape of the
curves is qualitatively different in GaAs.

4. Discussion

Although the relevant theoretical parameters for ion-implanted Si are not as
accurately known as for intrinsic Si, we have obtained a good fit between
theory and experiments by assuming a 100 fs hot carrier relaxation time, a
10 ps recombination time and a free-carrier absorption cross-section that is
twice that of crystalline Si. The agreement between theory and experiments
is sensitive to the relaxation time of the hot carriers even though τ_e is much
shorter than the pulsewidth, thanks to the extremely nonlinear behavior of
the system of coupled partial differential equations which describes the
physics of the processes. The recombination time is in qualitative agreement
with the recent results of BERGNER et al. [7], who measured the lifetime of
optically injected carriers (n $\leq 10^{20}$ cm^{-3}) in different disordered Si samples. The injected carrier densities achieved in our experiments are as high
as 10^{21} cm^{-3} [11]. When we repeat our experiments on more heavily damaged
Si layers, we note a reduced synergistic effect, which comes from the increased
band-to-band IR absorption. The reduced influence of free-carrier absorption
becomes then more difficult to evaluate.

Our results to date in intrinsic Si also indicate the presence in very
dense and hot plasmas of a stronger free-carrier absorption than at low excitation regimes. In fact, we have strong evidence that interconduction band
transitions become important in dense (> 10^{20} cm^{-3}) and hot (T_e > 2000K)
EHP's. For example, we have compared the calculated and measured single-
pulse melting thresholds at 1.06 μm for intrinsic Si as a function of pulse-
width. With a constant cross-section of either $\sigma = \sigma_o \equiv 5.1 \times 10^{-18}$ (T/300)
cm^{-2} [8], or $\sigma = 4 \times \sigma_o$, as was recently proposed [9], it was not possible
to obtain a good fit of these data [10], while a theory with $\sigma = \sigma (n, T_e)$ can
reproduce the data fairly well. Our two-color results also suggest a very
fast recovery in the IR fluence needed to melt, as a function of time delay
between the two laser pulses. We tentatively interpret this as due to
diffusion, a faster process in crystalline Si than in disordered Si. These
results will be discussed in detail in another publication.

131

Our GaAs results can be explained by the presence of efficient two-photon absorption (TPA) at 1.06 μm in addition to FCA. TPA is an "anti-synergistic" effect, since it becomes more efficient at higher $F_{1.06}$ and lower F_{532}. We have been able to obtain a qualitative fit using a model similar to that used for Si, after inclusion of TPA.

5. Conclusion

A novel two-pulse technique has been introduced to study the dynamics of very hot and dense EHP's in semiconductors. Our experimental results have allowed us to verify the conventional models for high intensity ultrashort laser excitation and melting, over a large range of excitation energies, including regions where conventional excite and probe techniques (using the same wavelengths) are rather insensitive.

6. Acknowledgements

This research was supported by the Air Force Office of Scientific Research. We thank D. Ress and D. Dale for technical assistance. P.M. Fauchet was supported by an IBM Postdoctoral Fellowship.

7. References

1. J.M. Liu, H. Kurz, and B. Bloembergen in Laser-Solid Interactions and Transient Thermal Processing of Materials, Narayan, Brown and Lemons, eds. (North-Holland, New York, 1983) pp. 3-12.
2. D. von der Linde and N. Fabricius, Appl. Phys. Lett. 41, 991 (1982).
3. C.V. Shank, R. Yen, and C. Hirlimann, Phys. Rev. Lett. 50, 454 (1983).
4. D.H. Auston, J.A. Golovchenko, and T.N.C. Venkatesan, Appl. Phys. Lett. 34, 558 (1979).
5. P.M. Fauchet, Zhou Guosheng and A.E. Siegman, in Ref. 1, pp. 205-210.
6. P.M. Fauchet and A.E. Siegman, Appl. Phys. Lett. 43, 1043 (1983).
7. H. Bergner, V. Bruckner, G. Kerstan and W. Norwick, Phys. Stat. Sol. (b)117, 603 (1983).
8. K.G. Svantesson and N.G. Nilsson, J. Phys. C. Sol. St. Phys. 12, 3838 (1979).
9. L.-A. Lompré, J.M. Liu, H. Kurz and B. Bloembergen, Appl. Phys. Lett, 44, 3 (1984).
10. I.W. Boyd, S.C. Moss, T.F. Boggess, and A.L. Smirl, to appear in Energy-Beam Solid Interactions and Transient Thermal Processing, Fan and Johnson, eds. (North-Holland, 1984); P.M. Fauchet, unpublished data.
11. P.M. Fauchet and A.E. Siegman, to appear in Ref. 10.

Picosecond Optical Excitation of Phonons in Amorphous As$_2$Te$_3$

C. Thomsen, J. Strait, Z. Vardeny[a], H.J. Maris, and J. Tauc

Department of Physics and Division of Engineering, Brown University
Providence, RI 02912, USA

J.J. Hauser

AT & T Bell Laboratories, Murray Hill, NJ 07974, USA

Introduction

We present a novel method of creating phonon excitations in a thin film of
a chalcogenide glass. Phonon generation and the detection is done
optically and thus allows measurement with picosecond time resolution.
Current methods of studying phonon propagation involve ultrasonics or heat-
pulse techniques. These use electronics, and the time resolution is
limited to nanoseconds. Our picosecond method is a promising tool for
studying propagation of phonons with extremely short mean free paths. As a
demonstration of this method we measured the sound velocity in a film of
a-SiO$_2$ and compare it with a computer simulation.

Experimental Method

Photoinduced transmission and reflection measurements were done using the
pump and probe method [1]. The source for both the pump and the probe
pulses was a passively modelocked dye laser producing 1 ps pulses with 2.0
eV photon energy, 0.5 MHz repetition rate, and 1 nJ pulse energy. The pump
and the probe were focused to a 50 μm diameter spot on a thin film of
amorphous As$_2$Te$_3$ sputtered onto a sapphire substrate. The samples had an
absorption coefficient $\alpha = 3.3 \times 10^5$ cm^{-1} for 2 eV photons. Thus the
absorption depth in the films $\zeta = \alpha^{-1}$ was about 300 Å.

Results

Figure 1 shows the transmission of the probe as a function of delay time
for samples of different thicknesses held at 295 K. The photoinduced
response is the superposition of two components: (1) a step-like change in
transmission ($-\Delta T/T \simeq 10^{-3}$) at time zero followed by a monotonic relaxation
and (2) a damped oscillation with a period that depends on sample
thickness. Response (1) is due to absorption by excited carriers and has
been studied extensively in other chalcogenide glasses and in amorphous
silicon [2].

Response (2) however is a new observation. The amplitude of the
oscillations in a-As$_2$Te$_3$ ranges from 3×10^{-5} to 3×10^{-4}. Their period
scales linearly with sample thickness. We have also observed oscillations
in cis-polyacetylene (Fig. 2), but in the trans-isomer transmission has
previously [3] been found to decay monotonically with time.

Interpretation

We interpret our results in terms of a stress wave generated by the laser
pulse. The bandgap is sensitive to strain and the transmission through the
sample is modulated by the stress wave. The oscillating transmission thus
results from an oscillating stress wave. The sound velocity in a-As$_2$Te$_3$
associated with the periods and sample thicknesses is v = 1.6 x 10^5 cm sec^{-1}.

(a)   Solid State Institute, Technion, Haifa, Israel

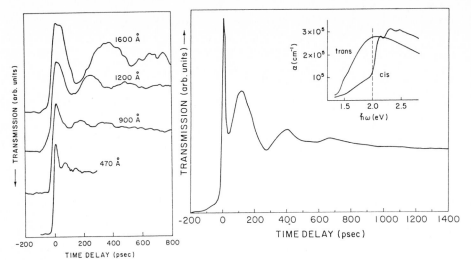

Figure 1: Photo-induced change in transmission of a-As$_2$Te$_3$ samples with various thicknesses

Figure 2: Photo-induced change in transmission of cis-polyacetylene at 80 K

This value is consistent with the measured transverse sound velocity [4] of 1 x 10^5 cm sec^{-1} in c-As$_2$Te$_3$ and the velocity of 2.5 x 10^5 cm sec^{-1} for longitudinal sound [5] in a-As$_2$S$_3$.

Strain decreases the absorption in chalcogenide glasses. For small strain η we can write the change in absorption coefficient α as:

$$\Delta\alpha = (d\alpha/dE_g)(dE_g/d\eta)\eta \tag{1}$$

where $E_g = 0.8$ eV is the bandgap and $dE_g/d\eta$ is the difference in deformation potentials of the conduction and valence bands ($dE_g/d\eta = dE_c/d\eta - dE_v/d\eta$). The fractional change in transmission for small $\Delta\alpha$ is, using (1),

$$\frac{\Delta T}{T} \simeq - \frac{d\alpha}{dE_g} \frac{dE_g}{d\eta} <\eta> d \tag{2}$$

where $<\eta>$ is the average strain in the sample and d is the thickness.

We will discuss now how the stress wave is generated and give an estimate of what amplitude to expect for transmission oscillations. Thermalization of excited carriers to the band edges provides a mechanism to cause a temperature rise $\Delta\Theta_{av}$ in the absorption region. The excess energy E_x from this process is $E_x = (\hbar\omega_{ph} - E_g)N$ where our photon energy is $\hbar\omega_{ph} = 2$ eV and N the number of absorbed photons. Thus

$$\Delta\Theta_{av} = E_x/\zeta AC \tag{3}$$

where A is the area of the beam and C is the specific heat per unit volume of the film. The sudden thermal expansion caused by the temperature rise will set the film into an oscillatory motion around a new equilibrium value with an amplitude of

$$<\eta> = \frac{1+\nu}{1-\nu} \frac{\zeta}{d} \beta\Delta\Theta_{av} \tag{4}$$

134

where ν is Poisson's ratio. Taking the values $E_x = 0.35$ nJ, $A = 2 \times 10^{-5}$cm^2, $C = 1.5$ JK^{-1}cm^{-3}, we find that for $\zeta = 300$ Å $\Delta\alpha_{av}$ is 4 K. Then for [6] $\beta = 1.9 \times 10^{-5}$ K^{-1}, and assuming a Poisson's ratio of 0.3, we find the initial amplitude of the oscillataions to be 1.3×10^{-4} for the 470 Å sample. Together with $d\alpha/dE_g = -3.5 \times 10^5$ cm^{-1}eV^{-1} estimated from an absorption edge measurement and a deformation potential of $dE_g/d\eta = 2.3$ eV [7] we find $\Delta T/T$ from equation (2) to be $\Delta T/T \simeq 3 \times 10^{-4}$. This calculated value of $\Delta T/T$ is well in agreement with the observed value.

Although the picture based on thermal expansion appears to give a sufficient explanation for the described experiment, we think that the real situation is more complicated. A treatment of phonon generation involving deformation potentials is published elsewhere [8].

In polyacetylene the strikingly different behaviour of the cis- and trans-isomers is understood by looking at their absorption coefficients as a function of photon energy (insert of Fig. 2). At 2 eV α increases rapidly with photon energy in cis-polyacetylene while in the trans-isomer it has a plateau which makes it much less sensitive to an absorption edge modulation process.

With this method of generating and detecting phonons room temperature measurements of phonon velocity and attenuation become possible in very highly attenuating materials. The frequency generated by our films is 15 GHz and it can be increased to ~ 100 GHz by using thinner films with higher phonon velocity. The excellent time resolution allows the study of phonon propagation in other thin films deposited on top of the As$_2$Te$_3$.

As a demonstration we have used As$_2$Te$_3$ as a transducer to launch a coherent acoustic wave into a film of a-SiO$_2$. (See insert of Fig. 3.) In a computer model we have simulated $\Delta T/T$ as a function of time delay. The model incorporates exponential absorption of light and various acoustic reflection coefficients from the interfaces. Figure 3 is a comparison of the simulation (a) and experimental data (b). The agreement is excellent. At point 1 the stress wave has left the a-As$_2$Te$_3$ because of the good

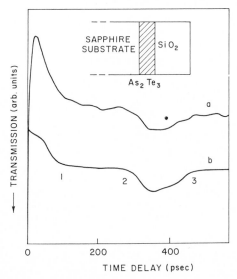

Figure 3: Experiment to measure sound velocity with a-As$_2$Te$_3$ as transducer. (a) Shows data for $\Delta T/T$, (b) is a computer fit

135

impedance match and the transmission remains on a constant value (1-2). Upon reflection at the free surface the acoustic wave inverts sign and returns into the absorbing film. The shape of the induced change in transmission (2-3) is a result of the non-uniform absorption of the light. Treating the velocity in a-SiO$_2$ as an adjustable parameter, we have obtained the best fit with $4.8 \pm 0.3 \times 10^5$ cm sec^{-1}, a value, which is in reasonable agreement with 5.8×10^5 cm sec^{-1} measured by ROTHENFUSSER et al. [9].

Acknowledgements

We thank H. T. Grahn and T. R. Kirst for technical assistance. This work was supported in part by the National Science Foundation through the Materials Research Laboratory at Brown University.

References

1. E. P. Ippen and C. V. Shank, Ultrashort Light Pulses, edited by S. L. Shapiro (Springer, New York, 1977), p. 102 ff.
2. Z. Vardeny, J. Strait and J. Tauc, Picosecond Phenomena III, edited by K. B. Eisenthal et al. (Springer, New York, 1982) p. 372.
3. Z. Vardeny J. Strait, D. Moses, T.-C. Chung and A. J. Heeger, Phys. Rev. Let. 49, 1657 (1982).
4. G. Cibuzar and C. Elbaum, private communication.
5. D. Gerlich, E. Litor and O. L. Anderson, Phys. Rev. B20, 2529 (1979).
6. We have used the value for As$_{0.45}$Te$_{0.66}$ given by J. Cornet, J. Schneider, 4th International Conference on the Physics of Non-Crystalline Solids, edited by G. H. Frishat, (Trans. Tech. Publications, 1976), p. 397.
7. J. M. Besson, J. Cernogora and R. Zallen, Phys. Rev. B22, 3866 (1980),
8. C. Thomsen, J. Strait, Z. Vardeny, H. J. Maris and J. Tauc, to be published.
9. Rothenfusser, W. Dietsche and H. Kinder, in Phonon Scattering in Condensed Matter, edited by W. Eisenmenger, K. Lassmann and S. Dottinger, (Springer, New York, 1984), p. 419.

Femtosecond Studies of Intraband Relaxation of Semiconductors and Molecules

A.J. Taylor, D.J. Erskine, and C.L. Tang

Materials Science Center, Cornell University, Ithaca, NY 14853, USA

In this talk we present a summary of our studies of the ultrafast relaxation dynamics of hot carriers in semiconductors [1] and of highly excited dye molecules in solution [2] using the equal pulse correlation technique [3]. These intraband processes occur on timescales comparable to the pulsewidths generated by current femtosecond (fs) technology, and hence present difficulties in extracting their relaxation times due to the presence of slow relaxation processes and of the coherent artifact.

1. Equal Pulse Correlation Technique

Our measurements are based on the saturation effect in the transmission of two 90 fs 612 nm laser pulses through a thin sample (dye jet or semiconductor) using the equal pulse correlation technique. The two pulses, which are orthogonally polarized, collinearly propagating and equal in energy, are focussed on the sample with an intensity such that saturable absorption occurs. Their combined, time-averaged, transmitted flux is measured as a function of temporal delay, τ. This transmitted flux reaches a peak at $\tau=0$, and reduces to a background value when τ exceeds the relaxation time of the absorption process. We call this peak a transmission correlation peak (TCP). In general, it consists of an incoherent portion whose shape yields information on the relaxation processes which depopulate the photo-excited state and a coherent artifact which contains no such information.

Using the formalism of Ref. 4, the TCP has the following form when the two input pulses are orthogonally polarized:

$$TCP_1(\tau) \propto \int_0^\infty dw \ A_{xxzz}(w) \ [AC(\tau+\omega) + AC(\tau-w)]$$

$$+ \int_0^\infty dw \ A_{xzzx}(w) \ [\xi(w,\tau) + \xi(w,-\tau)]. \tag{1a}$$

It has the following form when the input pulses have parallel polarization:

$$TCP_{11}(\tau) \propto \int_0^\infty dw \ A_{zzzz}(w) \ [AC(\tau-w) + AC(\tau+w)]$$

$$+ \int_0^\infty dw \ A_{zzzz}(w) \ [\xi(w,\tau) + \xi(w,-\tau)] \tag{1b}$$

where $AC(\tau)$ is the autocorrelation of the pulse intensity envelope and $\xi(w, \tau)$ is given by:

$$\xi(w,\tau) = \int_{-\infty}^\infty E^*(t) \ E(t+\tau) \ E^*(t+\tau-w) \ E(t-w) \ dt \tag{2}$$

where $E(t)$ is the electric field envelope. A_{ijkl} is the response function of the third-order susceptibility of the system for the combination of electric field polarizations $E_i^*E_jE_k^*E_l$:

$$A_{ijkl}(t) = [Y_{ij} \ Y_{kl} \ (1-\exp(-t/T_0)) + Y_{ijkl} \ \exp(-t/T_0)] \ G(t). \tag{3}$$

$G(t)$ is the impulse response of the population of the photoexcited state. T_O is the orientational diffusion time which characterizes the decay of the photoexcited anisotropy of the material. Y_{ij} and Y_{ijkl} are dimensionless coefficients which describe the symmetry of the saturable absorption process [4]. They are determined from the transition matrix elements $\hat{r}_{cv}$ in the following manner:

$$Y_{ij} = 1/4\pi \int d\Omega \, \hat{r}_{vc}^* \cdot \hat{\varepsilon}_i^* \quad \hat{r}_{cv} \cdot \hat{\varepsilon}_j \tag{4a}$$

$$Y_{ijkl} = 1/4\pi \int d\Omega \, \hat{r}_{vc}^* \cdot \hat{\varepsilon}_i^* \quad \hat{r}_{cv} \cdot \hat{\varepsilon}_j \quad \hat{r}_{vc}^* \cdot \hat{\varepsilon}_k^* \quad \hat{r}_{cv} \cdot \hat{\varepsilon}_l \quad . \tag{4b}$$

The first term in (1a) or (1b), which we denote TCP_i, represents the incoherent contribution to the saturation process. This is the term which provides information on the relaxation of the populations of the photoexcited state. It is proportional to the convolution of the AC and the response function of the population of the photoexcited state. The second term in (1a) and (1b), which we denote CA, is the coherent artifact contribution to the TCP. CA provides no information on the relaxation of the photoexcited state, as it is nonzero only when τ is less than the coherence time of the pulse. If the relevant relaxation process of the material has a decay constant much longer than the laser pulsewidth, then the CA will not present a problem, since it will simply appear as a spike at $\tau=0$. However, for processes which are faster than the pulsewidth, such as intraband relaxation, the amplitude of the CA must be determined and subtracted from the total TCP in order to extract a relaxation time from TCP_i.

For our experimental configuration of orthogonally polarized input beams, the fractional amount of CA in the TCP at $\tau=0$ is given by:

$$FRAC \quad CA = \frac{T_O}{T_R} \, Y_{xzzx}/(\frac{T_O}{T_R} \, (Y_{xzzx} + Y_{xxzz}) + Y_{zz}^2) \tag{5}$$

when the following assumptions are made: 1) The response function is of the form $G(t) \sim \exp(-t/T_r)$. 2) T_O and T_r are less than the pulsewidth. 3) The electric field envelope is real and temporally Gaussian. Equation 4 reveals that the coherent contribution is negligible if the absorption properties of the material are polarization-independent ($Y_{xzzx}=0$), or if T_O is much shorter than the pulsewidth or the fastest energy relaxation time, T_r. If, however, the input pulses have parallel polarizations, the coherent contribution will always be half of the TCP height at zero delay. Therefore, the experimentally determined ratio, R, of the total TCP height (at $\tau=0$) above the background (at $\tau=\infty$) when the input beams have parallel polarizations to the TCP height when the beams have orthogonal polarizations provides an estimate of T_O/T_r if the Y parameters are known:

$$\frac{T_O}{T_R} = Y_{zz}^2 \, (2-R)/(R(Y_{xzzx} + Y_{xxzz}) - 2Y_{zzzz}) \, . \tag{6}$$

Using (5) and (6) the CA can be estimated and subtracted from the total TCP, assuming that it has the same shape as the AC. The remaining TCP_i can then be analyzed to determine the relaxation times of the system. Independent of these assumptions, when $T_O=0$ R=2, and no CA is present. Conversely, when $T_O=\infty$ $R=2Y_{zzzz}/(Y_{xzzx}+Y_{xxzz})$, and there is the maximum fraction of CA of $Y_{xzzx}/(Y_{xzzx}+Y_{xxzz})$.

The equal-pulse correlation technique is particularly useful when studying processes which occur on time scales comparable to or faster than the pulsewidth (i.e., intraband relaxation) in the presence of much slower processes (such as interband relaxation), since the TCP corresponding to the fast decay process is then clearly revealed as a symmetrical, narrow peak on the

flat background which corresponds to the slow decay. After subtraction of
the CA, the remaining narrow peak is then proportional to a convolution of
the AC and a double-sided exponential with decay time T_r, of the fast
process. In contrast, when the conventional pump-probe technique is used,
fast processes are obscured by the step in transmission at $\tau=0$ and by the
long decay. A fast decay process is mainly revealed through a modification
of the rise of the initial step, making it difficult to determine T_r for
the fast process.

2. Hot Carrier Relaxation in Semiconductors

We have studied the relaxation of hot carriers excited by 2.02 eV photons [1]
in GaAs, $Al_{0.32}Ga_{0.68}As$ and an AlGaAs/GaAs multiple quantum well structure
(MQW) at room temperature. The band structures in Fig. 1 reveal that for
GaAs electrons are excited with energies 0.51, 0.45 and 0.15 eV above the
bottom of the conduction band by transitions from the heavy hole (H), light
hole (L) and split-off (S) valence bands, while in AlGaAs electrons are
excited from the heavy and light hole valence bands with energies of 0.17
and 0.11 eV. These photoexcited carriers leave the optically coupled region
via carrier-carrier scattering, polar optical phonon scattering within the
central valley and intervalley phonon scattering (electrons only). These
materials reveal a fast decay process out of the photoexcited state occurring
in <100 fs, as well as a ps process related to bandfilling.

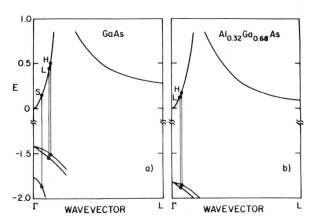

Fig. 1 Schematic of the
band structures for GaAs
and $Al_{0.32}Ga_{0.68}As$

Since the intraband relaxation time T_r is shorter than the pulsewidth,
we must determine the amplitude of the CA to analyze our data. In this
situation, the approximations described in the last section are appropriate,
particularly T_0,T_r<pulsewidth, so (5) and (6) are valid. The Y parameters,
calculated for the various transitions using the wavefunctions of Kane [5]
are displayed in Table 1. We measure R=1.75 which implies that $T_0/T_r=0.4$.
The CA is then ~20% of the TCP.

	Y_{zz}	Y_{zzzz}	Y_{xxzz}	Y_{xzzx}
GaAs				
Heavy Hole	0.28	0.091	0.069	0.069
Light Hole	0.28	0.087	0.074	0.056
Split Off	0.33	0.135	0.093	0.027
AlGaAs				
Heavy Hole	0.31	0.118	0.089	0.089
Light Hole	0.32	0.105	0.104	0.029

Table 1 Y parameters for
the 2.02 eV transitions in
GaAs and $Al_{0.32}Ga_{0.68}As$

The measured values of T_r for GaAs and AlGaAs are plotted versus photo-excited carrier density, n, in Fig. 2. The MQW exhibits qualitatively similar behavior, with T_r varying from 65 fs at $n=10^{18}$ cm^{-3} to 40 fs at $n=5\times10^{19}$ cm^{-3}. Calculated carrier-phonon scattering rates [1] are presented in Table 2, using a deformation potential of 10^9 eV/cm. From these rates we calculate T_r to be 45 fs for GaAs and 80 fs for AlGaAs, in good agreement with the low density data. We attribute the measured decrease in T_r versus n for AlGaAs to increasing carrier-carrier scattering. For GaAs, T_r may be independent of carrier density due to the large phonon-carrier scattering rates masking the increase in the carrier-carrier rates.

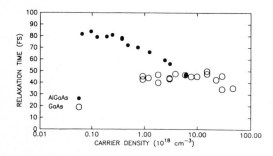

Fig. 2 Measured values of T_r for GaAs and AlGaAs versus carrier density

Table 2 Calculated polar optical phonon absorption, PO-ABS, polar optical emission, PO-EM, intervalley phonon, IV, and acoustic phonon, ACS, scattering rates for carriers generated by a 2.02 eV transition from the heavy hole valence band at 300[K]. Rates are in units of 10^{12}/sec

Material	GaAs		$Al_{0.32}Ga_{0.68}As$	
Carrier	Electron	Hole	Electron	Hole
PO-ABS	1.7	5.5	2.2	6.6
PO-EM	6.4	18.5	7.9	9.9
IV	22.	0.0	12.	0.0
ACS	0.6	5.5	0.5	9.8

At high carrier densities ($n>0.4\times10^{19}$ cm^{-3}) in GaAs and MQW (but not AlGaAs) a rising wing appears in the TCP, as shown in Fig. 3, which we believe is caused by the dynamic Burstein-Moss shift of the absorption edge due to the filling of the central valley by carriers draining from the outer valleys. The photoexcited levels in the conduction band connected to the split-off band transition then become saturated for large n. The 1.5 ps characteristic time of this rising wing reflects the time it takes the electrons excited with 0.5 eV of energy to establish a distribution with significant population at 0.15 eV. No rising wing is seen in AlGaAs since the split-off transition is not excited.

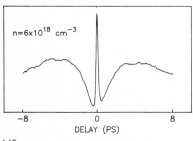

Fig. 3 TCP for GaAs with $n=6\times10^{18}$ cm^{-3}. The central peak corresponds to the rapid depopulation of the photoexcited state, while the rising wings are due to bandfilling effects

140

3. Vibrational Relaxation of Dye Molecules in Solution

We have also studied relaxation of highly excited vibrational states of dye
molecules such as Nile Blue, Rhodamine 640, DODC Iodide and Oxazine 725 in
ethylene glycol [2]. The absorption process for these dye molecules can be
modeled with the 3-level system shown in Fig. 4. For all of these dyes very
little broadening of the fast component of the TCP over the AC is seen, as
in Fig. 5, implying that t_{23} is faster than the laser pulsewidth.

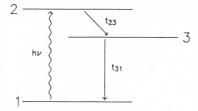

Fig. 4 Photoexcitation process for dye molecules. Absorption (1 to 2) populates high-
lying levels in the excited electronic state.
Vibrational relaxation (2 to 3) rapidly
follows in a time $t_{23}<1$ ps. Radiative decay
repopulates the ground state in a time
$t_{31}>100$ ps

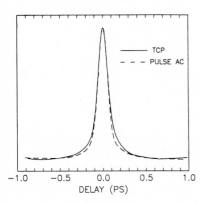

Fig. 5 TCP of Nile Blue in ethylene
glycol (solid line) with an autocorrelation of the laser pulse
(dotted line)

The Y parameters for dye molecules [4] are $Y_{zz}=1/3$ and $Y_{zzzz}/3=Y_{xxzz}=$
$Y_{xzzx}=1/15$. We have measured R=3, implying $T_0>>t_{23}$, according to (6).
This is in agreement with measured rotational diffusion times for large
molecules of >100 ps. The CA contribution to the TCP is then 50%. Since
we observe the incoherent portion of the long decay (t_{31}) to contribute
50±10% to the TCP, the fast component of the TCP appears to be almost
completely due to the CA. From the 10% uncertainty in the long decay contribution, we estimate the fast incoherent contribution to the TCP to be <10%.
Any process occurring in >20 fs that affects the saturation characteristics of
the photoexcited state would contribute more than 10% to the total TCP height,
which implies $t_{23}<20$ fs, assuming that the process corresponding to the transition from level 2 to 3 occurs. A lower bound for t_{23} of 6 fs is derived from
the absorption linewidth of 1800 cm^{-1}.

Experimentally we cannot distinguish between intramolecular and intermolecular relaxation processes. However, to understand our results and those
reported in Ref. 6, we suggest the following model: Because of the anharmonic
coupling to the many degrees of freedom of large molecules, the initial energy
is rapidly (<20 fs here) redistributed within the large molecule. This intramolecular energy distribution 'cools' down subsequently to that corresponding

to the solvent temperature through collisions with solvent molecules. This intermolecular relaxation, occurring on a ps time scale, is what we believe was observed in earlier ps experiments [6].

In conclusion, we have measured the initial relaxation time, T_r, characterizing the isotropic depopulation of carriers from their initially photoexcited levels in GaAs, AlGaAs and MQW as a function of carrier density. At low carrier densities, T_r=45, 85 and 65 fs, respectively, were observed. In GaAs and MQW we have observed a relaxation time of 1.5 ps corresponding to the redistribution of electrons from an energy of 0.5 eV above the bottom of the conduction band to an energy of 0.15 eV. For dye molecules in solution we have determined an upper limit of 20 fs for the initial relaxation out of the photo-excited state, assuming the fast process occurs.

This work was supported by NSF through the Materials Science Center of Cornell University and by the Joint Services Electronics Program.

References

1. D.J. Erskine, A.J. Taylor and C.L. Tang, Appl. Phys. Lett. (July 1984).
2. A.J. Taylor, D.J. Erskine and C.L. Tang, Chem. Phys. Lett. 103, 430 (1984).
3. A.J. Taylor, D.J. Erskine and C.L. Tang, Appl. Phys. Lett. 43, 989 (1983).
4. B.S. Wherrett, A.L. Smirl and T.F. Bogess, IEEE J. Quantum Elec. 19, 680 (19
5. E.O. Kane, J. Phys. Chem. Solids, 1, 249 (1957).
6. A. Laubereau and W. Kaiser, Rev. Mod. Phys. 50, 607 (1978).

Picosecond Laser Studies of Nonequilibrium Electron Heating in Copper

G.L. Eesley

Physics Department, General Motors Research Laboratories
Warren, MI 48090-9055, USA

The phenomenon of inequality between the electron and lattice temperatures in metals has been the subject of theoretical investigation for nearly thirty years [1,2]. The existence of such a nonequilibrium situation was postulated on the basis of the small specific heat of the electron gas, which is thermally insulated from the lattice for ultra-short time durations. Although electron-phonon (e-p) collision times are on the order of 0.01 psec at room temperature, several collisions between hot electrons ($\sim$1 eV) and phonons ($\sim$0.02 eV) are required for equilibration ($\sim$0.5 psec).

The advent of picosecond/femtosecond pulsed laser sources now permits the investigation of e-p relaxation in the nonequilibrium regime. Ultrashort light pulses can be used to extend the applicability of thermomodulation spectroscopy, which has conventionally been used to identify critical points in the interband optical absorption of solids [3]. In particular, it has been shown that at photon energies which probe the d-band to Fermi level transitions in noble metals, the thermoreflectance mechanism is dominated by the thermal smearing of occupied conduction electron states [4]. Thus, the thermally induced change in reflectivity at these photon energies is a measure of the conduction electron temperature.

Using picosecond laser pulses to sequentially heat a Cu film and probe the corresponding reflectivity transient in the vicinity of the d-band edge ($h\nu$ = 2.15 eV), we have observed a rapid heating and cooling transient ($\sim$4 psec) which is identified as the initial heating of electrons above the lattice temperature [5]. Thus, transient thermoreflectance spectroscopy (TTRS) provides the capability to temporally resolve electronic and lattice contributions to the reflectivity change, and to investigate their spectral behavior as well. In this work, we present results of TTRS experiments on single-crystal Cu at 100K and 300K, and we compare these results to calculations of energy relaxation under nonequilibrium conditions.

The TTRS apparatus consists of a modelocked argon-ion laser which synchronously pumps two dye lasers with a pulse repetition rate of 246 MHz. The heating dye laser is fixed in wavelength at 650 nm (1.9 eV), with an average output power of 145 mW and a pulse intensity autocorrelation of 7 psec full-width at half-maximum (FWHM). The probing laser is fixed at 594.5 nm (2.09 eV) for the measurements, and the average probe power at the sample is 1.5 mW. The probe pulse autocorrelation FWHM is 4 psec. The temporal crosscorrelation of the heating pulse with the probing pulse is 8.5 psec (FWHM) as measured by two-photon absorption in GaP.

An optical schematic of the TTRS configuration is shown in Fig. 1. The heating laser beam is polarized parallel to the plane of incidence and is focussed to a beam diameter of 20 μm. The probe laser beam is focussed to a diameter of 20 μm, and the retroreflected probe beam is detected by an avalanche photodiode. The heating induced modulation of the probe photo-

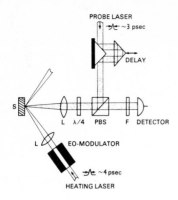

Fig. 1 Optical schematic of the transient thermoreflectance apparatus. L: lens; F: filter: PBS: polarization beamsplitter; λ/4: quarter-wave plate

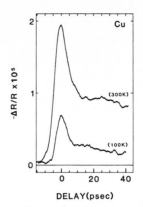

Fig. 2 Transient thermoreflectance signals from single crystal copper (110) at substrate temperatures of 300K and 100K

current is measured by a lock-in amplifier tuned to the 10 MHz modulation frequency of the heating laser. The lock-in output signal is measured as a function of the time delay of the probing pulse relative to the heating pulse, and the lock-in output is divided by the photodiode dc signal to yield the fractional change in reflectivity, $\Delta R/R$.

The sample is a 3 mm thick piece of single crystal Cu(110) mounted on a miniature refrigerator and contained in a vacuum of ∿0.1 μTorr. The Cu crystal was mechanically polished and then chemically polished in a solution of ammonium hydroxide, methanol and hydrogen peroxide.

The Cu transient thermoreflectance is shown in Fig. 2 for substrate temperatures of 300K and 100K. The rapid component of the data centered at zero time delay results from the heating of conduction electrons to a temperature above that of the lattice, followed by rapid equilibration and then the slow diffusion of heat out of the illuminated region [5]. At a probing photon energy of 2.09 eV, these TTRS signals are largely a result of the change in the d-band absorption due to the thermal smearing of final state densities near the Fermi level. Comparison of the thermoreflectance amplitudes with conventional thermoreflectance data indicates that the transient heating of the Cu lattice is ∿1K [4].

At a substrate temperature of 300K, the width of the rapid transient in Fig. 2 is essentially equal to the crosscorrelation width of the heating pulse and probing pulse. It was expected that at 100K the rapid transient would exhibit a slower decay time, indicative of a three-fold increase in the electron-phonon collision time [1]. Clearly this broadening is not exhibited in Fig. 2. In addition, we see that the 100K signal amplitude is approximately one-third that recorded at 300K.

A better understanding of these results can be gained by modelling the heating/cooling process with a two-temperature system of equations. That is, the electron heating proceeds by absorption of light and the lattice heating is driven by e-p collisions. For the case of heat diffusion out of the illuminated area being negligible compared to diffusion into the interior of the sample (z-direction), the macroscopic energy balance

144

equations take the form [2]

$$AT_e \, (\partial T_e/\partial t) = K(\partial^2 T_e/\partial z^2) - G(T_e - T_i) + I_a(\vec{r},t)$$

$$C_i \, (\partial T_i/\partial t) = G(T_e - T_i) \, , \quad \text{where} \tag{1}$$

A = electronic constant of heat capacity,

K = thermal conductivity,

G = electron-phonon coupling constant,

C_i = lattice heat capacity,

$I_a(\vec{r},t)$ = absorbed laser power per unit volume,

$T_e(\vec{r},t)$ = electron temperature profile,

$T_i(\vec{r},t)$ = lattice temperature profile.

Difference formulas can be written for Eq. (1) and solved numerically to yield the temperature profiles of the electrons and the lattice. The e-p coupling parameter G may be calculated from [1], $(T_e - T_i \ll T_i)$

$$G = \frac{\pi^2}{6} \, (mNv^2/\tau T_i) \, (T_e/T_0)^4 \int_0^{T_0/T_e} \left(\frac{x^4}{e^x - 1}\right) dx \, , \quad \text{where} \tag{2}$$

m = free electron mass,

N = conduction electron density,

v = velocity of sound in the metal,

τ = e-p collision time at T_i (from dc conductivity),

T_0 = Debye temperature.

For comparison with the data of Fig. 2, the surface electron temperature was calculated as a function of time, using the appropriate Cu physical constants and a heating pulsewidth of 4 psec. In addition, the results are multiplied by the temperature derivative of the Fermi distribution (df/dT) evaluated at the starting lattice temperature and the probe photon energy. The results are plotted in Fig. 3, where the broadening due to the finite probe pulsewidth has not been included.

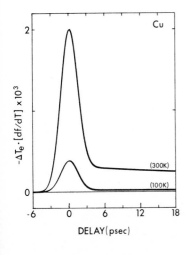

Fig. 3 Calculated surface electron heating multiplied by the Fermi smearing factor (df/dT) at substrate temperatures of 300K and 100K. At 300K the peak electron heating is 3.2K above the lattice, and at 100K the peak electron heating is 5.8K. The 4 psec wide heating pulse is centered at 0 psec time delay

145

Qualitatively we find that the features of the actual measurements (Fig. 2) are reproduced by the calculation. In Fig. 2, the peak $\Delta R/R$ at 100K is approximately one-third that at 300K. This trend is reproduced by the calculation shown in Fig. 3, and it is due to the decrease in df/dT at 100K for final states 0.06 eV below the Fermi level. Also, the ratio of the peak $\Delta R/R$ to that after the heating pulse is larger at 100K than the ratio at 300K, and this trend is also found in the calculations shown in Fig. 3. Most importantly, we note that Fig. 3 shows no significant broadening of the electron response at 100K relative to 300K. Our calculations indicate that at low substrate temperatures, the electron cooling rate is still dominated by electron-phonon coupling. Although the coupling is reduced by a factor of three at 100K, it is sufficient to cool the relatively small electron temperature rise (5.8K) during the 4 psec heating pulse.

In conclusion, we have used picosecond laser pulses to produce and probe the nonequilibrium heating of metal electrons above the lattice temperature. We have shown that macroscopic calculations of energy relaxation in the two-temperature system of electrons and phonons agree with initial measurements of Cu thermoreflectance transients at 100K and 300K.

Future experiments will investigate the effects of defects and alloying on the e-p coupling. In addition, sensitivity calculations are in progress to determine the conditions under which time-resolved measurements of e-p energy relaxation can be made.

1. M. I. Kaganov, I. M. Lifshitz and L. V. Tanatarov, Soviet Physics JETP $\underline{4}$, 173 (1957).
2. S. I. Anisimov, B. L. Kapeliovich and T. L. Perel'man, Soviet Physics JETP $\underline{39}$, 375 (1974).
3. M. Cardona, Modulation Spectroscopy, Ch. 5 (Academic, New York, 1969).
4. R. Rosei and D. W. Lynch, Phys. Rev. B $\underline{5}$, 3883 (1972).
5. G. L. Eesley, Phys. Rev. Lett. $\underline{51}$, 2140 (1983).

Picosecond Measurement of Hot Carrier Luminescence in $In_{0.53}Ga_{0.47}As$

K. Kash and J. Shah

AT & T Bell Laboratories, Crawford Corners Road; Holmdel, NJ 07733, USA

Time resolved optical measurements provide considerable information about carrier energy loss processes in semiconductors by measuring the time evolution of the carrier distribution function. Picosecond luminescence spectra using a streak camera or a Kerr shutter [1-3] and subpicosecond excite and probe absorption experiments [4,5] have been reported in GaAs. A comparable study in the smaller bandgap ternary and quaternary semiconductors would be of considerable interest from both the physics and the device viewpoint, but has not yet been reported.

We report here the first measurement of infrared luminescence from $In_{0.53}Ga_{0.47}As$ with 10 psec time resolution. We have used optical sum frequency, or up-conversion, as a light gate. While this technique has been used previously for picosecond studies of luminescence from CdSe [6] and emission from semiconductor lasers [7,8], our use of a sensitive photon counting technique for sum frequency photon detection and a precise, broadband calibration procedure has enabled us to measure photoluminescence spectra with a spectral resolution of <3meV, time resolution of 10 psec, and a dynamic range $>10^3$ over more than 200 meV spectral range. Our results show that the carrier distribution function has a Maxwellian high energy tail described by a carrier temperature T_C larger than the lattice temperature T_L. From the measurement of the time evolution of T_C, we deduce that the energy loss rate from the carriers to the lattice is approximately an order of magnitude smaller than calculated on the basis of a simple model applicable to the case of a low density Maxwellian distribution.

The measurements reported here were made on a 0.5 μm thick layer of moderate purity ($<10^{16}$ cm^{-3}) $In_{0.53}Ga_{0.47}As$ grown lattice matched to InP by liquid phase epitaxy. The sample was maintained at 10 K. Pulses from a synchronously pumped R6G dye laser, of 8 psec duration, at 610 nm wavelength and 4 MHz frequency, were split. One pulse train was focused to an 80 μm spot on the sample, while the other pulse train, with 12 nJ/pulse, was sent through a variable delay path. The luminescence from the epilayer and the delayed pulse train were then focused collinearly onto a 1 cm long crystal of $LiIO_3$. The time resolution of the up-conversion "gating" was approximately 10 psec in our case. The sum frequency signal was dispersed by a grating spectrometer with 3 meV resolution and detected with a cooled GaAs photomultiplier tube. To obtain a luminescence spectrum at a particular time delay, the angle of the crystal and the spectrometer wavelength were varied simultaneously. The up-conversion efficiency of the system versus infrared wavelength was measured by taking spectra with a calibrated tungsten lamp as the infrared source. The spectra presented below have been corrected for this spectral response.

We show in Fig. 1 the time evolution of luminescence at four different photon energies. We note that the luminescence rise and decay times become longer as the photon energy approaches the bandgap energy (~0.8 eV at 10 K).

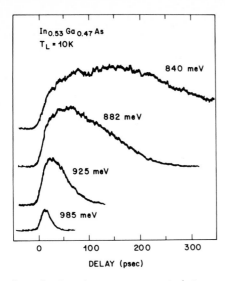

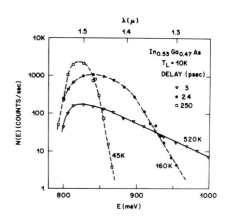

Fig. 1 Sum frequency signal (linear scale) versus time delay for various infrared photon energies above E_g (=813 meV).

Fig. 2 Infrared spectra at different time delays, measured after the middle of the 8 psec pump pulse.

In Fig. 2 we show the luminescence spectra obtained at three different delays for an incident energy/pulse of 2 nJ, corresponding to an estimated injected carrier density of 1.5×10^{18} cm^{-3}. We notice that at 3 psec delay the spectrum is broad and has a long high energy tail and that at longer delays the spectra become narrower with steeper high energy tails. These results can be qualitatively explained on the basis of a model of the cooling of the hot plasma by interaction with lattice modes, a model similar to that discussed previously for GaAs [9].

In order to obtain quantitative information from the spectra, a spectral lineshape analysis is required. Bandgap renormalization, plasma screening effects on the recombination matrix element, and the presence of optical gain complicate the spectra at low energies. However, for energies well above the chemical potential, the spectra are well described by the factor $E^2 \exp(-E/kT_c)$, where E is the photon energy and T_c the carrier temperature. We have deduced the carrier temperatures by fitting the high energy tails of the spectra at various time delays to this form and have plotted these T_c versus the time delay in Fig. 3.

The cooling of hot carriers is determined by the rate at which the hot plasma loses energy to the lattice. For the range of temperatures in Fig. 3, it is expected that the interaction of carriers with polar and non-polar optical phonons will dominate the energy loss processes in InGaAs just as in GaAs [10-12]. For the case of non-degenerate plasma, the cooling curve may be calculated in a straightforward manner following the case of GaAs [9], provided one takes into account the reduction of about a factor of three found in InGaAs compared to GaAs due to the smaller carrier mass and weaker electron-phonon coupling [13]. The result of this simple calculation, displaced by a factor of nine to longer times, is plotted in Fig. 3 as a solid curve. This curve fits the experimental points quite well, leading to the conclusion that the energy loss rate is about a factor of nine smaller than

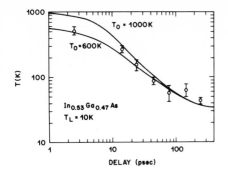

Fig. 3 Carrier temperature versus time for spectra taken with 2 nJ/pulse. The solid lines are the result of the simple calculation described in the text, shifted by a factor of nine to longer times, for initial carrier temperatures T_0 of 600 K and 1000 K.

expected on the basis of the simple model considered above. Similar discrepancies in earlier measurements in GaAs have been attributed [1-5] to the screening of the electron-phonon interaction at the high carrier densities created in these experiments. Estimates of the effect of screening and degeneracy at high carrier densities on the phonon emission rates are in fair agreement with experiments [12, 14]. However, while screening effects are undoubtedly present in both GaAs and InGaAs, the effect of the creation of a non-equilibrium phonon population on the cooling rate [15,16] must also be considered.

We thank A. E. DiGiovanni for excellent technical assistance and R. F. Leheny and P. A. Wolff for valuable discussions.

References

1. S. Tanaka, H. Kobayshi, H. Saito, and H. Shionaya, J. Phys. Soc., Japan 49, 1051 (1980).
2. E. O. Goebel, W. Graudzus, and P. H. Liang, J. Luminesc. 24, 573 (1981), and W. Graudzus and E. O. Goebel, Proceedings of the 16th International Conference on the Physics of Semiconductors I, 555 (Montpelier, 1982).
3. R. J. Seymour, M. R. Junnarkar, and R. R. Alfano, Solid State Commun. 41, 657 (1982).
4. C. V. Shank, R. L. Fork, R. Yen, J. Shah, B. I. Greene, A. C. Gossard, and C. Weisbuch, Solid State Commun. 47, 981 (1983).
5. R. F. Leheny, Jagdeep Shah, R. L. Fork, C. V. Shank, and A. Migus, Solid State Commun. 31, 809 (1979).
6. T. Daly and H. Mahr, Solid State Commun. 25, 323, (1978).
7. M. A. Duguay and T. C. Damen, Appl, Phys. Lett. 40, 667 (1982).
8. T. L. Koch, L. C. Chiu, Ch. Harder, and A. Yariv, Appl. Phys. Lett. 41, 6 (1982).
9. Jagdeep Shah, J. Phys. C7, 445 (1981).
10. E. M. Conwell, High Field Transport in Semiconductors (Academic Press, New York, 1967).
11. E. O. Goebel and O. Hildebrand, Phys. Stat. Sol. (b) 88, 645 (1978).
12. M. Pugnet, J. Collet, and A. Cornet, Solid State Commun. 38, 531 (1981).
13. Jagdeep Shah, R. F. Leheny, R. E. Nahory, and M. A. Pollack, Appl. Phys. Lett. 37, 475 (1980).
14. Ellen J. Yoffa, Phys. Rev. B23, 1909 (1981).
15. J. Collet, A. Cornet, M. Pugnet and T. Amand, Solid State Commun. 42, 883 (1982).
16. W. Pötz and P. Kocevar, Phys. Rev. B28, 7040 (1983).

Picosecond Carrier Dynamics in Semiconductors

E.O. Göbel, J. Kuhl, and R. Höger

Max-Planck-Institut für Festkörperforschung, Heisenbergstraße 1
D-7000 Stuttgart 80, Fed. Rep. of Germany

Picosecond and more recently femtosecond spectroscopy has established as a powerful tool for the direct investigation of the dynamics of nonequilibrium charge carriers and phonons in semiconductors [1]. In this paper we will discuss some recent results of the effect of localization of electronic states on carrier dynamics. Carrier dynamics of localized states are determined by trapping and detrapping mechanisms as well as by the modified transition probabilities of the nonequilibrium charge carriers with respect to delocalized states. Three different systems with different nature and degree of localization have been chosen, namely: GaAs/GaAlAs quantum wells, amorphous semiconductors, and the quaternary compound III-V semiconductor GaInAsP.

Picosecond Luminescence Studies of GaAs/GaAlAs Quantum Well Structures

Quantum well structures represent unique systems for the study of localization effects, because the degree of localization in the direction perpendicular to the quantum well layers can be continuously varied by changing the well thickness L_z and barrier height [2]. Inherent fluctuations in the quantum well thickness additionally result in random fluctuations of the bound energy levels and thus some localization within the plane of the quantum well layers.

We have investigated the kinetic of carrier trapping and recombination by means of picosecond luminescence spectroscopy. A synchronously mode-locked dye laser (Rh6G, 4.7 ps pulse width, 80 MHz rep. rate) is used for excitation and a synchroscan streak camera (S20 spectral response) together with a 0.25 m grating spectrometer for detection. Experimental results for a single GaAs/Ga$_{0.82}$Al$_{0.18}$As quantum well with L_z = 5 nm and 1 μm GaAlAs cladding layers at T = 10 K are depicted in Fig.1a. The upper part shows the time integrated luminescence spectrum for surface excitation of the top GaAlAs cladding layer revealing radiative transitions in the GaAlAs (1) as well as n=1 quantum well transitions including the light hole (2) and heavy hole (3) valence band subbands. The respective time behaviour of these three recombination processes is shown in the lower part of Fig.1a. The decay of the GaAlAs luminescence is extremely fast. The dependence of the GaAlAs luminescence decay on temperature and composition x is correlated to the respective dependence of the carrier mobility [3], i.e. the luminescence decay becomes faster as the mobility increases and vice versa. We therefore conclude that the decay of the GaAlAs luminescence is at least partly determined by the trapping of carriers into the GaAs quantum well. The onset of the quantum well luminescence (curve 2 and 3) corresponds to the decay of the GaAlAs luminescence reflecting again the trapping mechanism. The decay of the quantum well luminescence becomes faster with decreasing well thickness and amounts to about 350 ps for L_z= 5 nm at T = 10 K. The carrier lifetime in the quantum well is determined by radiative and nonradiative recombination transitions. Nonradiative recombination

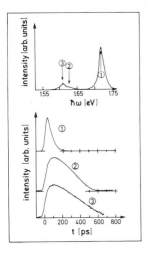

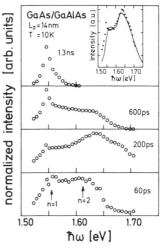

Fig.1a: Time integrated (upper part) and time resolved luminescence (lower part) of a GaAs/GaAlAs single quantum well with L_z = 5 nm

Fig.1b: Transient luminescence spectra of a GaAs/GaAlAs double quantum well with L_z = 14 nm. The inset shows a theoretical fit (full line) to a time resolved spectrum at t = 220 ps

is governed by interface recombination for the moderate carrier concentrations of the present experiments. Both interface as well as radiative recombination increase with decreasing well thickness, however, interface recombination is expected to decrease with temperature [4]. Radiative recombination thus should dominate the decay of the quantum well luminescence at low temperatures for samples with high interface quality even for well thicknesses as small as 5 nm [5]. We therefore attribute the faster luminescence decay of the quantum wells with respect to thick GaAs layers to an increase of electron-hole (exciton) recombination probability due to partial localization [4,6] in accordance,e.g.,with other time-resolved studies [7], earlier excitation spectroscopy results [8] as well as theoretical expectations [8,9,10].

The temporal change of the carrier distribution within the quantum well is more clearly seen in Fig.1b where transient spectra for a GaAs/Ga$_{0.79}$Al$_{0.21}$ double quantum well with L_z= 14 nm are depicted [3]. For this well thickness both the n=1 and n=2 subbands are bound within the quantum well. The respective energetic positions of the excitonic heavy hole transitions as determined by excitation spectroscopy are indicated by arrows. The halfwidth and intensity of the quantum well luminescence increases within the first 200 ps due to the delayed population by trapping of carriers out of the GaAlAs. Afterwards the halfwidth decreases because of relaxation, cooling and recombination of the electrons and holes. The transient spectra clearly reveal the appreciable filling of the quantum well states within the early time regime (t < 500 ps). The inset in Fig.1b shows a calculated lineshape (full line) for t = 220 ps taking into account the 2D joint density of states including a phenomenological damping of 10 meV as well as the 2D continuum exciton enhancement. The values for the density and effective temperature amount to n = 6.5·10^{12} cm^{-2}, and T_{eff} = 150 K, respectively. The effective carrier temperature is appreciably higher than in the 3D case, which may reflect the reduced cooling within the quantum well due to localization [11]. At carrier densities

151

below 10^{11} cm^{-2}, i.e. in particular for resonant excitation of the quantum well,the luminescence spectra are solely determined by 2D exciton recombination [3].

Femtosecond Excite and Probe Spectroscopy of Amorphous Semiconductors

Localized states in amorphous semiconductors are due to intrinsic and extrinsic defects and are statistically distributed within the amorphous network [12]. The picosecond carrier dynamics in amorphous semiconductors have been extensively studied by induced absorption experiments [13] as well as by optoelectronic correlation measurements [14]. Time resolution well below 1 ps is required, however, for a direct determination of carrier relaxation and trapping in high defect materials. We have used a colliding pulse mode locked ring laser (Rh6G and DODCI as laser and absorber dye, respectively) to investigate the induced absorption and reflectivity of amorphous semiconductors with a time resolution of 100 fs. Relative changes of the induced absorption and reflection as low as 10^{-6} can be detected. The experimental results of the amorphous materials (Si and GaAs) are also compared to data for the respective crystalline semiconductors.

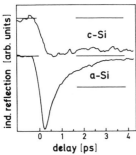

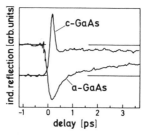

Fig.2a: Photoinduced reflection of amorphous and crystalline Si

Fig.2b: Photoinduced reflection of amorphous and crystalline GaAs

In Fig.2 results for the time response of the induced reflection of crystalline and amorphous Si (Fig.2a) and GaAs (Fig.2b) are shown for example. The a-Si and a-GaAs have been prepared by sputtering an about 0.1 μm thick layer on a transparent substrate without adding hydrogen. The material thus is characterized by a high defect density. The induced reflection of the a-Si decays with a time constant of 0.8 ± 0.2 ps to almost zero within a few picoseconds. The decay of the reflectivity signal in a-GaAs instead shows an initial fast component (time constant 0.6 ± 0.2 ps) followed by a slower decay with a time constant of about 7.5 ps. The time behaviour for the crystalline Si and GaAs is also different. A fast component is observed only for the c-GaAs but definitely is not present in c-Si. This fast component is only slightly broader than the crosscorrelation trace recorded in a KDP crystal. We therefore cannot exclude that this spike is a coherent artifact, which would require the existence of a polarization memory effect. On the other hand, this fast component may reflect changes in the dielectric function due to thermalization of the carriers [15]. Thermalization is indeed expected to occur on a subpicosecond time scale in GaAs because of the strong coupling of the charge carriers to LO phonons [16]. We therefore may conclude that in the a-GaAs both thermalization and trapping of carriers into localized states is observed and gives rise to the fast and slower component, respectively.

In contrast, in the c-Si and a-Si thermalization cannot be resolved explicitly and the decay of the induced reflectivity signal of the a-Si dominantly reflects the trapping of carriers. This preliminary interpretation of our results is further supported by the induced transmission data [17].

Picosecond Luminescence Spectroscopy of GaInAsP

Localization of electronic states in ternary and quaternary compound semiconductors may occur because of fluctuations of the band gap energy due to compositional fluctuations [18]. Various authors have pointed out that depending on growth conditions this can be particularly severe in GaInAsP [19].

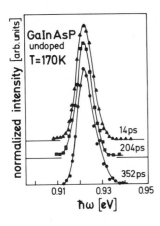

Fig.3: Transient luminescence spectra of a GaInAsP LPE-layer lattice matched on InP. Composition of the GaInAsP corresponds to an emission wavelength of 1.3 μm at T = 300 K

In Fig.3 we present picosecond luminescence spectra of GaInAsP, which indeed give additional support that localization may play an important role in the carrier dynamics of GaInAsP [20]. A frequency doubled passively mode locked Nd:YAG laser (25 ps pulse width) has been used for excitation of the GaInAsP LPE-layers grown lattice matched onto InP substrates. A CS_2 optical Kerr shutter together with a 0.5 m grating spectrometer and a cooled Ge-detector have been employed for measuring the time resolved luminescence spectra with a time resolution of 25 ps. The most striking result of Fig.3 is the narrow spectral halfwidth (~ 7.5 meV), which in addition is almost independent of time in spite of the rapid decrease of the luminescence intensity (roughly a factor of 50 within the first 300 ps). The expected linewidth for a simple band-to-band transition at these high excitation intensities (~ 10 MW/cm^2 which corresponds to an initial carrier concentration of about $8 \cdot 10^{18}$ cm^{-3}) should be in the range of 150-200 meV even if the spectra are dominated by stimulated emission. Furthermore, the linewidth should decrease with time because of the decrease of carrier concentration due to recombination. Similar results have been obtained also for various GaInAsP/InP double heterostructures. These experimental findings obviously cannot be explained by a simple band-to-band transition. We therefore conclude that the spectral behaviour and in particular the linewidth is governed by quasi localized states similar to,e.g.,heavily doped GaAs [21]. Compositional fluctuations, however, will result in an antiparallel (contravariant) modulation of the bands, opposite to the case of heavy doping [18]. Electrons and holes thus will be localized at the same positions,resulting in a enhanced recombination probability. As expected, this localization obviously is less important in the ternary compound InGaAs as can be concluded from recent picosecond luminescence results [22]. The above conclusions drawn on the base of the time resolved luminescence results are furthermore in accordance with recent absorption and stationary photoluminescence experiments [23].

In conclusion, we have shown that time resolved spectroscopy in the picosecond and subpicosecond time regime can be used as a powerful tool to study the effect of localization on the dynamics of nonequilibrium charge carriers. The respective trapping and detrapping mechanisms peculiar to a specific localized state as well as the modified recombination probabilities affect very sensitively the relaxation behaviour as demonstrated by the present results of various groups active in this field.

The results reported here have been obtained in collaboration with H.-U. Habermeier, H. Jung, J.C.V. Mattos (Univ. Campinas, Brasil), A. Mozer (Univ. Stuttgart), Th. Pfeiffer, and K. Ploog. Their various contributions are gratefully acknowledged.

References

1. Various examples can be found in: "Picosecond Phenomena III", K.B. Eisenthal, R.M. Hochstrasser, W. Kaiser, A. Laubereau (editors), Springer Series in Chemical Physics 23 (Springer, Berlin, 1982).
2. R. Dingle: "Confined Carrier Quantum States in Ultrathin Semiconductor Heterostructures" in Festkörperprobleme: Advances in Solid State Physics (Pergamon/Vieweg, Braunschweig, 1975), vol.15, p.21.
3. R. Höger, E.O. Göbel, J. Kuhl, K. Ploog: to be published.
4. R.J. Nelson: J.Vac.Sci.Technol. 15, 1475 (1978).
5. For a surface recombination velocity of 300 cm/s at T = 300 K (see, e.g., ref.[4]) and an interface barrier height of 10 meV the interface recombination lifetime at T = 10 K would amount to about 300 ns.
6. E.O. Göbel, H. Jung, J. Kuhl, K. Ploog: Phys.Rev.Lett. 51, 1588 (1983).
7. J. Christen, D. Bimberg, A. Steckborn, G. Weimann: Appl.Phys.Lett. 44, 84 (1984).
8. R.C. Miller, D.A. Kleinmann, W.T. Tsang, A.C. Gossard: Phys.Rev. B24, 1134 (1981).
9. G. Bastard, E.E. Mendes, L.L. Chang, L. Esaki: Phys.Rev. B26, 1974 (1982).
10. Y. Shinozuka, M. Matsuura: Phys.Rev. B28, 4878 (1983).
11. Y. Masumoto, S. Shionoya, H. Kawaguchi: Phys.Rev. B29, 2324 (1983).
12. See e.g.: "The Physics of Hydrogenated Amorphous Silicon II" (eds. J.D. Joannopoulos, G. Lucovsky, Springer, Berlin 1984).
13. J. Tauc: "Photoinduced Absorption in Armorphous Silicon" in Festkörperprobleme: Advances in Solid State Physics (Pergamon/Vieweg, Braunschweig 1982) vol.22, p.85.
14. D.H. Auston, P. Lavallard, N. Sol, D. Kaplan: Appl.Phys.Lett. 36, 66 (1980); D.H. Auston, A.M. Johnson, P.R. Smith, J.C. Bean: Appl.Phys.Lett. 37, 371 (1980).
15. D.H. Auston, S. McAfee, C.V. Shank, E.P. Ippen, O. Teschke: Solid State Electr. 21, 147 (1978).
16. D. von der Linde, R. Lambrich: Phys.Rev.Lett. 42, 1090 (1979).
17. J. Kuhl, E.O. Göbel, Th. Pfeiffer, A. Jonietz: Appl.Phys., A34,105 (1984).
18. S.G. Petrosyan, A. Ya. Shik: JETP Lett. 35, 437 (1982)
19. see e.g. in "GaInAsP Alloy Semiconductors" (ed. T.P. Pearsall, J. Wiley, Chichester 1982).
20. A. Mozer, M.H. Pilkuhn, J.C.V. Mattos, E.O. Göbel, to be published.

21. Y.I. Pankove: "Optical Processes in Semiconductors", Solid State Physical Electronics Series (ed. N. Holonyak, Jr., Prentice-Hall, Englewood Cliffs, N.Y. 1971) pp. 147.
22. K. Kash, J. Shah: Bull.Am.Phys.Soc. 29, 476 (1984) and to be published.
23. A. Mozer, PhD thesis, Stuttgart 1984, to be published.

Picosecond Dephasing and Energy Relaxation of Excitons in Semiconductors

Yasuaki Masumoto

The Institute for Solid State Physics, The University of Tokyo, Roppongi Minato-ku, Tokyo 106, Japan

1. Introduction

Dynamic relaxation processes of excitons have attracted a great interest of a number of researchers in solid state spectroscopy. By means of a variety of ultrafast spectroscopic techniques, it has now become possible to observe directly the dynamics of photo-excited excitons on the picosecond time scale from the various standpoints. This paper introduces the recent spectroscopic techniques used by us to study dynamic relaxation processes of excitons. One spectroscopic technique is the computer-aided picosecond induced absorption and luminescence spectroscopy to study energy relaxation processes of excitons. The other is the time-resolved non-degenerate four-wave mixing to study the dephasing relaxation processes of excitons. By using these techniques we have studied picosecond relaxation processes of excitons in three characteristic materials, that is, CuCl, CdSe and GaAs-AlAs multi-quantum-well structures (MQW).

2. Computer-Aided Picosecond Induced Absorption and Luminescence Spectroscopy

By observing the transient induced absorption and luminescence it is possible to derive the temporal evolution of the energy distribution of excitons. However, we require an enormous number of calculations to correlate the experimental data with the temporal evolution of the distribution function, the average energy and the number density of excitons. We can shorten this process and obtain precise information by using the computer data processing and graphics.

The Method of Transient Induced Absorption

The principle of the transient induced absorption (IA) is depicted in Fig.1. The initial and final states of the IA transition are the excitonic polariton and excitonic molecule states, respectively. As is shown in Fig.1, the dispersion of the polariton and the molecule is approximately expressed by $E_{ep}(k) = E_t + \hbar^2 k^2/(2M)$ and $E_m(k) = E_m + \hbar^2 k^2/(4M)$, respectively. Here k is the wavevector and M is the translational mass of exciton. Direct transition connects the same k state, so that the energy of absorbed photons E_{IA} is related to the energy of polaritons in the initial state by $E_{IA} = E_m(k) - E_{ep}(k) = E_m - E_t - \hbar^2 k^2/(4M)$. In this equation the polariton state is uniquely connected with the IA spectrum. This method has been applied to CuCl [1]. The IA band is far below exciton resonance (3.2025 eV) and is located at fairly transparent energy region (~3.17 eV). The dispersions of polaritons and molecules in CuCl are well known, so that we can precisely treat the experimental data to obtain the distribution function, average energy and number density of polaritons.

For the IA measurement, two kinds of 20 ps light pulses were generated by using a passively mode-locked Nd^{3+}:YAG laser system. Tunable excitation

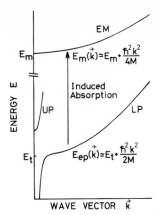

Fig.1 Schematic diagram of the induced absorption transition from the excitonic polariton state to the excitonic molecule state. The notations **EM**, **LP** and **UP** stand for the excitonic molecule, the lower branch excitonic polariton and the upper branch excitonic polariton, respectively

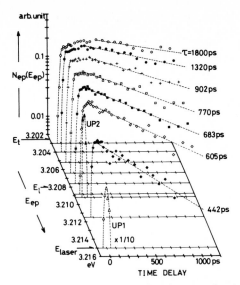

Fig.2 The energy- and time-resolved distribution of excitonic polaritons in CuCl at 4.2 K. E_t (= 3.2025 eV) and E_ℓ (= 3.2080 eV) stand for the transverse and longitudinal exciton energies, respectively. The excitation laser photon energy is 3.215 eV

pulses were second harmonics of parametric signals which were generated by a LiNbO$_3$ optical parametric oscillator (OPO) pumped by the second harmonics of the Nd^{3+}:YAG output. Broad band probing pulses were generated by a dye laser pumped by the third harmonics of the Nd^{3+}:YAG output. The energy of the excitation pulses was tuned to create the upper branch polaritons.

The energy distribution of polaritons is proportional to the product of the IA spectrum, the density of states of polaritons and the inverse of the joint density of states between polaritons and molecules. Therefore we must perform an enormous number of calculations of density of states to obtain the time-resolved energy distribution of polaritons. The data obtained by using an OMA system was transferred to a computer. Thus data processing was easily performed by using this computer. In Fig.2 computed results, that is, the energy- and time-resolved distribution of excitonic polaritons in CuCl, are shown. As is seen, dynamic energy relaxation of polaritons is vividly observed. The theoretical analysis indicates that this energy loss rate is well explained by the deformation-potential-type exciton-LA phonon interaction.

The Method of Transient Luminescence

Another method to probe the temporal evolution of the distribution of excitons is the computer-aided time-resolved luminescence spectroscopy. Two materials, CdSe single crystal and GaAs-AlAs MQW, have been studied by using this method [2,3]. Excitons in CdSe give an example of the typical Wannier excitons in II-VI semiconductors. On the other hand, excitons in GaAs-AlAs MQW give an excellent example of excitons in the two-dimensional disordered system.

Experimental set-up is illustrated in Fig.3. Light pulses given by a rhodamine 6G dye laser synchronously pumped by a mode-locked argon laser were used as the excitation source. Laser light has the pulse width of 1~2 ps and the output of 300 pJ/pulse. The lasing photon energy was 2.1 eV which corresponds to the band-to-band excitation of CdSe. For GaAs-AlAs MQW this energy is above the band gap of GaAs (well) and below that of AlAs (barrier). The spectrally resolved temporal response of the luminescence was analyzed by using a system consisting of a 25 cm-monochromator, a synchroscan streak camera (Hamamatsu-C1587), an SIT camera and computers. Another 50-cm monochromator and an ISIT camera were used to obtain the time-integrated spectra with the improved spectral resolution. The time-resolution of the combined system of the laser, the monochromator and the streak camera was 70 ps.

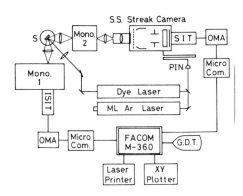

Fig.3 Experimental set-up for the transient luminescence. The meanings of the notations are as follows: **S** = sample; **Mono. 1** = 25-cm monochromator; **Mono. 2** = 50-cm monochromator; **S.S. Streak Camera** = synchroscan streak camera; **PIN** = PIN photodiode; **OMA** = optical multi-channel-analyzer; **Micro Com.** = microcomputer; **G.D.T.** = graphic display terminal

In ordinary semiconductor crystals, it is well known that the line shape of the LO-phonon Stokes sidebands of exciton luminescence directly reflects the energy distribution of excitons. On the other hand, the line shape of the zero-phonon band is related to it in a rather complicated manner because of the polariton effect. Therefore the analysis of the time-resolved LO sidebands is more favorable to derive the temporal development of the energy distribution of excitons than the analysis of the zero-phonon band. In our study of CdSe, therefore, we observed the energy- and time-resolved luminescence of the LO Stokes sidebands of the A exciton (A-LO and A-2LO bands).

In Fig.4 energy- and time-resolved A-LO luminescence is shown. This figure was obtained by means of computer graphics. Because of the flat dispersion of the LO phonon branch, the luminescence intensity at the energy E is proportional to the distribution of excitons at $E+E_{LO}$, where E_{LO} (= 26.3 meV) is the LO phonon energy. Thus we can see the energy relaxation of excitons from Fig.4. Excitons lose most of their energy at ~300 ps, which is much shorter than the lifetime of 2.8 ns. The analysis based on the energy- and time-resolved luminescence of the A-2LO band was found to derive the similar results. We have developed a new theory which gives the analytical expression for energy loss rates due to three types of exciton-phonon interactions, that is, the deformation potential type, the piezoelectric type and the Fröhlich type interactions. The observed energy relaxation rate of excitons is well explained by the sum of the three individual energy loss rates. Among these loss processes, the piezoelectric type interaction plays the predominant role in the energy relaxation of excitons described in Fig.4. This fact comes from the strong piezoelectricity of CdSe.

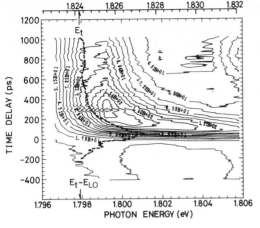

Fig.4 A contour map of energy- and time-resolved luminescence intensity of the A-LO band in CdSe at 4.2 K. The upper horizontal scale is shifted by E_{LO} (=26.3 meV). A bold dotted line shows the average energy of excitons drawn on the upper horizontal scale

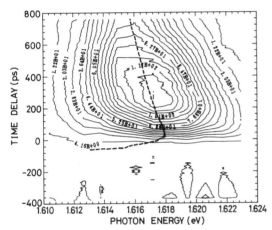

Fig.5 A contour map of energy- and time-resolved luminescence intensity of the excitons (n=1, e-hh) in GaAs-AlAs MQW (76 Å GaAs well, 33 Å AlAs barrier) at 4.2 K. A bold dotted line shows the average energy of excitons

In contrast to the case of CdSe, excitons in GaAs-AlAs MQW have two prominent characters, that is, two-dimensionality and localization in the disordered well. Due to the lateral fluctuation of the well thickness by half of the lattice constant, the resonance energy of exciton is spread. Therefore exciton absorption spectra are inhomogeneously broadened. We can neglect the polariton effect in this system and derive the temporal evolution of exciton energy distribution by observing the transient luminescence of the zero-phonon exciton band. In Fig.5, the energy- and time-resolved luminescence of the lowest exciton (n=1, e-hh) is shown. Population dynamics of excitons is directly visualized in the energy-time space. One can calculate the average energy and the number density of the exciton ensemble. Results indicate that excitons lose their energy in the exciton band at the rate of an order of 10^6 eV/s in all the samples whose well thickness ranges from 53 Å to 108 Å. This rate is much slower than the calculated kinetic-energy-loss rate. Slow rate is due to the localization of excitons in the disordered well. In fact, the theory developed on the basis of the localized exciton transfer model explains the observed relaxation rate.

3. Time-Resolved Non-Degenerate Four-Wave Mixing

The phase relaxation of excitons is considered to occur much faster than the energy relaxation. However, there has been no direct experimental information in the time domain about the phase relaxation of excitons. Dephasing of localized excitation can be measured in the time domain by means of time-resolved, degenerate four-wave mixing. In this technique, two light pulses, $(\vec{k}_1, E_1)$ and $(\vec{k}_2, E_1)$, which are resonant with some material excitation, are used, and the output of the $(2\vec{k}_2-\vec{k}_1, E_1)$ pulse is measured as a function of the time separation between the two pulses. This measurement is based on the principle that the third-order nonlinear polarization which generates the $(2\vec{k}_2-\vec{k}_1, E_1)$ pulse depends on the nondephased part of excitation generated by the first $(\vec{k}_1, E_1)$ pulse at the time when the second, delayed $(\vec{k}_2, E_1)$ pulse reaches the excitation. In the study of the phase relaxation of excitonic polaritons, however, one must note that polaritons are composite particles of excitons and photons. The polaritons are not localized but are propagated in the crystal being dephased with damping constant $\Gamma/2$. Incident photons, $(\vec{k}_1, E_1)$ and $(\vec{k}_2, E_1)$, are converted to polaritons inside the crystal. As they are propagated at the same group velocity $v_g(E_1)$, the first polariton pulse cannot be caught up with by the second, delayed pulse. If we suitably choose the second pulse $(\vec{k}_2, E_2)$ so that its group velocity is faster than that of the first E_1 polariton pulse, the second E_2 polariton pulse catches up with the first E_1 pulse, and the third-order nonlinear polarization which emits a $(2\vec{k}_2-\vec{k}_1, 2E_2-E_1)$ pulse is generated in proportion to the nondephased part of the first polariton pulse. By measuring the $2E_2-E_1$ output as a function of the relative time delay between the two pulses, we can measure the dephasing of the E_1 polariton pulse. To confirm this idea, we have attempted an experiment for excitonic polaritons in CuCl around the exciton resonance [4].

For the experiment, two tunable picosecond pulses are necessary. Two OPO's were pumped by the second-harmonic radiation of a mode-locked Nd^{3+}:YAG laser, and E_1 and E_2 pulses were obtained by taking the second harmonics of parametric signals. The temporal width of the E_1 and E_2 pulses were 20 ps. Two beams, $(\vec{k}_1, E_1)$ and $(\vec{k}_2, E_2)$, were focused on a CuCl crystal directly immersed in superfluid helium. When the temporal coincidence as well as the spatial overlap were optimum, a signal of $2E_2-E_1$ was clearly observed in the direction of $2\vec{k}_2-\vec{k}_1$.

In Fig.6, the output intensity of $2E_2-E_1$ pulses is shown as a function of the relative time delay t_2-t_1. The positive direction of t_2-t_1 means that the slower E_1 pulse is ahead of the faster E_2 pulse. In Fig.6, the calculated transit time of the excitonic polariton pulse through the sample of 14.15 μm thickness is also shown. The transit time of the E_2 pulse through the sample is 1.4∿1.5 ps, while that of the E_1 pulse varies from 3.88 ps to several hundred picoseconds. The traces **1** to **6** show asymmetry tailed toward $t_2-t_1>0$. This asymmetry grows with going from **1** to **3**, and then decreases from **3** to **8**. Then, the correlation traces **8** to **10** are sharp and almost symmetric. The asymmetric tail toward $t_2-t_1>0$ suggests that the phase of the E_1 polariton pulse survives for a while and then it is probed by the E_2 polariton pulse when the latter catches up with the E_1 polariton pulse. The reason why the asymmetry grows in going from **1** to **3** is that the transit time of the E_1 polariton pulse increases. Because the dephasing damping of the E_1 polariton pulse increases, the asymmetry decreases in going from **3** to **8**.

As above, we have demonstrated for the first time that the dephasing of excitonic polaritons in CuCl is directly measurable in the picosecond time

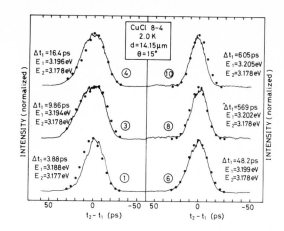

Fig.6 Intensity of the four-wave mixing $2E_2-E_1$ beam emitted from a CuCl crystal as a function of the relative time delay between the E_2 and E_1 pulses, t_2-t_1. The transit times of E_1 pulses are shown by Δt_1, while that of E_2 pulses is 1.4∼1.5 ps

domain by the time-resolved, non-degenerate four-wave mixing at 2.0 K. Theoretical analysis indicates that the observed dephasing constant $\Gamma/2$ is of the order of 0.01 meV and increases as the energy approaches the resonance energy of the transverse exciton from below. This energy dependence and the order of the value suggest that the optical dephasing of excitonic polaritons is attributable to the process of polariton-polariton scattering.

References:

1. Y. Masumoto and S. Shionoya: J. Phys. Soc. Jpn. 51, 181 (1982)
2. Y. Masumoto and S. Shionoya: To be published in Phys. Rev. B
3. Y. Masumoto, S. Shionoya and H. Kawaguchi: Phys. Rev. B 29, 2324 (1984)
4. Y. Masumoto, S. Shionoya and T. Takagahara: Phys. Rev. Letters 51, 923 (1983)

Femtosecond Dynamics of Nonequilibrium Correlated Electron-Hole Pair Distributions in Room-Temperature GaAs Multiple Quantum Well Structures

W.H. Knox, R.L. Fork, M.C. Downer, D.A.B. Miller, D.S. Chemla and C.V. Shank

AT & T Bell Laboratories, Crawford Corners Road, Holmdel, NJ 07733, USA

A.C. Gossard and W. Wiegmann

AT & T Bell Laboratories, Murray Hill, NJ 07974, USA

The GaAs multiple quantum well structure (MQWS) is a new and unique nonlinear optical material [1]. The reduced dimensionality of the electron-hole system in the ultra-thin (100 Å) layers gives rise to clear excitonic absorption peaks which are observed even at room temperature [2]. The GaAs MQWS may form the basis for future high speed optoelectronic devices operating at room temperature.

Here we investigate the dynamics of optical absorption in GaAs MQWS at room temperature under ultrashort excitation near the fundamental absorption edge. In particular, we are interested in the excitation of nonequilibrium distributions of excitons and carriers which have a mean energy of less than kT for the lattice. Our 150 fs optical pulses are shorter than the longitudinal optical phonon scattering time and hence we are able to observe such nonequilibrium distributions at room temperature for the first time. We observe partial relaxation of induced absorption changes in $\sim$ 300 fs, and interpret this as equilibration of the electron-hole system with the lattice.

Tunable excitation pulses in the 800-870 nm spectral range are obtained by focusing 100 µJ, 120 fs duration pulses at 620 nm from an amplified dye laser system operating at 10 Hz repetition rate [3] into a 1.5 mm thick flowing jet of ethylene glycol and selectively amplifying the resultant continuum pulse in a 1 cm cell of flowing LDS 821 laser dye in propylene carbonate. The amplifier cell is pumped with 30 mJ of 532 nm light from the system Q-switched Nd:YAG laser. A pair of diffraction gratings removes group velocity dispersion, a spatial filter rejects amplified spontaneous emission from the amplifier cell, and a 10 nm bandwidth interference filter selects the required spectral excitation distribution. Pulses of 150 fs duration are obtained with energy up to 50 nJ. The slight broadening of the IR pulse is a direct result of the amplification process. A beam splitter selects 10 percent of the continuum pulse before amplification for use as a probe pulse spectrum. The transmitted probe continuum pulse is detected and recorded with a spectrometer and optical multichannel analyzer.

Previously, Shank et al. [4] used non-tunable pulses at 752 nm to excite carriers high in the band in a GaAs MQWS at 77K and observed exciton screening, band filling and renormalization effects. In the present study, the generation of continuously tunable IR pulses of 150 fs duration allows us to selectively excite into excitonic or free carrier absorption regions with increased time resolution. In Figure 1 we show some measurements of absorption dynamics when pumping near the absorption edge. Figure 1a shows the unpumped absorption spectrum of a MQWS consisting of 65 periods of (96Å: GaAs – 98Å: Al$_{.3}$Ga$_{.7}$As). The absorption spectrum in the absence of excitation consists of peaks corresponding to heavy hole (HH) and light hole (LH) excitons in the n=1 quantum sub-band and the two-dimensional continuum [2]. The IR excitation pulse spectrum (also shown in Figure 1a) overlaps the lowest-lying

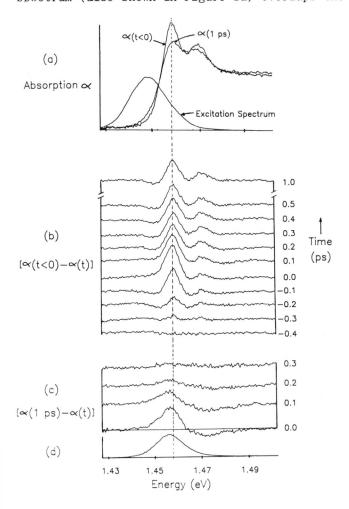

Figure 1.   (a) Absorption of MQWS before and after pumping with 150 fs IR pulse. IR pulse spectrum is also shown. (b) Differential absorption spectra at 0.1 ps intervals. (c) Differential spectra calculated relative to late absorption spectrum. (d) Product of IR pulse spectrum and absorbance, representing absorbed energy distribution function.

energy states of the system; heavy hole excitons or band-tail correlated electron-hole pairs. For this case, the excited carrier pair density is about 5×10^{10} cm^{-2} per layer. We find that after a transient evolution which lasts for about 1 ps the absorption spectrum approximates those obtained with picosecond and continuous wave sources [2]. In addition, the absorption spectrum at times t>1 ps is independent of whether the initial excitation was into excitonic or free carrier absorption regions. We interpret this feature for times t>1 ps as the absorption change introduced by the electron-hole gas once a quasi-equilibrium has been reached. The absorption eventually recovers when the electrons and holes recombine in a time scale of 30 ns [2]. Conversely, we find that the absorption spectrum in the transient regime (0<t<1 ps) is highly sensitive to the excitation spectrum. In Figure 1b we display the differential spectra calculated relative to the early, or unpumped absorption spectrum when pumped near the HH exciton position. The excitation pulse is centered at $\sim$ t=0. At the HH and LH exciton peaks we observe a decrease of the absorption in a time limited by the pump and probe pulsewidths. In the region of the low energy edge of the HH exciton we observe an additional feature which exists substantially for only several hundred fs, in the spectra at t=0, 0.1, 0.2 and 0.3 ps. This feature is highly reproducible and cannot be interpreted as a coherent coupling artifact since identical results are obtained with orthogonally polarized pump and probe beams and with pump and probe beams generated in separate continuum jets. It is the signature of a nonequilibrium distribution of correlated electron-hole pairs which occupy the lowest lying energy states of the system. In order to elucidate this feature more clearly we display the differential spectra in an alternate form which expresses the difference between the spectrum at time t and the quasi-equilibrium spectrum at late times, in the case 1 ps. In Figure 1c we display these unconventional differential spectra for the same data as in Figure 1b at four selected times. The vertical scale has been expanded by a factor of 1.6 relative to those in Figure 1b. The nonequilibrium feature occurs at nearly the peak of the absorbed energy distribution function, which is shown in Figure 1d. The quasi-equilibrium distribution is reached within approximately 300 fs. In contrast, when we pump well above the edge (820 nm) we observe similar late time behavior but no corresponding nonequilibrium feature, within experimental uncertainty.

In the case of excitation near the band edge, the nonequilibrium distribution can be considered as initially "cold" in comparision to the lattice, since the mean energy is less than kT for the lattice. The most probable mechanism for the equilibration of the distribution is absorption of optical phonons. Our measured response time is consistent with an estimate of the thermal ionization time of the HH exciton of 0.4 ps [2].

In the low excitation regime where the bandgap renormalization energy is small compared to the HH exciton binding energy (10 meV) the rapid transient can be interpreted as a saturation of the low-lying exciton states with subsequent heating to the lattice temperature by optical phonons. At higher excitation densities where the bandgap renormalization energy exceeds the HH exciton binding energy, the exciton peaks completely disappear leaving only a renormalized continuum. The nonequilibrium distribution is still observed but at an energy which is shifted below the HH exciton position. In this case, the rapid transient should be considered as a band-filling at the renormalized edge. At intermediate excitation densities where the distinction between excitons and correlated band-edge carriers is less clear, the rapid transient should be regarded as a filling of the lowest available energy states of the system, with approach to equilibrium resulting from absorption of optical phonons.

We have studied the dynamics of optical absorption in GaAs MQWS at room temperature with excitation from 150 fs pulses in the vicinity of the fundamental absorption edge. We observe partial relaxation of the absorption changes in $\sim$300 fs which we ascribe to equilibration of the created electron-hole distribution with the lattice. This result is consistent with estimates of optical phonon scattering rates.

[1]   D. A. B. Miller, Laser Focus 19, Vol. 7, p. 61 (1983).

[2]   D. S. Chemla, D. A. B. Miller, P. W. Smith, A. C. Gossard and W. Weigmann, IEEE J. Quantum Electron., QE-20, 265 (1984).

[3]   R. L. Fork, C. V. Shank, R. Yen and C. A. Hirlimann, IEEE J. Quantum Electron., QE-19, 500 (1983).

[4]   C. V. Shank, R. L. Fork, R. Yen, J. Shah, B. I. Greene, A. C. Gossard and C. Weisbuch, Solid State Comm. 47, 981 (1983).

Femtosecond Transient Anisotropy in the Absorption Saturation of GaAs

J.L. Oudar

Laboratoire de Bagneux, C.N.E.T. 196 rue de Paris, F-92220 Bagneux, France

A. Migus, D. Hulin , G. Grillon, J. Etchepare, and A. Antonetti

Laboratoire d'Optique Appliquée, Ecole Polytechnique, ENSTA
F-91120 Palaiseau, France

The dynamics of hot photoexcited carriers in semiconductors are currently receiving considerable interest[1], owing to the possibility of observing time-resolved luminescence spectra, absorption changes or other nonlinear effects, induced by picosecond, and now femtosecond laser pulses.

We report here the observation of a new ultrafast feature in the absorption saturation of GaAs, using a technique that measures directly the momentum relaxation of photoexcited electrons and holes (e-h). This is due to the fact that linearly polarized light slightly above the band gap generates carriers with an anisotropic momentum distribution. Correspondingly, the absorption saturation by intense pump pulses is anisotropic, which has enabled the observation of four-wave mixing due to orientational gratings [2,3]. Here we show that this anisotropy has a lifetime in the femtosecond range, much faster than the isotropic part of the absorption saturation, reflecting the orientational momentum relaxation of carriers.

This has been observed in a pump-and-test experiment using high power tunable femtosecond pulses [4], starting from the amplified output of a CW mode-locked dye ring laser. The anisotropy of the transmission changes induced in the sample by the pump pulse (linear dichroism $\alpha_\perp - \alpha_\parallel$) has been measured by detecting polarization variations of the transmitted test beam, as a function of the relative delay between pump and test pulses. A polarization-sensitive detection scheme (see Fig.1), similar to the one used in polarization spectroscopy, enables to detect a signal that varies linearly with the rotation ε of the test beam polarization. This rotation is related to the linear dichroism through the relation

$$\varepsilon = (1/4)\sin 2\chi \int_0^d (\alpha_\parallel - \alpha_\perp)\, dz \qquad (1)$$

where χ is the angle between pump and test input polarizations, and the integral is over the sample thickness d. The measured sign of ε corresponds to a rotation of the test polarization towards that of the pump, showing that $\alpha_\parallel < \alpha_\perp$.

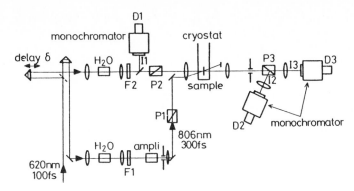

Fig.1 : Experimental set-up. P_1, P_2, P_3 are Glan polarizers. P_1 sets the pump polarization, P_2 the test one, P_3 analyzes the transmitted test polarization. Photodiodes D_1 to D_3 measure intensities I_1 to I_3 respectively.

The tunable excitation wavelength, obtained by dye-amplifying the spectrally filtered output of a femtosecond white light continuum, was set at 806 nm, i.e. very close to the band-edge (820 nm) of our GaAs sample at 77K. This creates electron-hole pairs with an excess energy less than the threshold for optical phonon emission. As a result, strongly inelastic LO-phonon scattering is minimized, which leaves the carrier-carrier interactions as the dominant scattering mechanism. The sample was a 1.5 μm thick GaAs layer obtained by molecular beam epitaxy. The pump energy density was adjusted to be about $100 \mu J/cm^2$, from which we estimated a free-carrier density of $6 \times 10^{17} cm^{-3}$. The test beam was generated in a distinct continuum water cell, to avoid any possible coherent artifact [5] and its wavelength was adjustable independently from that of the pump, by means of monochromators. Maximum polarization signal was obtained when the pump and test wavelength were identical. This signal is shown in Fig.2, where the dots are the experimental points, and the curves are calculated as the convolution of the instrument response (obtained by two independent methods) and exponential functions with time constant τ. From this we deduce that $\alpha_{\shortparallel} - \alpha_\perp$ has a decay time of 190±20fs. More details on this work will appear elsewhere [6].

While in molecular systems anisotropy decay [5] is discussed in terms of rotational diffusion of molecules in real space, our observations are obviously connected to another kind of orientational relaxation[6]. In terms of extended state Bloch functions with wave-vector $\vec{k}$, the valence-to-conduction band transition probability depends on the angle θ between $\vec{k}$ and the linearly polarized electric field $\vec{E}$. As a consequence, the e-h

167

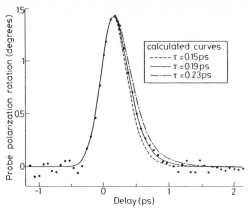

Fig. 2 : Temporal evolution of induced polarization rotation ε.

distribution is anisotropic and can be expanded in terms of the Legendre polynomials of $\cos\Theta$, leading to

$$f^{e,h}(\vec{k}) = f_0^{e,h}(k) + f_2^{e,h}(k) \ P_2(\cos\Theta) + f_4^{e,h}(k) \ P_4(\cos\Theta) + \ldots \qquad (2)$$

For the transition from the heavy-hole valence band (which represents 90 % of the band-to-band transitions) the squared transition matrix element is proportional to $1-P_2(\cos\Theta)=(3/2)\sin^2\Theta$. As a result f_2 in eq.(1) is negative, and it can be shown by proper averaging in $\vec{k}$ space that the measured dichroism is equal to :

$$\alpha_\perp - \alpha_\shortparallel = -(3/10) \ \alpha_0(f_2^e + f_2^h) \qquad (3)$$

i.e. directly proportional to $f_2 = f_2^e + f_2^h$, α_0 being the unsaturated absorption coefficient. Hence the rotation ε (see Eq.1) is a direct measure of the anisotropic part f_2 of e-h distribution. We note that the relaxation of f_2 depends on both elastic and inelastic collisions. In our case, where the test beam frequency is centered on the pump frequency, quasielastic collisions should dominate the observed relaxation. This is consistent with the fact that the observed relaxation of f_0, as measured by a conventional bleaching experiment, takes place on a time scale two orders of magnitude longer (i.e. 20 ps). On the other hand, the drift mobility relaxation time, deduced from Hall measurements at 77K, is calculated to be 1.7 ps, mostly due to ionized impurity scattering. From these considerations we conclude that carrier-carrier collisions are most likely the dominant mechanism for the observed relaxation. Further experiments at various pump intensities should give information on the dependance of orientational relaxation with carrier density.

In conclusion, we have developed a new technique for measuring the momentum relaxation time of photoexcited carriers, by purely optical means, in the femtosecond range.

References

(1) J. SHAH, Journ. de Phys. 42, C7-445 (1981) and references therein

(2) A.L. SMIRL, T.F. BOGGESS, B.S. WHERETT, G.P. PERRYMANN and A. MILLER, Phys. Rev. Lett.49, 933 (1982)

(3) J.L. OUDAR, I. ABRAM, C. MINOT, Appl. Phys. Lett. 44, 689 (1984)

(4) A. MIGUS, J.L. MARTIN, R. ASTIER, A. ANTONETTI and A. ORSZAG in Picosecond Phenomena III (Springer, N.Y. 1982) p.6

(5) C.V. SHANK and E.P. IPPEN, Appl. Phys. Lett. 26, 62 (1975)

(6) J.L. OUDAR, A. MIGUS, D. HULIN, G. GRILLON, J. ETCHEPARE and A. ANTONETTI, submitted to Phys. Rev. Lett.

Holographic Interferometry Using Twenty-Picosecond UV Pulses to Obtain Time Resolved Hydro Measurements of Selenium and Gold Plasmas*

G.E. Busch, R.R. Johnson, C.L. Shepard

KMS Fusion, Inc., P.O. Box 1567, Ann Arbor, MI 48106, USA

When high power laser pulses irradiate various targets in fusion research, high density plasmas are formed very rapidly. Characterizing and controlling these plasmas are primary goals in ICF (Inertially Confined Fusion) research. Holographic interferometry is an optical technique used routinely at KMS Fusion to visualize and delineate the plasma density gradients. Many types of targets are used in the experiments. Selenium and gold disks are compared in this paper.

In the selenium experiments, the laser driver Chroma delivers on-target about 10 Joules of green light (λ = 527 nm) in 100 to 200 psec. The spot size is 200 to 300 μm, giving a power density of 10^{14} - 10^{15} W/cm^2. Two sequential pulses in a YLF oscillator cavity are switched out. One feeds a regenerative oscillator cavity [1] to be temporally compressed 4:1 in 22 full passes. The other of the switched-out pulses feeds the Chroma amplifiers through a stacking device which permits a choice of 1 or 2 (to a maximum of 11) pulses to be serially propagated with 100 psec spacing through the Chroma amplifiers. The regenerated pulse - now about 25 psec long - is amplified, frequency doubled and quadrupled to UV (λ = 263 nm) and is used to probe the Se and Au plasmas. The non-linear frequency conversion further shortens this pulse to < 20 psec. Pulse durations of this order are necessary to prevent blurring of the plasma images obtained in these experiments because phase velocities of 2 - 3 x 10^8 cm/sec are generated.

The Chroma driver pulse and the compressed UV probing pulse have a precise time relation, both arising from the same cavity. By traversing nearly identical optical pathlengths to the target, the probe can illuminate and be used to record holographically the plasma at any relative driver time. The probing time is measured from the peak of the first Chroma pulse to the peak of the UV probe pulse and has a temporal resolution of 10 psec or better.

To generate holographic interferograms of plasmas, two holographic images are recorded sequentially on a single plate: one of the target before, and another during the arrival of the Chroma pulse. Since holographic images record phase, these two images interfere coherently when reconstructed with a visible laser yielding an image with fringes. These fringes can be inverted to produce electron iso-density contours within the plasma object. The method of holographic generation is represented in Fig. 1. Examples of the images are in other figures. The target-plasma object is imaged with a novel two-element standoff f/2 optic of a catadioptrict type with a resolution of 1000 lp/mm.

The holograms are reconstructed using a visible laser to produce 3D images. These are photographed and Abel-inverted [2,3] to obtain electron density gradients and distributions as a function of probe times.

170

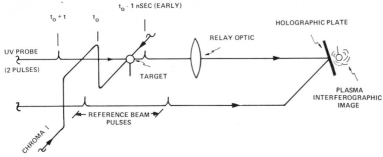

Fig. 1 Diagram showing the sequence of pulses from the main laser, CHROMA, and the UV probe through a fusion target to produce a plasma interferogram.

At $\lambda_{PROBE} = 0.26$ μm the range is from $10^{19} - 3 \times 10^{21}$ e/cm^3, for plasmas of these types. The time of the probe in relation to the driver is adjusted on different target shots to assess the plasma evolution dynamics. This is effective but is limited by shot-to-shot experimental scatter. A multiframe method has been developed which yields images which are spatially and temporally separated with frame rates between 5-25 billion per second which corresponds to frame separation times of 40 to 200 psec. The data shown here is from the single frame system.

The interferographic images in Fig. 2 a,b, and c depict plasmas produced from specially prepared disk targets. These have evaporated selenium (~ 2000 Å thickness) on a silicon substrate about 200 μm in diameter. Chroma enters right to left in the plane of the photos; the probe is in the orthogonal plane. From the inversions of such an image series, using known power densities, and probe times, many plasma hydrodynamic inferences have been made. Electron and phase velocities and accelerations in 3D, absorption boundaries, etc. are a few examples.

Probe time: + 90 psec	Probe time: + 140 psec	Probe time: + 40 psec

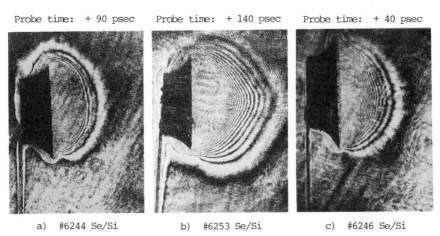

a) #6244 Se/Si	b) #6253 Se/Si	c) #6246 Se/Si

Fig.2 Holographic interferograms: 2000 angstroms selenium on silicon substrate.

These results are compared to hydro-simulation codes and analytical models. Figure 3 portrays the on-axis electron density profiles of the three Se plasmas. Figure 4 is an interferogram of a gold disk. Figure 5 depicts the phase velocity differences between typical Se and Au plasmas. Finally, Fig. 6 is a compilation of phase velocities inferred from a series of Au disk shots.

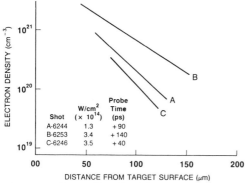

Fig. 3 Density profile of Se/Si targets along axis.

Probe time: + 200 psec

#5900 Au Disk

Figure 4

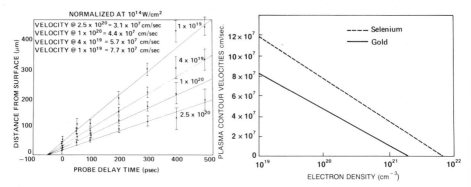

Fig. 5. Plasma velocities of Se and Au targets compared.

Fig. 6 Plasma velocities from disk surfaces for different electron densities.

*Work supported by the U.S. Dept. of Energy under Contract No. DE-ACO8-DP-40152.

1. J. E. Murray, "Temporal Compression of Mode-Locked Pulses for Laser Fusion Diagnostics," IEEE Journ. Quantum Elect., 17, 1713, 9/81.

2. R. N. Bracewell, Fourier Transform and Its Applications, (McGraw-Hill, New York, 1965).

3. D. W. Sweeney, J. Opt. Soc. Am. 64, 559 (1974).

Temporal Development of Absorption Spectra in Alkali Halide Crystals Subsequent to Band-Gap Excitation

W.L. Faust, R.T. Williams, and B.B. Craig

Naval Research Laboratory, Washington, DC 20375, USA

The electronic states of a semiconducting or of an insulating crystal often are treated as though the lattice were rigid, quite indifferent to the presence of carriers. However, departures from this picture are well known. In polar AB crystals, conduction-band electrons have enhanced mass, existing as polarons; and in many halide salts the hole is actually immobilized. Fields associated with an electron in the vicinity of such a hole, a self-trapped exciton (STE), can give rise to substantial displacements of nuclei and even to photochemistry. Transient lattice defects [1], with near-unity yields [2], can be recognized in optical absorption or in emission. Such absorption may cover much of the spectrum, from the near-infrared to the near-ultraviolet. For the alkali halides, there are minor yields of stable or long-lived defects; these populations have been studied for many years by conventional spectroscopic techniques. At low temperatures the ground-state $^3\Sigma_u^+$ STE is stable except for a spin-forbidden radiative transition to the electronic ground state of the crystal. However, certain higher states of the STE lead to generation of F-H defect pairs; this has been demonstrated in double-excitation experiments [3]. The stability of the ground-state STE against such defect creation is associated with only a small barrier -- about 110 meV in NaCl [4]. We have employed short-pulse white-light spectroscopy to record absorption of STE, F, and F-like species early after band-gap excitation by two 266 nm photons. We interpret the temporal evolution of the spectra in terms of thermal migration of Cl atoms over the above barrier [5,6]. In this system, photochemical defect production follows promotion to a potential sheet on which the barriers to halogen diffusion are small.

The apparatus has been described at a previous meeting of this series [2]. The Nd:YAG laser system provides 1064 nm pulses (30-50 mJ, 30 ps) in each of two beams, with control of the relative timing at the sample by a retroprism. The first beam was converted to 266 nm; about 8 mJ energy was available, but ca. 2 mJ sufficed for the current work. Focused to a 2 mm dia. spot on the crystal, the beam produced two-quantum excitation extending through crystal thicknesses of several mm; the consequent defect concentrations resulted in maximal optical density levels of about 2. White light was generated by focusing the second beam through a 1 m lens into a 15 cm cell containing D_2O and D_3PO_4. This beam was employed for dual-beam ratio-recording of absorption spectra between about 460 and 800 nm, with a vidicon polychromator system and a microcomputer for acquisition and processing of data. The spectra show some deficiency of recorded absorption near the edges of the field, where the primary detection sensitivity is reduced; this can be observed as the spectrometer is tuned to bring an absorption band toward an edge. Since measurements of absorption in excess of O.D. = 2 were not considered reliable, the spectra were truncated at this level before averaging. The original spectra (vs. wavelength) included some nonlinear spectral dispersion due to pincushion distortion in the electron optics of the vidicon. This was removed by use of a quadratic fit to nine lines of Hg and of Ne; the data then were converted to energy spectra. The data include uncompensated group velocity temporal

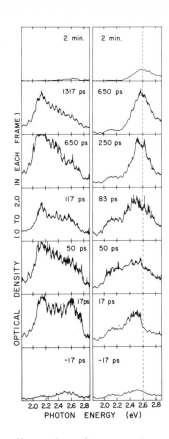

Fig. 1. Time series of absorption spectra for NaCl at 80 K (left) and 300 K (right); the probe pulse delay is given in each frame

dispersion; from streak camera records, the relative delay between probe wavelengths at the STE and F band peaks was estimated to be 15 ps.

Figure 1 presents time-series spectra of absorption in NaCl. Absorption toward the red, near 2.1 eV, is identified with an electron-transition of the STE. Blue absorption, near 2.6 eV, is due principally to F centers (there is a minor component due to a hole-transition of the STE). Note that for T = 80 K the entire spectrum of induced absorption is essentially transient. At 300 K, some F centers remain after several minutes. In the progressions of spectra there is suggested a flow conservative of oscillator strength. A broad initial spectrum, with features common to both temperatures, collapses into a band identified with the STE (low temperature) or with the F center (room temperature). This is consistent with the kinetic picture given previously in Ref. 5.

Several further points are developed in Refs. 5 and 6: (i) The STE (ground state) and F center populations are not independent. The sense is that of a molecular correlation diagram. The STE can be carried by adiabatic [7] nuclear displacements on a single electronic sheet into an F center (a vacancy occupied by an electron) plus an H center (a halogen atom in a split-interstitial configuration; Cl_2^- centered on a Cl^- site). Successive halogen replacements occur along [110] directions (face-diagonal), cumulatively amounting to diffusion of neutral atomic Cl; an extra Cl atom resides at Cl^- sites progressively further removed from the vacancy. Since Cl atoms must pass between Na^+

174

ions, there are potential maxima between minima corresponding to the successive halide sites. Barrier heights have been assigned on the basis of several types of data, as follows: temperature-dependence of the relative yields of STE's and F centers, the temperature-dependent STE lifetime [4], and early kinetics taken from the current data. (ii) Thermal activation is significant; the first two barrier heights are ca. 60 and 110 meV. At a temperature of 80 K the system does not escape the space of the central (STE) and nearest-neighbor (FHnn) minima. However, at 300 K the system may pass to FHnnn and to further-separated configurations. (iii) The spectroscopic properties of the system are such that the atomic diffusion can be visualized in the progressive modifications of the spectrum. The first displacement of Cl, in initial vacancy formation, transforms the STE into FHnn; a broadened and red-shifted F band reflects the neighboring interstitial. At FHnnn and at successive stages of further removal, the F center is less perturbed. The oscillator strengths are similar (near-unity) for the STE band, the F band, and presumably for the perturbed F band. [The H center also possesses a characteristic absorption band; it lies in the near-UV, however, and does not contribute to our spectra. Picosecond time-resolved UV absorption spectra for KI have been reported by SUZUKI and HIRAI [8]].

The kinetics which we have described for NaCl pertain specifically to thermally activated defect formation from relaxed STE's. In crystals such as KCl and KBr, one must invoke a second mechanism to account for very fast F center formation [1] proceeding even at very low lattice temperatures. This channel may have the character of a direct photochemical process; or it may depend upon a locally hot vibration associated with the nonradiative transition in which the STE reaches its ground state, with an excess energy of 1 or 2 eV.

1. J.N. Bradford, R.T. Williams, and W.L. Faust, Phys. Rev. Lett. 35, 300 (1975); R.T. Williams, J.N. Bradford, and W.L. Faust, Phys. Rev. B18, 7038 (1978).
2. W.L. Faust, L.S. Goldberg, T.R. Royt, J.N. Bradford, R.T. Williams, J.M. Schnur, P.G. Stone, and R.G. Weiss, Picosecond Phenomena, ed. C.V. Shank, E.P. Ippen, and S.L. Shapiro (Springer, Berlin 1978), p. 43.
3. R.T. Williams, Phys. Rev. Letters 36, 529 (1976); K. Tanimura and N. Itoh, Semiconductors and Insulators 5, 473 (1983).
4. Quantity from T. Karasawa and M. Hirai, J. Phys. Soc. Japan 39, 999 (1975); M. Ikezawa and T. Kosjima, J. Phys. Soc. Japan 27, 1551 (1969). The current interpretation was developed in Refs. 5 and 6.
5. R.T. Williams, Semiconductors and Insulators 5, 457 (1983).
6. R.T. Williams, B.B. Craig, and W.L. Faust, Phys. Rev. Letters 52, 1709 (1984).
7. The physical motion will involve relaxation of vicinal ions, in the course of a trajectory of minimal free energy-not readily represented in a diagram .
8. Y. Suzuki and M. Hirai, Semiconductors and Insulators 5, 445 (1983).

Kinetics of Free and Bound Excitons in Semiconductors

X.-C. Zhang, Y. Hefetz, and A.V. Nurmikko

Division of Engineering, Brown University, Providence, RI 02912, USA

With tunable short pulse laser sources it has now become possible to examine the kinetics of excitons in semiconductors with detail not possible previously. Here we illustrate the point in the case of excitonic phenomena at low densities for three compound semiconductors, Cu_2O, GaSe, and $Cd_{1-x}Mn_xTe$, the range of phenomena varying from energy-dependent free exciton collisions to thermalization processes characteristic of localized excitons.

Our experiments were performed with a pair of synchronously pumped, modelocked cw dye lasers in an excite-probe configuration. The apparatus is sensitive to small photoinduced modulation and has been used by us to study free electron and impurity bound exciton dynamics in several compound semiconductors (1).

(i) Exciton-Exciton Scattering in Cu_2O

In this work, scattering between nP and 1S excitons has been examined for the yellow exciton in Cu_2O. The experimental photomodulation spectra agrees well with a model where exciton-exciton scattering enters as an additional source of linewidth broadening. The time-dependent results show how an energy-dependent cross-section of this scattering for translationally cold nP excitons strongly favors cold 1S scattering partners. To our knowledge this is a first illustration where such a trend can be deduced for exciton scattering events directly in a time resolved experiment.

Figure 1 shows a photomodulated (probe) spectrum in the vicinity of the 2P resonance, observed at t_d=600 psec following the excitation of the crystal at energy 2.162 eV. Similar features, linear in excitation intensity, were observed up to the 7P transition. The quantity dT/T is equal to $(I_{p2} - I_{p1})/I_{p1}$, where I_{p1} and I_{p2} refer to the intensities of the probe beam transmitted through the sample in the absence and presence of excitation, respectively. A calculated, incrementally collision broadened lineshape is also included in Fig. 1 based on parameters from the measured absorption spectrum for the 2P transition.

Kinetic information is shown in Figure 2 where the signal amplitude dT/T is graphed with probe frequency fixed at the peak of the 2P resonance, for three different photon energies of excitation at t=0 at T=1.8 K. The energies correspond to absorption by the picosecond pulse of excitation in the 1S-phonon assisted tail below the 2P resonance, at approximately the 2P resonance, and in the vicinity of the 4P resonance, respectively. We see a lack of instantaneous response and a time resolved build-up, with variations of the lattice temperature up to 40 K having little effect on the observed risetime. The overall decay of the signals occurred on a scale >> 1 nsec and corresponds to orthoexciton recombination and ortho-para relaxation (2).

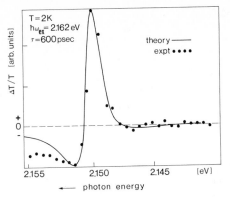
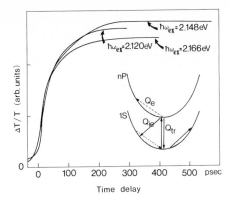

Figure 1: Photomodulation spectra near the 2P resonance in Cu₂O (left)
and initial dynamics interpreted as energy dependent exciton-
exciton scattering (right).

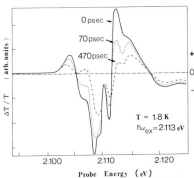
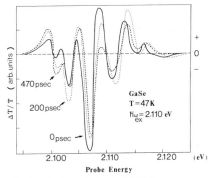

Figure 2: Spectra near n=1 free exciton in GaSe at two temperatures.

By considering the observed lineshapes, relative independence of the risetime on excitation energy, and the levels of overall excitation, we can
rule out contributions by simple lattice heating or by nonequilibrium optical
phonons to the photomodulation spectra. The observed dynamical features then
reflect the energy relaxation of the pump created excitons toward the 1S
orthoexciton minimum by phonon emission. When the mean 1S exciton energy is
lowered below that of the smallest optical phonon participating in the loss
process, the thermalization rate slows down notably, with long wavelength
acoustic phonons carrying the cooling process to completion. Our measurements
then imply that for an nP-1S exciton scattering process the dominant cross
section strongly favors a low energy, cold 1S exciton. The experiment selects nP
excitons as translationally cold because of the small probe photon momentum.
Recent measurements of transient photoluminescence indicate that full
thermalization occurs on a timescale of less than 500 psec (3). We estimate that
the 80 psec risetime measured in the photomodulation experiment implies the
cooling of the 1S gas to an average energy of less than 5 meV.

Those excitonic scattering processes which may show enhancement for small
relative energies (velocities) include both elastic and inelastic events. These
are sketched in the inset of Fig. 2. From recent calculations for
exciton-exciton scattering (4) we note that both elastic (Q_e) and inelastic

(Q_{ie}) collision cross-sections can peak for small values of $\bar{k}_r$, the relati
exciton momentum. A more detailed discussion of this for our case is present
elsewhere (5). In addition, we introduce the inelastic ´resonance transfe
scattering process, $2P(k_1=0)$, $1S(k_3) \rightarrow 1S(k_2=k_3)$, $2P(k_4=0)$, labelled "Q_{tr}".
is known for slow atomic collisions that such a process favors small relativ
velocities and reaches very large values ($Q_{tr} > Q_e, Q_{ie}$) when scattering involv
internal states which can be coupled through allowed dipole optical transition
Here, the nP-1S transitions are strongly electric dipole allowed and immediate
suggesting an exitonic analogy. Experimental sensitivity permitting, furthe
details of such slow collisions should be possible to obtain by using selecti
optical excitation in the energy range directly near the 1S orthoexciton energ
minimum (k=0), through the weakly absorbing quadrupole transition.

(ii) Exciton Confinement by Layer Stacking Faults in GaSe

There are suggestions that the rather fine but random structure which
often accompanies optical absorption or luminescence spectra in the n=1
direct free exciton region of GaSe has an origin in partial confinement
effects induced by stacking faults in this layered semiconductor (6). We
have used time-resolved spectroscopy to study these ideas in thin (1 to 5
micron) platelets of GaSe with emphasis on spectral diffusion within the
inhomogeneous n=1 transition as a function of temperature.

Figure 2 shows examples of time-resolved photomodulation spectra for a 5
μm thick GaSe sample at 2 and 47 K, following excitation within the free
exciton region. The spectra on several platelets is of random multicomponent
origin with different substructure but always characterized by the generic
lineshape for free excitons (including the telltale zero crossings in dT/T).
At 2 K we observe rather independent rates of amplitude decay over different
portions of the inhomogeneous spectra (typically over a range from 200 to
over 500 psec). The decay of the free exciton has a component from the
formation of impurity bound excitons which we have monitored. With
temperature increasing above 10K, distinctly different transient behavior
takes place (see Fig. 2 at T=47 K where the exciton photomodulated spectrum
has acquired also some additional width and structure). Now a homogeneous
lifetime over the spectrum is observed and clear evidence of spectral
diffusion is shown during an initial thermalization on the timescale of some
200 psec. Note how amplitudes of components both on the high energy and low
energy side of the approximate spectral center (where excitation occurred)
grow following the excitation while a monotonous decay occurs at line
center.

The absence of any significant spectral diffusion within the exciton
line at low temperatures is identified with spatial diffusion in a model
where random stacking faults partially localize the n=1 exciton in a
direction perpendicular to the layer planes. The stacking faults are taken
to be the origin for the stochastically inhomogeneous exciton line. In an
adiabatic approximation, the finite coupling of the relative (r) and
CM motion (R) gives a small modulation to the exciton internal energy
through the stacking disorder potential, whereas the confinement effects
follow from an equation of motion for the CM-envelope function. From the
observed absorption data an estimate can then be made for the typical
confinement distance in the c direction to be an average channel of some 15
unit cell layers thick (or approximately 240 A). The Bohr radius of a
spherical free exciton is estimated to be about 80 A in GaSe so that the
confinement only slightly modifies the three-dimensional nature of the
exciton. From the observed temperature trends we estimate that the
localization energy for the n=1 excitons is approximately 1-2 meV.

(iii) Exciton Localization in the Mixed Crystal $Cd_{1-x}Mn_xTe$

Crystal potential variations from alloy compositional fluctuations, inherent to a mixed semiconductor crystal, give opportunity for localization of free excitons (7). While cw luminescence spectra can provide several clues for such events, time resolved techniques should be of superior advantage. Recently, transient luminescence spectra on subnanosecond scale has provided new insight in this connection in $CdSe_{1-x}S_x$ (8). We have applied the excite-probe technique to begin detailed work on exciton localization in the diluted magnetic semiconductor $Cd_{1-x}Mn_xTe$ (9). One particularly useful aspect of our approach is the ability to extract optical signals under conditions of ´resonant excitation´, an instance of considerable difficulty in luminescence work because of Rayleigh scattering.

Briefly, a rough estimate can be made for the energy width below an ´effective exciton mobility edge´ where localized states dominate the spectrum. For $Cd_{.85}Mn_{.15}Te$ this band edge ´smearing´ should be about 10 meV for a statistical cation distribution. Then, in a time resolved experiment, one looks for the characteristics of exciton energy relaxation into these states as a function of initial energy of excitation and temperature. In our first experiments with x=.15 material we see distinctly different photomodulation spectra in the bandtail region at low temperatures, depending on whether the initial excitation is applied above or below a presumed mobility edge whose position is inferred from reflectance peaks (left and right panels of Figure 3, respectively). The overall spectral signature, however, is characteristic of a localized exciton (dT/T>0) as opposed to a free exciton. Superposed on the spectra are also contributions from longer living excitations, believed to be associated with shallow acceptors and donors in these nearly compensated samples.

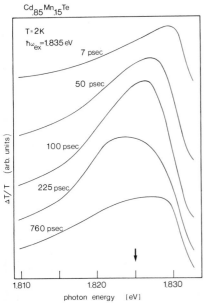
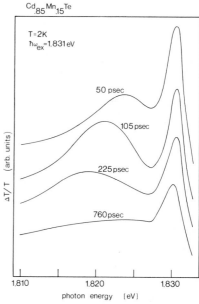

Figure 3: Transient spectra in CdMnTe (x=.15) at T=2 K in the free
          exciton tail for two energies of excitation. Arrow in left
          panel shows position of peak of cw photoluminescence.

The interpretation for the observed behavior is summarized as follows. In the case of injection to the initially delocalized states, exciton capture with comparable probability to the localized states takes place. This is followed by subsequent thermalization within the localized states as evidenced by the time dependent red shift on the left hand panel of Fig. 3. At 2 K, the thermalization is most likely occurring by phonon assisted tunneling. In strong contrast, excitation below the mobility edge generates a readily identifiable resonant response with a low frequency sideband which Stokes shifts with time. We observe that variation of the photon energy of excitation will cause a corresponding shift of the narrow peak. Qualitatively, the peak reflects direct generation of localized excitons which subsequently spectrally diffuse to lower energies, most likely by (energy dependent) phonon assisted tunneling (7). At temperatures approximately above 10 K, the ´resonant´ peaks begin to disappear, presumably showing the increased importance of thermal ionization to the extended states.

This work was supported by DOE through Office of Basic Energy Sciences and NSF/ECS.

References:

(1) J.H. Harris, S. Sugai, and A.V. Nurmikko, Appl. Phys. Lett. 40, 885 (1982); J.H. Harris and A.V. Nurmikko, Phys. Rev. Lett. 51, 1472 (1983), and references therein
(2) A. Mysyrowicz, D. Hulin, and A. Antonetti, Phys. Rev. Lett. 43, 1123 (1979)
(3) J.S. Weiner, N. Caswell, P.Y. Yu, and A. Mysyrowicz, Solid State Comm. 46, 105 (1983); J.S. Weiner, P.Y. Yu, and N. Caswell, Bull. Am. Phys. Soc. 29, 476 (1984)
(4) S.G. Elkomoss and G. Munschy, J. Chem. Phys. Solids 42, 1 (1981); S.G. Elkomoss and G. Munschy, J. Chem. Phys. Solids 45, 345 (1984)
(5) Y. Hefetz, X.-C. Zhang, and A.V. Nurmikko, Bull. Am. Phys. Soc. 29, 476 (1984), and to be published
(6) J.J. Fourney, K. Maschke, and E. Mooser, J. Phys. C 10, 1887 (1975), and references therein; Y. Sasaki and Y. Nishina, Physica 105B, 45 (1981)
(7) E. Cohen and M.D. Sturge, Phys. Rev. B25, 3828 (1982)
(8) J.A. Kash, A. Ron, and E. Cohen, Phys. Rev. B28, 6147 (1983)
(9) X.-C. Zhang and A.V. Nurmikko, Proc. 17th Int.Conf. Physics of Semic. (1984)

Determination of Surface Recombination Velocities for CdS Crystals Immersed in Electrolyte Solutions by a Picosecond Photoluminescence Technique

D. Huppert, S. Gottesfeld, Z. Harzion, and M. Evenor

Department of Chemistry, Tel-Aviv University, 69978 Ramat Aviv, Israel

S. Feldberg

Division of Chemical Sciences, Department of Energy and Environment
Brookhaven National Laboratory, Upton, NY 11973, USA

1. Introduction

The photoluminescence decay of an excited semiconductor depends on a number of factors: 1. The penetration depth of the exciting light. 2. Diffusion of photogenerated electron-hole (e-h) pairs into the semiconductor bulk. 3. Kinetics of band-to-band radiative recombination and nonradiative recombination of e-h pairs at bulk or surface centers. 4. Photoluminescence reabsorption.

In our experiments a semi-infinite uniform CdS semiconductor sample is irradiated with a laser pulse. In all the experiments performed, a high injection level is achieved and it is expected that the excess e-h pairs recombine in the bulk by a first order mechanism with a lifetime τ_B. The time-dependence of the local excess carrier concentration $\Delta n(x,t)$ is given by [1]

$$\frac{\partial}{\partial t} \Delta n = D^* \frac{\partial^2 \Delta n}{\partial x^2} - \frac{\Delta n}{\tau_0} + g(x,t) \qquad (1)$$

where $g(x,t) = \alpha \cdot g_0(t) \exp(-\alpha x)$. $g_0(t)$ is the laser pulse temporal shape, α is the absorption coefficient of CdS at the laser wavelength, and D^* is the ambipolar diffusion coefficient.

For the geometry employed, the boundary condition is:

$$\frac{\partial}{\partial x} \Delta n \Big|_{x=0} = (S/D^*) \cdot \Delta n(x = 0,t) \qquad (2)$$

where S is the surface recombination velocity, and $x=0$ corresponds to the front surface. The surface recombination process is assumed to be first order.

The instantaneous luminescence intensity is given by:

$$I(t) = k_r \int_0^\infty \Delta n^2(x,t) \exp(-\alpha' x) \, dx \qquad (3)$$

where k_r is the second order radiative rate constant and α' is the edge absorption coefficient.

We solved (1) and (3) numerically, including a numerical simulation for the generation function.

2. Experimental

Hexagonal n-type CdS single crystals (Cleveland Crystals) with a resistivity of 1-10 Ωcm were 1-2 mm thick, and faces 5x10 mm oriented perpendicular to the c-axis were used. The surfaces were polished using 0.05 μm alumina powder and subsequently etched in concentrated HCl, followed by rinsing with distilled water. The schematics of the time resolved picosecond photoluminescence apparatus were described elsewhere [2]. The samples were irradiated by a 352 nm 25 ps pulse (the third harmonic of a Nd/YAG laser). The photoluminescence was imaged onto the entrance slit of a C939 Hamamatsu streak camera, whose output was recorded and digitized on a PAR 1205D optical MCA interfaced to a microcomputer for data processing.

3. Results

Figure 1a shows the time resolved normalized photoluminescence intensity of the etched CdS crystal immersed in an aqueous solution of 2M NaOH. The solid line represents the experimental data and the dashed line is the simulated curve. In this case the surface recombination velocity was found to be very low ($S \lesssim 5 \times 10^3$cm/sec). Figure 1b shows the luminescence transient for the crystal immersed in a basic sulfide-polysulfide solution (2M NaOH, 0.1M S°, 0.5M Na$_2$S) after being previously immersed in the 2M NaOH solution for 2 hours. The surface recombination velocity evaluated in this case was 2×10^4cm/sec. Figure 1c shows the result for a crystal immersed immediately after the HCl etch in the sulfide-polysulfide solution. In this case the surface recombination velocity was much larger: 9×10^5cm/sec. The only other parameter evaluated by curve fitting, τ_B, was found practically the same in all cases: 2.4 nsec. For $g_0(t)$, a 25 psec FWHM Gaussian was used in the simulation. Other parameters for CdS were taken from the literature as follows: D*=0.8 cm^2/sec, $1/\alpha'$ (500 nm)=7.3x10^{-5}cm, $1/\alpha$(352 nm)=8x10^{-5}cm.

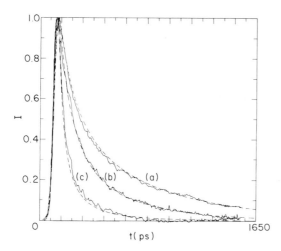

Fig.1. Experimental (solid lines) and simulated (dashed lines) photoluminescence curves for CdS single crystal: (1a) immersed in 2M NaOH solution; (1b) immersed in a solution: 2M NaOH, 0.1M S°, 0.5M Na$_2$S after immersion in 2M NaOH for 2h; (1c) immersed in the sulfide-polysulfide solution immediately after the HCl etch. Excitation wavelength: 352nm. All the plots are arbitrarily normalized.

4. Discussion

The measured decay times of the edge luminescence of the CdS semiconductor crystals were found before [3] to be very sensitive to the mechanical condition of the crystal surface. A sulfide polysulfide solution was used here in order to probe the effect of redox couples in solution on the rate of recombination processes occurring at the semiconductor-solution interface. This redox system is significant in the study of photoelectrochemical cells based on CdS photoanodes [4].

In the present work the large peak power of Nd-Yag laser pulse eliminates the (pre-existing) band bending, and thus the luminescence is controlled by the combined processes of the ambipolar diffusion of the photogenerated e-h pairs as well as by their bulk and surface recombination (1).
As the surface recombination rate increases, the decay of the photoluminescence becomes faster. From figure 1c it is obvious that adsorbed sulfide
strongly enhances the rate of the photoluminescence decay, which corresponds
to an increase of S to 10^6cm/sec. While both electrons and holes are con-
sumed at such a high rate at the surface under conditions of high injection,
this high surface reactivity will enhance the required photohole transfer
process under regular illumination conditions, when band bending lowers the
electron population near the surface.

When the CdS crystal is immersed in a NaOH solution prior to the immersion in the sulfide solution, the surface recombination velocity measured
in the sulfide solution becomes much lower (see Fig.1b). This phenomenon is
probably caused by the formation of an oxide or hydroxide layer on the CdS
crystal during the immersion in the NaOH solution, thus decreasing to some
extent the adsorption of the sulfide anions at the surface. Another possibility is a limited transport rate through the oxide film, which now lies
between the photo-carriers and the surface recombination centers.

The strong measured effect of the adsorbed layer of a redox system on
the surface recombination velocity is of interest. The involvement of
interfacial states, generated by adsorption from solution, in the process
of photocharge transfer has been invoked before [5]. It has been suggested
that inelastic charge transfer may occur from semiconductor electronic
bulk states to such surface states at rates expressed in terms of a surface
recombination velocity [5]. Our measurements show that the rate of this
charge transfer process in the CdS/S°, S^{-2} interface approaches the limit
set by the thermal velocity of the carriers to within an order of magnitude.

References

1. J. Vatikus, Phys. Stat. Sol. A34, 769 (1976).
2. D. Huppert and E. Kolodney, Chem. Phys. 63, 401 (1981).
3. D. Huppert, Z. Harzion, N. Croitoru and S. Gottesfeld in K.B. Eisenthal,
    R.M. Hochstrasser, W. Kaiser and A. Laubereau (Eds.) Picosecond Phenomena
    III, p. 360, Springer-Verlag, 1982.
4. a) G. Hodes, J. Manassen and D. Cahen, Nature, 261, 403 (1976).
    b) B. Miller and A. Heller, Nature 262, 680 (1976).
    c) A.B. Ellis, S.W. Kaiser and M.S. Wrighton, J. Am. Chem. Soc. 98,
       1635, (1976).
5. R.H. Wilson in "Photoeffects at Semiconductors-Electrolyte Interfaces",
    A.J. Nozik (Ed.), Washington, D.C.: ACS (1981), p. 103.

Pulsewidth Dependence of Various Bulk Phase Transitions and Morphological Changes of Crystalline Silicon Irradiated by 1 Micron Picosecond Pulses

S.C. Moss, I.W. Boyd, T.F. Boggess, and A.L. Smirl

Center for Applied Quantum Electronics, Department of Physics, Box 5368
North Texas State University, Denton, TX 76203, USA

We report the first pulsewidth study of the various bulk phase transitions of crystalline silicon (c-Si) induced by ultrashort pulses of λ = 1 μm laser radiation from 4 - 260 ps in duration. In particular, we find a continuous reduction in the single shot melting threshold, E_{TH}, for c-Si from 2.7 $\pm$ 0.3 J/cm^2 for 225 ps pulses to 0.6 $\pm$ 0.1 J/cm^2 for 6 ps pulses. Although pulses longer than 30 ps induce bulk melting followed by recrystallization, pulses of 10 ps duration or less actually produce an amorphized layer on the c-Si, contrary to published expectations [1]. We also find that where a phase change occurs, periodic ripple patterns are formed even for pulses as short as 4 ps. This reduces the proposed minimum irradiation time of 20 ps [2] required to form these structures on c-Si.

Single pulses were selected from the mode-locked output of either a Nd:phosphate glass (λ = 1.05 μm) or a Nd:YAG (λ = 1.06 μm) laser operating in the TEM$_{00}$ mode. A portion of each pulse was delivered to calibrated energy and pulsewidth monitors, in order that the incident energy and temporal duration could be determined on a shot-to-shot basis. Pinhole scans before and at focus indicated that the beams were Gaussian. The average pulsewidth from the Nd:glass laser was 7 $\pm$ 2 ps. Two pulsewidth ranges for the Nd:YAG laser (48 $\pm$ 12 ps and 169 $\pm$ 50 ps) were obtained by using different etalon output couplers. The radiation was focused to a spot size of 280 μm (fullwidth at 1/e^2 of the intensity) onto one of three high purity <111> orientation single crystal Si wafers. Two of these samples were 0.25 mm and 1 mm thick and were anti-reflection coated on the back side. The other was 1 mm thick and uncoated. All experiments were performed at room temperature. The irradiation procedure was strictly 1-on-1, in order to eliminate multishot damage phenomena [3]. The surface morphology was studied using High Contrast Optical Microscopy (HCOM).

For single shot irradiation events, a fluence threshold E_{TH} can be defined, above which a bulk phase change occurs in the material. Post-irradiation examinations of the morphology by HCOM enabled these bulk transitions to be easily distinguished. Figure 1 indicates the relationship between E_{TH} and pulsewidth for λ = 1 μm radiation. Clearly less energy is required by the shorter pulses to induce a permanent phase transition. For example, there is a factor of 4 reduction in E_{TH} for pulses of 6 ps duration compared to those 225 ps in duration. There was no significant difference in the measured values for E_{TH} for the three different samples. These data are in good general agreement with several isolated experimental points found in the open literature [4,5] and with theoretical predictions [6,7].

Close inspection of the morphology shows that for pulses longer than 30 ps the material recrystallizes to the original state, but with some point or line defects incorporated in the structure. However, for pulses shorter than 10 ps, the c-Si is transformed to an amorphous state at E_{TH}.

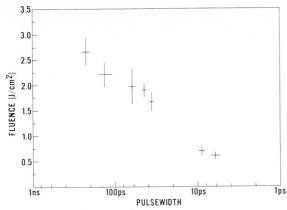

Fig. 1   Pulsewidth dependence of the single shot melting threshold of c-Si irradiated by 1 micron radiation.

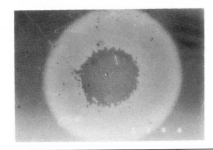

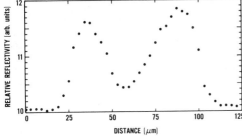

Fig. 2   (a) Surface morphology of c-Si following irradiation by a 6 ps pulse of 0.67 J/cm^2 at 1 micron, and (b) HeNe reflectivity profile of the surface, showing amorphous and crystalline regions.

At slightly higher fluences, only the perimeter of the irradiated area undergoes this transition while the central portion actually recrystallizes (see Fig. 2). At even higher energy densities, the recrystallized region becomes the dominant feature while the amorphous area appears only as a very thin surrounding ring. At the latter fluences a narrow ring of ripples always appears within the recrystallized material. Figure 2 shows a profile of the surface reflectivity at λ = 6328 Å monitored using a tightly focused HeNe laser beam and Si photodiode arrangement. The signal reflected from the edges of the profile in Fig. 2 obviously corresponds to that from conventional c-Si. A rise in reflectivity of 17% in the lighter ringed region is in very close agreement with the enhancement expected from a-Si over c-Si. In the center, the reflectivity corresponds to that measured at the edge, confirming that recrystallisation has taken place.

Laser-induced amorphization of c-Si has been reported previously only using nanosecond ultraviolet [8,9] and picosecond visible or uv radiation [10]. In these cases, the very short absorption depths and cooling times enable the liquid layer to freeze into the solid state before recrystallization can occur [10]. It had been assumed [1] that the absorption depth associated with 1 μm radiation was not small enough to provide sufficient quenching to form a-Si from the melt. Whether or not this is the determining criterion, we have clearly shown that the crystal-amorphous transformation can occur using 1 μm pulses of 10 ps duration, or less.

The phase transition kinetics are more complicated at 1 μm than in the visible or uv, primarily because of the nonlinear nature of the absorption that precedes melting at this wavelength. At high fluences, the absorption is strictly fluence-dependent, and the increase in the absorption coefficient with fluence can be attributed to free carrier absorption and a temperature dependence of the indirect absorption. Moreover, the decrease in melting threshold with decreasing pulsewidth shown in Fig. 1 goes roughly as the square root of the pulsewidth, suggesting that significant energy diffusion occurs during the pulse.

References

1. P. L. Liu, R. Yen, N. Bloembergen, and R. T. Hodgson, in "Laser and Electron Beam Processing of Materials" edited by C. W. White and P. S. Peercy (Academic, New York, 1980).
2. Z. Guosheng, P. M. Fauchet, and A. E. Siegman, Phys. Rev. B26, 5366 (1982).
3. I. W. Boyd, S. C. Moss, T. F. Boggess, and A. L. Smirl, at Materials Research Society Meeting in Boston, Nov. 1983, to be published.
4. K. Gamo, K. Murakami, M. Kawabe, S. Namba, and Y. Aoyagi, in "Laser and Electron Beam Solid Interactions and Materials Processing," edited by J. F. Gibbons, L. D. Hess, and T. W. Sigmon (Elsevier North Holland, 1981).
5. N. Bloembergen, H. Kurz, J. M. Liu, and R. Yen, in "Laser and Electron Beam Interactions with Solids", edited by B. R. Appleton and G. K. Celler (Elsevier, 1982).
6. A. Lietoila and J. F. Gibbons, J. Appl. Phys. 53, 3207 (1982).
7. P. M. Fauchet, A. E. Siegman, Appl. Phys. Lett. 43, 1043 (1983).
8. R. Tsu, R. T. Hodgson, T. Y. Tan, and J. E. Baglin, Phys. Rev. Lett. 42, 1536 (1979).
9. A. G. Cullis, H. C. Webber, N. G. Chew, J. M. Poate, and P. Baeri, Phys. Rev. Lett. 49, 219 (1982).
10. P. L. Liu, R. Yen, N. Bloembergen, and R. T. Hodgson, Appl. Phys. Lett. 34, 864 (1979).

Subthreshold Picosecond Laser Damage in Silicon Associated with Charge Emission

Y.K. Jhee, M.F. Becker, and R.M. Walser

Electronics Research Center, Electrical Engineering Department
The University of Texas at Austin, Austin, TX 78712, USA

Laser-induced surface damage observed during illumination of semiconductors by intense laser pulses has been much studied. The nature of damage has been established [1,2] by studying the nucleation and growth of damage near threshold intensities. The heterogeneous nucleation of damage in the early stage was also revealed. Furthermore, charge emission from a silicon surface induced by picosecond laser pulses at 532 nm was presented [3,4]. But the energy transfer mechanisms from the radiation field to the semiconductor in the multiple-pulse damage regime are not well understood. To contribute to further understanding of damage mechanisms induced by picosecond Nd:YAG laser pulses at 1.06 μm, damage at laser intensities below the one-shot damage threshold has been investigated. By recording charged particle emission which accompanies damage processes, some interesting information has been revealed. With a small capacitor and/or an electron multiplier tube, one can detect the charge emitted during small pit formation which may be considered as the initial damage morphology.

In this experiment, TEM$_{00}$ mode pulses were obtained by the use of a passively mode locked 1.06 μm Nd:YAG laser. The pulses had an average FWHM duration of 60 psec. A 330 μm thick (100) silicon sample was placed 8 mm behind the focus of 172 mm focal length lens where the beam spot diameter was approximately 100 μm. The silicon sample was chemically cleaned before insertion into the test chamber where a vacuum $< 10^{-5}$ torr was maintained during the experiment. The clean silicon sample was used as one electrode while a spiral wire placed 2mm before the sample served as the other electrode. A ±1300 V bias was applied between two electrodes.

The investigation of the incubation period [5] has been conducted by observing the first charge emission event which indicates its termination, and by measuring optical transmission during this period. Negative and positive charge is emitted at the same time and gives the same information. Nonlinear absorption is dominant and has an important role before damage is initiated. The incubation period, proportional to the number of pulses of fixed pulse repetition frequency (PRF), can be correlated to the lifetime of solids subject to a repeated load [6]. Under the assumption that the average temperature rise and the stress for each pulse during the incubation period is proportional to laser energy, the lifetime τ can be written

$$\tau = \tau_0 \exp \frac{U_0 - \gamma^2 E}{kT} = \tau_0 \exp \frac{U_0 - \gamma^2 E}{k^2 E} \tag{1}$$

where $U_0 - \gamma^2 E$ is the activation energy of excitation formation under laser irradiation, E is the irradiation energy, k^2, γ^2 and τ_0 are material dependent constants, and U_0 is the initial energy of activation. The experimental results plotted in Fig. 1 can be fitted well to this equation. This indicates that material excitations are accumulated due to laser irradiation. The incubation period as a function of the PRF was studied. From PRF experiments it was established that the incubation period is character-

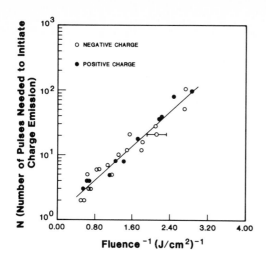

Figure 1: Number of pulses needed to initiate charge emission versus inverse fluence. (Error bars show the standard deviation of the laser energy fluctuations).

ized by irreversible phenomena (over at least 3 seconds), which suggests a mechanism such as long-lived excitation or permanent state accumulation. The heterogeneous nature of the nucleation of damage suggests that long-lived excitations or permanent states may act as nucleation seeds which are precursors to laser induced damage.

Charge emission after the incubation period was studied [7] to reveal further information. Positive and negative particles are emitted equally after damage initiation. The plot of the logarithm of the average charge emitted and inverse laser fluence shows a linear relation between the two quantities. This result suggests that the emission is due to thermal evaporation of silicon where the temperature of the emission site increases linearly with laser pulse energy [4]. In this case emitted charge, N, follows an Arrhenius relation [4],

$$N = N_0 \exp \frac{-U_0}{kT} = N_0 \exp \frac{-U_0}{k'E} \tag{2}$$

where E is the laser fluence, N_0 and k' are constants, and U_0 is the energy of activation. This result possibly indicates that the laser energy absorbed first by carriers is transferred eventually to the lattice. The local temperature is raised high enough to melt or evaporate silicon at pits, grooves, and holes; and some evaporated material is ionized. Thus charge emission depends on local temperature and follows the Arrhenius equation. Experimentally, this result was independent of PRF. This PRF independence of charge emission entirely excludes thermal accumulation pulse by pulse [8].

Damage morphology was compared with the charge emission event. The smallest damage observed after one charge emission event is pit formation at a low fluence. Damage propagates away from these pits with the formation of small ripples due to perturbation fields from the surface irregularity. While the ripple pattern propagates to undamaged areas, the ripple structure at the center of the damaged area is slowly destroyed due to redeposited materials transported from pits or grooves. Charge emission increases as damage grows until it covers most of the beam area. Additional laser pulses diminish the charge emission because of the destruction of grooves and holes at the center. These results suggest that charges are mainly emitted from pits, grooves, and holes within the damaged area.

In conclusion, nonlinear absorption is dominant before damage is initiated. Our data suggests that material excitations may be accumulated due to laser irradiation. The incubation period is characterized by phenomena that are irreversible for at least 3 seconds, and must, therefore, be due to some type of long-lived excitation, or permanent state accumulation. This long-lived excitation or permanent state acts as a nucleation seed, thereby creating its cluster which absorbs laser energy effectively. Evaporation of silicon takes place after a series of irradiations. A fraction of the evaporated particles are ionized and their flux satisfies the Arrhenius relation. Charges are mainly emitted from pits, grooves, and holes within the damaged area.

This research was supported by the DoD Joint Services Electronics Program through AFSOR contract F49620-82-C-0033.

References

1. R.M. Walser, M.F. Becker, D.Y. Sheng, and J.G. Ambrose, "Heterogeneous Nucleation of Spatially Coherent Damage Structures in Crystalline Silicon With Picosecond 1.06 μm and 0.533 μm Laser Pulses " in *Laser and Electron-Beam Solid Interactions and Materials Processing.*, T.J. Gibbons, W. Hess, and T. Sigmon, eds., New York, Elsevier, 1981, 177.
2. M.F. Becker, R.M. Walser, Y.K. Jhee, and D.Y. Sheng, "Picosecond Laser Damage Mechanism at Semiconductor Surfaces " in *Picosecond Lasers and Applications*, January 1982, Los Angeles, CA, SPIE Vol. 322, 93.
3. J.M. Liu, R. Yen, H. Kurz, and N. Bloemergen, Appl. Phys. Lett. 39, 755 (1981).
4. J.A. van Vechten, J. Appl. Phys. 53, 9202 (1982).
5. M.F. Becker, Y.K. Jhee, M. Bordelon, and R.M. Walser, "Observations of an Incubation Period for Multiple-Pulse Laser-Induced Damage " in proceedings of Conference on Lasers and Electro-Optics, May 1983, Baltimore, MD, 72.
6. G.P. Gusev, T.I. Musienko, G.T. Petrovsskii, L.R. Savanovich, and A.V. Shatilov, Sov. J. Opt. Technol. 48, 480 (1981).
7. M.F. Becker, Y.K. Jhee, M. Bordelon, and R.M. Walser, "Charged Particle Exoemission from Silicon during Multiple-Pulse Laser Induced Damage " in proceedings of the 14th Symposium on Optical Materials for High Power Lasers, NBS Special Publication #669, Boulder, CO, 1984.
8. M. Bordelon, R.M. Walser, M.F. Becker, and Y.K. Jhee, "A Study of the PRF Dependence of the Accumulation Effect in Multiple Pulse Laser Damage of Silicon " in proceedings of the 14th Symposium on Optical Materials for High Power Lasers, NBS Special Publication #699, Boulder, CO, 1984.

Third-Order Nonlinear Susceptibilities of Dye Solutions Determined by Non-Phasematched Third Harmonic Generation

A. Penzkofer and W. Leupacher

Naturwissenschaftliche Fakultät II-Physik, Universität Regensburg
D-8400 Regensburg, Fed. Rep. of Germany

The third-order nonlinear susceptibilities $\chi^{(3)}_{xxxx}(-\omega_3;\omega_1,\omega_1,\omega_1)$ of the dyes rhodamine 6G, fuchsin and methylene blue in methanol were determined versus concentration by non-phase-matched third harmonic generation. Picosecond light pulses from a mode-locked Nd-phosphate glass laser (wavelength 1.055μm, pulse duration 5 ps) were employed. For the quantitative measurements a special experimental arrangement was used which avoided disturbing light production by surrounding media [1]. The contributions from air were suppressed by slightly focusing the light beam to the sample cell in a vacuum chamber. The thicknesses of the cell windows were made equal to even integers of the coherence length $\pi/\Delta k$ so that the third harmonic signal was not changed by the windows. The output third harmonic signal originated only from the sample under investigation. For transparent media the third harmonic signal oscillates with cell thickness. This Maker fringe pattern was detected by using a slightly wedged cell and displacing it laterally. For substances absorbing at ω_3 or ω_1 the modulation disappears.

The mean energy conversion ratio $\bar{\eta}$ (averaged over a modulation period) of generated third harmonic signal $\bar{W}_3$ to input pulse energy W_1 is given by

$$\bar{\eta} = \frac{\bar{W}_3}{W_1} = \frac{\bar{\kappa}}{3^{3/2}} |\chi^{(3)}|^2 I_{10}^2 \tag{1}$$

with

$$\bar{\kappa} = \frac{4\omega_3^2[\exp(-3\alpha_1\ell)+\exp(-\alpha_3\ell)]}{n_3 n_1^3 c^4 \varepsilon_0^2 [(\alpha_3-3\alpha_1)^2+4\Delta k^2]} . \tag{2}$$

I_{10} is the peak intensity of the fundamental input pulse. A Gaussian shape is assumed. n_1 and n_3 are the refractive indices. α_1 and α_3 denote the absorption coefficients. c is the vacuum light velocity and ε_0 is the vacuum permittivity. The wave vector mismatch is $\Delta k = k_3-3k_1 = 3\omega_1(n_3-n_1)/c$.

The nonlinear susceptibility $|\chi^{(3)}|$ is determined by measuring all quantities entering (1) and (2). The refractive indices of the solvent methanol are measured with a Pellin-Broca prism method [2] while the refractive indices of the dye solu-

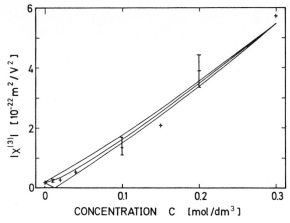

Fig.1 Nonlinear susceptibility of rhodamine 6G in methanol

tions were obtained by a normal-incidence reflection and transmission technique [3]. The absorption coefficients were determined with a spectrophotometer (cell thickness down to 10 μm).

The third-order susceptibilities of the dyes rhodamine 6G,
fuchsin and methylene blue in methanol were investigated (for
previous work see [4-8]). The concentration dependence of
$|\chi^{(3)}|$ is depicted in Fig.1 for rhodamine 6G dissolved in methanol. The susceptibility $\chi^{(3)}$ has contributions from the solvent (S) and the dye (D), i.e. $\chi^{(3)} = \chi_S^{(3)} + \chi_D^{(3)}$. The solute
susceptibility is complex, $\chi_D^{(3)} = |\chi_D^{(3)}| \exp(-i\varphi)$, due to reso-
nances at $2\omega_1$ and ω_3. The nonlinear dye susceptibility is related to the dye hyperpolarizability by [9]

$$\gamma_D^{(3)}(-\omega_3;\omega_1,\omega_1,\omega_1) = |\gamma_D^{(3)}| \exp(-i\varphi) = \frac{24\varepsilon_0 \chi_D^{(3)}}{N_D L^4} \qquad (3)$$

where N_D is the number density of the dye molecules and $L^4 =$
$(n_1^2+2)^3(n_3^2+2)/81$ is the Lorentz local field correction factor.

The curves in Fig.1 are fitted to the experimental $|\chi^{(3)}|$
point at the highest dye concentration. The upper curve belongs
to $\varphi=0$ ($\gamma_D^{(3)}$ real and positive), the middle curve is calculated
for $\varphi=\pi/2$ ($\gamma_D^{(3)}$ imaginary) and for the lower curve $\varphi=\pi$ is used
($\gamma_D^{(3)}$ real and negative). At low dye concentrations (C<0.04mol/l)
the real and imaginary part of $\chi^{(3)}$ may be deduced from the
concentration dependence of $|\chi^{(3)}|$. The data of Fig.1 indicate
a nearly complete imaginary hyperpolarizability due to resonances at $2\omega_1$ and ω_3 of the dye rhodamine 6G.

191

Table 1

Substance	$\lvert \gamma^{(3)} \rvert$ $[10^{-61} \ cm^4/V^3]$	φ $[^0]$
Methanol	1 ± 0.2	0
Rhodamine 6G	2200 ± 500	90 ± 45
Fuchsin	1300 ± 400	45 ± 45
Methylene blue	850 ± 250	0 ± 45

In table 1 the obtained hyperpolarizability values of the investigated substances are summarized. The enhancement of $\lvert \gamma_D^{(3)} \rvert$ compared to the solvent is clearly seen.

References

1. M. Thalhammer and A. Penzkofer, Appl. Phys. B32, 137 (1983).
2. K. Schmid and A. Penzkofer, Appl. Opt. 22, 1824 (1983).
3. W. Leupacher and A. Penzkofer, Appl. Opt. 23, 1554 (1984).
4. P.P. Bey, J.F. Giuliani and H. Rabin, IEEE J. QE-4, 932 (1968).
5. R.K. Chang and L.K. Galbraith, Phys. Rev. 171, 993 (1968).
6. P.P. Bey, J.I. Giuliani and H. Rabin, IEEE J. QE-7, 86 (1971).
7. J.C. Diels and F.P. Schäfer, Appl. Phys. 5, 197 (1974).
8. L.I. Al'perovich, T.B. Babaev and V.V. Shabalov, Sov. J. Appl. Spectrosc. 26, 196 (1977).
9. N.L. Boiling, A.J. Glass and A. Owyoung, IEEE J. QE-14, 601 (1978).

Excitation Transport and Trapping in a Two-Dimensional Disordered System: Cresyl Violet on Quartz

P. Anfinrud, R.L. Crackel, and W.S. Struve

Department of Chemistry and Ames Laboratory - USDOE, Iowa State University
Ames, IA 50011, USA

1. Introduction

Excitation transport and trapping processes involving dye molecules on surfaces are relevant to the efficiency of dye sensitizers on semiconductors in liquid-junction solar cells. Such processes have been treated theoretically for two-dimensional disordered systems [1,2], but have not yet been widely investigated experimentally [2,3]. We report here a picosecond photon counting study of cresyl violet (CV) submonolayers adsorbed onto fused quartz. The questions we address are: (i) What are the fluorescing CV species (CV monomers with varied adsorption geometries, CV dimers, etc.)? (ii) To what extent can a data analysis differentiate among different model functions for the time-dependent fluorescing state population? and (iii) How uniquely can the final decay parameters be determined from such an analysis alone? The last question is pertinent, because surface number densities of dyes in submonolayers, which can be difficult to ascertain directly, are important dynamical parameters in excitation transport.

2. Experimental

Coatings of CV on $\lambda/4$ fused quartz were prepared by treating substrates with ethylene glycol solutions of laser-grade dye, followed by centrifuging. Absorption spectra of the coatings exhibited single broad band maxima at $\sim$ 5300 Å. This indicates that CV dimers (and possibly higher oligomers) are the predominant CV species on quartz, since CV monomers in glycol and CV dimers in water exhibit band maxima at 6010 Å and $\sim$ 5200 Å, respectively. CV dimer surface number densities ranged from $\sim$ 1.9 x 10^{13} to < 2 x 10^{11} cm^{-2}.

A synchronous dye laser containing rhodamine 590 was pumped by a mode-locked argon ion laser to yield $\sim$ 10 ps fwhm, 5850 Å pulses with 47.6 MHz repetition rate. Sample substrates were excited at Brewster's angle, and fluorescence was detected with a Philips XP2020Q phototube. Fluorescence profiles were obtained by time-correlated photon counting using an Ortec 457 TAC with Canberra 30 MCA. Instrument functions evaluated with scattered fundamental were typically 280 ps fwhm. Profiles were analyzed using a convolute-and-compare algorithm based on the Marquardt nonlinear regression technique, which can fit data sets with arbitrary model functions.

As a measure of CV coating nonuniformity, fluorescence profiles were evaluated for randomly selected 0.1 mm dia areas on several of the samples. The resulting fluorescence lifetime parameters (see below) generally varied < 5% for lifetimes longer than the instrument function width.

3. Results

A typical fluorescence profile is given in Fig. 1 for a CV coating with surface density of 8.8×10^{12} cm^{-2}. The overall decay is accelerated at higher coverages due to excitation trapping by CV dimers. A number of model decay functions $N(t)$ were tested, including single exponential, biexponential, and $N(t) = A\exp[-t/\tau - 1.354\, C_T(t/\tau)^{1/3}]$. The latter is the exact decay function for excitation trapping by nonfluorescing trap molecules with reduced trap density C_T in a two-dimensional disordered system in the Förster limit of low fluorescing molecule density [2]. Of these functions, only the biexponential model yields fits with low χ^2, and the convolution of the optimized biexponential function is plotted in Fig. 1.

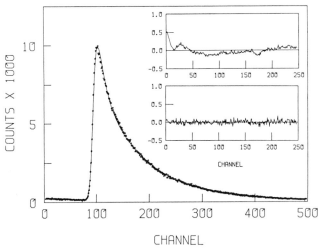

CHANNEL

Figure 1. Fluorescence profile from cresyl violet on quartz. Channel calibration is 25 ps/channel. Insets give autocorrelations for biexponential fit (upper) and single exponential plus Förster fit (lower).

Another model function which closely replicates the decay is $N(t) = A\exp(-t/\tau) + A'\exp[-t/\tau - 1.354\, C_T(t/\tau')^{1/3}]$. This would correspond to two emitting CV species: one decaying exponentially with lifetime τ, and the other decaying by excitation trapping as well as intramolecular relaxation with lifetime τ'. A comparison of autocorrelation functions [4] for the biexponential and exponential plus Förster fits (Fig. 1) shows that only the latter model yields an autocorrelation which is dominated by statistical noise. Final parameters for exponential plus Förster fits are given in Table I for a series of CV coatings with decreasing surface density. For all but the highest-density samples, τ' is comparable to CV monomer lifetimes in solution [5]. The C_T parameters, when combined with dimer (trap) number densities ρ from coating absorption spectra, yield estimates of the monomer-trap excitation transport Förster parameter R_o, since $C_T = \pi\rho R^2$ [1]. R_o values so obtained range from 15 to 21 Å for samples 3-6 in Table I, for which the derived C_T values are relatively insensitive to instrument function uncertainty. The fast-component lifetimes τ are very short compared to the instrument function width. They do not arise from CV dimer emission, because aqueous CV solution (in which CV exists primarily as dimers) emits only single-exponential isotropic monomer fluorescence. For all of the above model functions, the

194

Table I. Single exponential plus Förster fitting parameters

Sample	$\rho(10^{12}cm^{-2})$	A	τ(ps)	A'	τ' (ps)	C_T
1	19	0.282	15.7	0.0077	436	0.377
2	14	0.252	8.8	0.0879	2432	1.97
3	8.8	0.119	22.2	0.118	3231	0.980
4	5.9	0.133	21.3	0.0866	3280	0.418
5	-	0.0700	38.6	0.0832	3381	0.306
6	-	0.0329	90.8	0.0776	3425	0.226

final parameters correspond to well-defined convergences in the convolute-and-compare analyses.

The two-body approximation to the diagrammatic expansion solution of the transport-trapping master equation [1] was also tried as a decay model, with variable reduced monomer and trap densities as well as lifetimes. Severe convergence problems in this analysis indicated that independent knowledge of some of the decay parameters (e.g. C_T) is necessary for testing this model.

References

1. R. F. Loring, M. D. Fayer: Chem. Phys. 70, 139 (1982).
2. N. Nakashima, K. Yoshihara, F. Willig: J. Chem. Phys. 73, 3553 (1980).
3. K. Kemnitz, T. Murao, I. Yamazaki, N. Nakashima, K. Yoshihara: Chem. Phys. Lett. 101, 337 (1983).
4. A. Grinvald, I. Z. Steinberg: Analyt. Biochem. 59, 583 (1974).
5. G. S. Beddard, T. Doust, S. R. Meech, D. Phillips: J. Photochem. 17, 427 (1981).

Temporal Dependence of Third-Order Non-Linear Optical Susceptibilities of Fused Quartz and Liquid CCl$_4$

J. Etchepare, G. Grillon, I. Thomazeau, J.P. Chambaret, and A. Orszag

Laboratoire d'Optique Appliquée, Ecole Polytechnique - ENSTA
F-91120 Palaiseau, France

Temporal discriminations between non-linear effects involving mechanisms with different characteristic times have been, this last decade, the object of many works on organic liquids, supercooled liquids or plastics. Nevertheless, for materials with one or two orders of magnitude smaller non-linearities, the problem still subsists of directly measuring the relative contributions of electronic, nuclear or molecular reorientational origin. We present here the kinetics of the optical Kerr effect (OKE), taken on a sub-picosecond time scale, in two materials which belong to this class of small non-linearity media.

OKE is an efficient method of investigation of the non-linear suscepti-bilities : by using the pump (ω_p) and test (ω_t) technique, we get a direct measurement of a combination of $\chi^{(3)}_{ijkl}$ ($-\omega_t,\omega_t,\omega_p,-\omega_p$) coefficients. In the classical polarization configuration (pump and test beams linearly polarized and at 45° from each other), this combination is [1] :

$$\chi^{(3)}_{eff.} = \frac{1}{2} \left[\chi^{(3)}_{1212} (-\omega_t,\omega_t,\omega_p,-\omega_p) + \chi^{(3)}_{1221} (-\omega_t,\omega_t,\omega_p,-\omega_p) \right] .$$

Our experimental set up has been described earlier [2]. The laser system essentially consists of an intense pump pulse (several GW/cm^2) at 620 nm with a temporal width of $\sim$ 100 fsec, and a probe pulse selected at 650 nm, from a continuum generated in a 5 mm long water cell [3] . The high dynamical range obtained for the transmitted signal of the Kerr cell is strictly related to the probe pulse intensity and the high quality of the polarizers, allowing then to discriminate between effects with very different magnitudes. To deduce from the data the values of $\chi^{(3)}_{eff.}$ coefficients, we have fitted the measured absolute transmission curves to models involving exponential decay times : the laser pulses were described by biexponential temporal shapes and their dispersion effects coming from the passage through the optical elements and inside the Kerr cell accounted for ; liquid CS$_2$ was used as a calibration standard, keeping in mind that its overall well established d.c. Kerr constant [4] arises in fact from at least two different temporal processes [11] .

Figure 1 represents the pump-induced absolute transmission of a 4 mm fused quartz (suprasil) Kerr gate versus pump and test delay. We merely see an instantaneous process, the *overall* signal being well fitted by the third-order correlation of the pulses. This result confirms the indirect determination by HELLWARTH et al. [5] who, by comparing Raman data and intensity-induced polarization change measurements, have determined the fraction f_e of electronic origin to be of 79 ± 3 %. The corresponding value $\chi^{(3)}_{eff.}$ is in excellent agreement with the one reported by OWYOUNG et al. [6], (in the

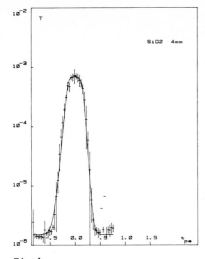
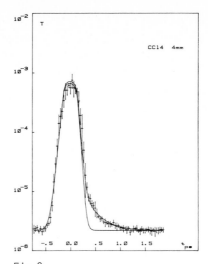

Fig.1.   Fig.2.

Absolute transmission of SiO2 (Fig. 1) and CCl4 (Fig. 2) versus the time
delay between the pump and test pulses (semi-logarithmic scale)

approximation $\chi^{(3)}_{1212} = \chi^{(3)}_{1221}$) ; it is also in good accordance with the value
of $\chi^{el.}_{1111}$ $(-\omega,\omega,\omega,\omega)$ obtained by third harmonic generation [7] . On the opposite to the SiO2 random network, liquid tetrachloride is composed of indivi-
dual molecules, but their spherical symmetry still prevents from a contribution of reorientational origin. In figure 2, we clearly discriminate a non-
instantaneous process characterized by a 0.5 psec. relaxation time : this
finding is well related to earlier light scattering and infrared absorption
measurements [8] which suggested the existence of two non-instantaneous processes with respectively 0.7 and 1.2 psec. response times. We found a $70 \pm 3\%$
ratio of electronic Kerr non-linearity, a proportion close to the 54 % value
of HELLWARTH et al. [9] determined by ellipse rotation measurement. The values we find of the effective non-linearities coefficients are again well
within the range of values deduced from the literature [7,10] .

Substance	CS2	Si O2 fused silica	CCl4
$\chi^{(3)}_{eff.} (-\omega_1,\omega_1,\omega_2,-\omega_2)$	161[b] (1.5 ps)[a]	1.1[a]	2.6[a]
	32[b] (0.20 ps)[a]		1.0[a] (0.5 ps)
$\chi^{(3)}_{1221}(-\omega,\omega,\omega,-\omega)$		1.5[e]	6.1[c]
$\chi^{(3)}_{1111}(-3\omega,\omega,\omega,\omega)$		2.6[d]	4.5[d]
$\chi^{(3)el.}_{1111} (-\omega_3,\omega_1,\omega_1,-\omega_2)$			11.0[f]

Values of third-order non-linear susceptibilities, relative to CS2 units: 10^{15}
esu. Measured relaxation times in parentheses a this work, b[4], c[9], d[7],
e[6], f[10].

In summary, we have demonstrated that OKE technique allows, on a subpico-second time scale, a direct discrimination between different temporal depen-dent non-linear phenomena. A very sensitive calibration standard with res-pect to CS_2 allows the determination of the respective importance of the associated coefficients.

References

1. See for example A. Owyoung - Ph. D. Thesis - California 1971.
2. J. Etchepare, G. Grillon, R. Astier, J.L. Martin, C. Bruneau and A. Antonetti - Picosecond Phenomena III. Springer Verlag (Berlin Heidelberg, New-York, 1982).
3. A. Migus, J.L. Martin, R. Astier, A. Antonetti, and A. Orszag Picosecond Phenomena III. Springer Verlag (Berlin, Heidelberg, New-York 1982).
4. See for example R.W. Hellwarth - Prog. Quant. Electron. 5, 1 (1977).
5. R.W. Hellwarth, J. Cherlow and Tien Tsai Yang - Phys. Rev. B, 11, 964 (1975).
6. A. Owyoung, R.W. Hellwarth and N. George - Phys. Rev. B, 5, 628 (1972).
7. M. Thalhammer and A. Penzkofer, Appl. Phys. B 32, 137 (1983).
8. J.A. Bucaro and T.A. Litovitz - J. Chem. Phys. 55, 3585, (1971)
9. R.W. Hellwarth, A. Owyoung and N. George - Phys. Rev. A, 4, 2342 (1971).
10. M.D. Levenson and N. Bloembergen, J. Chem. Phys. 60, 1323 (1974).
11. See for example J.M. Halbout and C.L. Tang, Appl. Phys. Lett. 40, 765 (1982).

High Excitation Electron Dynamics in GaInAsP

A. Miller, R.J. Manning, A.M. Fox

Royal Signals and Radar Establishment, Malvern, WR 14 3PS, United Kingdom

J.H. Marsh

Department of Electronic and Electrical Engineering, University of Sheffield
Sheffield, S1 3JD, United Kingdom

The quaternary alloy semiconductors such as GaInAsP are of interest for a number of optoelectronic applications at wavelengths compatible with low loss/minimum dispersion optical fibres. Here we report excite-probe absorption saturation measurements in 0.86eV band gap GaInAsP using high power, 5psec (fwhm) pulses at 1.176eV and demonstrate that the optical response of this alloy semiconductor can be extremely fast, $\sim$10psec for excited carrier concentrations in excess of 5×10^{19}cm^{-3}. This very short timescale is consistent with Auger carrier recombination. Evidence of induced intervalence band absorption is also apparent. Both of these mechanisms have been proposed as possible reasons for the strong temperature dependence of the threshold currents of 1.3 - 1.55μm wavelength injection lasers made from this alloy [1].

The 2.5μm thick $Ga_{0.35}In_{0.65}As_{0.78}P_{0.22}$ layer was grown on a semi-insulating InP substrate by liquid phase epitaxy. Single switched out, TEM$_{\infty}$ pulses from a Nd: phosphate glass laser were focussed to a spot size of 100μm (fwhm) on the room temperature epitaxial layer while delayed, attenuated probe pulses sampled the central part of this excited region. For the photon energies employed, the radiation is heavily absorbed in the alloy layer but the substrate is transparent. The low power transmission was $\sim$2% while at high carrier densities, the absorption was bleached due to band filling of the carriers up to and above the optically coupled states. For incident pulse energy densities between 1 and 15 mJ/cm^2, the transmission of excite pulses was found to increase by a factor of about 10. Surface damage occurred above 50mJ/cm^2. Figure 1 shows the transmission of probe pulses as a function of time delay after the arrival of excite pulses of average energy density in excess of 15mJ/cm^2. A 2psec wide coherence spike occurs at the mid-point of a more gradually rising transmission, which takes a time close to the overlap integration of the 5 psec pulses to reach a maximum value. The transmission subsequently falls to its original value in a further 10psec. This decay relates to the excess carrier lifetime but will also be affected by the dynamics of the electron and hole distributions within the bands.

At the onset of the increase in excite pulse transmission ($\sim$1mJ/cm^2), we can estimate that the generated carrier concentration just inside the front face is $\Delta N\sim5\times10^{19}$cm^3 assuming no recombination during the pulse. This should be the approximate density required to move a quasi-Fermi level to the energy of the optically coupled states. In experiments employing 120 psec optical pulses at 1.06μm, Sermage et al.[2] have determined that at carrier densities above 2×10^{18}cm^{-3}, the lifetime in 1.3μm GaInAsP is dominated by Auger recombination with coefficient, 2.3×10^{-29}cm^6/sec. This predicts a lifetime of 20psec at 5×10^{19}cm^{-3} and 10psec by 6.6×10^{19}cm^{-3}, in good agreement with the observed decay feature in fig. 1. We would thus expect significant recombination during the excite pulse at high excitation and the measured decay time may well be limited by the 5psec pulse duration.

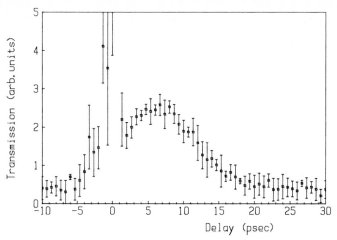

Fig.1 Probe pulse transmission versus delay at ∼15mJ/cm² excitation

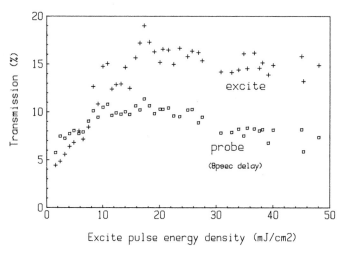

Fig.2 Excite and probe transmissions versus excite pulse fluence

The transmission of excite and probe pulses at a fixed delay of 8psec is shown as a function of excite pulse energy density in fig. 2. Initially, both excite and probe transmissions increase with energy densities up to ∼15mJ/cm² due to saturation of the interband absorption. At higher excitation levels, the transmission of both excite and probe pulses limit at about 16 and 10% respectively, well before the value expected for complete saturation (∼50%). This is reminiscent of results obtained in germanium using a similar laser system [3]. In that case, the limiting was determined to be due to induced absorption caused by transitions between the spin-orbit split and upper valence bands when the upper levels are depleted of electrons at carrier concentrations of ∼ 10^{20}cm^{-3}. Both Ge and the GaInAsP alloy used here have spin-orbit splittings of ∼0.3eV, and so the onset of intervalence band absorption would be expected to occur at about the same densities. We can obtain a rough estimate of the generated den-

200

sity at 15mJ/cm^2 excitation by assuming that the absorption saturation results in a fairly homogeneous distribution of carriers through the sample thickness. Again making the simplication that carrier recombination during the excite pulse is small, then the absorbed energy predicts $\Delta N \sim 3 \times 10^{20} \text{cm}^{-3}$. Thus, comparison with Ge which has been studied in some detail gives strong evidence that the limiting in fig. 2 is due to induced intervalence band absorption.

On the very short timescales encountered here, we would expect several other mechanisms to contribute to the results. For instance, (i) the electrons enter the conduction band with a large excess energy, 12kT, and can take a significant time to thermalise, (ii) cooling of these hot carriers by phonon emission will not be complete on the timescale of the observations, (iii) Auger recombination processes and free carrier absorption will result in further carrier heating, (iv) the high carrier densities will cause an appreciable band gap reduction through many-body interactions.

We have thus identified two mechanisms in GaInAsP which contribute to an ultrafast nonlinear optical response at 1.054μm. The carrier recombination time at high excitation approaches the response measured in amorphous silicon optoelectronic switches [4].

References

1. See e.g. 'Special issue on semiconductor lasers', IEEE J.Quantum Electron. QE-19, 897 (1983)
2. B. Sermage, H.J. Eichler, J.P. Heritage, R.J. Nelson, N.K. Dutta: Appl.Phys.Lett. 42, 259 (1983)
3. A. Miller, G.P. Perryman, A.L. Smirl: Optics Commun. 38, 289 (1981)
4. A.M. Johnson, D.H. Auston, P.R. Smith, J.C. Bean, J.B. Harbinson, A.C. Adams: Phys. Rev. B 23, 6816 (1981)

Picosecond Nonlinear-Optical Limiting in Silicon

T.F. Boggess, S.C. Moss, I.W. Boyd, and A.L. Smirl

Center for Applied Quantum Electronics, Department of Physics
North Texas State University, Denton, TX 76203, USA

We report a Si picosecond nonlinear-optical energy limiter for 1 μm radiation. This completely passive device has a high transmission for low input
energies, but it effectively clamps the output at a low value for high
input energies. The switching action is activated by the generation of
electron-hole pairs and therefore has a subpicosecond initiation time. The
recovery time is limited by the carrier lifetime in the material. This
device is attractive for pulsed applications where one wishes to restrict
the pulse energy incident upon sensitive optical components. Additional
potential applications include its use as an optical Zener diode and as a
fluence regulator for use in materials processing.

The device geometry is shown in the inset in Fig. 1 and is similar to
that used for optical power limiting with cw [1] and pulsed [2] lasers and
for intrinsic optically bistable devices [3]. A single 48 ps (FWHM) pulse
from a mode-locked Nd:YAG laser was focused onto an optically polished
1-mm-thick wafer of high-purity, undoped, single crystal Si of (1,1,1)
orientation. The transmitted beam was recollimated onto a 2-mm aperture
placed before a Si P-I-N photodiode. The response of this detector as a
function of energy incident on the Si is shown by the squares in Fig. 1.
The low energy response is linear with a transmission consistent with the
26% linear transmission of the Si and the 72% transmission of the pinhole.
When the incident energy exceeds 1 μJ, the response becomes nonlinear, and
above 10 μJ input energy the output is clamped at 1.25 ± 0.25 μJ.

The limiting action can be attributed to fluence dependent changes in
the refractive index and absorption coefficient caused by the absorption of
the near-band-gap 1.06 μm radiation. The contribution to the limiting
arising fron nonlinear absorption can be isolated by repeating the above

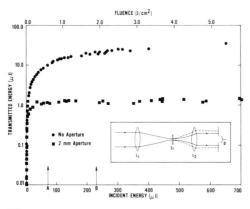

Fig. 1 Device response with
(squares) and without (circles)
the aperture in place. Energy
A marks the onset of observable
multishot damage and B the
single-shot melting threshold.
The device geometry is shown in
the inset

measurement with the aperture removed. The results are shown by the circles in Fig. 1. At the highest fluence, the nonlinear absorption has reduced the Si transmission by a factor of ~5. Preliminary analysis indicates that this nonlinear absorption can be attributed to free-carrier absorption and an increased indirect absorption caused by band-gap narrowing with increasing lattice temperature. Evidently, nonlinear absorption alone cannot account for the observed limiting action. This strong absorption, however, produces a large number of electron-hole pairs that in turn provide a large negative contribution to the refractive index. The resulting self-defocusing was verified by taking vidicon scans in the plane of the pinhole for various input energies. These studies indicate that self-defocusing provides the additional attenutaion observed when the pinhole is in place.

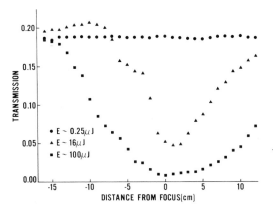

Fig. 2 Transmission of the device for three input energies as a function of Si position relative to L_1

An interesting feature, that is in contrast to previously demonstrated devices using thick liquid or gas filled cells [1-3], is the striking dependence of the present device on the position of the Si relative to the focus of L_1. This dependence is shown in Fig. 2, where the device transmission is displayed as a function of Si position for three input energies. For the low energy, the response is independent of the sample position, as would be expected. At the higher energies, however, the response becomes asymmetric in a manner that indicates a narrower spot at the pinhole with the Si at a given distance before focus than for the same distance beyond focus. Notice that for the intermediate energy, there is even a range of positions before focus where the beam at the aperture is apparently smaller than the undistorted beam. Similar position dependence of the self-defocusing has been reported in narrow-band-gap semiconductors using cw radiation [4]. Such behavior can be qualitatively described by considering the Si as a thin negative lens. For this simple model, the spot size in a fixed plane is unchanged when this negative lens is at the focus of L_1, it can narrow with the lens before focus, and broadens with the lens placed beyond focus. We have observed this effect by scanning the beam in the plane of the pinhole with the Si at focus and at 6 cm from focus. For these measurements the intermediate energy (16 μJ) was used because of the interesting structure seen in Fig. 2. This energy was increased when the Si was moved away from focus so as to maintain a constant fluence. The results are shown in Fig. 3. Notice that with the Si at focus the central portion of the beam is essentially unchanged, even though there is considerably more energy in the wings. When the Si is placed 6 cm before focus, again the energy in the wings increases, but the center of the beam

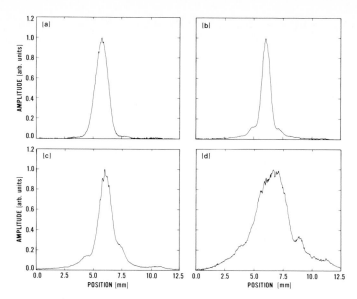

Fig. 3 Spatial beam profiles in the plane of the aperture with (a) the Si removed and with the Si at (b) -6 cm, (c) 0 cm, and (d) +6 cm from the focus of L_1. The fluence was fixed at ~130 mJ/cm^2 for scans (b)-(d)

obviously narrows. Finally, with the Si 6 cm beyond focus, we observe extreme overall broadening of the beam. If the limiter is operated with the Si in this latter position, it should be possible to decrease the energy at which limiting begins while simultaneously increasing the energy at which single-shot melting occurs.

References

1.   R. C. C. Leite, S. P. S. Porto, and T. C. Damen, Appl. Phys. Lett. 10, 100 (1967).
2.   M. J. Soileau, W. E. Williams, and E. W. Van Stryland, IEEE J. Quantum Electron. QE-19, 731 (1983).
3.   J. E. Bjorkholm, P. W. Smith, W. J. Tomlinson, and A. E. Kaplan, Opt. Lett. 6, 345 (1981).
4.   J. R. Hill, G. Parry, and A. Miller, Opt. Commun. 43, 151 (1982).

Nonlinear Absorption and Nonlinear Refraction Studies in MEBBA

M.J. Soileau, W.E. Williams, E.W. Van Stryland, S. Guha, and H. Vanherzeele

Center for Applied Quantum Electronics, Department of Physics,
North Texas State University, Denton, TX 76203, USA

J.L.W. Pohlmann, E.J. Sharp, and G.L. Wood

Night Vision and Electro-Optic Laboratories, Fort Belvoir, VA 22060, USA

In this paper we report measurements of nonlinear absorption and optical self-action (self-focusing) in 4-methyl benzylidene 4'-n-butylaniline (MEBBA) with picosecond pulses at 0.53 µm and 1.06 µm. P_2, the second critical power for self-focusing [1], was measured as a function of pulse-width (37 to 150 psec FWHM) and polarization state at 0.53 and 1.06 µm. This material, a new liquid crystal, was found to have a critical power lower than CS_2 for 125 psec pulses at 0.53 µm (indicating that n_2 is larger than that for CS_2).

Large nonlinearities have been previously reported in liquid crystals and such materials have been used in cw 4-wave mixing experiments [2]. These large n_2's are due to light induced reorientation of the large anisotropic liquid crystal molecules (a slow process). Roa and Jayaraman have reported measurements of n_2 in MBBA with 20 nsec ruby laser pulses [3]. Their value of n_2 was approximately 20 times that of CS_2 for similar conditions.

P_2 was measured using the technique shown in figure 1 and described in detail in reference 4. A lens (L_1) is used to focus light into the liquid crystal and a second lens (L_2) reimages the transmitted light through an aperture. Transmission through the aperture abruptly changes as P_2 is reached. The transmission is lowered by both self-focusing and the subsequent plasma formation. Figure 2 is an example of such laser induced switching for CS_2 and MEBBA at 0.53 µm with 125 psec pulses. The switching power was found to be independent of the focal length of L_1, at both wavelengths used, indicating that the dominant nonlinear process is self-focusing [4]. The P_2 for MEBBA at 0.53 µm is approximately 80 times smaller than at 1.06 µm for the same pulsewidth. Self-focusing theory predicts only a change of four between the two wavelengths. This large

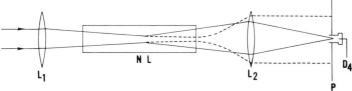

Fig. 1 Technique for measuring the onset of self-focusing. The solid lines schematically trace the input beam for low input power. The beam is focused into the nonlinear medium by lens L_1 and then imaged by L_2 through an aperture onto detector D_4. The transmission for low input powers can be near unity. As the input power is increased to approximately P_2, the critical power for self-focusing [1], the beam undergoes severe phase aberrations (i.e., nonlinear refraction) and the transmission through the aperture decreases. The high power situation is shown schematically by the dotted lines.

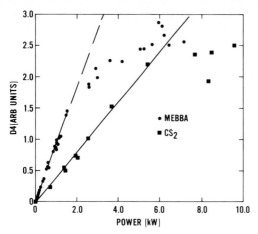

Fig. 2 Nonlinear optical switching in CS_2 and MEBBA (4-methyl benzylidene 4'-n-butylaniline). D_4 is the reading of a detector behind an aperture through which the beam transmitted through the nonlinear medium is imaged. D_4 is linear with input power until P_2 is reached. P_2 for the methoxy variation (MBBA) of this molecule is over an order of magnitude larger for similar conditions. The data shown are for 125 psec (FWHM), linearly polarized 0.53 μm pulses.

dispersion means that the mechanism responsible for nonlinear refraction with picosecond pulses is not simply molecular reorientation. The larger nonlinearity at 0.53 μm may be due to the onset of nonlinear absorption.

Nonlinear absorption in MEBBA was observed at 0.53 μm and the results are plotted in figure 3. This figure is a plot of the inverse transmission as a function of the incident irradiance (expected to be nearly linear for two-photon absorption). Within the uncertainties of the experiment no

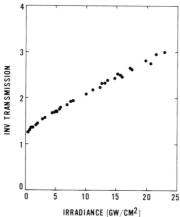

Fig. 3 The inverse of the transmission of a 0.5 cm pathlength cell filled with MEBBA is plotted as a function of the incident irradiance of 0.53 μm, 30 ps (FWHM) pulses. The nearly linear dependence is indicative of two-photon absorption.

pulsewidth dependence of the nonlinear transmission was observed. The two-photon absorption coefficient extracted from this data is 0.4 ± 0.1 cm/GW. No nonlinear absorption was observed at 1.06 μm up to irradiances where there was obvious breakdown of the material.

The relatively large and relatively fast nonlinearity in MEBBA indicates that this material could have important applications in areas of picosecond optical switching, phase conjugation and optical bistability.

1. J. H. Marburger, Progress in Quantum Electronics, J. H. Sanders and S. Stenholm, eds., New York: Pergamon, 1977, pp. 35-110.

2. Khoo, I. C., Phys. Rev. A, 25, 1040, 1982.

3. D. V. G. L. Narasimha Rao and S. Jayaraman, Appl. Phys. Lett., 23, 539, 1973.

4. M. J. Soileau, W. E. Williams, and E. W. Van Stryland, IEEE Jour. Quant. Elec., QE-19, 4, pp. 731-735, 1983.

High Density Carrier Generation in Indium Antimonide

M. Sheikbahae, P. Mukherjee, M. Hasselbeck, and H.S. Kwok

Department of Electrical and Computer Engineering, State University of
New York at Buffalo, Amherst, NY 14226, USA

1. Introduction

The generation of high density free carriers in intrinsic and doped indium
antimonide was studied with picosecond CO_2 laser pulses. It was found that
a very dense electron-hole plasma could be generated despite a large Burstein
bandgap shift. The interaction of this dense plasma with the laser pulse
was highly dynamic and nonlinear. In this paper, we shall present a consistent explanation of the observed nonlinear reflection and transmission
thresholds prior to the onset of melting and damage. The computation involved the interaction of dynamic Burstein shift, Keldysh tunneling, thermal
runaway heating and two-photon absorption.

2. Experimental

The experiments were performed with 75 ps CO_2 laser pulses. These pulses
were generated using the standard procedure of optical free induction decay.
Both transmission and reflection were measured as a function of the pulse
intensity. The transmission measurement is expected to be sensitive to the
generation of free carriers in the $>10^{15} cm^{-3}$ range because of the particular values of the free carrier absorption cross section ($10^{-15} cm^2$ for holes
and 10^{-16} for electrons). The sample thicknesses were 0.63 mm for the intrinsic sample and 0.37 mm for the intrinsic sample.

The reflectivity experiment was performed at a Brewsters angle. In such
an arrangement, any change of the refractive index could be sensitively
measured. [1] Since the critical density for the electron-hole plasma at
the CO_2 laser wavelength was $2.4 \times 10^{18} cm^{-3}$, therefore the reflectivity
measurement was sensitive to plasmas of this density also. Thus the combination of the reflectivity and transmissivity measurements should provide
information on the free carrier generation and interaction mechanisms at
concentrations of $10^{16} cm^{-3}$ to $10^{19} cm^{-3}$. Both intrinsic and n doped samples
were employed in the experiment. The n-InSb was doped to $1.5 \times 10^{18} cm^{-3}$.

The results can be summarized as follows: for the intrinsic sample, the
transmission remained relatively constant until an intensity of 2×10^5
W/cm^2, where it began to drop continuously until the sample was damaged at
4×10^9 W/cm^2. The reflectivity increased at 0.4×10^9 W/cm^2, and then decreased abruptly at 1.2×10^9 W/cm^2.

On the other hand, for the extrinsic sample, the transmission did not
change at all, until very close to the damage threshold. The reflectivity
likewise did not change until melting was achieved at 1.5×10^9 W/cm^2. In
both the intrinsic and extrinsic samples, the breakdown threshold, defined
by spark formation was at 4×10^9 W/cm^2.

3. Discussion

The behavior of the extrinsic sample can be understood by noting that two-photon absorption is disabled completely. The Burstein shifted bandgap is 0.23 eV at a carrier density of $0.7 \times 10^{18} cm^{-3}$. Densities higher than that will cause the bandgap to be larger than twice the CO_2 photon energy. Hence no plasma is generated except at intensities which are close to the melting threshold.

The qualititatively different behavior of the intrinsic sample can also be explained by TPA. The TPA coefficient is 10^{-6} cm/W. At an intensity of 2×10^5 W/cm^2, the carrier density generated is $4 \times 10^{15} cm^{-3}$. This is sufficient to cause the decrease in transmission due to free hole absorption, because of the large heavy to light hole absorption coefficient. This agrees well with the experimental observation.

The increase in reflectivity at 0.4×10^9 W/cm^2 must be due to the generation of a significant plasma density of $\sim 3 \times 10^{18} cm^{-3}$. However, TPA should be disabled at $0.7 \times 10^{18} cm^{-3}$. Therefore there is a discrepancy in the theoretical interpretation. This apparent contradiction can be resolved by using a combination of Keldysh tunneling carrier generation [2] and thermal runaway heating. Both of these two mechanisms are not disabled by the high carrier density.

4. Carrier Generation Calculations

Because of the high laser intensity, carriers continue to be generated even though TPA has been turned off. This rate is given by

$$(dN/dt)_K = 2\omega/9\pi (m^*\omega\eta/ha)^{3/2} Q(\eta, \Delta/h\omega) \tag{1}$$

$$\times \exp\left(-<\Delta/h\omega+1>\right)\{K(\eta a)-F(\eta a)\}/F(a)$$

where
$$\eta = eE/(m^* E_g)^{\frac{1}{2}}$$
$$a = (1+1/\eta^2)^{-\frac{1}{2}}$$
$$\Delta = (2/\pi)E_g(\eta a)F(a) .$$

K and F are complete elliptical integrals of the first and second kind, and Q is a sum of Dawson integrals representing the various multiphoton processes. Equation (1) has been evaluated and used to calculate heating of the lattice [3]. The bandgap is expected to vary as

$$E_g = 0.18 \text{ eV} + 1.5 \times 10^{-2}(\ln(N/10^{17}))^2 - C\Delta T, \tag{2}$$

where C is the temperature coefficient for bandgap shrinking. The effective mass depends also on the carrier concentration as

$$m/m^* = 1 + [5.67 \times 10^{-3} E_g^2 + 1.09 \times 10^{15} N^{2/3}]^{-\frac{1}{2}} . \tag{3}$$

Assuming no diffusion, the lattice temperature T can be calculated

$$dT/dt = N\sigma I(t)/C_p(T) \tag{4}$$

where C_p is the heat capacity. Finally, to include the effect of thermal runaway heating, the total carrier density generated is given by

$$N = \int_0^t (dN/dt)_K \, dt + N_i(T) \tag{5}$$

where $N_i(T)$ is the intrinsic carrier concentration at temperature T. It was found from the calculation that $N_i(T)$ was always much smaller than the contribution by Keldysh generation. Hence thermal runaway heating has a small effect on the observation. This is reasonable because the observed melting threshold for extrinsic InSb was higher than the intrinsic sample. This will not be true if thermal runaway heating is important because the extrinsic sample has a much higher initial carrier density.

The reflectivity and transmissivity is calculated using the usual formula for complex refractive index. The numerical result of the solution of equations (1)-(5) agrees very well with the experimental observation. No adjustable parameters were used in these calculations. It was found that for the intrinsic sample, reflectivity increases at 0.5×10^9 W/cm^2 and melting occurs at 1.2×10^9 W/cm^2. Without Keldysh generation this level of carrier generation and melting cannot be achieved except at unreasonably high intensities.

In summary, we have indirect evidence that Keldysh tunneling ionization has occurred in InSb at intensities higher than 10^9 W/cm^2. A very dense electron-hole plasma was generated which leads eventually to the melting of the sample.

This work was supported by the National Science Foundation.

References

1. M. Hasselbeck and H.S. Kwok, Appl. Phys. Lett. <u>41</u>, 1138 (1982).
2. L.V. Keldysh, Sov. Phys. JETP <u>20</u>, 1307 (1964).
3. M. Hasselbeck, M. Sheikbahae, P. Mukherjee and H.S. Kwok, to be published.

Measurement of Two-Photon Cross-Section in DABCO with the Use of Picosecond Pulses

G. Arjavalingam, J. H. Glownia and P. P. Sorokin

IBM Thomas J. Watson Research Center

P. O. Box 218

Yorktown Heights, New York 10598, USA

From a qualitative standpoint, the vapor phase molecule triethylenediamine (DABCO) has a disposition of energy levels[1] that seems favorable for a demonstration of gain by two-photon stimulated emission (TPSE).[2] Several material parameters, knowledge of which is required to assess the feasibility of the above experiment, have recently been determined. These include the population dynamics and absorption spectrum of the two-photon emitting $\tilde{A}$ (3s(+)) state.[2] In this talk some details are given about the measurement of yet another important parameter of the system, the transient regime two-photon cross-section B.

The two-photon excitation spectrum of DABCO[1] shows that the vibrational origin of the $\tilde{A} \leftarrow \tilde{X}$ two-photon transition is about as intense as the bands representing transitions $\tilde{A} \leftarrow \tilde{X}$ ($v' = 1 \leftarrow v'' = 0$), where v' is any one of three totally symmetric DABCO molecular vibrations. By the argument of mirror symmetry, one would thus expect the potential two-photon emitting transitions $\tilde{A} \rightarrow \tilde{X}$ ($v' = 0 \rightarrow v'' = 1$) to have equally large cross-sections. Thus a measurement of B at the vibrational origin is sufficient for purposes of estimating TPSE gain.

The vibrational origin of the $\tilde{A} \leftarrow \tilde{X}$ two-photon transition comprises several remarkably sharp ($\Delta\nu_D \sim 0.2\text{cm}^{-1}$) sequence bands ($m' \leftarrow m''$), arising from the thermal occupation of vibrational quantum number states of an a_1'' low-frequency torsional mode.[1] All the rotational substructure of the Q-branch transitions is contained within the $\sim 0.2\text{cm}^{-1}$ linewidth of each sequence band. Our procedure for measuring B was to excite DABCO molecules by an intense, well-characterized picosecond laser beam, tuned to the $0' \leftarrow 0''$ two-photon sequence band (558.7nm), and then measure the $\tilde{A}$ state population created by the picosecond laser pulse. The experiment was conducted at ~300 K, at which temperature the vapor concentration of DABCO is well known ($\sim 10^{16}/\text{cm}^3$).

The picosecond laser pulses originated from continuous trains of pulses produced by a Coherent CR599-4 dye laser, synchronously pumped by a mode-locked Coherent Innova 10 Ar$^+$-ion laser. A measured autocorrelation trace width of ~7psec implied a dye laser pulsewidth of ~4.5psec. A pulse averaged linewidth of ~4.5cm^{-1} was observed with a spectrograph and OMA combination. This is somewhat (~2x) larger than what would be expected for Fourier transform limited pulses. Amplification of selected pulses from the continuous trains was accomplished with the use of four prism dye cells[3], pumped by a XeCl excimer laser. With the use of this amplifier chain, nominally ~4.5psec pulses with maximum energy ~0.5mJ were produced. The ASE energy was ~5 percent of this value. The averaged linewidth of the amplified pulses was ~5.6cm^{-1}. The pulses were propagated with a 2mm beam diameter through a 50cm long DABCO cell.

The technique for recording time-resolved infrared absorption of spectra called TRISP[4] was used to measure the absorbance of the $\tilde{B} \leftarrow \tilde{A}$ ($3p_{x,y}(+) \leftarrow 3s(+)$) transition (origin at ~4045cm^{-1}) induced by the picosecond laser pulse. The $\tilde{B} \leftarrow \tilde{A}$ absorbance created by two-photon pumping could then be compared directly with the $\tilde{B} \leftarrow \tilde{A}$ absorbance created by optically pumping the DABCO vapor with a KrF laser. Since the $\tilde{A}$ state number density for the latter method of excitation has been earlier determined by a laser saturation method[2], the density created by two-photon excitation thus could also be obtained.

Some details involved in the comparison of $\tilde{B} \leftarrow \tilde{A}$ absorbances should be noted. When DABCO is optically pumped by a KrF laser the $\tilde{B}$ state is initially excited, but the molecules then undergo rapid non-radiative transitions to excited vibrational levels of the $\tilde{A}$ state. Two atmospheres of H$_2$ buffer gas were added to relax the vibrational distribution of $\tilde{A}$ state molecules resulting from $\tilde{B} \rightarrow \tilde{A}$ interconversions into a ~300 K thermal distribution within the $\tilde{A}$

state vibrational manifold. TRISP spectra taken under these conditions show the $\tilde{B} \leftarrow \tilde{A}$ transition again comprising a_1'' torsional mode sequence bands. For two-photon excitation of the $\tilde{A}$ state no buffer gas was employed, and the induced $\tilde{B} \leftarrow \tilde{A}$ absorbance was seen to comprise only one such sequence band, the a_1'' ($0' \leftarrow 0''$). The width of this sequence band was about twice as narrow in this case, compared to the KrF laser-pumped case, partly because there was no broadening due to buffer gas, and partly because the state selective method of excitation employed may have resulted in a non-thermal excited state rotational distribution. In any event, total areas under $\tilde{B} \leftarrow \tilde{A}$ absorbance curves were always used to compare $\tilde{A}$ state densities produced by the two methods of excitation. Proper account was taken of the decay of the $\tilde{A}$ state population in the case of KrF pumping by the time the ~10nsec long, ~2mm diameter TRISP broadband IR probe beam traversed the DABCO cell. The measured $\tilde{A}$ state density induced by two-photon pumping was $n_{\tilde{A}} \sim 4 \times 10^{12} cm^{-3}$. In deducing the value of B, the effective DABCO ground state population $n_{\tilde{X}}$ was assumed to be roughly one-third the DABCO vapor density of $\sim 10^{16} cm^{-3}$, since the picosecond laser beam was only resonant with the strongest a_1'' sequence band at the vibrational origin of the $\tilde{A} \leftarrow \tilde{X}$ two-photon transition. With B defined by the equation, $n_{\tilde{A}} = B\Phi^2 n_{\tilde{X}}$, where Φ is the fluence of the picosecond pulse, a value $B = 0.6 \times 10^{-36} cm^4$ is deduced.

The above value of B represents an effective two-photon cross-section for the ~ 4.5 picosecond pulses that were actually applied. Since $\Delta\nu_D \Delta t \sim 2.3 \times 10^{-2}$, the transient regime was involved. The so-called steady-state cross-section δ is given by $\sim 4(\ln 2/\pi)^{1/2} (\Delta\nu_D)^{-1} B$. Thus, $\delta \sim 1.88 \times 10^{-46} cm^4$ sec., but use of this quantity is inapplicable in our case. Since B is independent of the two-photon line width, the same value deduced above should be used in computing two-photon gain for the case that ~2atm of H_2 buffer gas are present, a requirement necessary for observing TPSE gain in this system[2]. At ~2atm of H_2 pressure the $\tilde{A} \leftarrow \tilde{X}$ two-photon sequence bands broaden to $\sim 1.5 cm^{-1}$. Calculation then shows that with ~4.5psec optical pulses, tuned to two-photon resonance with the a_1'' ($0' \rightarrow 0''$) sequence band corresponding to one of the $\tilde{A} \rightarrow \tilde{X}$ ($v' = 0 \rightarrow v'' = 1$) transitions, and having ~1mJ single pulse energy, one should be able to extract an additional millijoule of energy by TPSE in a ~22 pass ~300 K DABCO cell, pressurized to ~2atm of H_2, and end-pumped with a ~1J KrF laser beam. The Herriott-type resonator should be designed with a 0.7mm diameter average beam size.

References

1. D. H. Parker and Ph. Avouris, J. Chem. Phys. 71, 1241 (1979).
2. J. H. Glownia, G. Arjavalingam and P. P. Sorokin, paper (ThFF1) in XIII IQEC.
3. D. S. Bethune, Appl. Opt. 20, 1897 (1981).
4. D. S. Bethune, A. J. Schell-Sorokin, J. R. Lankard, M. M. T. Loy and P. P. Sorokin, in Advances in Laser Spectroscopy, Vol. 2, B. A. Garetz and J. R. Lombardi, editors (John Wiley and Sons, 1983).

Part III

Coherent Pulse Propagation

Coherence Effects in Pump-Probe Measurements with Collinear, Copropagating Beams

S.L. Palfrey and T.F. Heinz
IBM T.J. Watson Research Center, P. O. Box 218, Yorktown Heights, NY 10598

K.B. Eisenthal
Department of Chemistry, Columbia University, New York, NY 10027

When pump-probe measurements are performed using pulses originating from the same laser, it is necessary to consider the coherent interaction of the two pulses in determining material relaxation rates on the time scale of the pulse duration. In the usual geometry in which the beams cross at an angle the effects of the coherent coupling on the observed signal are well known [1-4]. In this case the coherent contribution to the signal may be viewed as a scattering of the pump radiation into the probe direction by a spatial grating induced when the pump and probe are simultaneously present in the sample. We consider here the influence of coherence effects on pump-probe measurements with collinear, copropagating beams of orthogonal polarization. Despite the fact that the grating vanishes in this geometry, we find, in contradiction with earlier predictions [4], that the coherent interaction still remains.

We first describe our experimental results. The sample, an optically thin jet of cresyl violet dye in ethylene glycol, was pumped and probed with 7 psec pulses obtained from a synchronously pumped dye laser tuned to resonance with the $S_0 \rightarrow S_1$ transition at 590 nm. With the pump beam chopped, we detected the synchronous modulation of the probe after the pump was blocked with a polarizer. The dotted points in Fig. 1(a,b) show the induced transmission of the probe beam as a function of delay with respect to the pump. Fig. 1(a) was obtained with the nearly transform-limited pulses coming directly from the mode-locked laser, while Fig. 1(b) was obtained with spectrally broadened pulses produced by sending the pulses through an optical fiber. Since the temporal broadening of the pulses in the fiber was minimal, the marked difference in the probe transmission reflects the importance of the coherence effects.

In Fig. 1, the increase in the probe transmission for long delay times arises from the bleaching by the pump of molecules in the solution with transition dipole moments that have projections along the polarization of the probe. The coherent contribution to the signal stems from this same bleaching mechanism. Intuitively, the coherence effects can be understood by

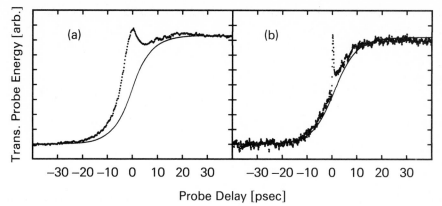

Figure 1. Induced probe transmission in cresyl violet obtained with (a) nearly transform-limited pulses and (b) pulses which were spectrally broadened in an optical fiber

noting that when the pump and probe are mutually coherent in the sample, they add to produce a total electric field with a well-defined polarization that is constant in time. As the pulses then propagate through the sample, they will induce a stronger anisotropy in the bleaching than if they had not been mutually coherent. Hence, fewer molecules will be excited and the pulses will pass through the sample with less total absorption.

We can describe the results of Fig. 1 analytically by treating the dye system as a collection of fixed, randomly oriented saturable absorbers with a slow recovery on the time scale of interest. We then find for an optically thin and weakly bleached sample that the change in the transmitted probe energy induced by the pump is given by [5]:

$$S(\tau) = \beta(\tau) + \beta'(\tau) + \gamma(\tau), \tag{1}$$

with

$$\beta(\tau) = \mathrm{Re}\left\{ \int_{-\infty}^{\infty} dt \int_{-\infty}^{\infty} dt' E^*(t-\tau)E(t)E^*(t')E(t'-\tau)\theta(t-t') \right\} \tag{1a}$$

$$= \frac{1}{2}\left| \int_{-\infty}^{\infty} dt E^*(t-\tau)E(t) \right|^2$$

$$\beta'(\tau) = \mathrm{Re}\left\{ e^{-2i\omega\tau} \int_{-\infty}^{\infty} dt \int_{-\infty}^{\infty} dt' E^*(t-\tau)E(t)E^*(t'-\tau)E(t')\theta(t-t') \right\} \tag{1b}$$

$$= \frac{1}{2}\mathrm{Re}\left\{ e^{-2i\omega\tau} \left[\int_{-\infty}^{\infty} dt E^*(t-\tau)E(t) \right]^2 \right\}$$

$$\gamma(\tau) = \mathrm{Re}\left\{ \int_{-\infty}^{\infty} dt \theta(\tau-t) \int_{-\infty}^{\infty} dt' |E(t')|^2 |E(t'-t)|^2 \right\}. \tag{1c}$$

217

Here, $E(t)$ is the complex electric field envelope of the pump (for an $e^{-i\omega t}$ time dependence) and τ is the delay time of the probe. The incoherent term, $\gamma(\tau)$, describes the familiar convolution of the response with the intensity autocorrelation. The terms $\beta(\tau)$ and $\beta'(\tau)$ represent the coherent contribution, and, in this limit of a slow recovery, are simply related to the *electric-field* autocorrelation. The contribution $\beta(\tau)$ is identical to that found for the crossed-beam geometry [2,3], while $\beta'(\tau)$ occurs only for collinear beams. Since $\beta'(\tau)$ oscillates as a function of delay time with a frequency 2ω, it will not contribute unless the delay is scanned very slowly.

The data in Fig. 1 can be explained well by Eqn. (1). The solid curves in Fig. 1 show the incoherent contribution $\gamma(\tau)$ as calculated from the experimental intensity autocorrelation function. The difference between these curves and the observed signal is given by $\beta(\tau)$. From Eqn. (1b), we predict that the coherent contribution will be largest when the pump and probe are coincident and zero for delays longer than the pulse coherence time. For nearly transform-limited pulses, we then expect the coherent contribution to be present whenever the pulses overlap, in agreement with Fig. 1(a). For the non-transform-limited pulses of Fig. 1(b), we see that the width of the coherent contribution is reduced substantially and is found to correspond quite well to the estimated 0.7 psec coherence time of the spectrally broadened pulse.

In order to observe the oscillatory behavior of the coherent contribution arising from $\beta'(\tau)$, we varied the path length of the probe beam by tilting a glass slide through which it passed. This delay was calibrated by leaking some pump light through the analyzer and monitoring the fringes resulting from the linear interference between the pump and probe beams. For delay times near $\tau=0$, we observed oscillations in the signal with a period of 2ω and a peak-to-peak height approximately equal to the average signal level, in accordance with Eqn. (1). Physically, the presence of these oscillations is reasonable since no cooperative bleaching will occur when the two beams are $90°$ out of phase. From the viewpoint of the four-wave mixing process through which the pulses interact, the oscillatory term arises from the interference of the forward-traveling phase-conjugate wave [3] with the probe beam, an effect possible only in the fully degenerate collinear geometry.

From the picture in which the coherent signal arises from a cooperative bleaching of the sample, it is evident that the effect of the mutual coherence of the two pulses should be to increase the transmitted energy of both the pump and probe. This behavior contrasts with what might be inferred from the grating picture of the coherent interaction: if one beam scatters off a grating into the other, only an exchange of energy between the two beams would be expected. We have checked directly whether or not the coherent interaction preserves the total transmitted energy. Experimentally, this was done by modulating both beams and then detecting the synchronous modula-

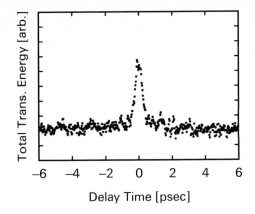

Figure 2. Total transmitted intensity in cresyl violet using spectrally broadened pulses of equal intensity; the pulse duration was 7 psec

tion in the combined transmitted energy at the sum frequency. Figure 2 shows the experimental results for the total transmitted energy obtained with the spectrally broadened pulses from the optical fiber. The feature centered at $\tau = 0$, being much narrower than the pulse duration, must be attributed to coherence effects. This peak and the flat response for delays greater than the pulse coherence time can be explained directly by Eqn. (1) applied separately to each beam.

In summary, we have shown that a coherent interaction between pump and probe pulses occurs even for collinear, copropagating beams. These coherence effects must in general be taken into account in pump-probe measurements performed in this geometry.

The authors gratefully acknowledge A.C. Balant, D.R. Grischkowsky and J. M. Halbout for valuable assistance during the course of this work.

References

1.   E. P. Ippen and C. V. Shank, in Ultrashort Light Pulses, ed. by S.L. Shapiro, Topics Appl. Phys. 18, (Springer, Berlin, 1977) p.83.
2.   Z. Vardeny and J. Tauc, Opt. Commun. 39, 396 (1981).
3.   B. S. Wherrett, A. L. Smirl, and T. F. Boggess, IEEE J. Quantum Electron. QE-19, 680 (1983).
4.   A. von Jena and H. E. Lessing, Appl. Phys. 19, 131 (1979).
5.   T. F. Heinz, S. L. Palfrey, K. B. Eisenthal, to be published in Optics Letters.

Observation of the 0π Pulse

Joshua E. Rothenberg, D. Grischkowsky and A. C. Balant

IBM Thomas J. Watson Research Center

P. O. Box 218

Yorktown Heights, New York 10598, USA

We have measured to 0.5 psec resolution the reshaping of a small area 7 psec pulse to a 0π pulse due to its resonant propagation through an optically thick sodium cell.

In coherent optics the pulse area θ is defined as the total angle the atomic state vector rotates around the electric field of the resonant driving pulse. The atomic excitation remaining after passage of the pulse is determined by this angle. For example, a $\pi/2$ pulse will put the atoms in a coherent superposition of the ground and excited states, a π pulse will put the atoms in the excited state, and a 2π pulse will take the atoms to the excited state and then back to the ground state. Another type of pulse which has generated a considerable amount of theoretical interest [1] but for which there has been only a few experimental investigations [2] is the 0π pulse. For this pulse, the initial part of the pulse excites the atoms and then due to a phase change of π in the electric field, the latter part of the pulse takes the atoms back to the ground state. Thus, just as in self-induced transparency, the atoms are completely in the ground state after passage of the pulse. There are two ways to produce such a 0π pulse. The first is to switch the phase of the second half of the pulse by electro-optic techniques. For this method it is difficult to get the 0π pulse area to the desired accuracy. The second method is to rely on the automatic reshaping to the 0π pulse which occurs when a weak pulse is propagated through a resonant vapor. This method will produce the 0π pulse area to any desired degree of accuracy. This fact is due to an area theorem for these weak pulses which states that the pulse area decays exponentially with the on-resonance absorption coefficient [1].

The area theorem is valid for any pulsewidth and applies to our experimental situation where we have short 7 psec pulses with spectral bandwidths as much as 30 times broader than the resonant linewidth of the atomic sodium vapor. Clearly, the energy of the pulse cannot decay exponentially. In fact, for an on-center absorption of $\exp(-400)$, we measure that the total energy of the output pulse from the sodium cell decreases by only 12%. The manner in which the pulse area decays is related to the reshaping of the pulse by its passage through the Na vapor. The electric field amplitude develops an oscillatory structure such that the first oscillation excites the atoms and then, due to a change of the phase angle by π, the second oscillation de-excites them and so on. This complicated excitation and de-excitation process remarkably leads to the simple exponential decay of the pulse area as the pulse propagates through the cell. For this reshaped pulse, the vapor is transparent to exponentially increasing accuracy. After the pulse passes by, essentially no excitation is left in the atomic system, even though the excitation oscillates during the pulse.

In our experiments the 7 psec input pulses to the sodium cell were from a synchronously pumped, mode-locked and cavity-dumped dye laser. The laser frequency was tuned to the 5890 Å transition of Na; the laser linewidth was 2 cm^{-1}. Assuming a $sech^2$ pulseshape, the transform limited value is 1.5 cm^{-1}. The peak input power to the cell was 100 W, and in the cell the beam diameter was 0.3 cm, corresponding to an input pulse area of $\pi/30$ radians. The 50 cm long, 2.5 cm diam Pyrex glass cell contained an excess of 99.95% pure Na metal transferred from a sealed ampoule. The cell was connected to a Varian Vac-Ion pump. The cell temperature was measured by thermocouples in contact with the glass. The absorption was calculated from the atomic number density corresponding to the cell temperature. A beamsplitter extracted part of the beam from the laser which was run through an optical pulse compressor [3] to produce a synchronized train of 0.5 psec probing pulses. The output pulse shapes from the sodium cell were measured with 0.5 psec resolution by cross-correlating with the compressed pulses by non-collinear generation of second harmonic light in a 0.3 mm long KDP crystal. The second harmonic light was monitored with a photomultiplier connected directly to a signal averager synchronized with the delay setting of the probing pulses as controlled by a stepping motor. A typical scan of the delay of the probing pulse took 100 seconds and usually 10 scans were averaged.

In Fig. 1, we show a series of our measurements (solid curves) of the output pulseshapes and the corresponding calculations (dashed curves) for different absorptions αl of the sodium cell (α is the absorption coefficient at the maximum of the 5890 Å line and l = 50 cm is the cell length). The input pulse is given by the top curve with αl = 0. The observed pulse reshaping is quite dramatic and shows time dependences faster than the 7 psec input pulses and significantly faster than the relaxation times of the Na atoms, T_1 = 16 nsec, T_2 = 32 nsec, and $T_2^* \cong 100$psec. The reshaping extends over time intervals very much longer than that of the input pulsewidths. As αl is increased by increasing the Na density, the number of oscillations in the output pulse envelope increases, the oscillations come closer together, and the duration of the pulse envelope increases. The theoretical results are obtained from numerical calculations based on linear dispersion theory following the procedure of Crisp [1] and using the Voigt profile for the absorption coefficients and the plasma dispersion function for the indices of refraction due to the two hyperfine components of the 5890 Å line.

The theoretical curves show general qualitative agreement with the experimental data. In particular, the theory correctly predicts the trends of the time positions of the envelope oscillations to come closer together, and in some cases the absolute positions of the envelope maxima and minima. It is seen, however, that the theory predicts these minima should go to zero which is not borne out by the data. These calculations assume transform limited pulses which have a center frequency that is exactly coincident with the Na absorption and include

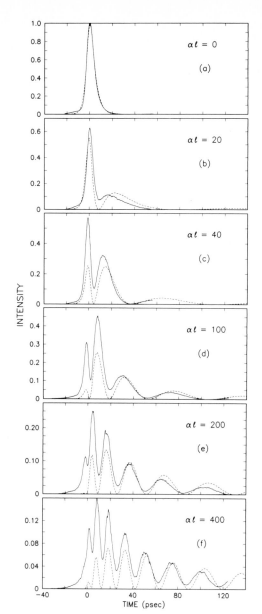

Figure 1 Observed (solid curves) and theoretical (dashed curves) reshaped pulse intensity after resonant propagation through sodium vapor of optical density αl. Theory is for transform limited pulses tuned exactly to the Na absorption line and involves no adjustable parameters

no adjustable parameters. As mentioned previously the bandwidth of the laser pulses exceeded the transform limit by ~1.5. In addition we found the center frequency of the laser drifted about the Na absorption line. In Fig. 2 we show theoretical fits which incorporate a linear chirp and frequency jitter in our input pulse. Nearly quantitative agreement is seen in these fits except for small delay. A possible explanation of this discrepancy may be a result of the unknown nature of the phase structure of the input pulses. The pulses thus have an

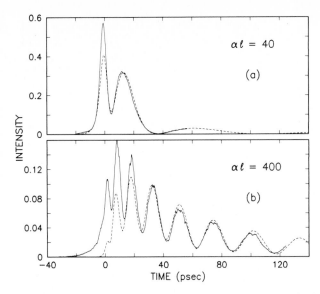

Figure 2   Reshaped pulse intensity, experiment (solid curves) and theory (dashed curves). Theory is for linearly chirped pulses with a triangular distribution of center frequencies consistent with the observed spectrum

excess of energy far from the Na resonance frequency, which travels faster, and thus contributes significantly to the signal at small delay.

In conclusion, we summarize by noting that this work is the first accurate subpicosecond observation of small area pulse reshaping, and that the cross-correlation technique employed here should be applicable to many similar ultrafast phenomena.

This work was partially supported by the Office of Naval Research.

1. M. D. Crisp, Phys. Rev. A1, 1605 (1970).
2. H. P. Grieneisen, J. Goldhar, N. A. Kurnit, A. Javan and H. R. Schlossberg, Appl. Phys. Lett. 21, 559 (1972).
3. B. Nikolaus and D. Grischkowsky, Appl. Phys. Lett. 43, 228 (1983).

Picosecond Two-Color Photon Echoes in Doped Molecular Solids

D.A. Wiersma, D.P. Weitekamp*, and K. Duppen

Picosecond Laser and Spectroscopy Laboratory, Department of Chemistry
University of Groningen, Nijenborgh 16, NL-9747 AG Groningen, The Netherlands

ABSTRACT

Two-dimensional picosecond photon echo spectroscopy is used to study vibrational deactivation in doped molecular solids. Evidence for intermediate levels in the relaxation pathway is presented.

The three pulse stimulated echo, long known in magnetic resonance (1), has more recently been demonstrated on optical transitions (2). An interesting way of looking at the stimulated-echo formation relies on the fact that the first two excitation pulses create an "ordered" population as a function of detuning. This ordered population can also be looked upon as a "grating" in frequency space from which a probe pulse can "reflect" an echo. Since a grating in only one of the levels suffices, a corollary is that the echo may be stimulated at some entirely different frequency ω_2 than the one ω_1, at which the grating was prepared. We report the observation of such a connected two-color stimulated echo (C2CSE) in a solid and use it to confirm the relaxation out of the 747 cm^{-1} vibrational state of the S_1 manifold of pentacene in naphthalene. It has also been realized (3) that, after redistribution of the initially excited population grating, the echo may be stimulated from *any* state into which the grating has relaxed. We report here the first observation of such a relaxed two-color stimulated echo (R2CSE) and note that this echo can be seen as the optical analogue of the cross-relaxation peak in a two-dimensional nmr spectrum. The rise and fall of this relaxed echo can be used as a probe for the vibrational deactivation pathway(s) subsequent to excitation of various vibrational levels in the S_1 manifold. Prior to discussing briefly some of the results we note that a necessary prerequisite to observation of such two-color photon echoes is that the inhomogeneous broadening on the selected transitions is correlated. Furthermore we note that in these echo effects interference with stimulated emission and super-fluorescence is avoided by using small-angle excitation pulses. Some of the results are shown in Fig. 1. In (A) the decay of the C2CSE is presented with the solid line being a fit to a population decay parameter of the 747 cm^{-1} level of 33 psec. In (B) the rise of the R2CSE is shown where the probe wavelength is at the pure electronic transition. The dashed line is the expected echo-intensity rise if the initially excited vibration would directly decay into the vibrationless state. The mismatch between the dashed curve and the experimental one shows that intermediate levels in the relaxation pathway must exist. The solid line in (B) is a fit to a model where one intermediate bottleneck level is assumed with a lifetime of 16 ± 2 psec.

---

* Present address: Department of Chemistry, University of California at Berkeley, Berkeley, California 94720, U.S.A.

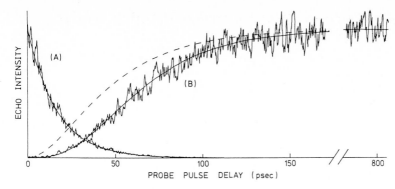

Fig. 1. Two-color stimulated picosecond echo signals. In (A) the vibronic
transition at 17335 cm^{-1} (747 cm^{-1} above the origin) is excited with
the first two excitation pulses. The probe pulse is applied to the
transition at 16579 cm^{-1}. In this connected two-color stimulated
echo C2CSE the excitation pulses and probe pulse have one level
in common. In (B) the excitation pulses are at the same frequency
as in (A). The probe pulse is now applied to the transition at
15832 cm^{-1} which is unconnected to the transition excited. This
relaxed two-color stimulated echo (R2CSE) intercepts the incoming
population in the vibrationless state.

REFERENCES

1. E.L. Hahn: Phys. Rev. 80, 580 (1950); W.B. Mims, K. Nassau and
    J.D. McGee: Phys. Rev. 123, 2059 (1961)
2. N.A. Kurnit and S.R. Hartmann: Bull. Am. Phys. Soc. 11, 112 (1966)
3. P. Ye and Y.R. Shen: Phys. Rev. A 25, 2183 (1982)
    W.H. Hesselink and D.A. Wiersma in: Spectroscopy and Excitation
    Dynamics of Condensed Molecular Systems (Eds. V.M. Agranovich and
    R.M. Hochstrasser, North-Holland Publishing Company, Amsterdam).

Subpicosecond Accumulated Photon Echoes with Incoherent Light in Nd³⁺-Doped Silicate Glass

H. Nakatsuka, S. Asaka, M. Fujiwara, and M. Matsuoka

Department of Physics, Faculty of Science, Kyoto University, Kyoto 606, Japan

1. Introduction

Since the first observation of accumulated photon echoes by HESSELINK and WIERSMA [1], nearly transform-limited mode-locked pulse trains have been used as the excitation source, and the resolution times have been limited by the widths of the pulses. In this paper we show theoretically and experimentally that the time resolution of the accumulated photon echoes is determined not by the pulse width but by the correlation time of the excitation field [2]. This indicates that it is possible to obtain a subpicosecond resolution in accumulated photon echoes even by using a cw laser (non-mode-locked), if the laser output has a sufficiently broad and smooth spectrum.

2. Theory

In an accumulated photon echo experiment a sample is irradiated by two excitation beams which originated from a single laser. The two beams are noncollinear, and one beam (2nd beam) is delayed with respect to the other (1st beam). The excitation of the sample by the two beams results in a periodic distribution of the population in space and in frequency within the inhomogeneous spectral width of the sample. Accumulated photon echoes can be regarded as free induction decays induced by the 2nd beam from this population modulation or accumulated population grating [3]. Accumulated photon echoes are effective in a system where the lifetime (T_1) of the population grating is sufficiently long. It is known that as far as the excitation intensity is within the range where the saturation broadening is negligible, the formation of the population grating can be described by a rate equation if $T_1 \gg T_2$, where T_2 is the homogeneous transverse relaxation time between the two resonant levels. In this situation the shape of the steady-state population grating is determined simply by the power spectrum of the excitation light and the relaxation times of the sample [3]. We express the fields of the excitation beams as

$$E_1(t, \vec{r}) = E(t - \vec{n}_1 \cdot \vec{r} / c) \quad , \tag{1}$$

$$E_2(t, \vec{r}) = E(t - \tau_{12} - \vec{n}_2 \cdot \vec{r} / c) \quad , \tag{2}$$

where $\vec{n}_1$ and $\vec{n}_2$ are unit vectors representing the beam directions of the 1st and 2nd beams, respectively, and are assumed to be nearly parallel, and τ_{12} is the delay time of the 2nd beam with respect to the 1st beam at $\vec{r}=0$. The shape of the steady-state accumulated grating $H(\omega,\vec{r})$ in the ground (or excited) level is given as

$$H(\omega, \vec{r}) \propto \int_{-\infty}^{\infty} d\omega' \, [\gamma^2 + (\omega - \omega')^2]^{-1} S(\omega', \vec{r}) \quad , \tag{3}$$

where $\gamma = 1/T_2$, and $S(\omega,\vec{r})$ is the power spectrum of the excitation field and is expressed as

$$S(\omega, \vec{r}) = \lim_{T\to\infty} T^{-1} \left| \int_{-T/2}^{T/2} dt \{E_1(t, \vec{r}) + E_2(t, \vec{r})\} \exp(-i\omega t) \right|^2 . \tag{4}$$

In (3) the inhomogeneous width of the sample is assumed to be much larger than the spectral width of the excitation field, and the saturation of the depth of the grating is neglected. Echoes are induced by $E_2(t,\vec{r})$ from the grating $H(\omega,\vec{r})$, and thus the echo polarization is given as

$$P(t, \vec{r}) \propto \int_{-\infty}^{t'} d\tau \, G(t'-\tau-\tau_{12}) \exp[-2\gamma(t'-\tau)] E(\tau-\tau_{12}) \quad , \tag{5}$$

where $t' = t - \vec{n}_2 \cdot \vec{r}/c$, and $G(\tau)$ is the autocorrelation function of the excitation field, and is given as

$$G(\tau) = 2\pi \lim_{T\to\infty} T^{-1} \int_{-T/2}^{T/2} dt \, E(t) E^*(t-\tau) \quad . \tag{6}$$

From (5) the averaged echo intensity observed with a slow detector is given as a function of τ_{12} as

$$I(\tau_{12}) \propto \int_{-\tau_{12}}^{\infty} d\tau \int_{-\tau_{12}}^{\infty} d\tau' \, G(\tau) G^*(\tau') G(\tau'-\tau)$$
$$\times \exp\left[-2(2\tau_{12} + \tau + \tau')/T_2\right] \quad . \tag{7}$$

If the width of $G(\tau)$, or the correlation time τ_c of the excitation field, is much smaller than T_2, then

$$I(\tau_{12}) \propto \exp(-4\tau_{12}/T_2) \quad . \tag{8}$$

From (6) and (7) we see that τ_c becomes the resolution time in measuring T_2. Since the correlation time τ_c is related to the spectral width $\Delta\omega$ of the excitation field as $\tau_c \approx \Delta\omega^{-1}$, one only needs to use an excitation light with a broad and smooth spectrum to obtain a high time resolution.

3. Experimental Results and Discussion

A schematic diagram of the experiment is shown in Fig.1. The excitation source was an Ar^+-laser-pumped dye laser which was operated in either mode-locked or cw (non-mode-locked) oscillation. The average input power of each beam at the sample was 10 mW, and the beam diameter at the focus was 100 µm. The angle between the two beams was 20 mrad. The sample was a 3 mm thick piece of 3 % Nd^{3+}-doped silicate glass which was kept at about 20 K. The dye laser was resonant with the $^4I_{9/2} \leftrightarrow {}^2G_{7/2}, {}^4G_{5/2}$ transition of Nd^{3+}, and the center frequency was 5910 Å. The excitation beams E_1 and E_2 are chopped at frequencies f_1 (210 Hz) and f_2 (360 Hz), respectively, and the sum frequency (f_1+f_2) component of the echo signal was

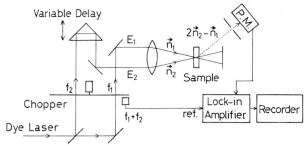

Fig.1
Schematic diagram
of the experiment

picked up by a lock-in amplifier. The application of this
cross-modulation technique in the accumulated photon echo
experiment resulted in a great improvement in the signal to
noise ratio compared to the usual single beam modulation. In
Fig.2(a) the decay curve of the accumulated photon echo is
shown where the excitation light was a nearly transform-limited
mode-locked pulse train with pulse widths of about 3 psec. On
the other hand, Fig.2(b) is the echo decay curve when a
broad-spectrum cw laser (non-mode-locked) was used. The
spectral width of the cw dye laser was 14 A corresponding to
the correlation time τ_c of 0.7 psec. The two figures 2(a) and
(b) give similar values of T_2, 60 and 57 psec, respectively,
and these values are consistent with the result by SHELBY [4].
The above result clearly shows that short pulses are not always
necessary in the accumulated photon echo experiment.

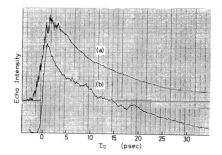

Fig.2
The decay curves of the
accumulated photon echo in
Nd^{3+}-doped silicate glass
where the excitation is (a)
by a nearly transform-limited
mode-locked pulse train, and
(b) by a broad-spectrum cw
laser (non-mode-locked)

For a higher time resolution a laser with a broader spectrum
is needed. From a practical point of view one of the most
useful broad-spectrum lasers for the accumulated photon echo
experiment is an incompletely mode-locked (far from transform-
limited) dye laser, where the dye laser is pumped by a
mode-locked Ar^+ laser, and their cavity lengths are mismatched
by a few μm. In this way we obtained a broad-spectrum dye
laser with a spectral width of 44 A, which corresponds to the
correlation time τ_c of 220 fsec. The echo decay curve obtained
with this broad-spectrum laser is shown in Fig.3. The peak in
the curve near $\tau_{12}=0$ was observed for the first time in this
time resolution. The width of the peak may be still limited by
the resolution time of the present experiment. The decay of
the accumulated photon echoes depends not only on the
transverse relaxation but also on slower relaxations up to T_1.

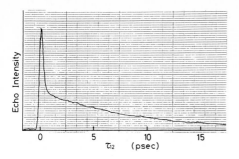

Fig.3
The decay curve of the accumulated photon echo in Nd^{3+}-doped silicate glass by an incompletely mode-locked pulse train

Hence, a possible explanation of the sharp peak would be a spectral diffusion of the population grating or a thermal grating effect in the glass. The shorter T_2 obtained from the wing of the decay curve in Fig.3 is due to a somewhat higher temperature of the sample than in the cases of Figs.2(a) and (b).

The use of an incoherent light in accumulated photon echoes enables us to measure T_2 in femtosecond resolution without difficulties of ultrashort pulse generation, and this method will be a useful tool in ultrafast coherent transient spectroscopy.

References

1. W.H. Hesselink and D.A. Wiersma: Phys. Rev. Letters <u>43</u>, 1991 (1979).
2. This fact was first predicted by T. Yajima and co-workers at the Meeting of the Physical Society of Japan (autumn, 1982).
3. H. Nakatsuka, S. Asaka, M. Tomita and M. Matsuoka: Opt. Commun. 47, 65 (1983).
4. R.M. Shelby: Opt. Letters <u>8</u>, 88 (1983).

Femtosecond Dephasing Measurements Using Transient Induced Gratings

A.M. Weiner, S. De Silvestri*, and E.P. Ippen

Department of Electrical Engineering and Computer Science and
Research Laboratory of Electronics, Massachusetts Institute of Technology
Cambridge, MA 02139, USA

Transient four-wave mixing techniques have been developed in the past several years for the investigation of ultrafast dephasing of electronic transitions in condensed matter [1,2]. These experiments involve self-diffraction of two noncollinear pulses from an optically induced absorption grating. We have recently reported an improved scheme utilizing three separate input pulses [3]. Similar three pulse grating experiments [4] can be used to study processes such as spatial diffusion, orientational relaxation and excited state relaxation; but the applicability to dephasing studies had not been previously recognized. Compared to the two pulse scattering, our three pulse experiment provides several important advantages, including resolution below the pulsewidth, clear demarcation between homogeneous and inhomogeneous broadening, and separation of energy relaxation (T_1) effects from the phase relaxation (T_2) effects.

The experimental geometry is shown in Fig. 1. The method relies on an optically induced grating, formed by the interference of pulses #1 and #2. When the two pulses are separated temporally, a grating can still be formed provided that the dephasing time T_2 is sufficiently long. Therefore, by measuring the grating amplitude as a function of the delay τ, one can measure the dephasing time. This is accomplished using pulse #3 as a delayed probe to scatter off the grating into directions $\bar{k}_3 \pm (\bar{k}_1 - \bar{k}_2)$.

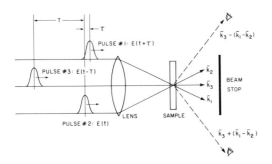

Fig. 1

Experimental geometry for dephasing measurements by the three pulse scattering technique

For homogeneously broadened transitions the scattered energy is always a symmetric function of the delay τ. The Fourier transform $G(\Omega)$ of the grating amplitude is written as the product of the absorption spectrum $\alpha(\omega)$ with the laser power spectrum $\phi(\omega)$:

$$G(\Omega) \sim \alpha(\omega_L + \Omega)\, \phi(\omega_L + \Omega) \quad , \tag{1}$$

*Permanent Address: Centro di Elettronica Quantistica e Strumentazione Elettronica del CNR, Istituto di Fisica del Politecnico, Milano, Italy

where ω_L is the laser carrier frequency and $\Omega = \omega-\omega_L$. To obtain the scattered energy, one calculates the inverse Fourier transform of (1) with respect to τ and squares its amplitude.

Inhomogeneous broadening can lead to asymmetric traces; this asymmetry provides a simple criterion for differentiating between the two types of line broadening. Note that in the case of wide inhomogeneous broadening our three pulse scattering technique is formally equivalent to the stimulated photon echo [5] in the low field limit. We wish though to emphasize the generality of the three pulse scattering technique, which includes both the homogeneous and inhomogeneous limits within its framework.

We have used 70 fsec pulses from a CPM ring dye laser [6] to perform three pulse scattering experiments on a variety of organic dye solutions in thin cells. In initial experiments using parallel polarization for all pulses, scattering was strongly enhanced due to the formation of a cumulative thermal grating. When thermal gratings dominate, the experimental results will depend only on the total absorption spectrum according to (1), independent of any inhomogeneous broadening.

We have checked the validity of (1) by comparing parallel polarization data with calculated curves. Three pulse scattering data for Nile blue in methanol are shown in Fig. 2; the data are almost indistinguishable from the squared electric field autocorrelation function, appropriate for $T_2=0$. This $T_2=0$ limit is expected because of the very shallow curvature of the Nile blue absorption spectrum at the laser wavelength. Scattering data for rhodamine 640 in methanol as well as for Nile blue are plotted logarithmically in Fig. 3. The excellent agreement between experimental and calculated curves demonstrates the validity of (1) even when excitation is not at line center and material coherence cannot be neglected.

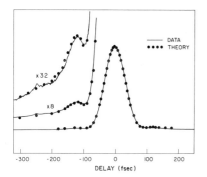

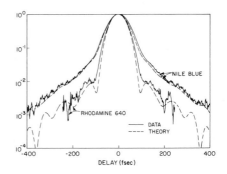

Fig. 2 Scattering data for Nile blue dye in methanol, using parallel polarization

Fig. 3 Scattering data for rhodamine 640 dye and for Nile blue in methanol, using parallel polarization

In order to obtain true dephasing information, the thermal grating effect can be eliminated using orthogonal polarizations for pulses #1 and #2. Both parallel and orthogonal polarization scattering was measured for solutions of Nile blue and of oxazine 725. The close correspondence between the scattering data of the two polarization conditions and the good agreement with the curves calculated assuming homogeneous broadening indicate that the dephasing times of these solutions are less than the 20 fsec experimental resolution.

We have also studied the dephasing of cresyl violet dye in solid poly-(methyl methacrylate) using amplified [7] CPM laser pulses. Scattering curves taken at a temperature of 15 K are shown in Fig. 4 for both scattering directions. The asymmetry of the curves, illustrated most dramatically by the 60 fsec peak shift, show that inhomogeneous broadening is present. At room temperature the asymmetry and peak shifts are no longer evident; homogeneous broadening apparently dominates. Detailed interpretation is complicated by the multilevel character of the transition; nevertheless, our data confirms that asymmetry of the three pulse scattering curves can serve as a sensitive indicator of inhomogeneous broadening.

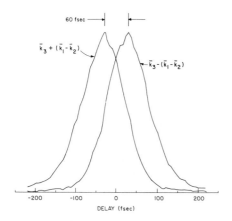

Fig. 4

Scattering data for cresyl violet dye in solid poly (methyl methacrylate) at 15 K

This work was supported in part by the Joint Services Electronics Program under contract DAAG-29-83-K-003. A.M. Weiner was a Fannie and John Hertz Foundation Fellow. S. De Silvestri is a NATO Science Fellow.

References

1. T. Yajima, Y. Ishida and Y. Taira, Picosecond Phenomena II, eds., R.M. Hochstrasser, W. Kaiser and C.V. Shank (Springer-Verlag, New York 1980).
2. J.G. Fujimoto and E.P. Ippen, Opt. Lett. 8, 446 (1983).
3. A.M. Weiner and E.P. Ippen, Opt. Lett. 9, 53 (1984).
4. D.W. Phillion, D.J. Kuizenga and A.E. Siegman, Appl. Phys. Lett. 27, 85 (1975).
5. T. Mossberg, A. Flusberg, R. Kachru and S.R. Hartmann, Phys. Rev. Lett. 42, 1665 (1979).
6. R.L. Fork, B.I. Greene and C.V. Shank, Appl. Phys. Lett. 38, 671 (1981).
7. R.L. Fork, C.V. Shank and R.T. Yen, Appl. Phys. Lett. 41, 223 (1982).

Picosecond Pulse Multiphoton Coherent Propagation in Vapors

H. Vanherzeele* and J.-C. Diels

Center for Applied Quantum Electronics, Department of Physics
North Texas State University, P.O. Box 5368, NT Station, Denton, TX 76203, USA

We present a method to study multiphoton excitation by intense picosecond pulses in vapors. The technique, based on interferometric cross
correlations of the multiphoton stimulated reradiation with the pulse
itself, requires an interferometric delay line to generate accurately
phased pulse sequences. Experimental data for two-photon coherent propagation in lithium vapor agree with our theoretical model.

The experimental setup consists in a high gain picosecond dye laser
system, coupled to an interferometer and a computerized data acquisition
system as presented schematically in Fig. 1. Data acquisition and processing are performed on line at the 10 Hz repetition rate of the laser
source. The latter consists of a synchronously pumped dye laser, followed
by a 3-stage amplifier which itself is pumped by a frequency doubled, Q-
switched Nd:YAG laser. The dye laser pulses are tuned to resonance with
the 2s-4s two photon transition of lithium at 571.2 nm with a 3-plate
birefringent filter. The oscillator-amplifier, on which we have reported
in more detail elsewhere [1], delivers 1.5 mJ pulses with a distortion free
wavefront and Gaussian spatial profile.

To determine the temporal properties of the pulses, we use the interferometric delay line required for the target experiment. This delay line
splits the main pulse into two pairs of pulses, and continuously varies
their delay. (The use of stepping motors, piezo-electric drivers or a
combination of both of them has to be ruled out because of the subwave-
length accuracy required over a wide range of delays.) The relative phase
between the pulses (slowly increasing at a minimum rate of approximately
2 every 4s), as well as the absolute delay between the pulses, is determined for each laser shot from a He-Ne laser interferogram (channel 4 in
Fig. 1). One of the pair of pulses is used to analyze the waveform and the
phase modulation of the pulse. The ratio of the second harmonic after the
interferometer (channel 3) to that before the interferometer (channel 1)
provides both the interferometric (8:1) autocorrelation and (after
averaging), the second order intensity (3:1) functions. From these data,
we determined the temporal pulse shape to be [1]:

$$E = \exp[-(t/5.27)^2 + i\phi], \quad \phi = 1.43\ E^2 \quad, \text{ where t is in ps.}$$

The other pair of pulses (channel 2) is sent into a lithium heat pipe.
The construction and operation of the heat pipe are very similar to the
simple heat pipe described in [2]. The ratio of the second harmonic of the
transmitted signal to that of the reference signal (channel 1) is recorded
as a function of the delay and the relative phase between the pulses. The

---

*On leave from the Vrije Universiteit Brussel, Pleinlaan 2, B-1050
Brussels.

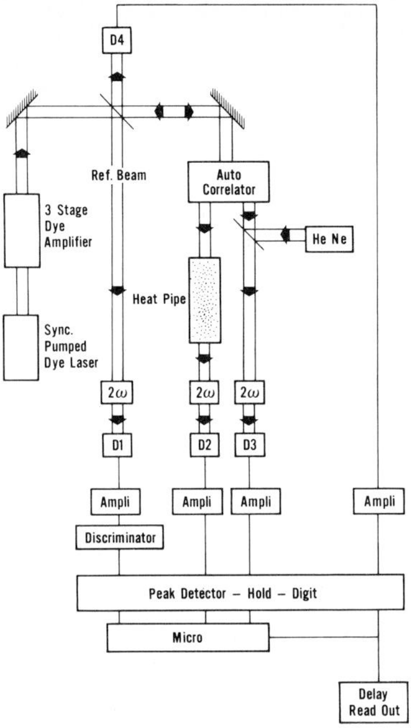

Fig. 1 Experimental setup

periodicity in phase of this signal is a direct measurement of the order of the process ($2\pi/n$ for n photon resonance). The shape and decay of the signal provide simultaneously information about coherence times as well as spectroscopic information [3]. It should be noted that because of our accurate knowledge of the pulse shape, our temporal resolution is at least one order of magnitude smaller than the pulse duration (6.2 ps FWHM).

In order to generate an accurate display of signal (or transmission) versus phase, the software of our data acquisition system averages data over several periods for a given phase and given intensity levels. At the same time, phase averaged values of the data are also computed. Experimental results for both types of averaging processes are shown in Fig. 2. Figure 2a represents the phase independent data, which in the absence of lithium vapor correspond to the conventional intensity autocorrelation (upper trace). An example of the phase dependent data, taken near zero delay, is shown in Fig. 2b (the upper trace corresponds again to the absence of vapor).

One might question whether or not the second harmonic measurements correspond exactly to $\int |E^2|^2 \, dt$ in the presence of phase modulation induced by the interferometric delay line and the propagation through the nonlinear medium. Therefore, we have verified that the spectrum of the second harmonic signal is indeed proportional to the square of the spectrum of the fundamental for bandwidths larger than 10 THz, far in excess of that of the pulses used in this experiment.

The evolution of the two-photon resonant system is described by a set of three differential equations for the components ρ_{11}, ρ_{22}, and ρ_{12} of the density matrix [4]. The dominant terms in the polarization of the medium are a resonant term proportional to $\rho_{12}E^*$ and a term proportional to $\rho_{22}E$

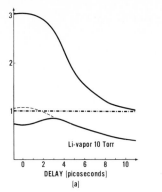

Fig. 2 Second harmonic intensity before (upper traces) and after (lower traces) propagation through Li vapor, as a function of pulse delay (a) and relative phase between the pulses (b)

(a)   (b)

which describes the ionization losses from the upper 4s level of lithium. In contrast to the long medium considered in [4], we linearized the propagation equations, consistent with a maximum measured attenuation factor of approximately 30%. The linear equations enabled us to calculate the transmitted signal (and the second harmonic energy) as a function of delay for all phase differences, thus providing the same large set of data as obtained experimentally. The agreement between theory and experiment is excellent, with the only adjustable parameter being the exact detuning from resonance. Of particular interest is the region of enhanced transmission for delays of the order of the pulse width, where the complicated phase dependence reflects the phase dependence of the input intensity, the stimulated emission, as well as reduced photo-ionization for the 90° out of phase tail.

Our method can be easily extended to the study of different order processes. Dye vapors (single photon) and mercury vapor (four photons) are also being investigated.

This work was supported by AFOSR under grant no. AFOSR 820044.

References

1.   H. Vanherzeele, H. J. Mackey, and J.-C. Diels, Appl. Opt. 23, No. 11, 000 (1984).

2.   L. A. Melton and P. H. Wine, J. Appl. Phys. 51, 4059 (1980).

3.   S. Besnainou, J.-C. Diels and J. Stone, J. Chem. Phys., July 15 (1984).

4.   J.-C. Diels and A. T. Georges, Phys. Rev. A19, 1589 (1979).

Stimulated Photon Echo for Elastic and Depolarizing Collison Studies

J.-C. Keller and J.-L. Le Gouët

Laboratoire Aimé Cotton*, C.N.R.S. II, Bâtiment 505
F-91405 Orsay Cedex, France

The stimulated photon echo technique has been applied previously to the study of small angle elastic scattering [1] but the method may also prove attractive to investigate quasi-resonant inelastic processes such as depolarizing, state-changing or energy pooling collisions. For these processes, total cross section have been obtained in the past using different experimental methods. However, they deserve to be reexamined in the perspective of their angular dependence. We present here the first angular analysis of a collisional depolarizing process, together with the determination of elastic scattering cross section in an excited state.

Our measurements are performed on ytterbium vapour, using levels $6s^2\ {}^1S_0$, $6s6p\ {}^3P_1$ and $6s7s\ {}^3S_1$ as levels a , b and c respectively (Fig. 1). The sample is excited by a sequence of three co-propagating laser pulses. Due to the interaction with the first two pulses, the population ρ_{bb} is modulated in the longitudinal-velocity space and the period of this modulation is $(k_1 t_{12})^{-1}$. The dipoles ρ_{bc} and thus the echo are produced by the third pulse from the velocity modulated part of level b population [1,2]. During the time interval t_{23} between the second and the third laser pulses, echo ralaxation results not only from the depopulation of level b but also from the destruction of the modulation in the velocity-space by elastic (velocity changing) collisions. Laser beams are linearly polarized; the polarization of the first two pulses is either perpendicular (Fig. 1c) or parallel (Fig. 1d) to that of the third pulse. In the latter case the signal appears only if depolarizing (m state changing) collisions occur.

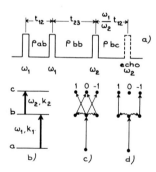

Figure 1 :

Excitation sequence and level scheme.

---

* Laboratoire associé à l'Université Paris-Sud.

The echo signals $I_{\parallel}$ and $I_{\perp}$, respectively for parallel and for perpendicular polarizations, are given by :

$$\sqrt{\frac{I_{\perp}}{I_0}} = \exp - \left(\gamma_{ab} + \frac{\omega_1}{\omega_2}\gamma_{bc}\right)t_{12} \cdot \exp - (\gamma_{e\ell} - \tilde{W})t_{23}$$

$$\cdot \left[2 + \exp - (\tilde{W} - \tilde{W}^{a\ell})t_{23}\right]$$

$$\sqrt{\frac{I_{\parallel}}{I_{\perp}}} = \frac{2 - 2\exp - [\tilde{W} - \tilde{W}^{a\ell}]t_{23}}{2 + \exp - [\tilde{W} - \tilde{W}^{a\ell}]t_{23}}$$

where γ_{ab} and γ_{bc} are the relaxation rates of optical coherences ρ_{ab} and ρ_{bc} ; $\gamma_{e\ell}$ is the total rate for elastic collisions in level b ; $W(\Delta v_z)$ and $W^{a\ell}(\Delta v_z)$ are the collision kernels corresponding respectively to the restitution of population and alignment in level b ; $\tilde{W}(k_1 t_{12})$ and $\tilde{W}^{a\ell}(k_1 t_{12})$ are their Fourier transforms :

$$\tilde{W}(k_1 t_{12}) = \int W(\Delta v_z) \cos k_1 \Delta v_z t_{12} \, d\Delta v_z \quad ; \quad \gamma^{e\ell} = \tilde{W}(0) \quad .$$

All the information on the angular dependence of the collisions is contained in the collision kernel which is expressed in terms of the differential cross section $d\sigma/d\theta$.

Using the experimental set-up described in [2], we have measured the quantities $d(\gamma^{e\ell} - \tilde{W})/dp$ and $d(\tilde{W} - \tilde{W}^{a\ell})/dp$ (p : perturber pressure) for different values of t_{12} and for different noble gas perturbers (He, Ne and Xe). In our case we are still in a scattering domain correctly described by classical mechanics and the data concerning elastic collisions in Yb $(6\,^3P_1)$ are interpreted within the frame of classical small angle scattering using a Van der Waals potential for the long range interaction. This models leads to : $d(\gamma_{e\ell} - \tilde{W})/dp = g(m,m_p,T)\cdot(t_{12})^{1/3}\sigma^{5/6}$. σ is the total elastic cross section ; m and m_p are respectively the active atom and the perturber masses and T is the sample temperature. From the experimental data (Fig. 2) and using the above expression, we have obtained the

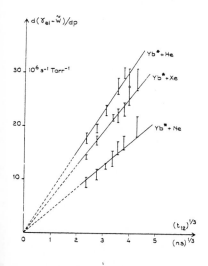

Figure 2 :

Experimental points and fitted curve for elastic collisions.

following cross sections for elastic scattering of Yb ($6s6p\ ^3P_1$) :
$$\sigma_{He} = 220(20)\ \mathring{A}^2\ ;\quad \sigma_{Ne} = 320(20)\ \mathring{A}^2\ ;\quad \sigma_{Xe} = 1240(50)\ \mathring{A}^2\ .$$

Data concerning level b depolarization have been completed by the measurement of the collision rate for alignment destruction $\gamma^{al} = \tilde{W}(0) - \tilde{W}^{al}(0)$. For this purpose, we used two-pulse photon echo produced on bc transitions after some delay of the laser excitation and with convenient laser polarizations (Fig. 1c,d) . The corresponding cross sections are : $\sigma^{al}_{He} = 94(3)\ \mathring{A}^2\ ;\quad \sigma^{al}_{Ne} = 123(7)\ \mathring{A}^2\ ;\quad \sigma^{al}_{Xe} = 300(20)\ \mathring{A}^2$.
In the frame of Anderson's model [3] we can use these values to determine the long range branch of the anisotropic component of the interatomic potential. As its isotropic component can be derived from elastic cross-section values, we dispose of the two parameters which are needed for the calculation of the small angle differential cross section for depolarizing collisions and of the corresponding collision kernel. A reasonable agreement with experimental data is obtained (Fig. 3a,b).

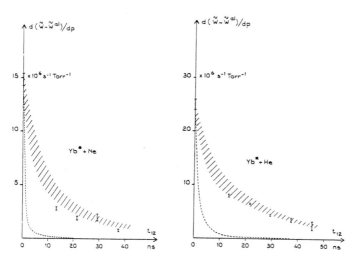

Figure 3 : Experimental points and calculated curve (dashed area) for depolarizing collisions.

[1] T. W. Mossberg, R. Kachru and S. R. Hartmann, Phys. Rev. A 20, 1976 (1979) and references therein.
[2] F. de Rougemont, J.-C. Keller and J.-L. Le Gouët in Laser Spectroscopy VI (1983) p. 53.
[3] P. W. Anderson, Phys. Rev. 76, 647 (1949) ; 86, 809 (1952).

Coherent Transient Spectroscopy with Ultra-High Time Resolution Using Incoherent Light

Norio Morita, Tatsuo Yajima, and Yuzo Ishida

Institute for Solid State Physics, University of Tokyo
Roppongi, Minato-ku, Tokyo 106, Japan

In transient spectroscopy, the time resolution is usually limited by the pulse width or the risetime of light sources or equivalent perturbers. In the extremely short time region, however, its improvement by shortening the pulse width is not an easy way even with the recent development of femto-second pulse technique. We propose here a new method of transient spectros-copy where the time resolution is not limited by the light duration t_p but limited only by the correlation time τ_c of light sources, being generally much shorter than t_p. A temporally incoherent light has a short τ_c corre-sponding to the reciprocal bandwidth, and appears like a single pulse of duration τ_c in the auto-correlation measurement. This kind of light is therefore expected to play essentially the same role as a short pulse in nonlinear spectroscopy utilizing the auto-correlation technique. We present here a basic theory and an experiment which confirm this idea.

A nonlinear optical process relevant to examine and realize the basic idea is the resonant degenerate four-wave mixing, where two incident light beams with wave-vectors $\vec{k}_1$ and $\vec{k}_2$ at frequency ω produce output light with $\vec{k}_3=2\vec{k}_2-\vec{k}_1$ at ω through the third-order nonlinearity of a resonant material (Fig.1). This process is known to include various nonlinear coherent tran-sient processes in the low field limit [1,2]. Consider a model of two-level atomic system with longitudinal (T_1) and transverse (T_2) relaxation times. When two input coherent pulses of duration t_p and mutual time delay τ are used, the correlation trace, i.e., the output light energy versus τ, be-comes the functions $\exp(-2\tau/T_2)$ and $\exp(-4\tau/T_2)$ for homogeneously and in-homogeneously broadened transitions, respectively, if $t_p \ll T_2$. These re-sults are independent of T_1, and reflect the features of free induction decay and photon echo. We have further calculated the correlation traces for the same nonlinear process with incoherent incident beams (including cw light) in place of short pulses, in order to see whether the trace shows relaxation behaviors even for $T_1,T_2 \ll t_p$. A density matrix equation is solved for the incident light field with Gaussian random complex amplitude.

Figs.2(i) and 2(ii) show the calculated correlation traces with incoher-ent incident light in the limits $\tau_c \to 0$ and $t_p \gg T_1,T_2$.

Fig.3 illustrates typical features of the correlation traces for incoher-ent light together with those for short pulses ($t_p \to 0$) shown for compari-

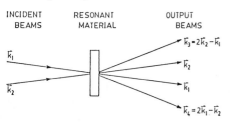

INCIDENT   RESONANT   OUTPUT
BEAMS   MATERIAL   BEAMS

Fig.1   Configuration of the four-wavemixing with two input beams

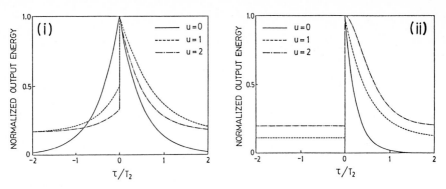

Fig.2 Calculated correlation traces of the transient four-wave mixing with incoherent light for various values of $u = T_2/T_1$ in the limits $\tau_c \to 0$ and $t_p \gg T_1, T_2$. (i) homogeneously broadened transition, (ii) inhomogeneously broadened transition

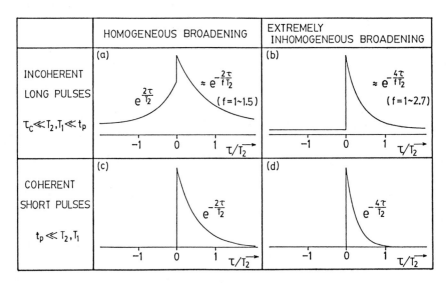

Fig.3 Comparison between the calculated correlation traces with incoherent light and with coherent short pulses. Though the decay curves at $\tau > 0$ in (a) and (b) are not always expressed by simple functions, they can be approximated to be single-exponential functions of the forms shown in the figures, where f varies with T_2/T_1. The curves in (a) and (b) are drawn for $T_2/T_1 = 0.5$ to show clearly the differences between the cases of two kinds of light. However, for most range of T_2/T_1, the shape and rate of decay curves are almost the same for the two cases

son. It is revealed that the traces for incoherent light are essentially of the same form as those for short pulses, but with the resolution limited by τ_c. There are, however, some features peculiar to the case of incoherent light : (1) Relaxation profiles appear for both signs of τ in the homogeneously broadened case. (2) The profile is not exactly exponential but

slightly deformed by the T_1 effect.   (3) A small background appears.   The
features (2) and (3) disappear in the limit $T_2/T_1 \to 0$. In the determination
of relaxation times by the present method the feature (2) brings about some
complications.   However, detailed examination showed that the trace is
dominantly governed by T_2 for most possible values of T_1, and that the error
of T_2 determined by an effective decay time of the trace is not so significant.

All these properties can well be explained physically and even mathemati-
cally by regarding the incoherent light as a series of random ultrashort
pulses. The resulting trace can be interpreted as arising from the integration of numerous transient four-wave mixing processes caused by various
combinations of these pulses.

The incoherent light for this purpose must have poor temporal coherence,
but have good transverse spatial coherence (directivity) and sufficient
intensity.   As practical light sources, the amplified spontaneous emission
or the broad-band laser far from transform-limited are suitable.   We performed an experiment demonstrating the essential point of the new method for
the Na 3S-3P transition by using the imperfectly mode-locked dye laser
pulses with $t_p \simeq 6$ ps and $\tau_c \simeq 0.5$ ps.   Although the observed correlation traces were too complicated to compare directly with the simple theory,
a clear quantum beat between the D_1 and D_2 transitions with the period of
1.9 ps has been observed even in the time scale much shorter than t_p.   This
can be regarded as a direct evidence of the basic idea.   Recently, Asaka et
al. [3] also demonstrated the same kind of facts by an experiment for Nd^{3+} :
glass using a broad-band cw dye laser source.

A coherently phase-modulated pulse such as self-phase-modulated one has a
reciprocal bandwidth τ_B much shorter than t_p, and is expected to play a
similar role to incoherent light.   Calculations for the same nonlinear
process revealed that τ_B plays the same role as τ_c of incoherent light and
that T_2 decay profile in the correlation traces with the resolution of τ_B
can readily be obtained even under the condition $T_2 << t_p$.

These results provide a new approach of studying ultrafast phenomena in
the subpicosecond and femtosecond regions without requiring ultrashort light
pulses.

References

1. T. Yajima and Y. Taira : J. Phys. Soc. Japan 47, 1620 (1979)
2. P. Ye and Y.R. Shen : Phys. Rev. A 25, 2183 (1982)
3. S. Asaka, H. Nakatsuka, M. Fujiwara and M. Matsuoka : Phys. Rev. A 29,
    2286 (1984)

Part IV

Stimulated Scattering

Interaction-Induced, Subpicosecond Phenomena in Liquids

P.A.Madden

Royal Signals and Radar Establishment
St.Andrews Road, Great Malvern, Worcester WR14 3PS, UK

1 Introduction

The non-linear susceptibility $\chi^{(3)}$ of condensed samples is, in part, determined by interaction-induced (I-I) effects; that is, by changes in molecular polarizabilities which result from intermolecular interactions. Here the way in which these effects may influence time-domain experiments with subpicosecond pulses on $\chi^{(3)}$ (such as Kerr transient studies or 4-wave mixing) will be considered. Quite analogous I-I effects have been extensively studied by conventional frequency domain spectroscopy, particularly light scattering, and the intention of this article is to review the information so gained in the context of the time domain experiments, where the I-I effects have only recently been resolved.

The time domain experiments measure some property of the non-linear polarization $(\underline{P})$ induced by the electric field $(\underline{E})$ of a laser pulse. The polarization is made up of an instantaneous electronic or hyperpolarization **part,** described by $^{el}\chi^{(3)}$, and a part which lags behind the laser field. It may be written

$$P_\alpha(t) = {}^{el}\chi^{(3)}_{\alpha\beta\gamma\delta} E_\beta(t)E_\gamma(t)E_\delta(t) + E_\beta(t) \int_{-\infty}^{t} d\tau \, C_{\alpha\beta\gamma\delta}(t-\tau)E_\gamma(\tau)E_\delta(\tau) \ . \quad (1)$$

Provided that the measured polarization is truly third-order in the field, then C(t) is simply the correlation function of the total polarizability of the sample [1]

$$C_{\alpha\beta\gamma\delta}(t) = \langle\alpha_{\alpha\beta}(t) \, \alpha_{\gamma\delta}\rangle \ . \quad (2)$$

Although one normally thinks of the time domain experiments in terms of pumping and probing this result shows that the dynamical processes observed are just those of the spontaneous thermal fluctuations in the sample, in the absence of the laser field. Precisely the same correlation functions may be observed by light scattering, the depolarised light scattering spectrum is just [2]

$$I(\omega) \propto \text{Re} \int_{0}^{\infty} dt \, e^{i\omega t} \, C_{xzxz}(t) \ . \quad (3)$$

In what may be called the "conventional picture" the polarizability of the sample is regarded as the sum of the gas-phase molecular polarizabilities, i.e.

$$\alpha_{ab} = \sum_{i} \alpha^i_{ab} \ . \quad (4)$$

It is now well recognised [3], in the spectroscopic field, that this viewpoint is inadequate for condensed matter and that the contributions of the
polarizabilities due to the interactions between pairs, triplets, etc. of
molecules must also be considered, i.e.

$$\alpha_{ab} = \sum_i \alpha_{ab}^i + \sum_{ij}' \alpha_{ab}^{ij} + \dots \quad , \tag{5}$$

these are the interaction-induced effects of the title. The presence of
such terms may readily be inferred from the shape of light scattering spectra,
such as the depolarised Rayleigh spectrum of CS_2 shown in fig. 1a. The single
molecule (i.e. the α_{xz}^i terms of (5)) contributions to α_{xz} are expected to relax
by molecular reorientation; for CS_2 this motion is diffusional down to times
of order 150fs [4] and the correlation time is about 1.5 ps [5]. In the conventional picture then, the Rayleigh spectrum should be lorentzian out to
about 35cm^{-1} with a width of 3.5cm^{-1}. Such a lorentzian is shown in the
figure (dotted line), it is clear that there is substantially more intensity
at the higher frequency; the crosses show the difference between the total
and the lorentzian. This is attributed to the I-I effects; these relax
through intermolecular motions which change the relative position and orientation of molecules. For CS_2 such motions are more rapid than molecular
reorientation in space.

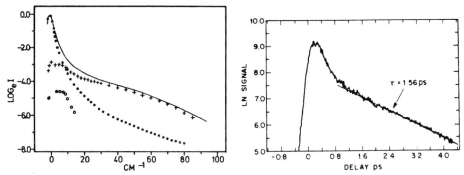

Figure 1   Rayleigh spectrum [5] and Kerr transient induced by a 100fs
pulse [6] for CS_2 ..... - a lorentzian for the expected reorientational
spectrum, +++ the difference between it and the total.

To resolve the corresponding difference between the predictions of the
conventional picture and observation in the time domain requires a time
resolution of about 300fs, which has only been achieved quite recently.
Figure 1b shows the Kerr transient of CS_2 obtained with a 100fs laser pulse
[6]. This clearly shows a non-instantaneous (and therefore non-electronic)
component which is too rapid to be associated with reorientation; it was
recognised by Greene and Farrow as corresponding to the interaction-
induced Rayleigh wing [6].

The contribution to the Kerr transient from I-I effects could be quite
large. The static, or d.c., Kerr effect of CS_2 is 10% electronic, 50%
reorientation and 40% I-I [7]. However, for pulses shorter than the reorientation time the relative importance of the rapidly relaxing I-I and
instantaneous electronic terms will be enhanced.

2 Nature of the Interaction-Induced Polarizability

In order to understand the factors which affect the shape and amplitude of I-I transients it is necessary to establish the way in which the I-I polarizabilities depend upon the intermolecular coordinates. Much work has been done to understand this issue, primarily on "forbidden spectra", that is on spectra which are forbidden by symmetry for the isolated molecule (or in the conventional picture). Such spectra are purely interaction-induced. An example of a forbidden spectrum is the depolarised Rayleigh spectrum (DRS) of a fluid of spherical molecules or atoms, since the molecular polarizability has no off-diagonal components. Other forbidden spectra are the Raman spectra of the u-symmetry vibrations of a centrosymmetric molecule or the infra-red spectra of the g-symmetry ones. Whilst the latter do not correspond to contributions to $\chi^{(3)}$ their study has provided a general framework to which studies on the I-I polarizability belong [3].

The central conclusion of this work is that the dominant contributions to the I-I properties are the result of electrodynamic interactions between the otherwise undistorted charge distributions of molecules, that is the effects of overlap between the electron clouds of different molecules is not an important effect. This conclusion pertains to normal "Van der Waals" materials, but not to strongly interacting materials like ionic melts.

If overlap effects are neglected the interaction-induced properties may be predicted from simple multipole expansions [3,8]. For example, the most important two-particle term responsible for DRS in an atomic fluid becomes the dipole-induced dipole (DID) polarizability

$$\alpha_{xz}^{ij} = \alpha \; \underline{T} \; (\underline{r}^{ij}) \; \alpha \tag{6}$$

where $\underline{T}$ is the dipole field tensor

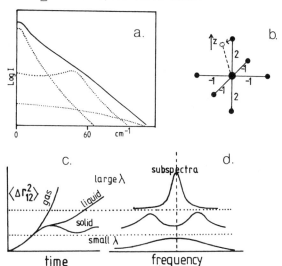

Figure 2 a) Solid line is the DRS of argon [14]. b) Illustrates the cancellation of the DID polarizability in atomic fluids, numbers give $r^3 \times T_{zz}$. c) Mean square displacement of a pair of atoms to indicate the origin of the subspectra shown in d) also shown as dotted lines in a).

$$T_{\alpha\beta}(\underline{r}) = (3r_\alpha r_\beta - r^2 \delta_{\alpha\beta}) \; r^{-5} \qquad (7)$$

and α is the atomic polarizability. Similar explicit forms for other I-I effects may also be deduced [3].

The validity of this central conclusion has been established through careful quantitative gas-phase studies (ref 9 & 10) and the applicability of the multipole mechanisms in the condensed phase has been checked by the comparison of spectral data generated by computer simulation with experimental results [11]. The multipole models have also proven useful in semi-quantitative comparisons of the properties of different I-I spectra of the same fluid (see e.g. 12).

3 I-I Effects and the Dynamics of Intermolecular Encounters

Since the I-I polarizability depends upon the relative position and orientation of molecules the decay of its correlation function directly reflects the dynamics of intermolecular encounters. In forbidden spectra (e.g. the DRS of argon) or the analogous purely I-I time domain phenomena (e.g. Kerr effect in argon) these fundamental events are directly observed, without interference from the molecular α^1 terms. Theories of the appropriate correlation functions have been developed and the model of the inter-molecular dynamics which underlies them has been applied [13] to the description of transport and relaxation rates in liquids, which are determined by the same fundamental events.

The lineshape of the DRS in liquid argon [14] is shown in fig 2a, the analogous Kerr transient may be found in ref [15]. The lineshape is best described as piecewise exponential, with a shoulder at 60cm^{-1} and an additional low frequency feature. The high frequency portion of the line is only slowly dependent upon density and temperature but the amplitude and width of the low frequency feature are quite sensitive (like the viscosity or diffusion coefficient). The decay of the I-I polarizability is therefore quite complex; most liquid state relaxations give exponential correlation functions (or lorentzian spectra) corresponding to simple Markov processes.

The lineshape in the dense fluid is best understood as a consequence of the relaxation of <u>cooperative</u> local density fluctuations. The density of a fluid at the triple point is rather close to that of the solid (typically 10% less) so that the molecules are locally well ordered. Consider the total DID polarizability ($\sum_j \alpha^2 \, T_{xz}(\underline{r}^{1j})$) of atom 1 when its first shell of neighbours is perfectly ordered, as in fig 2b. The total polarizability vanishes, as pairs of neighbours give contributions of equal magnitude and opposite sign, therefore the DID intensity for the perfectly ordered configuration also vanishes. However, the perfect cancellation is destroyed when the position of a molecule fluctuates with respect to the others (i.e. cooperatively), as in the dotted line of the figure. The observed intensity should therefore be associated with the fluctuations which destroy the local order, rather than the mere presence of particles close to each other. (Note that the sum of the <u>pair</u> intensities, i.e. $\sum_j (\alpha^2 T_{xz}(\underline{r}^{1j}))^2$, is large even for the ordered configuration.) Experimentally it is known that the DID intensity per atom falls rapidly with increasing density [16] to about 1% of the sum of pair intensities at the triple point. The consequence of this cooperativity for the relaxation is that the slow separation of pairs of particles over large distances does not contribute to the observation; this is beautifully illustrated by the computer simulation of Ladd et al.[17].

To interpret the lineshape [18,3] the local density fluctuations are resolved into fourier components and the spectrum described as a super-position of _subspectra_, each associated with the relaxation of a density fourier component of a particular wavelength (λ). To relax such a fourier component requires molecular motion over a distance comparable to a wavelength (i.e. from a peak to a trough of a density wave). Because of the cancellation effects described above the important range of wavelengths is from about an intermolecular spacing downwards. The relaxation of the subspectra of different wavelengths then reflects distinctive characteristics of the inter-molecular motion, this is illustrated in figs.2c and 2d. The mean-square displacement $<\Delta r^2(t)>$ in a liquid is gas-like ($<\Delta r^2(t)> \propto t^2$) at short times, where such small distances are travelled that no collisions occur, and diffusive at long times ($<\Delta r^2(t)> \propto t$) when displacements over an inter-molecular spacing or more occur. At intermediate times (the domain of correlated collisions) the motion has an oscillatory solid-like character. The long wavelength subspectra are therefore diffusive in character, as they are relaxed by motion over an intermolecular spacing or so. They will have a lorentzian shape and a width which is sensitive to the thermodynamic state (like the diffusion coefficient). The very short wavelength subspectra are gas-like, broad and insensitive to temperature. At intermediate wavelengths the subspectra have shifted peaks reflecting the oscillatory solid-like motion. The subspectra are illustrated in figure 2d; their superposition, as in fig 2a gives the full spectrum and it is in this way that the spectrum's complexity may be understood.

4 The Interference Between Molecular and Interaction-Induced Polarizabilities

In many cases of practical significance both molecular (i.e. the α^i terms of (5)) and I-I terms contribute. Examples are the Rayleigh spectrum and Kerr transient of CS_2, shown in fig 1. In such cases we may wish to decompose the observation into molecular and I-I parts, so as to determine a reorientation time for example. In this section the possibility of effecting such a separation are discussed and the limited applicability of the separation procedure highlighted.

To simplify notation (5) will be re-written

$$\alpha_{ab}(t) = R(t) + Q(t) \quad , \tag{8}$$

where R is the sum of all molecular polarizabilities, which relax by reorientation, and Q is all the I-I terms. The total polarizability correlation function is then

$$C_{abab}(t) = <R(t)R> + <Q(t)Q> + <R(t)Q> + <Q(t)R> \quad . \tag{9}$$

As discussed in the introduction the Rayleigh spectrum and Kerr transient in CS_2 appear to show relaxations on two distinct timescales. The naive interpretation of this observation is that the slower corresponds to $<R(t)R>$ and the faster to $<Q(t)Q>$ with the cross-terms being negligible.

The polarizability correlation function of CS_2 has been studied by computer simulation with the intermolecular variable Q given by the pair DID term. The three contributions to the r.h.s of (9) are illustrated in fig. 3a. These show that the naive interpretation is quite false:- all three terms decay on the same timescale and the cross-correlation between Q and R is quite large.

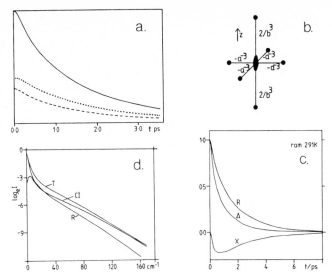

Figure 3   a) Components of C_{xzxz} for CS_2 [21] $-\langle R(t)R\rangle$ - - - $\langle Q(t)Q\rangle$, ...
$\langle R(t)Q\rangle$ equn. 9.   b) The imperfect cancellation of the DID for a non-
spherical molecule, numbers give T_{zz}.   c) Components of C_{xzxz} from equn. (15)
R & $\triangle$ - autocorrelation functions, X - cross-term.   d) Spectra from equn.
(15), T - total, R - reorientation, CI - collision-induced, c.f. fig. Ia .

A more tenable interpretation is obtained through a different decomposition of the polarizability; instead of (8) write

$$\alpha_{ab} = (1 + f)R(t) + \Delta(t) \tag{10}$$

where
$$\Delta = Q - fR \tag{11}$$
and
$$f = <QR>/<RR> . \tag{12}$$

f is called the projection of the I-I polarizability along the molecular one,
it may be thought of as the average I-I polarizability for a given molecular
orientation. As illustrated in fig.3b the DID I-I polarizability of a non-
spherical molecule with its neighbours in their average configuration is
non-zero, as illustrated in the figure the equatorial neighbours are closer
to the central molecule than the axial one and therefore have a larger DID
polarizability contribution (contrast the spherical case in fig 2a where
the I-I polarizability vanishes except during a fluctuation). fR represents
this non-zero average value and Δ is the part of the I-I polarizability
which is purely fluctuation-induced (and thereby analogous to the total I-I
polarizability in the spherical case). Notice that Δ and R are orthogonal
in the sense that
$$<\Delta(0) R(0)> = 0 \tag{13}$$
so that the zero time polarizability correlation function (i.e. the total
Rayleigh intensity or Kerr constant) may be written as the sum of two
amplitudes
$$C_{abab}(0) = (1 + f)^2 <R(0)^2> + <\Delta(0)^2> , \tag{14}$$

249

one associated with reorientation and the other the fluctuating part of the interaction-induced polarizability.

This decomposition is not useful unless the two amplitudes can be recognised from their distinctive time (or frequency) signatures; it is argued [19,3] that this should be possible if a timescale separation exists between the reorientational and intermolecular motion. In this circumstance the relaxation of the neighbours to their equilibrium position following a change in the orientation of the central molecule is rapid compared to the overall rate of molecular reorientation. The mean I-I polarizability will then follow the reorientation of the central molecule and all the slow reorientational time dependence in Q should be removed when Δ is formed. The polarizability correlation function, now written

$$C_{abab}(t) = (1+f)^2 <R(t)R> + [<\Delta(t)\Delta> + 2(1+f) <\Delta(t)R>] \quad , \qquad (15)$$

should then contain a slow, reorientational part, given by the first term in (15), and a rapidly relaxing part given by the square bracketed term, which may be called fluctuation- or "collision-" induced. Note that the amplitude of the reorientational term is not that of the conventional picture, it is modified by a "local-field factor", given by $(1+f)^2$ [20]. The amplitude of the square-bracketed term is just $<\Delta^2>$, by virtue of (13).

This separation may also be tested in the computer simulation [21], the results are shown in Fig 3c. Whilst $<\Delta(t)\Delta>$ is appreciably faster than $<R(t)R>$ it still contains some slow character, as does the cross-term. However, as the figure shows, the slow parts of these functions are of opposite sign and when they are combined in the square-bracketed, collision-induced term of (15) the result relaxes quickly.

The Rayleigh spectrum calculated in the simulation is shown in fig. 3d (solid line), it is very similar in shape to the experimental one (fig 1). The spectrum of the reorientational term $(1+f)^2 <R(t)R>$ (crosses) and of the square-bracketed collision-induced term (dotted) are also shown. The cancellation of the slow tails is witnessed by the fact that the collision-induced term does not contain a large low frequency feature. The reorientational term dominates the low frequency spectrum and the collision-induced one becomes most important at high frequencies. The two calculated

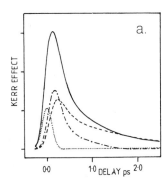
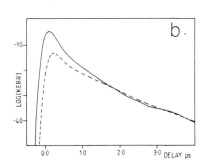

Figure 4   a) The Kerr transient of CS_2 (-) calculated in a computer simulation for a 100fs pulse and its electronic (....), reorientational (- - -) and C.I. (- -) components (eq. 15) b) A semilog plot of the total and the reorientational (- - -) component, for comparison with fig 1.

spectra coincide rather well with the separation of the experimental
spectrum [5] made on purely empirical grounds, which is displayed in
fig. 1a. It should be noted that this type of decomposition fails completely in fluids like N_2 in which the timescale separation is not found
[22].

The same simulation data may be used to calculate a Kerr transient, to
compare with fig 1b. In order to do this we must add an instantaneously
relaxing electronic term. The weight of the latter is known from D C Kerr
studies to be 11% of the Kerr constant [7]. Figure 4a shows the Kerr
transient calculated from the simulation, under conditions chosen to
reproduce those of the experimental study. The separate contributions to
the transient (as from eq (15)) are also shown. The collision-induced
component is the largest contributor to the transient at short times. In
fig 3f the total transient and the reorientational component (from
$(1+f)^2 <R(t)R>$) are plotted semi logarithmically. The total agrees rather
well with experiment (fig 1b). Part of the short-time transient can be
seen to be due to short-time, non-diffusional characteristics of the
molecular motion, though these are swamped by non-reorientational effects.
The reorientational correlation function only dominates the transient at
quite long times, so that reorientation times should be extracted from
such data with care.

References

1 R.W. Hellwarth: Prog. Quant. Electr. 5, 1 (1977)
2 B.J. Berne and R.W. Pecora: Dynamic Light Scattering (Wiley, N.Y. 1976)
3 P.A. Madden: "Interaction-Induced Effects", in Molecular Liquids,
     Ed. A.J. Barnes (Reidel 1984)
4 D.J. Tildesley and P.A. Madden: Mol. Phys. 48, 129 (1983)
5 T.I. Cox, M.R. Battaglia and P.A. Madden: Mol. Phys. 38, 1539 (1979)
6 B.I. Greene and R.C. Farrow: Chem. Phys. Lett., 98, 273 (1983)
7 M.R. Battaglia, T.I. Cox and P.A. Madden: Mol. Phys. 37, 1413 (1979)
8 A. Buckingham: Adv. Chem. Phys. 12, 107 (1967)
9 L. Frommhold: Adv. Chem. Phys. 46, 1 (1981)
10 J.D. Poll and J.L. Hunt: Can. J. Phys. 59, 1448 (1981)
11 B.J. Alder and E.L. Pollock: Ann. Rev. Phys. Chem. 32, 311 (1982).
     P.A. Madden and D.J. Tildesley: Mol. Phys. 49, 193 (1983)
12 P.A. Madden and T.I. Cox: Mol. Phys. 43, 287 and 43, 307 (1981)
13 A.J. Masters and P.A. Madden: J. Chem. Phys. 74, 2450, 2460 (1981).
     D. Kivelson and P.A. Madden: J. Phys. Chem. 86, 4244 (1982)
14 S.-C. An, C.J. Montrose and T.A. Litovitz: J. Chem. Phys. 64, 3717 (1976)
15 B.I. Greene, P.A. Fleury, H.L. Carter and R.C. Farrow: Phys. Rev. A29,
     271 (1984)
16 M. Zoppi and F. Barocchi: Can. J. Phys. 59, 1475 (1981)
17 A.J.C. Ladd, T.A. Litovitz and C.J. Montrose: J. Chem. Phys. 71, 4242
     (1979)
18 P.A. Madden: Mol. Phys. 36, 365 (1978)
19 T.F. Keyes, D. Kivelson and J.P. McTague: J. Chem. Phys. 55, 4096 (1971)
20 T.F. Keyes and B.M. Ladanyi: Mol. Phys., 33, 1063; 33, 1099; 34, 765
     (1977)
21 P.A. Madden and D.J. Tildesley: Mol. Phys. - to be published
22 D. Frenkel and J.P. McTague: J. Chem. Phys. 72, 2801 (1980)

Transient Infrared Spectroscopy on the Picosecond and Sub-Picosecond Time Scale

H.-J. Hartmann and A. Laubereau

Physikalisches Institut, University of Bayreuth
D-8580 Bayreuth, Fed. Rep. of Germany

In recent years coherent and spontaneous Raman scattering of ultrashort pulses has been used for numerous studies of molecular dynamics in the electronic ground state. Application of these techniques is restricted by the selection rules and also weakness of the Raman interaction. Corresponding infrared methods taking advantage of the different selection rules and higher efficiency of electric dipole coupling have been lacking.

Recent progress in the generation of widely tunable ultrashort pulses in the long wavelength range [1] has opened the field of ultrafast infrared spectroscopy. In this brief article first examples of this kind are demonstrated applying coherent pulse propagation at low intensity level (small area pulses). Drastic pulse reshaping is reported for resonant infrared pulses of a few ps duration in optically thick samples, allowing for the first time to subtract quantitative information from this physical effect.

For moderately thick samples, novel features are found for the coherent propagation of low-intensity pulses. Our theoretical propagating pulse develops an approximately exponential slope which deviates from the dephasing time T_2 of the molecular transition. This time behaviour is termed nearly-free induction decay (NFID) in contrast to the well-known effect of free induction decay (FID) which occurs for very short samples.

Some numerical results are depicted in Figs. 1a and b [2]. A specific experimental situation is considered with resonant, weak input pulses of Gaussian shape and duration t_p (FWHH) and two values of propagation length, $\alpha\ell = 0.2$ and $\alpha\ell = 1$. α denotes the conventional absorption coefficient at the maximum of the absorption line. The intensity of the transmitted pulse is evaluated solving the Maxwell-Bloch equations for the homogeneous case of a single molecular transition frequency (Fig. 1a) and of two neighbouring transitions (Fig. 1b). Experimentally, the short pulse intensity is detected by means of an ultrafast light gate using parametric up-conversion [3]. This technique provides the convolution of the IR pulse of interest with the probing pulse (Gaussian shape, duration t_p) operating the parametric device. The resulting conversion signal $C(t_D)$ is plotted in Fig. 1 on a logarithmic scale versus delay time between investigated and probing pulse. For small values of $t_D/t_p < 2$ the shape and peak amplitude of the transmitted pulse (solid line) is practicly unchanged compared to the signal curve of the input pulse (dotted line). For larger values of t_D, the $C(t_D)$ curve in Fig. 1a displays an extended, weak wing with approximately exponen-

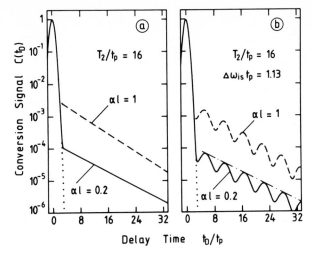

Fig. 1 : Calculated conversion signal $C(t_D)$ of the transmitted pulse vs. delay time for $T_2/t_p = 16$ and two values of absorption length, $\alpha \ell = 0.2$ and 1.0.
a) single transition frequency;
b) two transition frequencies leading to a beating phenomenon. Signal curves of the Gaussian input pulse (dotted lines).

tial decay, which represents nearly free induction decay (NFID) of the radiating transition dipoles. The effective decay time τ of the signal is found to differ from $T_2/2$ by several per cent for the example $\alpha \ell = 0.2$, while a larger deviation of $\simeq 25\ \%$ occurs for $\alpha \ell = 1$ (broken line). It can be shown that for a single transition frequency the decay time is given by [4]:

$$\tau \simeq \frac{T_2}{2}\ (1 + \frac{\alpha \ell}{4})^{-1} \tag{1}$$

if $\alpha \ell \leq 1 \ll T_2/t_p$. We note that T_2 may be readily determined from measurements of the trailing pulse wing. Fig. 1 also shows that the intensity of the exponential slope increases with $\alpha \ell$; this makes NFID attractive in comparison with FID necessitating $\alpha \ell \ll 1$.

Similar results are presented in Fig. 1b for the case of isotopic line splitting. Two species with frequency difference $\Delta \omega_{is} \times t_p = 1.13$, equal dephasing time T_2 and the ratio of $\alpha_1/\alpha_2 = 3$ are considered ($\alpha = \alpha_1 + \alpha_2$). Most important is the oscillatory time behaviour of the decaying pulse wing in Fig. 1b. A novel beating phenomenon occurs which is superimposed on an approximately exponential slope and reflects the coherent superposition of the adjacent molecular transitions.

As a consequence of the low intensity level with negligible population changes considered in Fig. 1, intensity variations of the input pulse do not change the shape of the signal transient $C(t_D)$. This property favours quantitative applications of the coherent propagation effect investigated here. Calculated data on the propagation through thick samples ($\alpha \ell \gg 1$) will be discussed below in context with experimental data.

We have substantiated our theoretical results by an experimental study of vibration-rotation transitions of pure HCl gas and in mixtures with argon buffer gas. Data will be also presented on liquid CH_3I. Our measuring system is described as follows

[3]. Single picosecond pulses, generated by a Nd:glass laser system, pass through a saturable absorber for minor pulse compression and enter a double-pass parametric generator.

Tunable infrared pulses are efficiently produced at the "signal" (ω_s) and "idler" (ω_{IR}) frequency positions by stimulated parametric amplification in a 3 cm $LiNbO_3$ specimen. Values of t_p = 2 ps to 4 ps and a high pulse quality are adjusted ($t_p \times \Delta\nu \simeq 0.7$). The "idler" pulse is resonantly tuned to the molecular transition and serves for the study of the propagation effect; it passes the sample cell and enters a parametric light gate designed for high-sensitivity pulse analysis. The "signal" component is used as probing pulse and operates the gate device. Parametric up-conversion is efficiently generated in a 1 mm $LiNbO_3$ and subsequently detected by a photomultiplier. Our infrared detection scheme achieves sub-picosecond time resolution (0.3 ps) and a large measuring range of 10^6. Using a second infrared light gate in a simultaneous reference experiment, the shape and amplitude of the input infrared pulse are also measured.

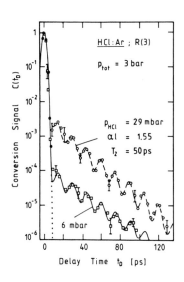

Fig. 2: Time-resolved IR spectroscopy of HCl:Ar (R(3) line) at 3 bar total pressure: conversion signal of the transmitted pulse versus delay time; open squares, full curve: p_{HCl} = 6 mbar; open circles, broken line: p_{HCl} = 29 mbar; the input pulse (full points, broken curve) refers to the same ordinate scale; theoretical curves. A novel beating effect and exponential dephasing are observed.

Experimental data for the system HCl:Ar with natural isotope abundance are presented in Fig. 2 for a total pressure of 3 bar and two HCl concentrations [4]. The infrared pulse is tuned to the maximum of the R(3) line of HCl^{35} at 2963 cm^{-1} [5]. The corresponding line of the HCl^{37} is shifted by $\simeq$ 2 cm^{-1} and also interacts with the incident radiation. The conversion signal $C(t_D)$ is plotted in the Fig. versus delay time between the infrared and probing pulse. $C(t_D=0)=1$ marks the maximum signal observed with an evacuated sample cell. Around $t_D=0$, the conversion signal of the transmitted pulse (open squares, solid line for p_{HCl} = 6 mbar) closely follows the rapid build-up and immediate decay of the input signal (full points, dotted line). For larger time values, the transmitted pulse wing displays a beating phenomenon, which is superimposed on an approximately exponential slope. From the oscillations of the signal curve the beating period 15.0 + 0.3 ps is directly measured, which corresponds to $\Delta\omega_{is}/2\pi c = \bar{2}.22 + 0.04$ cm^{-1}, in good agreement with spectroscopic data [5]. From the exponential signal slopes values of $T_2 = 52 + 5$ ps and $T_2 = 50 + 4$ ps are obtained for $p_{HCl} = 6$ mbar and $\bar{2}9$ mbar, respectively. Varying the pressure values of HCl and buffer gas, a linear dependence of $1/T_2$ on pressure has béen established as expected for pressure-broadened molecular transitions [2-4].

Results on coherent propagation in a thick sample ($\alpha\ell = 19.5$) are presented in Fig. 3. The R(3) transition of pure HCl is investigated at p = 3 bar and 1.3 cm sample length (open circles, solid curve). The drastic change of pulse shape compared to the input pulse (full points, dotted line) and the large peak transmission of $\simeq 0.2$ should be noted. For large $\alpha\ell$, repetitive absorption and coherent re-emission of the interacting molecules leads to a complex signal decay. Quantitative agreement between theory and experimental points is obtained. Our result for the dephasing time of the R(3) line is $T_2 \times p = 21.5 \pm 1$ ps bar (self-broadening).

We have investigated various vibration-rotation transitions. Results for the foreign-gas broadening of the R branch of HCl: Ar (J=0 to 8) are shown in Fig. 4. The experimental values of $1/T_2$ (full points) refer to an argon pressure of 1 bar and vanishing concentration of HCl ($p_{HCl} \rightarrow 0$). The corresponding Lorentzian linewidth scale is also shown. The open rectangles in the Fig. indicate the scatter of reported spectroscopic linewidth data [6]. The solid curve represents the theoretical results of Ref. 7 considering only rotational relaxation mechanisms. The

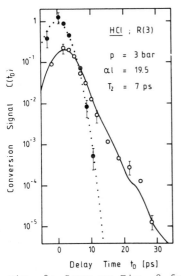

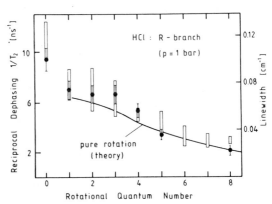

Fig. 4: Foreign gas broadening of HCl : Ar: the reciprocal dephasing time $1/T_2$ (normalized to p=1 bar) vs. rotational quantum number of the vibration-rotation transitions (full points); corresponding linewidth data (FWHH).

Fig. 3: Same as Fig. 2 for neat HCl at 3 bar and large absorption ($\alpha\ell = 19.5$). The drastic change of the pulse wing allows the determination of $T_2 = 7 \pm 0.6$ ps.

good agreement with the time-resolved data supports the conclusion that vibrational relaxation gives only negligible contribution to the dephasing times. Comparison with picosecond CARS
data on HCl suggests that apart from rotational population decay pure rotational dephasing gives a significant contribution
to the observed R(J) time constants [2].

The time resolution supplied by our tunable infrared pulses
also allows investigations on the sub-picosecond time scale. As
an example the symmetric CH_3-stretching mode of neat CH_3I at
room temperature is studied in Fig. 5. For a cell length of
100 μm, a value of $\alpha \ell \approx 6$ is adjusted. This corresponds to the
peak transmission 2×10^{-3} of the absorption band under stationary
condition. In transient IR spectroscopy, however, the maximum
transmission observed with pulses of 2.1 ps is considerably
larger; a value of approximately 0.07 is shown in the Fig. Most
important, due to the transient character of the interaction,
the slope of the transmitted pulse (open circles, solid curve)
decays with a time constant of 0.5 ps notably more slowly than
the input pulse (full points, dotted line, time resolution 0.3
ps). The result of the dephasing time is $T_2 = 1.0 \pm 0.1$ ps in
excellent agreement with linewidth data. Since the vibrational
dephasing time has been determined to be $T_2(vib) = 2.0 \pm 0.2$ ps
in an independent experiment [8], a significant contribution to
the measured dephasing time originates from rotational dynamics
in liquid CH_3I.

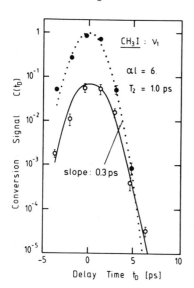

Fig. 5: Time-resolved IR spectroscopy of liquid CH_3I at 295
K; the ν_1 mode at 2948 cm^{-1} is
studied; input pulse (full
points), transmitted pulse
(open circles) and theoretical
curves; coherent pulse reshaping on the sub-picosecond time
scale yields $T_2 = 1.0 \pm 0.1$ ps.

In summary we emphasize that we have applied low-intensity
picosecond pulses for a novel time-resolved infrared spectroscopy. The coherent propagation effect investigated here has a
linear intensity dependence which facilitates quantitative applications. The low intensity level also avoids competing nonlinear effects. In comparison with Raman techniques the different selection rules of our infrared spectroscopy should be

noted. In this way information on additional molecular transitions becomes experimentally accessible. The dynamics of a vibration-rotation transition in the gas phase has been studied on the picosecond time scale, verifying for the first time the exponential decay times of these systems predicted theoretically. Our technique is also well applicable to condensed matter where coherent pulse reshaping has been observed in sub-picosecond time domain.

References:

1. A. Laubereau, L. Greiter, and W. Kaiser: Appl. Phys. Lett. 25, 87 (1974);
   A. Fendt, W. Kranitzky, A. Laubereau, and W. Kaiser: Optics Comm. 28, 142 (1979);
   T. Elsaesser, A. Seilmeier, W. Kaiser, P. Koidl, and G. Brandt: Appl. Phys. Lett. 44, 383 (1984).
2. H.-J. Hartmann and A. Laubereau: J. Chem. Phys., July 1984.
3. H.-J. Hartmann and A. Laubereau: Optics Comm. 47, 117 (1983).
4. H.-J. Hartmann, K. Bratengeier, and A. Laubereau: Chem. Phys. Lett., in print.
5. G. Herzberg: Molecular Spectra and Molecular Structure, Vol. I (van Nostrand, Princeton 1950).
6. J. Wormhoudt, J.I. Steinfeld, and I. Oppenheim: J. Chem. Phys. 66, 3121 (1977); see literature quoted in Ref. 7.
7. E.W. Smith and M. Giraud: J. Chem. Phys. 66, 1762 (1977).
8. H.R. Telle and A. Laubereau: Chem. Phys. Lett. 94, 476 (1983).

Quantum Fluctuations in Picosecond Transient Stimulated Raman Scattering

N.Fabricius, K.Nattermann, and D.von der Linde

Universität-GHS-Essen, Fachbereich Physik,
D-4300 Essen, Fed. Rep. of Germany

1. Introduction

When an intense light beam interacts with a Raman-active medium, stimulated Raman scattering[1] (SRS) gives rise to optical gain at a Stokes shifted frequency $\omega_S = \omega_L - \omega_V$, corresponding to the difference between the frequency of the pump wave, ω_L, and the frequency of some Raman-active material excitation, ω_V. For example, an input wave at frequency ω_S entering the Raman medium at the front face (z=0) of a pencil-shaped pumped volume is amplified such that at the exit face (z = L) the intensity of the Stokes wave is $I_S(L) = I_S(0) \ e^G$, where e^G is the Stokes gain.

Amplification of Stokes light can be very well described by a semiclassical theory in which the light waves are treated as classical fields, whereas the response of the material is accounted for by taking the quantum expectation value of the physical quantities of interest. Many aspects of SRS can be quantitatively explained with the use of this semiclassical theory. However, in the semiclassical picture, no Stokes light is generated by the stimulated Raman process unless an input wave is supplied. On the other hand, it is clear that such an input is always provided by ordinary spontaneous Stokes Raman scattering. A proper description of the initiation of the stimulated process is obviously beyond the scope of the semiclassical theory.

Raymer and Mostowski[2] have recently presented a detailed quantum theoretical treatment of the spontaneous initiation of SRS. An interesting new aspect of their work is the prediction that in pulsed SRS large, macroscopic fluctuations of the energy of the Stokes pulses should be observed. For highly transient conditions these fluctuations were shown to result from the fundamental quantum uncertainties of the initial state of the system [3].

In this report we present measurements of the probability distribution of the Stokes pulse energy in highly transient SRS. After a brief outline of the theoretical background closely following the work of Raymer et al. we describe the experiment in which picosecond excitation pulses are used to observe transient Stokes pulse generation in molecular hydrogen gas at high pressures[4].

2. Theoretical Background

In the semiclassical approach the stimulated Raman process is usually described by a set of coupled wave equations[5]:

$$\left(\frac{\partial}{\partial z} + \frac{1}{v} \frac{\partial}{\partial t}\right) E_S = -i \ \kappa_2 \ Q^* \ E_L$$

$$\left(\frac{\partial}{\partial t} + \frac{1}{T_2}\right) Q^* = i \kappa_1 E_S E_L^*.$$

E_S and E_L represent the classical complex wave amplitudes of the Stokes wave and the laser pump wave, respectively, both propagating in the z direction with approximately the same group velocity v (E_L is assumed to be constant). Similarly, Q is the classical wave amplitude describing the material excitation, typically a molecular vibration or an optical phonon mode. T_2 is the dephasing time which accounts for the damping of Q, κ_1 and κ_2 are coupling constants related to the stationary Stokes gain e^{G_S}, i. e., $G_S = 2\kappa_1\kappa_2 T_2 |E_L|^2$.

In the quantum description the classical wave amplitudes are replaced by Heisenberg field operators, e.g., following the notation of Raymer et al. [2]:

$$E_S, E_S^* \longrightarrow \hat{E}_S^{(-)}, \hat{E}_S^{(+)}$$
$$Q, Q^* \longrightarrow \hat{Q}, \hat{Q}^+$$

E_L can still be regarded as a classical amplitude, because the pump wave is very strong and constant.

Here, we are interested in a transient situation in which the duration t_p of the pump pulses is much shorter than the dephasing time T_2, $t_p \ll T_2$. In this case the Stokes field operator describing the Stokes wave at the exit face at z=L is given by

$$\hat{E}_S^{(-)}(L,\tau) = -i\kappa_2 E_L(\tau) \int_0^L dz\ \hat{Q}^+(z,0)\ I_0((4\kappa_1\kappa_2(L-z)\int_0^\tau |E_L(\tau')|^2 d\tau')^{1/2})$$

where $\tau = t - z/v$, and I_0 is the modified Bessel function of order zero. An additional term proportional to $\hat{E}_S^{(-)}(0,\tau)$ has been neglected, because we assume that no input wave is present. This operator solution is completely analogous to the well-known solutions of the classical coupled wave equations. Note that the Stokes field is proportional to the field operator $\hat{Q}^+(z,0)$, describing the <u>initial state</u> of the material, before the laser pulse has been turned on.

The quantity measured experimentally is the Stokes pulse energy, which, in the quantum picture, is given by the expectation value of the operator

$$\hat{W}_S = \int d\tau\ \hat{E}_S^{(-)}(L,\tau)\ \hat{E}_S^{(+)}(L,\tau).$$

Using the solutions for $\hat{E}_S^{(-)}$ and $\hat{E}_S^{(+)}$ the Stokes energy can be written

$$\langle\hat{W}_S\rangle \sim \int_0^L \int_0^L dz'\ dz''\ K(z',z'') \langle\hat{Q}^+(z',0)\hat{Q}(z'',0)\rangle,$$

where $K(z',z'')$ is essentially a convolution of the Bessel functions. The important term in the expression for $\langle\hat{W}_S\rangle$ is the expectation value of the operator product for $\tau = 0$,

$$\gamma = \langle\hat{Q}^+(z',0)\ \hat{Q}(z'',0)\rangle.$$

Returning to a classical physics interpretation for a moment, one sees that γ can be made arbitrarily small, i.e., γ —> 0 for t —> 0, or, in other words, there is no Stokes output in a classical system initially at rest. On the other hand, according to quantum theory, γ has a well-defined, finite value even if the system is in the ground state (t=0). This is a manifestation of the zero point motion of the quantum system. From the expression for the Stokes energy this zero point motion can thus be identified as the basic trigger mechanism responsible for the generation of Stokes light in the absence of an input. Yet another aspect of the quantum theory of SRS -which has been pointed out only quite recently[3]- is the fact that the Stokes energy W_S should exhibit characteristic fluctuations around the average value $<\hat{W}_S>$. For transient SRS the probability distribution of the Stokes energy has been predicted to be an exponential function [3]:

$$P(W_S) = \frac{1}{<\hat{W}_S>} \exp(-W_S/<\hat{W}_S>) .$$

Thus large, macroscopic energy fluctuations with a standard deviation $\Delta W_S = <\hat{W}_S>$ are expected, as a result of the quantum nature of the initiation of SRS. In the transient case these fluctuations are caused by fundamental quantum uncertainties associated with the Heisenberg operator $\hat{Q}(z,0)$ describing the initial state of the system.

3. Experimental

In our experiments we used the $Q_{01}(1)$ vibrational transition of molecular hydrogen with a frequency corresponding to 4155 cm^{-1}. For sufficiently high gas pressures the Raman line width is determined by homogeneous pressure broadening. Our measurements were done with gas pressures ranging from 12 bars to 25 bars, corresponding to a variation of T_2 from 526 ps to 253 ps. With excitation pulses of t_p=30 ps we are thus safely in the transient regime, $t_p \ll T_2$. The excitation pulses were generated by a Nd:YAG laser system consisting of an actively and passively mode-locked oscillator, a single pulse selector, and two amplifier stages. The amplified single pulses were frequency doubled using a KD*P crystal. Streak-camera measurements showed that the pulses were approximately Gaussian with a half-width of 30 ps.

The second harmonic beam was collimated to a diameter of 175 μm (half-width of the Gaussian beam) and passed through a 16-cm-long pressure cell filled with hydrogen. The Stokes light at the exit of the cell was separated from the pump beam by a dispersing optical prism and suitable optical filters. The energy of the Stokes pulses was measured with the use of a calibrated photomultiplier tube followed by suitable signal digitizing and processing equipment. Care was exercised to ensure that possible distortions of the Stokes signal distribution due to photomultiplier statistics and the finite resolution of the digitizers were negligible.

In the experiment we simultaneously measured the energy of each excitation pulse and of the corresponding Stokes pulse. For the determination of the Stokes energy distribution belonging to a certain excitation energy we accept all events within ± 1 % of the average excitation energy. As a safeguard against possible fluctuations of the pulse width or changes of the spatial energy distribution of the excitation beam -which would escape detection if only the total energy was measured- we also monitor the unsaturated conversion efficiency in a separate second harmonic generator. Events were rejected if the measured conversion efficiency deviated by more than 20 % from the average value.

4. Results and Discussion

To avoid depletion of the pump and competition with other nonlinear optical processes we work with pump-to-Stokes conversion efficiencies of only a few times 10^{-10}. Nevertheless, the high gain limit of SRS holds, with a transient gain coefficient $G_T = 2(2G_S t_p/T_2)^{1/2}$ of 20 to 30. Thus the Stokes pulse energies are well above the level of spontaneous scattering.

The full circles in Fig. 1 represent the measured average Stokes energy versus pump energy for different gas pressures. The solid curves are calculated using the theory of transient SRS in a one-dimensional approximation, which is justified because the experimental geometry corresponds to a Fresnel number of approximately 0.4. We fitted curve (a) in Fig. 1 by treating the effective area of the pump beam and the absolute Stokes energy scale as adjustable parameters, which were then used to calculate the Stokes energy for the other gas pressures. From the excellent agreement with experimental data we conclude that the transient theory of SRS provides a sound basis for the study of the statistics of the Stokes energy.

Each data point in Fig. 1 represents the average value of the Stokes energies of a few thousand individual events. Four examples of measured distributions are depicted in Fig.2 as histograms. The height of the bars corresponds to the number of Stokes pulses whose energy falls in the interval given by the width of the bars (15 fJ), divided by the total number of events. The shaded areas at the tops represent the statistical error.

We find that the most frequent events correspond to Stokes energies near zero. The probability of higher Stokes energies falls off exponentially, as shown by the straight line connecting the tops of the bars. We have thus established that the energy distribution of the Stokes pulses is indeed given by an exponential probability law, in agreement with the theory

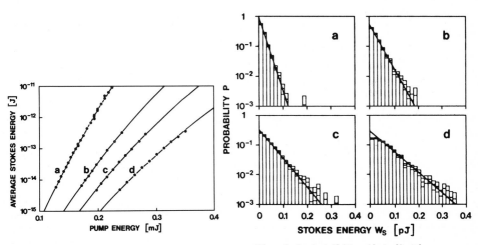

Fig.1 Average Stokes energy versus the energy of the excitation pulses. The gas pressures and the dephasing times are:
a) 25 bars-253 ps; b) 18 bars-353 ps;
c 15 bars-423 ps; d) 12 bars-526 ps.

Fig. 2 Probability distributions of the Stokes pulse energy at 18 bars for different excitation energy.
a) 187 μJ; b) 196 μJ;
c) 204 μJ; d) 214 μJ.

which predicts that such macroscopic energy fluctuations should be observed in transient SRS as a result of the fundamental quantum fluctuations of the system which are responsible for the initiation of the stimulated process. The remarkable conclusion is that even in a hypothetical situation in which the excitation pulses are absolutely reproducible the Stokes pulses exhibit large energy fluctuations with a standard deviation ΔW_S equal to the average value $\langle \hat{W}_S \rangle$, i.e., 100 % random fluctuations.

It should be emphasized that these large fluctuations are characteristic of transient SRS in a single mode situation with a Fresnel number of the order of unity. For multimode conditions in a geometry corresponding to large Fresnel numbers the fluctuations associated with different spatial modes are statistically independent, and from the central limit theorem the probability distribution of the total Stokes energy is expected to be a much narrower Gaussian distribution. We note also that in SRS under steady-state-like conditions, $t_p \gg T_2$, the memory of the initial state of the system is wiped out by the damping process. In this case collision-induced fluctuations associated with the damping mechanism determine the fluctuaions of the Stokes energy, which can be much less than those in the transient case[6].

References

[1]   N. Bloembergen,
      Am. J. Phys. 35, 989 (1967)

[2]   M. G. Raymer, and J. Mostowski,
      Phys. Rev. A24, 1980 (1981)

[3]   M. G. Raymer, K. Rzazewski, and J. Mostowski,
      Opt. Letters 7, 71 (1982)

[4]   N. Fabricius, K. Nattermann, and D. von der Linde,
      Phys. Rev. Lett. 52, 113 (1984)

[5]   R. L. Carman, F. Shimizu, C. S. Wang, and N. Bloembergen,
      Phys. Rev. A2, 60 (1970)

[6]   K. Rzazewski, M. Lewenstein, and M. G. Raymer,
      Opt. Comm. 43, 451 (1982)

Transient Coherent Raman Spectroscopy: Two Novel Ways of Line Narrowing

W. Zinth, M.C. Nuss, and W. Kaiser

Physik Department der Technischen Universität München
D-8000 München, Fed. Rep. of Germany

Recently, a new Raman technique was introduced, which allows to improve the spectral resolution beyond the limitations of spontaneous Raman spectroscopy /1/. Here we present the first data of the short excitation and prolonged interrogation technique using a continuously tunable excitation from a cw mode-locked dye laser system. In addition, we show a related but different line narrowing method, which uses prolonged excitation and delayed probing /2/. The smooth tunability in both techniques allows the quantitative determination of frequency positions and amplitudes of individual transitions. In congested spectral regions new information, not available from spontaneous Raman data, is obtained.

The principle of the two line-narrowing techniques is the following. Spectral resolution is determined by the inverse of the observation time τ_{obs}. In standard steady-state experiments the observation time equals the intrinsic time constants such as energy and/or phase-relaxation times. Transient coherent techniques, however, allow to increase the observation time under special experimental conditions.

1. Short Excitation and Prolonged Interrogation Spectroscopy (SEPI)

During the transient coherent Raman excitation molecules are driven at the excitation frequency $\nu_E = \nu_1 - \nu_2$ by two light pulses of frequency ν_1 and ν_2. A Raman transition of frequency ν_V close to ν_E is coherently excited. At the end of the exciting process the coherent excitation oscillates at ν_V and decays with the dephasing time T_2. A third delayed probe pulse (frequency ν_3) interacts with the coherently vibrating molecules and generates an anti-Stokes spectrum. The crucial point is the following. A narrow anti-Stokes spectrum is observed at late delay times, provided a long Gaussian shaped probing pulse is used. For pulse durations $t_p > 1.4\ T_2$ the observed SEPI spectrum shows narrower bands than the spontaneous Raman spectrum.

The experiments are performed using two synchronously pumped dye lasers (ν_1, ν_2) excited by a mode-locked Ar^+ ion laser. The pulses from the first laser have a fixed frequency ν_1 and a duration of $t_p \approx 12$ ps. They are used in the excitation and probing process, i.e. $\nu_1 = \nu_3$. The second exciting pulse at the tunable frequency ν_2 is shorter by a factor of two. The coherent spectrum is detected in a geometry for anti-Stokes phase-matching.

The tunable SEPI technique is applied to a mixture of pyridine and methanol (50% by volume), where strong hydrogen bonds are present. In the spontaneous Raman spectrum the hydrogen bonded complexes show a broad and featureless band around 1000 cm^{-1} (see Fig. 1a), a frequency close to the skeletal vibration of pyridine at 991 cm^{-1}. In Fig. 1b we show a set of SEPI spectra taken with excitation frequencies between 995 and 1002 cm^{-1}. In clear contrast to the broad spontaneous Raman band several narrower lines are resolved.

From SEPI spectra measured at different pyridine:methanol compositions we obtain the following results: (i) The broad spontaneous band around 1000 cm^{-1} is made up by three transitions at 997.3, 1000, and 1001 cm^{-1}. (ii) These frequency positions do not depend on concentration. (iii) The amplitudes of the components vary with concentration. In this way we may assign the different transitions to specific hydrogen bonded complexes.

2. Prolonged Transient Excitation Spectroscopy

Two long laser pulses ($t_p > T_2$) at frequencies ν_1 and ν_2 with amplitudes $E_1(t)$, $E_2(t)$ excite the molecular vibration at the difference frequency $\nu_E = \nu_1 - \nu_2$. At the end of the driving process a third pulse probes the relaxing material excitation by

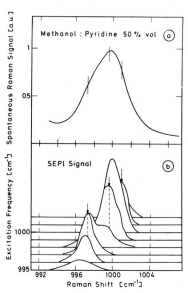

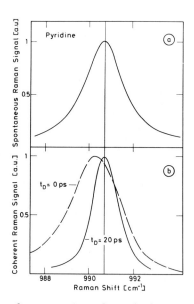

Fig. 1 Example of technique one: Short excitation and prolonged probing. Mixture of pyridine:methanol (50% by volume. (a) Spontaneous Raman spectrum of the hydrogen bonded complexes. (b) Transient coherent Raman spectra for different frequencies of excitation $\nu_E = \nu_1 - \nu_2$.

Fig. 2 Example of technique two: Prolonged excitation and delayed probing. Pure pyridine. (a) Spontaneous Raman spectrum. (b) The coherent anti-Stokes Raman signal is plotted versus the tunable frequency at $\nu_E = \nu_1 - \nu_2$. Without delay: Standard CARS (broken curve. With delayed probing (solid lin

coherent Raman scattering. The energy of the coherently produced anti-Stokes light around $2\nu_1 - \nu_2$ is measured as a function of the frequency difference ν_E of the exciting pulses. The salient feature of this technique is /2/: the observed transient coherent Raman band will be narrower than the spontaneous spectrum when sufficiently long exciting pulses of suitable shape are used (the shape of the driving force $E_1(t) \times E_2(t)$ should be close to Gaussian).

In the experiment we use pulses (ν_1, ν_2) of duration $t_p \simeq 10$ ps and frequency width $\Delta\bar{\nu} \simeq 1$ cm^{-1}. Part of the pulse at frequency ν_1 is shortened in a saturable absorber and used as the delayed probing pulse. The total coherent anti-Stokes signal is recorded with a broad band filter as a function of the detuning of ν_E.

Fig. 2 gives an example of the prolonged excitation technique using pure pyridine. Fig. 2a shows the spontaneous Raman band around 991 cm^{-1} (spectral resolution 0.5 cm^{-1}). A standard coherent anti-Stokes (CARS) spectrum is obtained with zero time delay between excitation and probing pulse (Fig. 2b, broken curve). The spectrum is asymmetric and broader than the spontaneous Raman line. The asymmetry and the slight shift of the peak position to smaller frequencies is due to the nonresonant susceptibility $\chi_{NR}^{(3)}$. The solid curve in Fig. 2b shows the transient coherent Raman spectrum obtained with delayed probing. Now, the asymmetry and the frequency shift have disappeared /3,4/. Of importance is the prolonged excitation process: the observed spectrum is narrower than the spontaneous Raman band. In congested spectral regions an improved resolution - similar to Fig. 1b - has recently been achieved.

References

1   W. Zinth, M.C. Nuss, W. Kaiser, Chem. Phys. Lett. 88 (1982) 257
2   W. Zinth, M.C. Nuss, W. Kaiser, Optics Commun. 44 (1983) 262
3   W. Zinth, A. Laubereau, W. Kaiser, Optics Commun. 26 (1978) 457
4   F.M. Kamga, M.G. Sceats, Optics Lett. 5 (1980) 126

Picosecond Transient Raman Spectroscopy: The Excited State Structure of Diphenylpolyenes

T.L. Gustafson, D.A. Chernoff, J.F. Palmer, and D.M. Roberts

The Standard Oil Company (Ohio), Corporate Research Center
4440 Warrensville Center Road, Warrensville Heights, OH 44128, USA

The spectroscopy of diphenylpolyenes, particularly the ordering and structure of the electronic excited states, is of interest because of the role that polyenes have in various photochemical and biological processes. These molecules have been studied vigorously for many years, both by steady-state measurements[1] and, more recently, by direct observation of the dynamics with high time resolution[2].

For the present work we have concentrated on the transient Raman technique to provide information on the excited state structure of the "diphenylpolyenes", trans-stilbene and diphenylbutadiene(DPB). The vibrational resolution of Raman spectroscopy is a powerful adjunct to fluorescence and absorption measurements. The observed vibrational frequencies of the excited state provide considerable information about its potential energy surface. This is important in considering mechanisms for energy transfer from the optical modes to the reaction coordinate. Also, the intensities in the resonance Raman spectra reflect the character of a higher electronic state.

Figure 1 is a diagram of our picosecond transient Raman spectrometer, which we have described in detail elsewhere [3,4]. To pump the sample and to probe its Raman spectrum, we used the second harmonic and fundamental frequencies, respectively, of a high repetition rate, amplified, synchronously pumped, cavity dumped dye laser[3-6]. Our choice of laser differs from that of other workers, who use low repetition rate, high peak power lasers[7]. We note that the observed Raman signal depends on both the peak and average powers of the laser. The average power of the high repetition rate laser is quite high. The moderate peak power of the laser permits tight focusing at the sample, with consequent efficient illumination of the spectrograph. We used a single stage monochromator coupled to an optical multichannel analyzer for detecting the Raman scatter. We were able to eliminate the baseline of Rayleigh scatter from the transient difference spectra by using our own OMA software for rapid sequencing of the four illumination conditions (both pump and probe incident on the sample; pump only; probe only; dark)[4].

The transient Raman spectra of trans-stilbene over the region from 170 cm^{-1} to 1800 cm^{-1} obtained at delays of -100, 0, 25, and 75 picoseconds are displayed in Fig. 2. These spectra are unsmoothed. It was necessary to use two spectrograph settings to obtain the complete spectrum; the overlap from 930 cm^{-1} to 1130 cm^{-1} indicates the reproducibility between separate settings of the spectrograph. There are 18 vibrational bands discernible. The relative intensities of all the bands follow the 70 picosecond lifetime of the excited state.

We have discussed in detail our assignments for S_1 trans-stilbene Raman bands[4]. Fifteen bands can be assigned as a_g fundamentals and three as a_u fundamentals in the C_{2h} point group. The most interesting structural information comes from the two broad, asymmetric bands at 1242 and 1567 cm^{-1}. We assign these to the

266

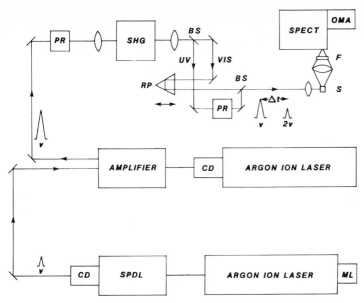

Fig. 1: Schematic diagram of the picosecond transient Raman apparatus: ML, mode locker; SPDL, synchronously pumped dye laser; CD, cavity dumper; PR, polarization rotator; SHG, second harmonic generator; BS, dichroic beam splitter; RP, retroprism; S, sample; F, filter; SPECT, spectrograph; OMA, optical multichannel analyzer.

C_o-phenyl single and C_o-C_o double bond stretch, respectively. These bands show little resolvable sub-structure even when observed at higher resolution. The nature of the asymmetry, with the higher energy band tailing to lower energy and the lower energy band tailing to higher energy, suggests a distribution of conformations in S_1 having different amounts of phenyl twist[3]. Hochstrasser and coworkers first suggested the existence of such conformers to account for excitation energy effects observed in the transient absorption spectra in solution[8]. Our results support the claim that different conformers exist, but we have not succeeded in measuring the rate of their interconversion.

The band intensities in the S_1 t-stilbene Raman spectrum permit us to infer that the high lying electronic state, S_n, which provides the resonance enhancement, has an electron distribution quite different from S_1. The strongest band at 284 cm^{-1} can be approximately assigned as a C_o-C-C bending motion. Other strong features include the C_o-phenyl single and the C_e-C_e double bond stretching bands at 1242 and 1567 cm^{-1}, respectively, the C-H deformation mode at 1181 cm^{-1}, and ring motions at 621 and 845 cm^{-1}. These last three modes involve primarily the ortho and meta positions of the phenyl rings. We conclude that electron promotion from S_1 to S_n creates a significant displacement of electron density (i.e. change in polarizability) in the olefinic portion of trans-stilbene and in the region of the ortho and meta carbon atoms on the phenyl rings.

Building on the understanding we obtain from our study of trans-stilbene we looked at the excited state structure of diphenylbutadiene(DPB). The photophysics of DPB has been studied extensively, but some ambiguity remains as to the ordering

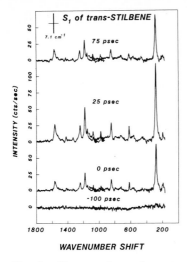

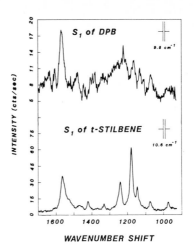

Fig. 2: Picosecond transient resonance Raman spectra of t-stilbene in hexane at delays of -100, 0, 25, and 75 picoseconds: 296.3 nm pump; 592.7 nm probe.

Fig. 3: Picosecond transient resonance Raman spectra of DPB in THF (20 picosecond delay; 308.8 nm pump; 617.7 nm probe) and t-stilbene in hexane (25 picosecond delay; 296.3 nm pump; 592.7 nm probe).

of the electronic excited states in fluid solution at room temperature[9]. Early two photon absorption spectra suggest that the lowest excited single state is ^1A_g[10], but a recent study of radiative and nonradiative decay rates of DPB suggests that the lowest excited state is "stilbene like", ^1B_u[9].

We compare the excited state spectra of DPB and trans-stilbene over the region from 930 cm^{-1} to 1800 cm^{-1} in Fig. 3. We note that the S/N of the DPB spectrum is reduced owing to a low excited state absorption at our probe wavelength 618 nm; the maximum of the excited state absorption is at ~670nm [11]. There are 8 vibrational bands that are observable; the two most prominent bands are at 1567 and 1225 cm^{-1}. These two bands compare with the bands at 1567 and 1242 cm^{-1} in t-stilbene. We assign the DPB bands as the C_o-C_o double bond stretch, and C_o-phenyl and C_o-C_o single bond stretch, accordingly. We note that the band at 1567 cm^{-1} in DPB does not appear to be asymmetric, whereas the comparable band in t-stilbene is. And the 1225 cm^{-1} band in DPB is broader than the 1242 cm^{-1} band in t-stilbene, suggesting a broader range of conformations that affect the single bond stretch in DPB. This may be a consequence of the fact that the center of inversion in DPB is a single bond, but in t-stilbene it is a double bond.

The relative intensities of the bands vary between the S_1 spectrum of DPB and t-stilbene, but there appears to be a one-to-one correspondence in many of the peak positions. The intensities are derived from a higher electronic state and these states may be quite different in DPB and t-stilbene. Since the peak positions are indicative of the S_1 structure, the similarity between the DPB and t-stilbene spectra suggests that the excited state structure of DPB in fluid solution is very similar to that of t-stilbene. We will report these results elsewhere in greater detail [12].

Acknowledgments

We wish to thank R. L. Swofford, K.-J. Choi and G. R. Fleming for many helpful discussions.

REFERENCES

1. J. Saltiel, J. D'Agostino, E. D. Megarity, L. Metts, K. R. Newberger, M. Wrighton, and O. C. Zafirou, Org. Photochem. 3, 1(1971).
2. R. M. Hochstrasser, Pure Appl. Chem. 52, 2683(1980).
3. T. L. Gustafson, D. M. Roberts and D. A. Chernoff, J. Chem. Phys. 79, 1559(1983).
4. T. L. Gustafson, D. M. Roberts and D. A. Chernoff, J. Chem. Phys. 81, XXX(1984).
5. D. H. Waldeck, W. T. Lotshaw, D. B. McDonald, and G. R. Fleming, Chem. Phys. Lett. 88, 297(1982).
6. T. L. Gustafson and D. M. Roberts, Opt. Commun. 43, 141(1982).
7. H. Graener and A. Laubereau, Chem. Phys. Lett., 102, 100(1983) and references therein.
8. F. E. Doany, B. I. Greene and R. M. Hochstrasser, Chem. Phys. Lett., 75, 206(1980).
9. S. P. Velsko and G. R. Fleming, J. Chem. Phys. 76, 3553(1982) and references therein.
10. R. L. Swofford and W. M. McClain, J. Chem. Phys., 59, 5740(1973).
11. R. A. Goldbeck, A. J. Twarowski, E. L. Russell, J. K. Rice, R.R. Birge, E. Switkes, and D. S. Kliger, J. Chem. Phys., 77, 3319(1982).
12. T. L. Gustafson, J. F. Palmer and D. M. Roberts, unpublished.

Time-Resolved Nonlinear Spectroscopy of Vibrational Overtones and Two-Phonon States

G.M. Gale, M.L. Geirnaert, P. Guyot-Sionnest, and C. Flytzanis

Laboratoire d'Optique Quantique du C.N.R.S., Ecole Polytechnique
F-91128 Palaiseau Cédex, France

Vibrational overtones and multiphonon states are related to important processes in dense molecular systems. Recently, we have proposed [1] and for the first time demonstrated [2] the use of nonlinear optical techniques to study these states. In particular, it was shown there that coherent excitation and probing of the time evolution of overtones and two-phonon states is feasible.

We anticipate that this technique and similar ones open the path to a direct study of the dynamics of cooperative excitation of pairs of molecules, two-phonon states (free or bound) and large wave vector phonons. Furthermore they allow one to create such states in an off-equilibrium situation. Here we summarize the results on CS_2 [2] and present some preliminary ones on CO_2 using the CAHORS technique (Coherent AntiStokes Higher Order Raman Scattering).

We recall that the spectrum in the overtone region of a given mode arises [3,4] from the competition between the anharmonicity within a single molecule and the simultaneous excitation of the intramolecular vibration on pairs of interacting molecules. In molecular crystals the anharmonic potential within a molecule may overbalance the delocalizing effect of the intermolecular interaction and give rise as well as to the free two-phonon quasicontinuum to a bound two-phonon state [5,7] with drastically different spectral features. These states may be coherently driven with high coupling efficiency by laser excitation of appropriate frequency and their subsequent evolution and decay process observed.

These techniques contain two stages, a) an excitation stage where the molecules are selectively prepared in the desired state by a second-order Raman process and b) a probing stage where a second weak delayed pulse monitors the instantaneous state of the excited system. In the first demonstration of this technique we studied the first overtone and two phonon states of a Raman-inactive mode: the ν_2 mode in CS_2 and the coupled $\{2\nu_2,\nu_1\}$ states in CO_2 [5].

The effective Hamiltonian for the coupling of the vibrational motion with the light is [1,2].

$$h''_{\sigma\sigma} = -\frac{1}{4} \alpha^{(2)}_{\sigma\sigma} Q_{\sigma\sigma} E^2$$

where $E = E_1 \cos\omega_1 t + E_2 \cos\omega_2 t$, $\omega_1 - \omega_2 \approx \Omega_0$ the overtone frequency, $\alpha^{(2)}_{\sigma\sigma} = \partial^2\alpha /\partial q^2_\sigma$ the second-order Raman tensor of mode σ and $Q_{\sigma\sigma} = q_\sigma q_\sigma$ the overtone amplitude. As a consequence of the interaction (1) the expectation value $<Q_{\sigma\sigma}>$ grows resonantly and modulates the polarisation set up by E. In the probing stage, delayed by t_D with respect to the excitation stage, the electromagnetic field of frequency ω_p is scattered off this modulation and shifted by the overtone frequency, $\omega_s = \omega_p - \Omega_0$ and

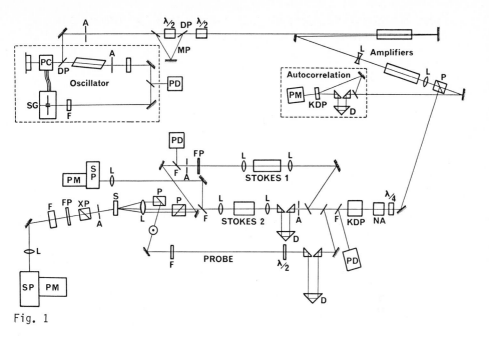

Fig. 1

$\omega_A = \omega_p + \Omega_o$, the Stokes and anti-Stokes frequencies respectively. The coherent scattering amplitude $S(t_D)$ at the anti-Stokes frequency gives then
the desired information on the dynamical evolution of the coherence of the
overtone.

The experimental system used for the CO_2 observations is shown in Fig. 1
(for CS_2 see [2]). The system is driven by a single 5 ps average width pulse
produced by a cavity dumped passively mode-locked Nd^{3+}/ phosphate glass oscillator and subsequently amplified to 2 GW power. This pulse,after traversing a nonlinear absorber cell with a small signal transmission of 10^{-3}, is
doubled in frequency to 527 nm. This green beam is then split into three parts,
the first of which is rotated in polarisation by 90° and serves as a probe
of the vibrational overtone coherence. The two other parts pump two Raman
generators ; Stokes 1 and Stokes 2, employing deuterated methanol and water,
respectively, to provide a suitable frequency difference for excitation of
the relevant CO_2 vibrations in a Raman amplifier set up $\omega_v = \omega_{s1} - \omega_{s2}$. This
method has the advantage of pushing excitation related noise into the red
and allowing additional frequency selective noise rejection.

The experimental results for the bound two-phonon state associated with
the v_2 internal vibration (at 801 cm^{-1}) in crystalline CS_2 (T = 160 K) are
shown in Fig. 2 where the coherent anti-Stokes signal is plotted on a logarithmic scale vs probe delay t_D. Exponential decay is observed but with a
dramatic increase of T_2 to 14 ps in the crystalline phase over that obtained
[2] in the liquid for the $2v_2$ overtone (T_2 = 1.8 ps).

In CO_2 because of the interaction of the $2v_2$ overtone with the neighboring v_1 mode the two-phonon spectrum contains two mixed states of frequencies v^- and v^+ at 1276 cm^{-1} and 1383 cm^{-1} respectively. The CO_2 crystals,
which were grown in a specially adapted cryostat, were of sufficiently high

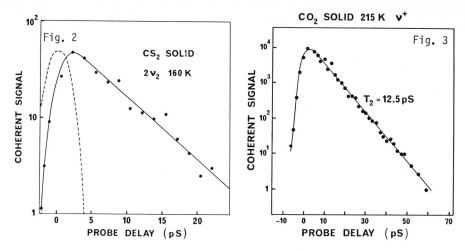

Fig. 2 CS₂ SOLID 2ν₂ 160 K

CO₂ SOLID 215 K ν⁺

Fig. 3 $T_2 = 12.5\,pS$

quality to allow us to cool down to at least 77 K without severe optical degradation of the sample.

Fig. 3 shows the experimental results for the v^+ line of CO_2 at 215 K in the solid. We note the dynamic measurement range of 10^4 and the observed exponential decay yielding a $T_2 = 12.5$ ps. Exponential decay is maintained for both components over the whole temperature range considered from 77 K to 250 K where CO_2 is liquid.

Fig. 4 depicts the temperature dependence of T_2 for both components up to and across the solid/liquid phase transition, where we plot the decay rate $2/T_2$ as a function of T°K. Both lines exhibit a strong temperature variation of more than a factor of 2 with v^- varying more rapidly. Another striking feature to remark is that the v^+ and v^- curves appear to be parallel.

The observed behavior can be understood with the assumption [1,2,8] that the primary mechanism of loss of coherence is the coupling and overlap of the local bound two-phonon state with the delocalized free two-phonon states into which the former eventually disintegrates. It can be shown [1,8] under certain conditions that the loss of coherence evolves exponentially

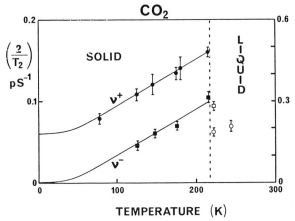

CO_2

$\left(\dfrac{2}{T_2}\right)$ pS^{-1}

SOLID

LIQUID

v^+

v^-

TEMPERATURE (K)

Fig. 4

in time with a time $T_2 \sim 1/\nu_0$, where ν_0 is the density of the free two-phonon states calculated where the overlap with the bound state is maximal ; consequently T_2 will be strongly sensitive to structural phase transitions and show strong temperature dependence there. This exponential decay is confirmed with CS_2 and CO_2 although in the latter case the occurrence of Fermi resonance [9] must be explicitly taken into account. In a more detailed treatment the polariton character of mode ν_2 which is infrared active, must also be considered.

In conclusion we have shown that improved picosecond coherent techniques can be extended to the study of the dynamics of vibrational overtones in liquids and two-phonon states in solids. Many variations of the above techniques and combination with the one-vibron time resolved studies [10] can be conceived to obtain important information about the pathways of the coherence and energy transfer of large amplitude vibrational motion in condensed matter. They can be used [8] as a starting point for the observation of parametric instabilities, two-phonon echoe and other nonlinear processes.

References

1 - C. Flytzanis, G. Gale and M.L. Geirnaert in Picosecond Spectroscopy on Chemistry, edited by K. Eisenthal (Plenum, N.Y. 1983)

2 - M.L. Geirnaert, G.M. Gale and C. Flytzanis, Phys. Rev. Lett. 52, 815 (1984)

3 - A. Ron and D.F. Hornig, J. Chem. Phys. 39, 1129 (1963)

4 - J. Chesnoy, D. Ricard and C. Flytzanis, Chem. Phys. Lett. 42, 337 (1979)

5 - F. Bogani, J. Phys. C11, 1283, 1297 (1978)

6 - M. H. Cohen and J. Runalds, Phys. Rev. Lett. 23, 1378 (1969)

7 - J.C. Kimball, C.Y. Fong and Y.R. Shen, Phys. Rev. B23, 4946 (1981)

8 - to be published

9 - F. Bogani and P.R. Salin, to be published (the authors are indebted for receiving this preprint)

10 - A. Laubereau and W. Kaiser, Rev. Mod. Phys. 50, 608 (1978)

Direct Picosecond Determination of the Character of Vibrational Line Broadening in Liquids

G.M. Gale, P. Guyot-Sionnest, and W.Q. Zheng

Laboratoire d'Optique Quantique du C.N.R.S., Ecole Polytechnique
F-91128 Palaiseau Cédex, France

Vibrational transitions in liquids and solids may be broadened by many different physical processes. The diverse contributions to a given band cannot always be unambiguously separated in conventional spectroscopy, especially if the band is perturbed by neighbouring overlapping lines such as vibrational overtones and hot bands, and it is often particularly difficult to determine if the band contour is Gaussian, Lorentzian or of intermediate shape.

We show in this letter that very high dynamic range (10^9), short time resolution (<200 fs), coherent excitation and probe techniques allow a definitive assignment of the character of the line-broadening mechanism by the direct observation of the evolution of the coherent vibrational amplitude over very many decay times. In a primary application, the coherent amplitude of the ν_1 mode of acetonitrile at 2943 cm^{-1} is observed to relax exponentially over 20 decay times and is thus demonstrated to be very predominantly homogeneous with an associated phase relaxation time $T_2 = 1.65 \pm 0.07$ ps in excellent accord with spontaneous Raman scattering data[1] ($\Delta\tilde{\nu}_{1/2} = 6.5$ cm^{-1}, $T_2 = 1.63$ ps). This result is relevant to recent proposals and attempts to apply special picosecond techniques (selective k-matching[2,3], high laser depletion[4] to measure directly the homogeneous component of an inhomogeneously broadened Raman line.

The experimental system shown in figure 1 is driven by a single 5 ps average width pulse produced by a cavity dumped passively mode-locked Nd^{3+}/phospate glass oscillator. To avoid possible self-focussing problems in the amplifier chain this single pulse is split into two pulses with a time separation of 500 ps and crossed polarisations, which are subsequently amplified to 1 GW and 2 GW powers respectively. The 1 GW pulse is at present used to monitor the autocorrelation function of the amplified 1.054 µ beam. The 2 GW pulse, after traversing a non-linear absorber cell with a small signal transmission of 10^{-3}, is doubled in frequency to 527 nm. This green beam is then split into three parts, the first of which is rotated in polarisation by 90° and serves as a probe of the vibrational coherence ; the second drives a stimulated Raman generator using ethanol to provide a Stokes pulse, displaced by 2930 ± 10 cm^{-1} from the laser frequency, for excitation of the vibration in a Raman amplifier set-up and the remaining energy at 527 nm is used as a pump pulse for Raman excitation. These three pulses are directed into the 2 mm long sample cell containing CH$_3$CN, with a non-collinear k matching geometr. The high spatial quality of the beams produced by this system allow very severe angular and polarisation selection of the collimated off-axis coherent anti-Stokes signal and consequent excellent noise rejection.

The experimental results on the ν_1 mode of acetonitrile at 20°C are shown in Figure 2 (full circles) where each point represents an average of at least 20 acceptable laser shots. We plot the coherent anti-Stokes signal on a lo-

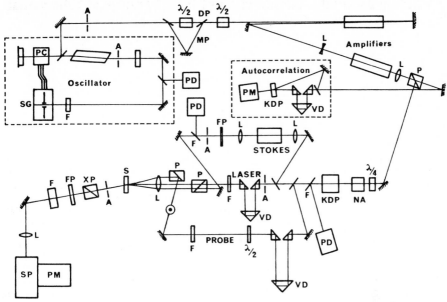

Fig. 1 - Experimental set-up : PC Pockel's cell, SG spark gap, F filter, A aperture, λ/2 half-wave plate, MP multiplexer, DP dielectric polariser, L lens, P polariser, VD variable delay, KDP frequency doubler, PM photomultiplier, FP Fabry-Pérot, λ/4 quarter wave plate, NA non-linear absorber, PD photocell, S sample cell, XP crossed polariser, SP monochromator.

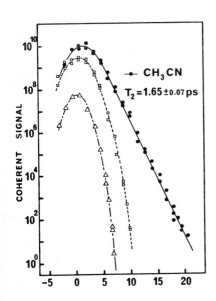

Fig. 2 - Coherent anti-Stokes signals plotted on a logarithmic scale vs probe delay in ps for the ν_1 mode of CH_3CN (full circles), for CH_3CH_2OH (system response function open squares) and for H_2O (open triangles). The curves (full, dashed and dot-dashed lines) are calculated.

garithmic scale as a function of probe delay (t_D) in ps. It is noteworthy that the experimental points extend over a dynamic measurement range of 10^9. The point at $t_D > 5$ ps follow an exponential decay curve (full line) over eight orders of magnitude yielding a decay time $\tau = 0.82_5$ ps and a $T_2 = 2\tau$ = 1.65 ±0.07 ps in excellent agreement with spectroscopic data (T_2 = 1.63ps). The simultaneously measured system response function is shown by open squares and, as expected, decays very rapidly, giving response function signal levels more than 10^5 below the coherent response of CH_3CN at t_D = 10 ps. The final slope of the response function yields a decay time $\tau \sim 250$ fs which is the limiting time resolution of our system in this experiment. This extremely clear distinction between the coherent evolution and the system response argues powerfully in favour of the present measurement.

We can now consider the situation for a Raman line which is predominantly *inhomogeneous* as reported for the ν_1 line of acetonitrile in reference (4). In this case, the decay of the coherent vibrational amplitude is not uniformly exponential but decreases rapidly at small delays as $\exp -\Delta^2 t^2$, where Δ^2 is the second moment of the inhomogeneous spectral distribution. Even if the correlation time t_c of the inhomogeneous frequency autocorrelation function is relatively small ($t_c \approx 4 - 5$ ps) as suggested by the authors of ref (4), quasi-Gaussian decay is maintained over several orders of magnitude, due to destructive interference of the different inhomogeneous components, and only at long time ($t_D > 10$ ps) becomes exponential but with a final decay rate $\tau^{-1} \approx 2\Delta^2 t_c$ *must faster* than the experimentally observed slope. Note also in this case τ is unrelated to any dephasing time of the system under investigation.

One can attempt to reproduce the observed final decay time by choosing an unrealistically small t_c for inhomogeneous broadening. This fiction, however contains an intrinsic instability as the calculated modulation speed moves from slow to moderate ($\Delta\, t_c \approx 1$) the second moment increases, and the line collapses to a Lorentzian.

In conclusion, we have shown using high dynamic picosecond techniques with simultaneous supervision of time resolution that the ν_1 line of acetonitrile is predominantly homogeneous with a phase relaxation time T_2 = 1.65 ± 0.07 ps at 20°C in excellent accord with spectroscopic data. This definitive measurement is in clear contradiction with a recent report[4] where large laser pulse width (t_L = 6-8 ps) and small dynamic range (<10^3) severely limit the system time resolution necessary to resolve the fast decay of this homogeneous Raman line.

This high dynamic technique should be applicable to many other systems having Raman linewidths up to 20 cm^{-1}, to provide an unambiguous assignment of the nature of the line-broadening mechanism. The high sensitivity of our experimental system also means that previously inaccessible weak Raman features may now be studied directly[5].

We further conclude, in view of the negative report of ref. 5 on selective k-matching, that no convincing demonstration of a time-resolved determination of the homogeneous isotropic Raman linewidth of an inhomogeneously broadened system has yet been made and hence this field, potentially information rich, remains open.

References

1 - J. Yarwood, R. Arndt and C. Doge, Chem. Phys. 25, 387 (1977)

2 - A. Laubereau and W. Kaiser, Rev. Mod. Phys. 50, 607 (1978)

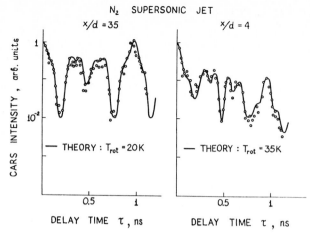

Fig. 1. The determination of the gas rotational temperature using data on the pulse response of the nitrogen jet

of the Q-band ($\lesssim$ 0,5 cm^{-1}) is much smaller than the value of $(C\tau_\rho)^{-1}$, so that the coherent excitation of molecules by pulses with a duration of $\tau_\rho \approx 30$ ps can be viewed as an impact excitation.

We have experimentally measured and theoretically calculated the pulse response, the energy of the anti-Stokes pulse $W_a(\tau)$, as a function of the delay time τ between excitation and probing. The typical beating patterns of the pulse response of the free nitrogen flow are shown in Fig. 1 for distances from the nozzle of x = 0.4 and 3.5 mm. In the same figure the solid lines represent the calculated dependence using the value shown for both T_{rot} and T_2, which give the best fit to the experimental results.

Introducing the dephasing collisional cross-section,

$$\pi\sigma_d^2 = (T_2 n\bar{v})^{-1},$$

where n is the molecular number density, $\bar{v}$ is the mean thermal velocity of molecules in the jet and T_2 is the dephasing time, one can determine the $\pi\sigma_d^2$ in any point inside the flow. We calculate n and $\bar{v}$ using the following formulae for the isentropic gas expansion:

$$n = n_o (T/T_o)^{1/(\gamma-1)} ,$$

$$\bar{v} = \sqrt{16KT/\pi m},$$

where T_o = 300K, γ = 1.4 for N_2, m is the molecular mass of N_2, and T is the translational temperature of the flow. The last value we determine using formulae from [4] (see also [5]). The values of $\pi\sigma_d^2$ calculated in such a manner are represented in Fig. 2 as a function of

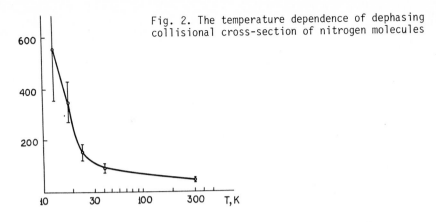

Fig. 2. The temperature dependence of dephasing collisional cross-section of nitrogen molecules

translational temperature. At room temperature, $T = 300$ K, $\pi\sigma_d^2$ is calculated from the data obtained from the cell measurement of static N_2 gas: $\pi\sigma_d^2 = 43$ $\mathring{A}^2$. This value is remarkably close to the gas-kinetic one at $T = 300$ K. The cross-section $\pi\sigma_d^2$ at $T \approx 20$K is equal to 330 $\mathring{A}^2$.

The results obtained indicate the broad feasibility of picosecond time-domain CARS techniques in the study of free gas flows. These possibilities will further increase if Fourier-limited pulses are used ($\Delta\nu = 0.6$ cm^{-1} for $\tau = 30$ ps); the sensitivity will increase too.

In our experiments we obtained close to transform-limited pulses from OPO by using the injection of external radiation from the tunable GaAs laser ($\lambda = 0.85$ μ).

The OPO consisted of two LiNbO$_3$ crystals. The first stage was collinearly pumped by the SH of the Nd:YAG laser to produce a tunable output near λ $\lambda_2 = 1.4$ μm. The second stage, pumped by the fundamental frequency of a Nd:YAG laser, was used as an amplifier. In this case, OPO generated highly reproducible pulses with a pulse length of $\tau_\rho = 20$ ps and FWHM $\Delta\nu = 1.2$ cm^{-1}; the product of $\Delta\nu \cdot \tau_\rho$ was thus 0.7.

In conclusion, the "quantum beats" of individual J-components of the Q-band of nitrogen molecules cooled in a supersonic jet are detected and numerically interpreted. The data obtained allow the rotational temperature distribution along the jet axis to be measured. The high "spectral resolution" determined by the long delay times available allows us to establish, for the first time, the substantial increase in the collisional dephasing cross-section as the temperature decreases down to 20 K.

Part V

Transient Laser Photochemistry

Picosecond Chemistry of Collisionless Molecules in Supersonic Beams

A.H. Zewail *

Arthur Amos Noyes Laboratory of Chemical Physics, California Institute of
Technology, Pasadena, CA 91125, USA

I. Introductory Remarks

The dynamics of collisionless intramolecular vibrational-energy redistribution
(IVR), bond breakage, and bond formation in large isolated molecules is currently a very important and challenging problem. A fruitful approach to this
problem involves the study, via direct measurements in the time domain, of the
decay parameters of energetically excited molecules. Using such an approach
one can determine unimolecular rate constants as well as study any quantum
mechanical coherence phenomena that may be involved in a decay process. Furthermore, in cold beams these rates and coherence effects can be studied as
a function of the *energy* and *character* of individual vibrational modes in a
molecule. The results of such studies are important to understanding the
nature of energy flow within molecules and to assessing the possibility of
vibrational mode-selective laser chemistry [1].

Over the past five years we have been examining vibrational level-selec-
tive rate processes and coherence effects using the technique of picosecond-
jet spectroscopy [2]. In this technique a picosecond laser selectively excites molecular vibronic levels and allows for time domain studies of rates
and coherence effects. The measurements are made on molecules that have
been cooled vibrationally and rotationally in a supersonic jet expansion.
Such cooling greatly reduces the spectral congestion that arises from thermal effects and allows for the single vibronic level excitation of even
very large molecules. With the picosecond-jet technique we have studied
several problems. These include:

1) Coherence and energy redistribution in isolated large molecules
(anthracene, stilbene) [3-6].
2) Quantum beats and Zeeman effect probing of radiationless processes in
pyrazine [7].
3) Intramolecular trans-cis isomerization of t-stilbene [6], 1,4-diphe-
nylbutadiene [8], and styrene [9].
4) Intramolecular proton transfer in methyl salicylate [10].
5) Chromophore-selective excitation and photochemistry of intramolecular
charge transfer formation in anthracene-$(CH_2)_3$-N,N-dimethyl aniline [11].
6) Bond breakage and IVR of partially solvated molecules in the jet [12].
7) Bond rotation dynamics and IVR in bianthryl [13].

Here, we review the techniques used in this laboratory. The work published
in the above-mentioned areas will not be discussed here (see references
cited); instead we shall highlight some recent advances.

---

*Camille & Henry Dreyfus Foundation Teacher-Scholar

II. The Picosecond Laser-Jet Beam Appartus

The arrangement we have used to interface the picosecond laser excitation source with the molecular beam involves two laser systems. The first is a synchronously pumped dye laser system, whose coherence width, coherence time, and pulse duration have been characterized [14] by the second harmonic generation (SHG) autocorrelation technique. The pulse width of these lasers is typically 1-2 ps, or, when a cavity dumper is used, 15 ps. The bandwidths of the lasers can be varied using intracavity filters and etalons. Typical bandwidths of the frequency doubled laser output are between 2 and 0.4Å. The second laser system used a synchronously pumped dye laser (oscillator) in connection with a 3-stage amplifier to increase the pulse energy to the millijoules range.

For detection we use three techniques:

1) <u>Time-correlated single photon counting</u> (detecting dispersed or total fluorescence) using a fast photomultiplier (or microchannel plate). This method can achieve a resolution of 70 ps without deconvolution. With deconvolution one can obtain the time constants of single exponential decays having lifetimes of $\sim$ 10 ps.

2) <u>Pump-probe multiphoton ionization detection.</u> This technique has been developed to measure short-time transient behavior [2,15,16]. Resolution is limited by the pulse width of the laser ($\sim$ 2 ps).

3) <u>Pump-probe picosecond/mass spectrometry.</u> We use this arrangement to resolve the picosecond dynamics of molecular fragmentation on the picosecond time scale in the beam.

III. Recent Advances

As mentioned in the introduction, molecular beam-picosecond techniques have been applied to a variety of problems which are discussed in the cited references. More recently, however, we extended the method to allow for (i) direct picosecond (pump-probe) multiphoton ionization in the supersonic expansion, (ii) polarization anisotropy measurements of intramolecular dynamics, and (iii) detection of ultrafast quantum beats(and their Fourier-transform spectra). Here we briefly discuss these studies.

A. Picosecond pump-probe and polarization techniques

In a recent communication [16], we reported on the use of pump-probe and polarization techniques in supersonic beams. We obtained data on stilbene and aniline at vibrational and rotational temperatures of less than [10K] and [5K], respectively. In these experiments the pump laser pulse excites an optically active mode in S_1, the first excited singlet state surface, and a delayed pulse probes the population of the excited mode as a function of the delay time (τ). This is achieved by monitoring the ion current induced by the probe as a function of τ. The high sensitivity of the method in the molecular beam, where the density and optical pathlength are very small, is evident from the quality of the transients S/N obtained (see Fig. 1). To probe the effect of rotations and coherence we polarized the probe ‖ or ⊥ with respect to the pump. Our time resolution is $\sim$ 1 ps. As shown in Fig. 1, the results for stilbene exhibit nonexponential behavior for modes excited at high excess vibrational energy. Preliminary analysis of the data in relation to *intramolecular dephasing* and polarization anisotropy (Fig. 2) is presented in our communication, and more work is now in progress by N. Scherer in this laboratory.

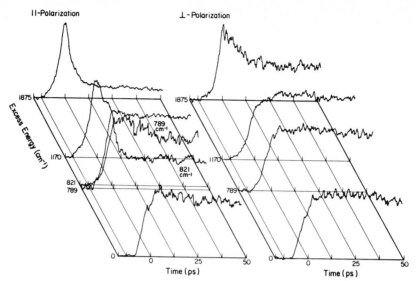

Fig. 1. Picosecond pump-probe transients of stilbene in a supersonic jet expansion. The carrier gas is He (20 to 30 psig), and x/D is 75. The ordinate refers to the excess vibrational energy (in cm⁻¹). The data presented here are for parallel (‖) polarization and ⊥ polarizations. Note the sensitivity of the transient shape to the excess energy (esp. 789 cm⁻¹ vs. 821 cm⁻¹), and the absence of the biexponentiality for zero-excess energy.

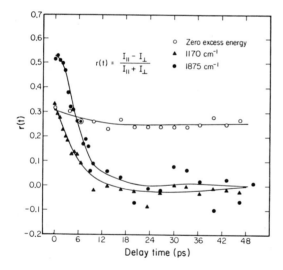

Fig. 2. The polarization anisotropy, $r(t)$, as a function of the delay time and excess vibrational energy. Note that for zero excess energy $r(t)$ is essentially flat over the 50 ps time scale, and holds at a value of 0.25. At higher excess energy and long times this $r(t)$ drops to a value of zero within our error.

In the figure: $r(t) = \dfrac{I_{\parallel} - I_{\perp}}{I_{\parallel} + I_{\perp}}$

○ Zero excess energy
▲ 1170 cm⁻¹
● 1875 cm⁻¹

B. Picosecond time-resolution of IVR and quantum beats

In continuation of our efforts in the area of quantum beats in large isolated molecules (see Section I), we have recently made two new advances. First, it was shown that nonchaotic *multilevel* vibrational energy flow is present in large polyatomic molecules with excess vibrational energy. The

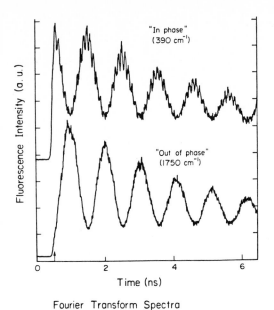

Fig. 3. Quantum beats observed in jet-cooled anthracene excited to $E_{vib}= 1420$ cm^{-1}. The arrow marks the temporal position of the excitation pulse. Note the *in phase* and *out-of-phase* behavior observed when different vibrational modes are detected (see text for cited references).

Fourier Transform Spectra
$\overline{\nu}_{ex}$: 1380 cm^{-1}

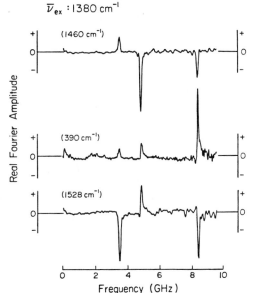

Fig. 4. Fourier transform spectra of the quantum beat residuals of three bands in the fluorescence spectrum of anthracene excited to $E_{vib} = 1380$ cm^{-1}. Note the relative phases of the bands which map out the selective coupling among vibrational states (Ref. 17). In this case, the number of levels involved is three.

molecule studied is anthracene (at a rotational temperature of [< 5K]). The results of this work, reported recently [17], are summarized in Fig. 3 and 4. From these results we obtained matrix elements for vibrational coupling and, more importantly, established the *number* and nature of modes involved in the intramolecular vibrational-energy redistribution (IVR).

Second, at higher excess vibrational energies we observed a "transition" in the behavior of IVR; instead of quantum beats among few levels ($\sim$ 3) at

287

moderate excess energies, a fast decay of $\sim$ 75 ps followed by a modulated long time decay was observed at higher excess energy. These observations [18] were interpreted as a manifestation of *dissipative* IVR where the initial deposited vibrational energy at 1792 cm^{-1} irreversibly flows to other modes. (We estimate the total number of levels to be $\sim$ 10.) The method gives a direct real-time view of this collisionless (beam) dissipative IVR process by picosecond-fluorescence techniques. The relevance of these studies to mode coupling and selectivity are discussed in a recent review article by BLOEMBERGEN and ZEWAIL [19].

ACKNOWLEDGMENTS

We acknowledge support by the National Science Foundation under Grants DMR-8105034 and CHE-8211356. This is Contribution No. 7047 from the Arthur Amos Noyes Laboratory of Chemical Physics, California Institute of Technology.

References

1. See, e.g., A. H. Zewail: Physics Today 33, 27 (1980)
2. For a recent review see, A. H. Zewail: Faraday Discuss. Chem. Soc. 75, 315 (1983)
3. Wm. R. Lambert, P. M. Felker and A. H. Zewail: J. Chem. Phys. 75, 5958 (1981)
4. P. M. Felker and A. H. Zewail: Chem. Phys. Lett. 102, 113 (1983)
5. W. R. Lambert, P. M. Felker and A. H. Zewail: J. Chem. Phys., Sept.(1984)
6. J. A. Syage, Wm. R. Lambert, P. M. Felker, A. H. Zewail and R. M. Hochstrasser: Chem. Phys. Lett. 88, 266 (1982)
7. P. M. Felker, Wm. R. Lambert and A. H. Zewail: Chem. Phys. Lett. 89, 309 (1982)
8. J. F. Shepanski, B. W. Keelan and A. H. Zewail: Chem. Phys. Lett. 103, 9 (1983)
9. J. A. Syage, F. Al Adel and A. H. Zewail: Chem. Phys. Lett. 103, 15 (1983)
10. P. M. Felker, Wm. R. Lambert and A. H. Zewail: J. Chem. Phys. 77, 1603 (1982)
11. P. M. Felker, J. A. Syage, Wm. R. Lambert and A. H. Zewail: Chem. Phys. Lett. 92, 1 (1982); J. A. Syage, P. M. Felker and A. H. Zewail: J. Chem. Phys., Sept. (1984)
12. P. M. Felker and A. H. Zewail: Chem. Phys. Lett. 94, 448 (1983); ibid. p. 454; P. M. Felker and A. H. Zewail: J. Chem. Phys. 78, 5266 (1983).
13. L. R. Khundkar, P. M. Felker and A. H. Zewail: work in progress.
14. D. M. Millar and A. H. Zewail: Chem. Phys. 72, 381 (1982)
15. J. W. Perry, N. F. Scherer and A. H. Zewail: Chem. Phys. Lett. 103, 1 (1983)
16. N. F. Scherer, J. F. Shepanski and A. H. Zewail: J. Chem. Phys., accepted for publication
17. P. M. Felker and A. H. Zewail: Phys. Rev. Lett., submitted for publication
18. P. M. Felker and A. H. Zewail: Chem. Phys. Lett., in press (1984)
19. N. Bloembergen and A. H. Zewail: Feature Article in J. Phys. Chem. (in press)

Energy Transfer in Picosecond Laser Generated Compressional Shock Waves

A.J. Campillo, L.S. Goldberg, and P.E. Schoen

Optical Sciences Division, Naval Research Laboratory
Washington, DC 20375, USA

We are employing picosecond fluorescence spectroscopy and laser driven compressional shocks to study the energy transfer from macroscopic mechanical disturbance to intramolecular modes in condensed media. By compressional shock we mean a large amplitude material disturbance propagating at supersonic speed, across which the pressure, density, temperature and internal state of molecules change in a nearly stepwise fashion. It is believed that a molecule in the path of such a disturbance witnesses a change from ambient to high pressure in times as short as a picosecond as the front passes. In general, the shock arises from an amplitude-dependent wave velocity which causes the leading edge of a high pressure compressional wave to steepen as it propagates, eventually sharpening to a near discontinuity.

Shock waves subject matter to extraordinary conditions that test our understanding of fundamental physical processes. Indeed, unexplained behavior is evident in many materials at rather modest shock pressures (ca. 10 to 100 kbar). Anomalous electrical and optical effects [1] and nonthermal chemistry [2-4] have been observed. Consequently, there has recently been a growing interest in the use of lasers as in situ, time resolved probes of intramolecular energy transfer processes and subsequent chemistry occurring in condensed media under moderate shock loading [5-6]. In this paper we summarize our initial efforts towards the characterization of shocks driven by ultrafast laser pulses and understanding the energy transfer processes that occur.

Multi-kbar shocks were produced in thin (ca. 20 to 200μ) confined foils of copper by heating with 10ps duration 1.054μm laser pulses (1 to 10J/cm^2) which generate plasmas in the region between the front surface of the foil and an overlaying glass window. As the trapped plasma/metal vapor attempts to expand, an intense compressional wave is launched into and through the foil. We measure the magnitude of shock pressure in an experiment in which the rear of the copper foil forms one mirror of a Michelson interferometer, illuminated by a single frequency Argon ion laser. This approach yields particle velocity histories from which shock pressures are estimated (see Figure 1) using the known equation of state. Depending on laser energy, rear surface motions in the range 10^{-2} to $10^{-1}\mu$m/ns are measured, corresponding to pressures of 2 to 20 kbar (1 bar = 1 atm). Both the magnitude and temporal characteristics of measured pressure profiles are consistent with a simple model describing the shock generation process [7]. In this model, the hot plasma/metal vapor is approximated by an ideal gas confined to a volume, V, which is expanding at a particle velocity u_p which is proportional to the pressure, P. The resulting compressional wave, after traversing the foil, is of the form

$$P(t) = P_{Max} (1-e^{-t/\tau_r}) (\frac{T}{T-t})^{1/2}$$

where τ_r is a risetime and T is ca. 2 to 4ns, both dependent upon the

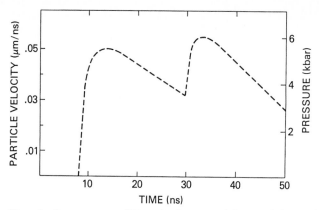

Fig. 1. Pressure profile of picosecond laser driven shock at rear of 40μm foil as determined using a Michelson interferometer. The 2ns risetime is an artifact reflecting pressure "ring-up" in a thin layer of fluid contacting the foil to the witness plate. The second peak in the profile is due to a reflected shock wave

geometry. Maximum pressure vs laser fluence was observed to follow a linear dependence.

An interesting problem concerns the details by which a macroscopic shock couples energy to intramolecular modes of a molecule. Are certain modes selectively populated upon shock loading? This is a difficult question to answer because the molecule almost certainly thermalizes within picoseconds of shock front passage. Because shocks travel relatively slowly (ca. 5μm/ns) it is necessary to limit the sample thickness to at most a few molecular layers to maintain sufficient time resolution to follow subsequently the grow-in and decay of intramolecular vibrational population. While we are actively working towards this temporal/spatial regime we report here preliminary results of an experiment to measure the effective shock temperature of a liquid several nanoseconds after the front arrives.

A schematic of the experiment is shown in Figure 2. Upon passing through the foil (8ns transit time) the shock, following the profile shown in

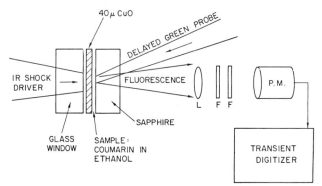

Fig. 2. Schematic of experiment to measure the shock-induced temperature rise in a submicron thick sample of coumarin in ethanol

290

Figure 1, heats a submicron layer of sample, 10^{-3} M coumarin 540A in ethanol, sandwiched between the rear of the foil and a second window. Temperature is estimated using the laser-induced fluorescence enhancement scheme of SEILMEIER et al.[8] with the coumarin serving as the "molecular thermometer." At ca. 10 kbar we observe a fluorescence enhancement of ca. 20% corresponding to a shock-induced temperature rise of 20°. We suspect that this preliminary estimate is conservative due to a probe intensity dependent coumarin fluorescence which reduces the observed enhancement. The initial results, however, illustrate the potential of this approach to tackle previously unanswerable questions in shock science.

1. G.E. Duvall, K.M. Ogilvie, R. Wilson, P.M. Bellamy and P.S. Wei, Nature 296, 846 (1982).
2. L.V. Barabe, A.N. Dremin, S.V. Pershin and V.V. Jakovly, Comb. Exp. Shock Waves (USSR) 5, 528 (1969).
3. G.A. Adadurov, V.V. Gustov and P.A. Jampolskii, Comb. Exp. Shock Waves (USSR) 7, 243 (1973).
4. D.W. Dodson and R.A. Graham in Shock Waves in Condensed Matter 1981, W.J. Nellis, L. Seaman and R.A. Graham (Editors), Am. Inst. of Phys. New York (1982) p. 62.
5. S.C. Schmidt, D.S. Moore, D. Schiferl and J.W. Shaner, Phys. Rev. Lett. 50, 661 (1983).
6. A.J. Campillo and P.E. Schoen, Proceedings of the APS 1983 Topical Conference on Shock Waves, Am. Inst. of Phys. New York (1983).
7. A.J. Campillo, L.S. Goldberg and P.E. Schoen, Bull. Am. Phys. Soc. 29, No. 4, p. 724 (1984).
8. A. Seilmeier, P.O.J. Scherer and W. Kaiser, Chem. Phys. Lett. 105, 140 (1984).

The Role of A and A' States in the Geminate Recombination of Molecular Iodine

D.F. Kelley and N.A. Abul-Haj

Department of Chemistry and Biochemistry, University of California
Los Angeles, Los Angeles, CA 90024, USA

Introduction

The photodissociation and recombination dynamics of solution phase molecular iodine have been extensively studied both theoretically and experimentally. The simplicity of the reaction, $I_2 \rightarrow I+I \rightarrow I_2$, along with iodine's well understood spectroscopy [1], have resulted in numerous picosecond laser experiments [2-6] as well as molecular dynamics[5], generalized Langevin[7] and other theoretical studies.[8-10] Despite these efforts, several fundamental questions have remained unanswered. For example, the rate of vibrational relaxation in the ground electronic state has not been directly determined. However, recent experimental[5] and theoretical[5,7,8] studies have suggested that vibrational relaxation may take ~100ps. Furthermore, recent rare gas matrix isolation studies by FLYNN[11] and theoretical studies by MILLER[9] have suggested that recombination into excited electronic states may be an important relaxation mechanism. The results presented here will further address these questions of electronic and vibrational relaxation.

Experimental

The experimental apparatus was based on an active/passive mode locked Nd^{+3}: YAG laser. Sample excitation was with the second harmonic (~1mJ, 532nm, 25ps) Some samples were excited with ~.5mJ 683nm (532 Raman shifted with H_2) pulses. Following second harmonic generation the 1064 fundamental was passed through an H_2O/D_2O (3:1) cell to produce the visible continuum (450-775nm) probe pulse. This light was split into reference and interrogation components, then recombined and focused through a .25M SPEX spectrometer. The output of the spectrometer was imaged onto a PAR 1254E SIT vidicon, which was interfaced into a DEC LSI 11/02 minicomputer.

Results and Discussion

Time resolved difference spectra [$\log(I/I_0)_{\text{no excitation}}$ - $\log(I/I_0)_{\text{with excitation}}$] have been taken following excitation at 532 or 683nm. Several different solvents (CCl_4, $CHCl_3$, CH_2Cl_2 and hydrocarbons) and iodine concentrations (5×10^{-2}-2×10^{-3}M) were examined. 532nm light promotes a transition to the $v \cong 33$ level of the $B(^3\Pi_0)$ state (see Fig. 1). Following excitation the molecule predissociates by crossing onto the repulsive $^1\Pi_{1u}$ state. The solvent rapidly dissipates the recoil energy of the atoms. Following equilibration among the 11 electronic states that correlate to ground state atoms, recombination on the $X(^1\Sigma)$, $A(^3\Pi_1)$, and $A'(^3\Pi_2)$ surfaces can occur. The degree of equilibration among these surfaces is largely determined by the separation obtained by the recoiling atoms, and therefore depends upon solvent viscosity. As MILLER[9] has pointed out, level hopping occurs while diffusion is taking place. Incomplete equilibration among the electronic states as the atoms recombine results in a frac-

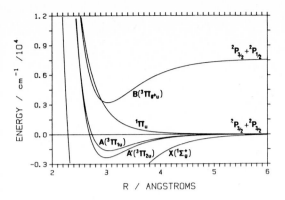

Fig. 1. Potential energy curves for some of the relevant electronic states of molecular iodine. There are 11 curves that correlate to ground state atoms, and 9 that correlate to one excited atom.

tion of the molecules being "frozen out" into the $A(^3\Pi_1)$ and $A'(^3\Pi_2)$ states. Excitation at 683nm results in similar dynamics, except that dissociation occurs directly on the $A(^3\Pi_1)$ surface.

In all cases excitation causes a significant ground state depletion, which results in a negative absorbance or bleach in the 450-575nm region. A positive absorption in the 575-775nm region was also observed (see Fig. 2). The kinetics of the 450-575nm bleach (in CCl_4) are presented in Fig. 3. The ground state recovery follows a biphasic decay to its residual value. The first component relaxes with a 140ps time constant and corresponds to recombination directly into the ground electronic state. WILSON et al.[5a] have calculated that an absorption maximum in the red should relax back to the static absorption maximum at ~510nm as vibrational relaxation occurs. No evidence for this behavior is observed in Fig 2, and we therefore conclude that vibrational relaxation is complete within ~40ps. The 140ps time constant therefore represents the time required for surface crossing and recombination. Non-geminate recombination occurs on the microsecond timescale and the amount of bleach present at long times (>10ns) is a measure of the fraction of atoms that escapes the solvent cage.

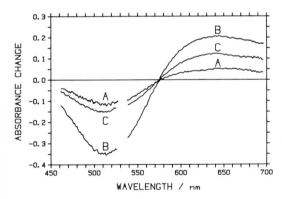

Fig. 2. Transient difference spectra of I_2 in CCl_4 taken at different delays following excitation. A:10ps, B:120ps, C:1.8ns.

The second component of the ground state recovery follows a 2.7ns exponential decay (in CCl_4) as does the 575-775nm absorption band. This band cannot be assigned to a vibrationally unrelaxed state or to an iodine-solvent complex; nearly identical spectra are observed in a wide variety of solvents. It must therefore be assigned to an excited electronic state. The only

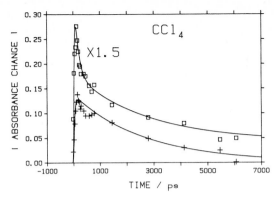

Fig. 3. Experimental plots of the absolute value of the absorbance change at 510nm (□) and 640nm(+) in CCl₄. The 510nm change is negative and that at 640nm is positive.

bound excited states that are energetically accessible are the $A(^3\Pi_1)$ and $A'(^3\Pi_2)$ states. The A' state lies $700cm^{-1}$ lower and probably has almost all of the excited population. From the amount of red absorption and the amount of bleach recovery which proceeds with a 2.7ns time constant, we estimate the extinction coefficient of this transition to be $\sim 1200 \, \ell \, mole^{-1}cm^{-1}$. Selection rules suggest that the upper state of this transition is a $^3\Sigma_g$ repulsive state that correlates with $^2P_{\frac{1}{2}}$, $^2P_{3/2}$ atoms. The 2.7ns decay time corresponds to the time required for intersystem crossing from the A' to the ground electronic state.

Chlorinated hydrocarbons and alkanes show nearly identical spectra, but different kinetics. In less viscous solvents the recoiling atoms obtain larger separations, thereby facilitating surface crossing to the ground state. This same effect also causes a higher probability for cage escape in lower viscosity solvents. Quantum yields for recombination into the A' and ground states and for cage escape are given in Table I.

Table I

Solvent	Viscosity/cp	$\Phi_{A'}$	Φ_X	Φ_{escape}	$T_{A'}/ns$
CCl₄	0.97	0.38	0.5	0.12	2700
CHCl₃	0.58	0.35	0.52	0.13	980
CH₂Cl₂	0.43	0.34	0.45	0.21	500
Hexane	0.336	0.23	~0.40	~0.37	~100
Hexadecane	3.34	0.42	~0.50	~0.08	~ 90

The lifetime of the A' state is also affected by the nature of the solvent. Charge transfer interactions with low ionization potential solvents stabilize the separated atoms with respect to the bound A' state molecule. This lessens the well depth at the A' state and thereby shortens its lifetime. This effect is particularly pronounced in hydrocarbons, which are known to form charge transfer complexes with iodine atoms. The observed A' state lifetimes are also given in table I.

At high concentrations and long delay times another absorption band was observed in the 580-775nm region (see Fig. 4). The band is present in all solvents examined but largest in those which have a high quantum yeild for cage escape, such as hexane. Furthermore, its appearance was at the diffusion limited rate over the 8×10^{-3}- 5×10^{-2}M concentration range studied, suggesting that this band may be assigned to I_3. The values of the observed I_3 and A' state absorption intensities allow us to estimate an I_3 extinction coefficient of about $2000 \, \ell \, mole^{-1}cm^{-1}$ the 600-750nm range. Despite this large extinction coefficient the low repetition rate of these experiments (~ 1 Hz) prevents buildup of these products from being a problem.

294

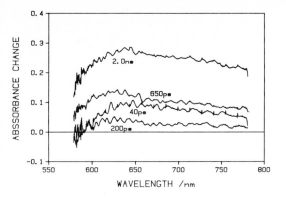

Fig. 4. Transient difference spectra of I_2 (5×10^{-2}M) in hexane taken at the indicated times following 532nm excitation.

Acknowledgment is made to the donors of The Petroleum Research Fund, administered by the ACS for partial support of this work. We also acknowledge the National Science Foundation and the Research Corporation for partial support of this work.

DFK would also like to thank IBM for a Faculty Development Award.

References

1. R.S.Mulliken,J.Chem Phys. 55,288(1971), and references therein.
2. T.J.Chuang,G.W.Hoffman, and K.B.Eisenthal, Chem.Phys.Lett.25, 201(1974).
3. (a)C.A.Langhoff,B.Moore, and M.DeMeuse, J.Am.Chem.Soc.104,3576(1982); (b) J.Chem.Phys.78,1191(1983).
4. D.F.Kelley and P.M.Rentzepis, Chem.Phys.Lett. 85,85 (1982)
5. (a)P.Bado,P.H.Berens,J.P.Bergsma,S.B.Wilson, and K.R.Wilson, in *Picosecond Phenomena III,*(Springer, Berlin, 1982),p.260.(b)P.Bado and K.R.Wilson,J.Phys.Chem.88,655(1984).
6. C.Harris,private communication.
7. C.L.Brooks III,M.W.Balk, and S.A.Adelman,J.Chem.Phys.79.784,804(1983).
8. D.J.Nesbitt and J.T.Hynes,J.Chem.Phys.77.2130(1982).
9. (a)D.P.Ali, and W.H.Miller,J.Chem.Phys.78,6640(1983);(b)Chem.Phys.Lett. 105,501(1984).
10. J.T.Hynes et.al.J.Chem.Phys. 72,177(1980),and references therein.
11. P.B.Beeken,E.A.Hanson, and G.W.Flynn,J.Chem.Phys.78,5892(1983).

Molecular Dynamics of I_2 Photodissociation in Cyclohexane: Experimental Picosecond Transient Electronic Absorption

Philippe Bado*, Charles G. Dupuy, John P. Bergsma, and Kent R. Wilson

Department of Chemistry,
University of California San Diego,
La Jolla, California 92093

Experimental and theoretical studies of the wavelength dependence of the transient electronic absorption of I_2 in cyclohexane following photodissociation at 680 nm indicate a role for the vibrational decay of the already recombined I_2 molecules.

1. Introduction

We have recently studied the molecular dynamics of I_2 photodissociation in thirteen different solvents following excitation into the A $1_u(^3\Pi)$ state. [1,2] As shown in Fig. 1, for a given solvent, the time scale of the transient electronic spectral response varies strongly with the probe photon wavelength. We have interpreted this as an indication that the response times arise from the time needed for vibrational or perhaps electronic decay of the already recombined I_2 molecules. Our experimental results correlate at least partially with the theoretical picture developed recently by our group,[3] and by the HYNES[4] and ADELMAN groups.[5]

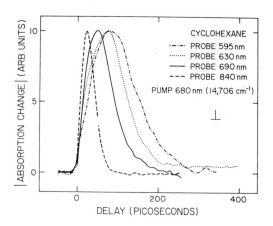

FIG. 1. *Experimental transient electronic absorption for I_2 photodissociation in cyclohexane at four different probe wavelengths following excitation at 680 nm. The vertical scale is the absolute value of the absorption change, computed by taking the square root of the output of the quadratic detection system.*

Our interpretation has been questioned by KELLEY and co-workers who have obtained in a recent study, under different experimental conditions, non-probe wavelength dependent spectra following excitation at 532 nm.[6] In light of these new experimental data we reexamine our results for the case of I_2 photodissociation in cyclohexane, for which a large set of experimental and theoretical data is available.[1,2]

---

* Present address: Laboratory for Laser Energetics, University of Rochester, 250 East River Road, Rochester, New York 14623

2. Experiment

Our experimental apparatus has been described in detail elsewhere[2, 7] Briefly, it consists of a mode-locked Argon ion laser synchronously pumping in parallel two independently tunable dye lasers. A multiple radio and audio modulation system provides a very sensitive detection technique for extracting weak signals from noise.[7] As indicated in the table below, this light source differs considerably from the frequency doubled Nd laser source used by other investigators.[6, 8-10]

Table 1. Comparison of physical parameters of two different experimental techniques.

	Dual synch-pumped dye lasers (this work)	second harmonic of Nd lasers [6, 8-10]
Pump wavelength	680 nm	530 nm
Probe wavelength	540-850 nm	530 nm
		400-850 nm [a]
Pump pulse energy	100 pJ	1-10 mJ
Pulse length	6-12 ps	6-30 ps
Repetition rate	80 or 240 MHz	1-10 Hz
Optical density	$\sim 0.2^b$	$\sim 1.0^b$
Ratio photons absorbed per pulse to number of I_2 molecules [c]	$\sim 10^{-6}$	~ 1

[a] with continuum generation.

[b] decreasing the concentration by a factor of five does no change the observed transient spectral response.

[c] calculated for a single pulse in the sampled volume, assuming Beer's Law remains valid.

Our tunable source offers several convenient features: (i) I_2 is photodissociated directly into the A $1_u(^3\Pi)$ state; [11] (ii) little excess energy is given to the recoiling I radical, thus the faction of atoms escaping the cage is reduced;[12] (iii) multiphoton processes are unlikely, due to the very low energy per pulse (less than 1 in 10^6 I_2 molecules are dissociated per pulse). The major drawback of our laser source is its high repetition rate. Even at high flow rates, the same sample volume is excited ~ 100 times before being replaced by fresh solution.

In contrast, the 530 nm excitation in the Nd laser experiments promotes the ground state X $0_g^+ (^1\Sigma)$ molecules either to the B $0_u^+ (^3\Pi)$ state,[11] (which presumably predissociates at some later time, but may absorb another photon in the meantime) or to the dissociative B''$1_u(^1\Pi)$ state.[11] As shown in Fig. 2, our excitation at 680 nm eliminates the ambiguities concerning the initially excited state, and the low power level minimizes the possibility of multiphoton phenomena due to high pulse energies.

3. Results and Discussion

We have examined the recombination process at a variety of probe wavelengths from 585 nm to 840 nm. Observation at higher probe photon energy gave longer spectral response times, as shown in Fig. 1. If the time needed for the iodine atoms to recombine is the principal cause for the time scale of the electronic absorption, then the observed time delay should be approximately independent of the probe wavelength. Since there is instead a large change in time scale as a function of the probe wavelength, our results indicate that recombination time is not the dominant mechanism. In light of this result , a more likely explanation would be vibrational relaxation of the already recombined molecules. This interpretation is reinforced by the presence of transient absorption in the near infrared where vibrationally excited molecules would be

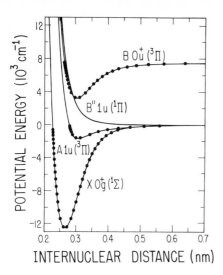

FIG. 2. *Potential energy curves used in the molecular dynamics calculation of the electronic absorption spectrum of I_2. The curves are derived from gas phase experimental data* [11] *and the dots show the RKR turning points for every fifth vibrational level.*

expected to absorb. The connection with the vibrational decay is further supported by the correlation between the observed transient response times and the presence of solvent vibrational modes in the frequency range needed to vibrationally relax hot I_2.[1,2] Furthermore the trend of the experimental wavelength dependence shown in Fig. 1 (shorter wavelength probe photons giving probe responses with larger half widths) corresponds with the *theoretical* wavelength dependence, shown in Fig. 3, which was calculated using molecular dynamics simulation.[3]

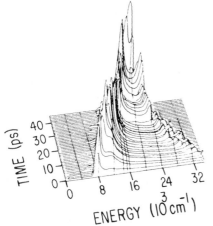

FIG. 3. *Theoretical molecular dynamics calculation of the transient electronic absorption spectra for I_2 photodissociation in cyclohexane following excitation to the $A\ 1_u(^3\Pi)$ state at time zero. At 45 ps. the spectrum has relaxed to the equilibrium distribution.*

There are still some uncertainties in our interpretations. The iodine atoms are born on an excited electronic state. The process whereby they go from the initial A state to final X state involves at least one electronic state transition whose details are not well known, and may involve other intermediate states as well. Our molecular dynamics simulations treat the electronic relaxation in a simplified way. After excitation to the A state the I_2 molecule is allowed to return to the ground X state when the internuclear separation reaches 0.425 nm, a point where the excited A state is very close in energy to the ground X state. If the iodine atoms collide with solvent molecules, causing them to recombine, they will vibrationally relax in the ground X state and give rise to the transient spectra shown in Fig. 3. However, collisions with solvent molecules can occur before the excited I_2 returns to the ground X state, causing I_2 to become trapped in the A state, as hypothesized by Kelley and co-workers. The absolute time scale of the calculated vibrational relaxation is quite sensitive to the choice of the potential parameters. For those parameters tested so far, the theoretical values are shorter than our observed spectra. The trends in our calculated wavelength dependence obtained with this simplified approach agree with our experimental results and are consistent with the calculations of HYNES[4] and ADELMAN.[5] Thus the dynamics calculations support but do not prove that vibrational relaxation is an important part of the recombination dynamics.

With our high repetition rate system there is the possibility of complications due to a quasi-steady state concentration of intermediates with lifetimes greater than a few nanoseconds. These are more likely in reactive solvents like the halogenated methanes and the alcohols. This might explain why our experimental results show agreement for the alkanes, but not the halogenated solvents, when compared with the results which the KELLEY[6] and HARRIS[13] groups obtained at low repetition rates. For alkane solvents, both our results and those of the HARRIS group[13] exhibit a probe wavelength dependence, which is consistent with a role for vibrational decay.

The I_2 photodissociation and recombination reaction is one of the simplest possible solution reactions, beginning and ending with the same homonuclear diatomic. Our experimental and theoretical results indicate that the role of vibrational relaxation must be considered in the description of the basic steps of this solution phase reaction.

We thank NSF, Chemistry and ONR, Chemistry for the financial support which made this work possible.

References

1. P. Bado and K. R. Wilson, *J. Phys. Chem.* **88**, 655 (1984).

2. P. Bado, C. Dupuy, D. Magde, K. R. Wilson, and M. M. Malley, *J. Chem. Phys.* **80**, 5531 (1984).

3. J. P. Bergsma, P. H. Berens, K. R. Wilson, D. R. Fredkin, and E. J. Heller, *J. Phys. Chem.* **88**, 612 (1984).

4. D. J. Nesbitt and J. T. Hynes, *J. Chem. Phys.* **77**, 2130 (1982).

5. S. A. Adelman, *Adv. Chem. Phys.* **44**, 143 (1980).

6. D. F. Kelley, N. Alan Abul-Haj, and Du-Jeon Jang, *J. Chem. Phys.* **80**, 4105 (1984).

7. P. Bado, S. B. Wilson, and K. R. Wilson, *Rev. Sci. Instrum.* **53**, 706 (1982).

8. T. J. Chuang, G. W. Hoffman, and K. B. Eisenthal, *Chem. Phys. Lett.* **25**, 201 (1974).

9. D. F. Kelley and P. M. Rentzepis, *Chem. Phys. Lett.* **85**, 85 (1982).

10. C. A. Langhoff, B. Moore, and M. DeMeuse, *J. Am. Chem. Soc.* **104**, 3576 (1982).

11. J. Tellinghuisen, *J. Chem. Phys.* **76**, 4736 (1982).

12. L. F. Meadows and R. M. Noyes, *J. Am. Chem. Soc.* **82**, 1872 (1960).

13. M. Berg, A. L. Harris, J. K. Brown, and C. B. Harris, in *Ultrafast Phenomena*, edited by D. Auston and K. Eisenthal (Springer-Verlag, Berlin, 1984). In press.

Iodine Photodissociation in Solution: New Transient Absorptions

M. Berg, A.L. Harris, J.K. Brown, and C.B. Harris

Department of Chemistry, University of California at Berkeley, and
Materials and Molecular Research Division of Lawrence Berkeley Laboratory
Berkeley, CA 94720, USA

We report new visible transient absorption measurements on I_2 photodissociation and recombination in solution, with improved time resolution and greatly enhanced signal-to-noise ratio. We also report the first picosecond absorption studies on I_2 in the ultraviolet. New short-time transients are found in both the red and ultraviolet regions.

Five millimolar iodine solutions were photodissociated by 1-2 ps light pulses at 590 nm from a 10-Hz, amplified synchronously-pumped dye laser[1]. Probe pulses at 295 nm were obtained by frequency doubling the 590 nm pulses. Other wavelengths were obtained using continuum generation and bandpass filters. Transmission changes were typically 1-5% in the visible and 20-30% in the ultraviolet. Table 1 summarizes the results.

Fig. 1 shows the transient bleaching at 500 nm. The delay in reaching the maximum bleach has been previously reported[2-5]. It has been hypothesized that the initially populated B state absorbs at 500 nm, so that the bleaching is not complete until the B state predissociates[2]. This model implies that the bleach has an instantaneous component due to excitation from the ground state to the more weakly absorbing B state, and a slower component due to B state predissociation. For the first time, our data clearly resolve two rise times, one at the expected experimental time resolution, thereby confirming this model and allowing accurate values of the predissociation time to be determined.

In CCl_4 (Fig. 1C) and the other chlorinated solvents, the decay of the bleach is clearly resolved into two components, in agreement with a

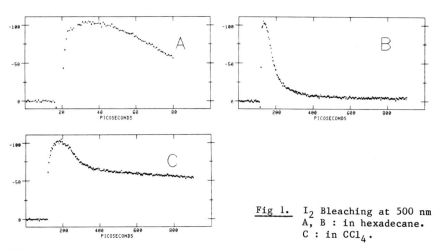

Fig 1. I_2 Bleaching at 500 nm
A, B : in hexadecane.
C : in CCl_4.

Table 1 : Recovery Times (ps)

	710 nm Absorb.		635 nm Absorb.		500 nm Bleach		350 nm Absorb.		295 nm Absorb.	
Hexadecane	62	±21	61	±24	52	±11	*		*	
	12	± 4	23	±13			*		*	
Nonane	66	±25	64	±21	71	±17	*		*	
	10	± 6	19	±16			*		*	
Hexane	67	±23	65	±10	83	±20	65	±15	67	±20
	12	± 7							6	± 2
Cyclohexane	41	± 3	68	±10	92	±10	72	±17	61	± 8
CCL_4	2900	±980	3000	±1300	*		2500	±670	*	
	34	± 6	70	±40	62	±20	8.5	±1.5	10	
$CHCL_3$	1200	±190	1100	±110	*		*		1000	
	35	± 7	46	±44	75	±40	*		10	
CH_2Cl_2	520	±40	501	±15	*		590	±160	640	±380
	11	± 7	20	±11	85	±85	5.4	±2.2	5	

* data not available

recent report[5], but in contrast to earlier interpretations[2-4]. The long decay has been attributed to relaxation from recombined molecules trapped in the excited A and A' states[5]. The transient bleach is not resolved into two decay components in the alkane solvents. However, further data discussed below suggest that the same two components are present, but have similiar time constants.

Fig. 2 shows examples of a newly observed, induced absorption in the ultraviolet. Iodine-atom/solvent contact charge-transfer transitions in the ultraviolet are known from nanosecond experiments[6]. In alkanes the atoms absorb at both 350 nm and 295 nm. However, in chlorinated solvents the absorption should be significant only at 295 nm. In the alkanes at both 350 nm and 295 nm, the absorptions show a constant offset out to nearly 900 ps, long after all geminate recombination dynamics are expected to be finished. In methylene chloride, the decays extrapolate to a 22% offset at 295 nm and a 0% offset at 350 nm. On this basis we attribute the long time transmission offsets in the ultraviolet to atom/solvent contact charge-transfer absorption.

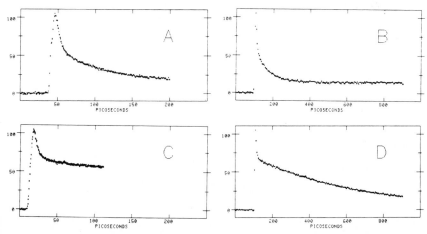

Fig. 2. I_2 absorption at 350 nm. A, B : in hexane. C, D : in CH_2Cl_2.

On the same basis, the time-dependent part of the ultraviolet absorption cannot be due to iodine atoms, since it is strong at both 295 nm and 350 nm in chlorinated solvents. Rather it must result from unrelaxed states of the I_2 molecule. In all the solvents two decays are present, one of 5-10 ps and another of 60 ps in alkanes and 500-3000 ps in chlorinated solvents. The short ultraviolet decay times agree with the rise times seen in the green and in the red (see below). This absorption can be reasonably, although tentatively, assigned to the B state and the decay time to the predissociation time. The longer decay in chlorinated solvents matches the decay of the A state as measured in the green and the red (see below) to within experimental accuracy. In all solvents, we attribute the longer decay to an ultraviolet absorption of the A state, either intrinsic or charge-transfer. Since the A state absorption stabilizes immediately after the B state predissociation, this model raises the possibility that the partitioning between separated atoms and bound, though unrelaxed, molecular states occurs on a time scale of ≤ 10 ps. This is an order of magnitude faster than has been suggested by previous experimental interpretations[2-4].

Two reports of an induced absorption beyond 600 nm following I_2 photodissociation have been made, but the reports directly conflict as to whether the transient decays are independent of wavelength[5,7]. In the chlorinated solvents, our data resolve two decay components in the red absorption, one wavelength independent and one which is faster at longer wavelengths (Fig. 3). The fast component was not resolved in either of the previous studies[5,7].

The longer, wavelength-independent decay time matches the time measured for A state relaxation in the green and ultraviolet, and so can be attributed to A state absorption. The short decay time is a factor of two shorter at 710 nm than at 635 nm. This type of behavior has been predicted as a result of vibrational relaxation on the ground state potential, since different vibrational levels are preferentially probed at different wavelengths[7]. The short time behavior may also be complicated by decay between the similiar A' and A states, although this cannot account for the wavelength dependence of the decay rates[8].

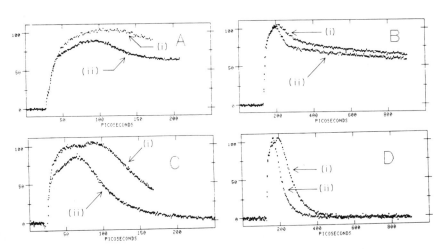

Fig. 3. I_2 Absorption at i) 635 nm and ii) 710 nm. A, B : in CCl_4. C, D : in cyclohexane.

In the straight chain alkanes, the decays do not separate into two components as clearly as in the chlorinated solvents. However, curve fitting of 710 nm decays definitely suggests a decomposition into two decay times. Significantly, the longer decay time matches the A state decay time measured at 350 nm. Thus, it appears that A state and vibrational decay are also present in the alkanes, but are similiar in timescale. Under this assumption, the 635 nm decays can also be decomposed into two components, although the timescales are even closer. By a reasonable extrapolation, the vibrational and A state decay time would be too close to resolve in the bleach recovery at 500 nm.

In summary, no <u>unique</u> model can be proposed with certainty from the data currently available. However, the new data appear to be internally self-consistent with B state predissociation in 5-10 ps with rapid ($<$10 ps) partitioning between separated atoms, A and A' excited state molecules and vibrationally excited ground state atoms. This is followed by slower excited state electronic relaxation and ground state vibrational relaxation. This model can be tested by exciting the I_2 molecules to the directly dissociative A states using longer wavelength pump light[7]. With carefully designed experiments and the higher time resolution and significantly improved signal-to-noise reported here, one can hope to understand fully the early partitioning among the various intermediate states following photodissociation in this and other systems.

This research was supported by the National Science Foundation.

References

1. M. Berg, A. L. Harris, J. K. Brown, and C. B. Harris, Optics Lett. 9, 50 (1984).
2. T. J. Chuang, G. W. Hoffman, and K. B. Eisenthal, Chem. Phys. Lett. 25, 201 (1974).
3. C. A. Langhoff, B. Moore, and M. DeMeuse, J. Chem. Phys. 78, 1191 (1983).
4. D. F. Kelley and P. M. Rentzepis, Chem. Phys. Lett. 85, 85 (1982).
5. D. F. Kelley, N. A. Abul-Haj, and D.-J. Jang, J. Chem. Phys. 80, 4105 (1984).
6. S. R. Logan, R. Bonneau, J. Joussot-Dubien, and P. Fornier de Violet, J. Chem. Soc., Faraday 1 71 (1975).
7. P. Bado, C. Dupuy, D. Magde, K. R. Wilson, and M. M. Malley, J. Chem. Phys., In press (1984).
8. D. F. Kelley, Private communication.

Photodissociation of Triarylmethanes

L. Manring and K. Peters

Department of Chemistry, Harvard University, 12 Oxford Street
Cambridge, MA 02138, USA

Introduction

Recently there have appeared several reports of picosecond absorption studies for the photodissociation of triarylmethanes(1,2,3). This class of molecules is of interest as photoexcitation leads to the formation of ion pairs, the kinetics of which are highly solvent dependent. The 300 nm photolysis of malachite green leucocyanide (MGCN) results in the formation of malachite green cation (MG^+) whose appearance, when monitored at 610 nm, is non-exponential. In order to develop a more comprehensive understanding of the molecular events that give rise to this behavior we have examined the picosecond absorption kinetics of a series of TAMs.

Experimental Technique

The laser flash photolysis apparatus has been described in detail previously (4). Briefly, the picosecond absorption spectrometer consists of a 10 Hz Nd^{+3}:YAG laser (Quantel International) with a pulse width of 25 psec. The detector is an OMA II Vidicon (PAR-1215-1216-1217) interfaced to a 200 mm spectrograph (JY-UFS-200). The photolysis energy at 266 nm is 0.1 mj per pulse and each absorption spectrum is an average of at least 200 laser pulses.

Results

The following series of triarylmethanes (TAM) were examined in both polar and non-polar solvents by picosecond absorption spectroscopy.

$$X = H$$
$$OH$$
$$OCH_3$$

Laser excitation of 1 mM malachite green leucohydride (MGH) in acetonitrile produced a species with an absorption λ_{max} = 610 nm within 25 psec after excitation, Figure 1a, which decays with a $\tau_{1/2}$ = 2.1±0.4 nsec. This spectrum is assigned to the $S_n^* - S_1^*$ transition of MGH based on its similarity to the species observed after photolysis of N,N-dimethyl-p-toluidene (DMT) in acetonitrile, Fig. 1b.

Photolysis of malachite green hydroxide (MGOH) in acetonitrile produces a transient whose absorption spectrum has a λ_{max} = 610 nm and is formed with-

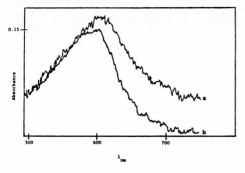

Fig. 1. a. 266 nm excitation of 1 mM
MGH in acetonitrile.
b. 266 nm excitation of 1 mM
DMT in acetonitrile.

in 25 psec of excitation, Fig. 2a. The somewhat broad spectrum initially
observed sharpens at later times and by 2 nsec takes on the characteristic
shape of MG^+, Fig. 2b.

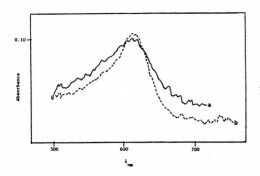

Fig. 2. Spectra generated by 266 nm
excitation of 1mM MGOH in
acetonitrile.
a. ——— 50 psec
b. - - - 2 nsec.

Excitation of either MGH or MGOH in cyclohexane gives a transient within
25 psec which has a broad absorption from 500 nm to 700 nm with a λ_{max} =
610 nm, Fig. 3. The transient from MGH decays with a $\tau_{1/2}$ = 6.3± 1.0 nsec
and the transient from MGOH decays with a $\tau_{1/2}$ = 1.6 ± 0.2 nsec.

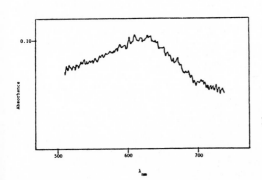

Fig. 3. Spectrum generated by 266 nm
excitation of 1 mM MGOH in
cyclohexane at 100 psec.

Excitation of malachite green methyl ether ($MGOCH_3$) in 90%acetonitrile/
10% methanol produces an initial broad absorption within 25 psec (λ_{max} =
610 nm, A = 0.03) which within 2 nsec grows into the distinct MG^+ spectrum
(λ_{max} = 610 nm, A = 0.25), Fig. 4. The $\tau_{1/2}$ for the appearance of MG^+
from $MGOCH_3$ is 1.0 ± 0.1 nsec. In 100% methanol, the rate of MG^+ formation
is greatly accelerated with a $\tau_{1/2}$ = 400 ± 50 psec.

305

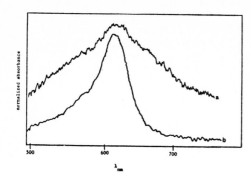

Fig. 4. Spectra generated by 266 nm
excitation of 1mM MGOCH$_3$ in
90% CH$_3$CN/10% CH$_3$OH at
a. 25 psec.
b. 2 nsec.

Discussion

The results of the picosecond absorption studies of MGH, MGOH and MGOCH$_3$ are
consistent with initial formation of the first excited singlet state which
has a visible absorption with λ_{max} = 610 nm for all three species based on
the similarity with the first excited state absorption spectrum of DMT.
Subsequent to S$_1^*$ formation for the malachite green derivatives,
the excited species may decay to the ground state or heterolytically cleave
forming the ion pair.

The results indicate that the rate for formation of MG$^+$ is both dependent upon the solvent and leaving group. In cyclohexane, only S$_1^*$ is observed,suggesting that the ion pair in non-polar solvents is energetically
higher than S$_1^*$. The importance of the stability of the resulting ions is
further demonstrated by the fact that S$_1^*$ for MGOCH$_3$ decays to form the ion
pair with a half-life of 400 psec in methanol and slows to 1 nsec in 90%
acetonitrile/10% methanol. The methoxide ion formed is more stabilized by
hydrogen bonding in methanol than in acetonitrile-methanol mixture.

Recently, it has been reported that malachite green leucocyanide undergoes very fast cleavage in ethanol or glycerol to give MG$^+$ initially in a
pyramidal conformation which then via a solvent restricted motion forms the
more stable propeller conformation (2). The proposed intermediacy of a
pyramidal cation was based on the observation of a fast rise in absorbance
at 610 nm ($\tau_{1/2}$ 50 psec) followed by a slower rise (120 psec to complete
formation of MG$^+$) after 266 nm excitation of MGCN. This result contrasted
with that in glycerol where the very fast appearance at 610 nm ($\tau_{1/2}$ =
50 psec) is followed by a much slower rise (2 nsec to complete formation
of MG$^+$). It was suggested that the immediately observed absorbance at
610 nm is the result of some overlap between the aryl group π-electron
systems immediately after photoionization, with MG$^+$ still in the pyramidal
configuration characteristic of MGCN. The increased viscous drag of the
glycerol (relative to ethanol) was suggested to cause slower attainment of
the propeller conformation; hence, the slower rise in the 610 nm absorption.
Based on our results, its seems likely that the initially formed absorbance
at 610 nm is due to the first excited singlet state absorption and the
subsequent rise in absorption is due to the cleavage of the first excited
singlet state to yield the ion pair. The effect of viscosity on the MG$^+$
appearance could just as easily be attributed to a viscosity effect on the
activation barrier to cleavage.

306

Acknowledgements

This work is supported by a grant from the National Science Foundation, CHE-8117519. KSP acknowledges support from the Henry and Camille Dreyfus Foundation for a teacher-scholar grant and the Alfred P. Sloan Foundation.

References

1.K.G. Spears, T.H.Gray and D. Huang, "Picosecond Phenomena III", Eds. K. Eisenthal, R. Hochstrasser, W. Kaiser and A. Laubereau, Springer-Verlag, Berlin, 1982, p.278.

2.D.A.Cremers and T.L.Cremers, Chem. Phys. Letts., 94, 102 (1983).

3.L. Manring and K. Peters, J. Phys. Chem., in press.

4.J. Simon and K. Peters, J. Am. Chem. Soc., 105, 4875 (1983).

Femtosecond Time Resolved Multiphoton Ionization: Techniques and Applications

B.I. Greene

AT & T Bell Laboratories, Murray Hill, NJ 07974, USA

Ionization spectroscopies have long been known to be extremely sensitive methods for detecting small number densities of molecular or atomic species. In this paper, we report on the first extension of a well-known nanosecond laser ionization technique into the picosecond and femtosecond time regimes [1, 2]. We present below data exemplifying the utility of this type of measurement for the study of ultrafast intramolecular relaxation processes. Finally, we discuss special interpretative considerations relevant to work in this area [1-4].

The demonstration by LETOKHOV et al. [5] of the two step nanosecond time delayed ionization of H_2CO out of its excited $^1A_2$ state suggested the general utility of the method [6,7]. In this general scheme, one laser pulse is used to excite a sample of vapor phase molecules, either by direct one-photon, or simultaneous multiphoton absorption. Subsequently, a second laser pulse will ionize these excited molecules via a one- or multiphoton optical transition from the initially prepared state into the ionization continuum. An arbitrary time delay between the preparation and ionization pulses can easily be inserted. Clearly then, in the absence of collisions, time delay dependent ionization yields will reflect upon the quantum mechanical evolution of the excited state.

The appeal of using picosecond and femtosecond laser pulses for these measurements is immediate. For intermediate and large size molecules, the femtosecond and picosecond timescale is one in which intramolecular vibrational energy redistribution (IVR), in the most general sense, is expected to occur. Additionally, dynamics associated with highly excited, ultrashort lived electronic states of molecules can be directly probed with femtosecond time resolution.

We present time resolved ionization results utilizing two different excitation schemes. In Fig. 1a, a single uv photon excites a set of vibronic levels in the lowest excited singlet state S_1. A second time delayed pulse then ionizes the molecule via a S_N resonantly enhanced three-photon (or 2-photon, for the case of stilbene below) absorption. Any unimolecular processes occurring between excitation and ionization that effect the cross section for ionization will result in a time delay dependent ionization yield. Such possible processes include dissociations, conformational changes, intramolecular vibrational energy redistribution on the excited S_1 surface, intersystem crossing or internal conversion.

Figure 1b depicts a two-photon uv excitation into a vibronic level of a molecular Rydberg state. A single visible laser photon then ionizes the molecule, and as the delay between pulses is scanned, the lifetime of the state is directly measured.

Time resolution in these experiments is limited only by the pulsewidth, and requires no high speed electronics or detectors of any kind. Additionally, picosecond and certainly subpicosecond timescale dynamics can be

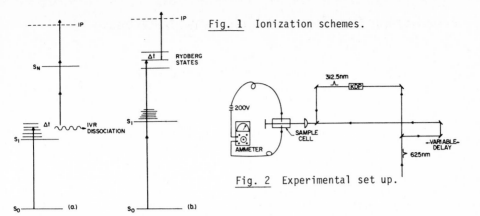

Fig. 1   Ionization schemes.

Fig. 2   Experimental set up.

safely assumed to occur without collisions. Even for sample pressures of 100 torr, mean hard sphere collision periods are roughly 1 nanosecond.

The experimental set up utilized is quite simple (Fig. 2). Amplified pulses from a colliding pulse modelocked ring dye laser are obtained as described in detail elsewhere [8, 9]. Pulses are divided at a beamsplitter, with one path frequency doubled (312.5 nm) and delayed with respect to the other path (625 nm). The two beams are attenuated and recombined at a dichroic beamsplitter. A 1 m quartz lens is used to focus the beams to a roughly 0.5 mm spot inside the ionization cell. Visible and uv pulse energies are typically 25×10^{-6}J and 1×10^{-6}J respectively. Samples consist of room temperature molecular vapors at equilibrium pressures above the solid or neat liquid.

As a first demonstration of the utility of the technique in elucidating dynamics associated with ultrafast unimolecular processes, we chose to study excited state conformational dynamics in cis-stilbene [1]. There have been numerous experimental papers published on picosecond gas phase stilbene photophysics (see Perry et al. and references cited therein). To date, only our study focuses on cis-stilbene, as the dynamics manifest are decidedly subpicosecond and consequently, in real time, must be probed with a subpicosecond laser. Fig. 3 depicts schematically the relevant energy levels and coordinates.

Fig. 3   Stilbene energy level diagram.

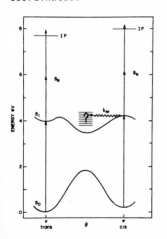

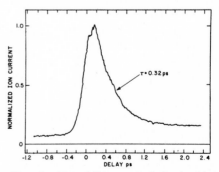

Fig. 4   Cis-stilbene photoionization current vs delay time.

The success of the measurement relies on the fact that while in the initially prepared planar (with respect to the ground state ethyenic bond) excited state, a strongly one-photon resonant enhanced two-photon ionization can occur. As the initial state evolves into newly accessible regions of phase space (presumably corresponding to a twisted form of the molecule) the excited state transition responsible for the resonant enhanced ionization disappears. Fig. 4 shows the ionization current as a function of delay time between uv preparation and visible ionization. We measure k_{nr}, corresponding to the rate of twisting about the central c-c bond in this picture, to be 3.1×10^{13} sec^{-1}. To our knowledge, this is the first and only real time resolution of a unimolecular reaction occurring in one vibrational period along a reaction coordinate [1].

In another set of experiments, the ultrafast relaxation of molecular Rydberg states has been investigated [2]. Fig. 1b shows the scheme whereby vibronic levels of the 3Rg and 4d Rydberg states in benzene and toluene respectively are populated via two-photon absorption. A single time delayed visible photon then raises the molecule into the ionization continuum.

Typical data are shown for perdeutero-toluene in Fig. 5. Lifetimes directly obtained in such a fashion are: benzene (70 ± 20 fs), d-benzene (110 ± 20 fs), toluene (170 ± 20 fs), and d-toluene (190 ± 20 fs). These lifetimes are consistent with linewidths recently measured by Whetten et al. using two-photon resonant, multiphoton ionization spectroscopy [10]. Our data are consistent with the notion that the highly excited valence states in benzene and toluene are extremely short-lived, even in comparison to the Rydberg states. We see no evidence of bottlenecking in the $^1E_{2g}$ state in benzene.

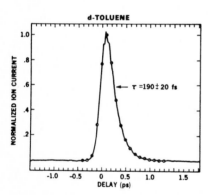

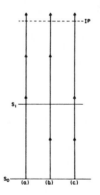

Fig. 5 Decay of 4 d Rydberg state in d-toluene.

Fig. 6 Simultaneous MPI possibilities.

We turn finally to a discussion of potential artifactual effects. A major problem can arise when the excitation pulse and ionization pulse are overlapping in time. Consider the one uv photon excitation, two visible photon (delayed) ionization scheme depicted in Fig. 6a. Simultaneous 3-photon absorption events 6a, b and c are also possible. In general, cross sections for the simultaneous absorptions cannot be assumed to be negligible. Depending on the specific molecule, light intensities, wavelengths and polarizations, the sum cross section of several different simultaneous two color absorption processes may or may not be comparable to that of a single sequential, time delayed process.

Data to be particularly cautious with contains an enhanced ionization signal in and around the time of pulse overlap (a prompt signal) followed by a sizable long lived signal. The long lived signal could be attributable to ionization out of an electronically excited state. Fig. 7 suggests how the "prompt" signal can be artifactually constructed out of instantaneous simultaneous multiphoton ionization processes superimposed on the normal instrumental rise of the long lived signal. A more detailed analysis appears in Fig. 8. The curve labeled "Artifact" is the construct shown in Fig. 7. The outer trace "$\tau = 0.44\ \Gamma$" is derived by numerical convolution of two Gaussian pulses (FWHM = Γ) with a 0.44 Γ single exponential decay sample response. Except for a small difference in the position of absolute zero time delay, these curves are very similar. In fact, it takes only an 8% under-estimation of the actual experimental pulsewidth,eg.,if one models the data assuming 4.0 ps pulses when in reality the pulses are 4.3 ps, and the two traces in Fig. 8 become virtually identical in shape.

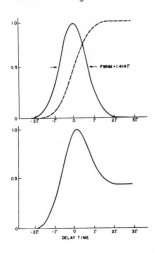

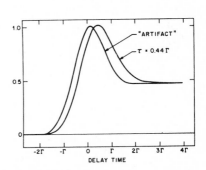

Fig. 7 Top: Autocorrelation of pulse FWHM = Γ with con-volved instrument response. Bottom: Normalized weighted algebraic sum of above.

Fig. 8 Comparison of artifactual signal discussed in text with a real sample response $\tau = 0.44\ \Gamma$.

Experimentally we have studied several molecules which display this prompt signal phenomena, where a single exponential fit to the apparent decay yields a value of roughly one half the pulsewidth. These molecules include trans-stilbene, naphthalene, quinoxaline, iron pentacarbonyl, and a variety of halo substituted naphthalenes. We originally omitted mention of this effect because we believed it to be artifactual in origin [1]. In light of this discussion and subsequent experimental work in the area [4], we maintain that it is imperative to address this issue. Before attributing experimental responses such as those depicted in Fig. 8 to anything other than instantaneous effects, a precise determination of the t = o point must be made.

A most exciting direction for picosecond and femtosecond time resolved MPI has recently been demonstrated by GOBELI et al. [3]. With tunable lasers, and most importantly, mass spectral or photoelectron analysis, an added dimension of intramolecular dynamic information will become available.

Even without this instrumentation however, our work has demonstrated the utility of the time resolved technique. By providing hitherto unobtainable realtime information on subpicosecond intramolecular relaxation events, we hope to encourage still more ambitious work in this area in the future.

References

1. B. I. Greene and R. C. Farrow, J. Chem. Phys. 78, 3336, (1983).
2. J. M. Wiesenfeld and B. I. Greene, Phys. Rev. Lett. 51, 1745, (1983).
3. D. A. Gobeli, J. R. Morgan, R. J. St. Pierre, and M. A. El-Sayed, J. Phys. Chem. 88, 178, (1984).
4. J. W. Perry, N. F. Scherer, and A. H. Zewail, Chem. Phys. Lett. 103, 1, (1983).
5. S. V. Andreyev, V. S. Antonov, I. N. Knyazev, and V. S. Letoknov, Chem. Phys. Lett. 45, 166, (1977).
6. D. H. Parker and M. A. El-Sayed, Chem. Phys. Lett. 42, 379, (1979).
7. T. G. Dietz, M. A. Duncan, M. G. Liverman, and R. E. Smalley, J. Chem. Phys. 73, 4816, (1980).
8. R. L. Fork, B. I. Greene, and C. V. Shank, Appl. Phys. Lett 38, 671, (1981).
9. R. L. Fork, C. V. Shank, and R. T. Yen, Appl. Phys. Lett. 41, 223, (1982).
10. R. L. Whetten, K. J. Fu, and E. R. Grant, J. Chem. Phys. 79, 2626, (1983).

Threshold Ionization in Liquids

G.W. Robinson, J. Lee, and R.A. Moore
Picosecond and Quantum Radiation Laboratory, Texas Tech University
Lubbock, TX 79409, USA

1. Introduction

One should not underestimate the importance of condensed phase reaction dynamics on ultrafast timescales, even when one's primary interest is in reactions and synthetic schemes that occur in much slower times. In the primordial phases of a chemical reaction, the reaction must usually "make a choice" among a number of possible pathways. The choice favors pathways having the fastest reaction rates, and this choice determines overall product yields.

Among the fastest reactions in solution are charge transfer reactions. The photoejection of electrons or protons into the surrounding solvent is perhaps the simplest of charge transfer reactions, yet the most important solvent, water, has always presented somewhat of an enigma when involved in these processes. Qualitative differences between water and the closely related alcohols occur that extend far beyond any consideration of dielectric constant.

Mobility measurements [1] and spectroscopic studies [2] of excess electrons in liquids indicate an increasing tendency for localization in the more polar solvents. The spectroscopic studies also show that in polar liquids, solvent clusters, rather than single solvent molecules, are important in electron localization. Studies of both electron and proton rates in charge transfer reactions have suggested that water clusters may also play an important role in chemical reaction mechanisms [3,4]. However, no quantitative methods for analyzing such data have been available. In this paper such a method is presented.

2. Threshold Ionization

Threshold ionization (TI), that is, ionization near the energy threshold, can sometimes be initiated by one-photon (1-P) absorption into a low lying excited state of a dissolved molecule. This process is qualitatively different from photoionization using ionizing radiation or multiphoton absorption, on which most intuition about these matters has been based. Electrons produced by TI can not form ordinary "solvated electrons", since they do not have sufficient kinetic energy to escape the local cationic environment.

The indole molecule, which is an important chromophoric moiety in proteins, provides a good example of the 1-P TI process when excited to its lowest excited singlet state with ~350nm radiation. The extreme solvent dependence of TI for indole is illustrated by the fact that 1-P TI takes place on picosecond timescales in liquid water, but occurs with effectively zero rate in methanol. In fact, studies can be carried out in water/methanol mixtures using methanol as the "inert" solvent.

313

3. Experimental

The photophysics of indole has been investigated in seven different water/methanol mixtures (0-100%) at a series of temperatures ($\pm 1^\circ$ accuracy) between 258K and 348K. Both picosecond rates and steady-state quantum yields were measured in order to obtain a comprehensive description of the TI process. Full details of the experimental methods and results will be published elsewhere [5].

4. Theory

The theoretical description of the TI photophysics is based on a generalization of the following kinetics model,

$$I \cdot W + h\nu_1 \rightarrow I^* \cdot W$$
$$I \cdot M + h\nu_2 \rightarrow I^* \cdot M$$
$$I^* \cdot W + M \rightleftharpoons I^* \cdot M + W$$
$$I^* \cdot W \rightarrow I^+ \cdot W^-$$
$$I^* \cdot W \rightarrow I \cdot W$$
$$I^* \cdot M \rightarrow I \cdot M \quad ,$$

where I labels the indole ground state, I* the indole lowest excited singlet state; W = H_2O, and M = methanol.

The generalization of the kinetics model uses probability matrix methods similar to the Markov "gambler's ruin" problem [6] to take account of the fact that more than one H_2O molecule is involved cooperatively in the TI process. The rate matrix for a critical cluster size of four H_2O molecules (N=4) is

$$\underset{\approx}{R} = \begin{pmatrix} A_{11} & J_{12} & 0 & 0 & 0 & K_1 \\ J_{21} & A_{22} & J_{23} & 0 & 0 & K_2 \\ 0 & J_{32} & A_{33} & J_{34} & 0 & K_3 \\ 0 & 0 & J_{43} & A_{44} & J_{45} & K_4 \\ 0 & 0 & 0 & J_{54} & A_{55} & K_5 \\ 0 & 0 & 0 & 0 & 0 & 1 \end{pmatrix} \quad . \tag{1}$$

A row matrix $\underset{-}{C}(0)$ having elements

$$C_n(0) = \frac{N!}{(n-1)!(N-n+1)!} [M]^{n-1}[W]^{N-n+1} \quad , \tag{2}$$

with n = 1,N+1, represents the initial concentrations. The first five rows of $\underset{\approx}{R}$ denote consecutively an I* molecule having an H_2O cluster of 4,3,2,1,0 molecules associated with it. The sixth row of $\underset{\approx}{R}$, which represents the ground state of indole, acts like a trap. A_{nn}(n=1,N+1) is the probability that the configuration $I^* \cdot W_{N+1-n} M_{n-1}$ does not change during an incremental time step Δt. $J_{n,n+1}$ is the probability that an H_2O cluster loses one H_2O molecule during Δt, while $J_{n,n-1}$ is the probability that the cluster increases by one during the incremental time. The probabilities K_n are for

314

spontaneous decay (radiative + nonradiative). The following relationships,

$$J_{n,n+1} \propto \{(N-n+1)/N\}P[M]$$

$$J_{n,n-1} \propto \{(n-1)/N\}P[W] \,, \tag{3}$$

are used, where $P = pn(T)$ is an elementary rate for solvent exchange, $n(T)$ is the viscosity, and $[\ ]$ represents volume fractions. In addition, independent experimental evidence provides the following form for $K_n(n>1)$,

$$K_n \propto [(N-n+1)/N][k_r(W)+k_{nr}^0(W)] + [(n-1)/N][k_r(M)+k_{nr}(M)], \tag{4}$$

a weighted average of the spontaneous rates. The form of K_1 is

$$K_1 \propto k_r(W) + k_{nr}^0(W) + k_e\exp(-\beta E_1), \tag{5}$$

the spontaneous rate in pure H_2O. We are thus assuming that the temperature dependent TI process does not occur for cluster sizes having $<N$ H_2O molecules. In the above expressions, $k_r(W)$ and $k_r(M)$ are the radiative rates in pure H_2O and methanol; $k_{nr}(M)$ is the nonradiative rate (internal conversion plus intersystem crossing) in pure methanol; $k_{nr}^0(W)$ is the temperature independent nonradiative rate and $k_e\exp(-\beta E_1)$ is the temperature dependent TI rate in pure water, with $\beta=(k_B T)^{-1}$ and E_1=43.4 Kj mol^{-1}.

5. Results and Conclusions

The dynamics are reproduced by successive matrix multiplications on a VAX 11/730. The quantity $\underset{\sim}{C}(m)=\underset{\sim}{C}(0)\underset{\sim}{R}^m$ gives the concentrations of the various species after time $m\Delta t$, where Δt is chosen sufficiently small that it behaves as an infinitesimal. Steady state quantum yields are reproduced by adding a small initial concentration $\delta\underset{\sim}{C}(0)$ $(\delta<<1)$ to $\underset{\sim}{C}(m)$ after each Δt and examining the relative rates (probabilities) when the steady state is reached. A comparison of the theoretical rates and quantum yields, containing only the two adjustable parameters p and N, with the diverse experimental data at various temperatures and water concentrations shows good agreement throughout. See Fig. 1.

The best value of N is 4±1, a value that has already been implied [7] on the basis of energetic arguements. The value of $P(293K) = pn(293K)$ is 10^{10} s^{-1}, which is of the magnitude expected from self-diffusion coefficients in the solvent. The presence of the large activation energy for the indole TI process suggests that the water structure responsible for electron trapping is far different from the normal water structure.

To investigate the generality of these methods, we have applied them to photophysical data, rates and yields, of certain anilino-naphthalene-sulfonate (ANS) derivatives, where the TI process is suspected to occur. The case of 6-p-toluidinyl-2-naphthalenesulfonate (TNS) is interesting, since the overall photophysics appears quite different than it does in the indole derivatives. This difference is mainly caused by the absence of an activation barrier in TNS, a phenomenon that we have attributed [3] to the preformation of the requisite water structure caused by the intramolecular charge transfer characteristics of post-excited TNS. In spite of the large external differences in the photophysics of these two types of molecules, the experimental data for TNS are well fit [8] to a TI mechanism, again with a water cluster size of N=4±1.

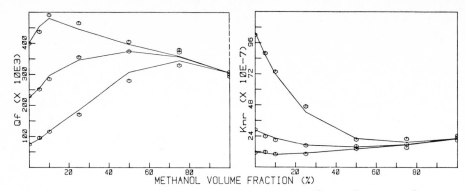

<u>Fig. 1</u>   Quantum yields and nonradiative rates at $5^{\circ}C$, $35^{\circ}C$ and $65^{\circ}C$ for indole.   Raising the temperature decreases Qf and increases Knr.   The hexagons are the experimental points.   For clarity, straight line segments have been drawn between the corresponding theoretical points (N=4)

In conclusion, we feel that these experiments and their parallel interpretation for the dissimilar appearing photophysics of indole and TNS are starting to provide a clearer picture of the way that near-threshold electrons behave in the presence of liquid water. Most remarkable is the growing possibility of a similar behavior of protons in liquid water [4,9]. The influence of water clusters on long range electron and proton charge transfer reactions in aqueous media must surely have to come under consideration in the future analysis of chemical and biological reaction mechanisms.

6. Acknowledgements

This work is supported by the National Science Foundation (CHE-8215447) and the National Institutes of Health (GM-23765).

7. References

1. G.R. Freeman, Ann. Rev. Phys. Chem. <u>34</u>, 463 (1983).
2. G.A. Kenney-Wallace, in <u>Photoselective Chemistry Part 2</u>, ed. J. Jortner (John Wiley, New York, 1981).
3. R.A. Auerbach, J.A. Synowiec and G.W. Robinson, in <u>Picosecond Phenomena</u>, ed. by C.V. Shank, E.P. Ippen, S.L. Shapiro, Springer Series in Chemical Physics, Vol. 4 (Springer, Berlin, Heidelberg, New York, 1978).
4. D. Huppert, E. Kolodney and M. Gutman, in <u>Picosecond</u> Phenomena, ed. by C.V. Shank, E.P. Ippen, S.L. Shapiro, Springer Series in Chemical Physics, Vol. 4 (Springer, Berlin, Heidelberg, New York, 1978).
5. J. Lee and G.W. Robinson, J. Chem. Phys., August 1, 1984.
6. L. Takács, <u>Stochastic Processes - Problems and Solutions</u> (Methuen, London, 1960); E. Parzan, <u>Modern Probability Theory and Its Applications</u> (John Wiley, New York, 1960).
7. R. Klein and I. Tatischeff, Chem. Phys. Lett. <u>51</u>, 333 (1977).
8. R.A. Moore, J. Lee and G.W. Robinson, J. Am. Chem. Soc., submitted.
9. D. Huppert, E. Kolodney, M. Gutman and E. Nachliel, J. Am. Chem. Soc. <u>104</u>, 6949 (1982).

Picosecond Multiphoton Laser Photolysis and Spectroscopy of Liquid Benzenes

H. Miyasaka, H. Masuhara, and N. Mataga

Department of Chemistry, Faculty of Engineering Science, Osaka University
Toyonaka Osaka 560, Japan

The high peak power of the ps laser pulse easily induces nonlinear phenomena such as multiphoton excitation as well as multiphoton decomposition and ionization of molecules. We have already reported excimer formation of liquid benzene by the nonlinear excitation of the 355 nm ps laser pulse [1] and the production of solvated electron by 266 nm ps multiphoton excitation of water and alcohols [2]. This characteristic behavior under the excitation of powerful ps laser is similar to that induced by radiolysis and comparison of the results obtained by ps radiolysis with that by multiphoton laser photolysis will be an interesting problem. In this report, results of investigation on the mechanism of excimer formation in neat benzene and its alkyl derivatives induced by multiphoton excitation will be given.

A microcomputer-controlled ps laser photolysis system with repetitive mode-locked Nd^{3+}:YAG laser [3,4] was used to measure transient absorption spectra. The third harmonic (355 nm, 22 ps FWHM) was used for excitation. Ps continuum generated by focussing the fundamental pulse into D_2O liquid was used as monitoring light. A double beam optical arrangement was adopted and spectral range of 380 nm can be covered with a multi-channel photodiode array by one shot.

Fig. 1(a) shows the time-resolved transient absorption spectra of liquid benzene. A tail-like absorption at 0 ps which may be due to precursor species is gradually replaced by a broad band with a peak around 505 nm which is due to excimer [4,5]. Fig. 1(b) shows the plot of benzene excimer

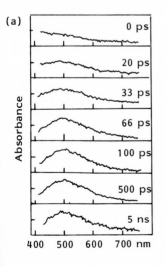

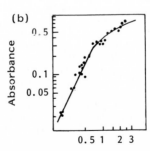

Fig. 1 (a) Time-resolved absorption spectra of degassed neat benzene. (b) Benzene excimer absorbance (at 100 ps) vs. excitation pulse intensity relation.

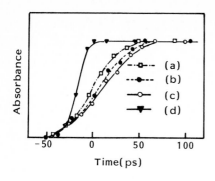

Fig. 2 Absorbance rise curves of excimer and hydrated electron. The rise times were obtained to be 57 ps, 65 ps, 73 ps and 27 ps for benzene (a), toluene (b), t-butyl-benzene (c) excimers and hydrated electron (d), respectively (see text).

absorbance at 100 ps after excitation against the excitation intensity. A linear relation of the slope of 1.93 was obtained, which indicates that the monomer S_1 and excimer state are produced by simultaneous two photon absorption because benzene has no absorption at 355 nm.

In order to obtain excimer formation rates of benzene and alkyl benzenes in pure liquid state, we have measured the excimer absorbance rise curves as shown in Fig. 2 by this two-photon absorption method. In this figure, the rise curve of the absorbance of hydrated electron in water produced by multi-photon excitation of the 266 nm ps laser pulse represents the time responce of the apparatus. The rise times of absorbances from 10% to 90% of their plateau values were obtained as given in Fig. 2. These rise times should be given as a convolution of excitation pulse, monitoring one and true formation time (t_0) of excimer. Assuming simple reaction scheme, M*+M→Ex, for excimer formation process, deconvolutional analysis by computer simulation method was performed. According to this result, the true excimer formation time was estimated to be about 20~30 ps, and observed rise time of the hydrated electron was found to correspond to t_0=0. On the other hand, in the case of the 266 nm ps laser pulse employed as excitation light, about 10 ps was obtained as the excimer formation time of liquid benzene. Further, from the bimolecular rate constant of excimer formation of benzene, (2±1) ×10^{10} M^{-1} s^{-1}, which was obtained by transient absorption measurement upon 0.4 M cyclohexane solution excited with the 266 nm ps pulse, the excimer formation time can be estimated to be less than 5~10 ps. These values are smaller than that obtained by means of multiphoton excitation of 355 nm. Possible mechanism for this delayed excimer formation may be as follows. In gas phase, benzene is nonradiatively deactivated when excited to higher levels than 5.1 eV (the so-called third channel). However, in liquid phase, weak fluorescence can be observed even when excited at 5.1 eV, and fluorescence yield decreases with the increase of the excitation energy [6]. Moreover, when excited to higher states than 6.5 eV, it has been concluded from the experiments using electron scavenger that liquid benzene undergoes geminate ion-pair formation or photoionization [6]. From this sort of studies and the measurements of excitation spectra of two-photon fluorescence of liquid benzene, 6.8~7.6 eV has been proposed as the ionization potential of liquid benzene, and moreover it has been concluded that benzene→benzene charge transfer (CT) state exsists around 6.5 eV [7,8].

As indicated in Fig. 3, the 355 nm two-photon excitation energy is close to the ionization potential in liquid state. Therefore, this two-photon excitation will produce geminate ion-pair which is a probable precursor of excimer. In order to confirm this possibility, we have examined the effect of addition of other liquids upon the excimer rise times of benzene solution. The rise time was obtained to be 45 ps with 20% cyclohexane, 75 ps with 20%

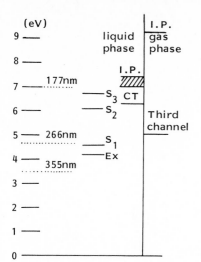

(eV)

Fig. 3　Energy levels of benzene

ethanol, and 89 ps with 20% acetonitrile compared with 57 ps in neat benzene, which shows the excimer rise time increases with the increase of the solvent polarity. This result indicates strongly that the ionized or geminate ion-pair state is produced by two-photon excitation with 355 nm pulse. The life-time of this state is a few 10 ps before recombination leading to the forma-tion of excimer state as well as benzene monomer S_1 state. The lifetime of the ion-pair state will become longer with the increase of the liquid polarity owing to the stabilization of the ion-pair state. Conversely, it might be possible to study the behaviors of the ion-pair state by observing the excimer formation dynamics. The present results of ps multiphoton excitation of liquid benzenes clearly show the potentiality of this method for the study of the excited state of liquid, in general.

References

1. H. Masuhara, H. Miyasaka, N. Ikeda, and N. Mataga, Chem. Phys. Lett. 82, 59 (1981).
2. H. Miyasaka, H. Masuhara, and N. Mataga, Chem. Phys. Lett. 98, 277 (1983).
3. H. Miyasaka, H. Masuhara, and N. Mataga, Laser Chem. 1, 357 (1983).
4. N. Nakashima, M. Sumitani, I. Ohmine, and K. Yoshihara, J. Chem. Phys. 72, 2226 (1980).
5. (a) R. Cooper and J. K. Thomas, J. Chem. Phys. 48, 5097 (1968). (b) R. Bonneau, J. Joussot-Dubien, and R. Bensasson, Chem. Phys. Lett. 3, 353 (1969). (c) J. K. Thomas and I. Mani, J. Chem. Phys. 51, 1834 (1969).
6. F. P. Schwarz and M. Mautner, Chem. Phys. Lett. 85, 239 (1982).
7. T. W. Scott, and A. C. Albrecht, J. Chem. Phys. 74, 3807 (1981).
8. T. W. Scott, C. L. Braun, and A. C. Albrecht, J. Chem. Phys. 76, 5195 (1982).

Chemical Reactions in Condensed Media

D. Statman, W.A. Jalenak, and G.W. Robinson

Picosecond and Quantum Radiation Laboratory, Texas Tech University
Lubbock, TX 79409, USA

The generalized Langevin equation (GLE) is given by

$$\dot{\vec{p}}(t) = \vec{F}[\vec{x}(t)] - \int_0^t d\tau \; \beta(\tau)\vec{p}(t-\tau) + \vec{f}(t) \tag{1}$$

where $\vec{F}[\vec{x}(t)]$ is the position dependent force, $\vec{f}(t)$ is a stochastic force, and $\beta(t)$ is a memory function $(k_B T \mu)^{-1}\langle\vec{f}(t)\cdot\vec{f}(0)\rangle$, where μ is the mass and k_B is Boltzmann's constant. The zero-frequency friction is $\int_0^\infty dt\beta(t) \equiv \beta_0$.

In the case of chemical reactions $\vec{F}[\vec{x}(t)]$ can be approximated to be $\mu\omega_b^2\vec{x}(t)$, ω_b representing the harmonic curvature at the top of the potential barrier [1]. For fluids, one can think of two limiting cases:

(A) $\beta(t) = \beta_0\delta(t)$ $\qquad\qquad\qquad\qquad\qquad\qquad\qquad\qquad$ (2)

(B) $\beta(t) = \beta(0)e^{-\gamma t}\cos\omega t$. $\qquad\qquad\qquad\qquad\qquad\qquad$ (3)

Laplace transformation of $\beta(t)$ gives the "frequency dependent friction"

(A) $\hat{\beta}(s) = \beta_0$ $\qquad\qquad\qquad\qquad\qquad\qquad\qquad\qquad\qquad$ (4)

(B) $\hat{\beta}(s) = \beta(0)(s+\gamma)/[(s+\gamma)^2+\omega^2]$. $\qquad\qquad\qquad$ (5)

Grote and Hynes [2] have shown that the positive solution of

$$s^2 - \omega_b^2 + s\hat{\beta}(s) = 0 \tag{6}$$

is the reactive frequency λ determining the rate constant, $k=(\lambda/\omega_b)k^{TST}$. Case (A), with the delta-function form for $\beta(t)$, is the Kramers limit. In case (B), the quartic equation

$$s^4 + 2s^3\gamma + s^2[\beta(0)+\gamma^2+\omega^2-\omega_b^2] + s\gamma[\beta(0)-2\omega_b^2] - \omega_b^2(\omega^2+\gamma^2) = 0 \tag{7}$$

must be solved.

For cis-trans isomerization, the reaction barrier is characterized by a periodic potential function. We choose the form

$$V(\theta) = V_0[\sin^2\theta + 4q\sin^2{\tfrac{1}{2}}\theta] \tag{8}$$

where θ is the dihedral angle. The barrier frequency for $|q|<0.25$ is

$$\omega_b \approx \left\{2V_0[1-\tfrac{1}{2}q\sin^{-1}2q]/Y\right\}^{\frac{1}{2}} \tag{9}$$

320

where Y is a "reduced" moment of inertia [3]. Henceforth, as a first approximation, the "disymmetry factor" q will be assumed zero.

The barrier frequency must be consistent with the barrier height. For example, in the case of stilbene [4,5], where $Y=4.95\times10^{-38}$ gcm^2 and $V_0=3.54$ kcal mol^{-1}, ω_b from (9) is 3.15×10^{12} s^{-1}. Experimentally, ω_b has been independently determined [5] to be about 1.0×10^{13} s^{-1}. The difference may be caused by $q \neq 0$, uncertainties in Y or V_0, or entropic factors that elevate the experimental rates.

Molecular dynamics calculations show that for a Lennard-Jones fluid at various temperatures and densities, the force auto-correlation function is like case (B), a damped cosine, whose power spectrum allows determination of ω and γ. Knowing $\beta(0)$, γ, ω, ω_b, and using (7), one may readily solve for λ and compare these results with the rate constants obtained directly from the molecular dynamics calculations.

Figure 1 shows the force autocorrelation function along the reaction coordinate for a zero barrier. Figure 2 compares the power spectrum of the force along the reaction coordinate to that for bulk solvent. Note that β_0 is ~50× larger for the solvent than for the reaction coordinate. This shows that, at least in this system, there is a distinct difference between the bulk friction and friction along the reaction coordinate.

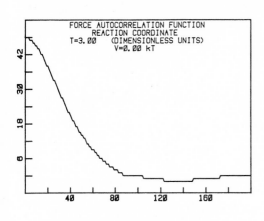

Fig. 1 Force auto-correlation function along reaction coordinate; dimensionless units ($\tau=\sigma\sqrt{m/\epsilon}\div1000$). Note tendency for damped oscillatory behavior

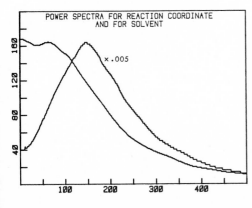

Fig. 2 Power spectra (Fourier transforms) of force auto-correlation functions along cis-trans reaction coordinate and for a solvent molecule; dimensionless units ($\omega=2\pi/32768\tau$)

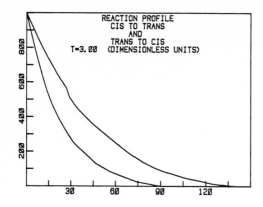

Fig. 3 Reaction profile comparing cis→trans with trans→cis rates; dimensionless units ($\tau = \sigma\sqrt{m/\varepsilon} \div 3000$)

It is also interesting to note that the cis→trans rate is slower than the trans→cis rate (Fig. 3). This preference for cis over trans, in the absence of an energy difference, is probably caused by entropic factors.

The authors are indebted to the National Science Foundation (CHE-8215447) for financial support of this research. David Statman acknowledges the Robert A. Welch Foundation (K-099e) for his Fellowship.

References

1. H.A. Kramers, Physica 7, 284 (1940).
2. R.F. Grote and J.T. Hynes, J. Chem. Phys. 73, 2715 (1980).
3. G. Herzberg, Infrared and Raman Spectra, pp. 225-227, D. Van Nostrand Co. Inc. Princeton, N.J., 1945.
4. G. Rothenberger, D.K. Negus, and R.M. Hochstrasser, J. Chem. Phys. 79, 5360 (1983).
5. G.R. Fleming, private communication.

Subpicosecond and Picosecond Time Resolved Laser Photoionization of Phenothiazine in Micellar Models

Y. Gauduel, A. Migus, J.L. Martin, J.M. Lemaître, and A. Antonetti

Laboratoire d'Optique Appliquée, INSERM, Ecole Polytechnique-ENSTA
F-91120 Palaiseau, France

1. Introduction

Aqueous surfactant solution permits the organization of the reactants at a molecular level and can be used as simple model of electron transfer (1) (2) (3). Ionic micellar solutions which contain a diffuse double layer (Gouy-Chapman region) may be typical of interfacial systems such as membranes. These agglomerates are useful to investigate the effect of the electrostatic potential between the interior and the bulk aqueous phase in the process of electron solvation.

This paper reports an experimental work on the process of photo-ejection of electrons in different micellar solutions (aqueous or reversed micelle), studied by subpicosecond laser spectroscopy.

2. Experimental

Purified anionic micelles are prepared by dissolving phenothiazine (PTH) in sodium lauryl sulfate solution (SLS) and by gently stirring the mixture at 60°C for several hours. SLS is purified by several recrystallizations from methanol. In other experiments, micellar aggregates of aerosol-OT in apolar solvent are used. AOT, supplied by the Fluka Chemical Co, is purified by dissolving it in methanol with active charcoal, filtering and drying in vacuo. Reversed micelles are prepared by dissolving AOT in heptane and adding a precise volume of water to the mixture. PTH is added after the preparation of reversed micelles. Care is taken to exclude light. Water is deionized and doubly distilled. Experiments are performed at ambient temperature (20°C) in fixed volume quartz cuvette of 0.2 cm light path length, continuously moved horizontally.

Laser photoionization of PTH is performed starting from the 100 fs amplified output of a CW passively mode locked dye ring laser (4). Half of the energy is used to generate a broad band femtosecond continuum, while the other half is focussed into a KDP crystal to induce an excitation pulse of 20 μJ at 310 nm. The continuum beam is divided in two parts and used as double beam spectrophotometer. Kinetics at different wavelengths are obtained by two diodes located at the exit slit of a 0.25 m spectrometer. The two signals (probe and reference pulses) are directed through an electronic chain to a computer for treatment of data.

3. Results

Subpicosecond photolysis of PTH ($1.5 \ 10^{-4}$M) incorporated into SLS micelle (0.1 M) leads to the appearance of an absorption at 750 nm (Figure 1). Such an effect in the red spectral region may be attributed to hydrated electron (e_{aq}^{-}) because addition of acetone reduces the amplitude of the signal and decreases the lifetime of this transient absorption with a rate constant of $0.91 \ 10^{10}$M^{-1} s^{-1}. This result agrees with the values found by other authors (5,6).

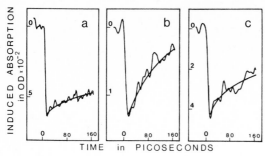

Figure 1 : Transient absorption at 750 nm showing the decay
of solvated electron in micellar solution
(a) PTH $(1.5 \ 10^{-4}M)$, SLS (0.1M) (b) PTH $(1.5 \ 10^{-4}M)$,
SLS (0.1M) + acetone (1M) (c) : PTH $(1.0 \ 10^{-4}M)$, SLS (0.025 M).

In *anionic micelles*, the risetime of hydrated electron absorption at
750 nm is not instantaneous and the best fit is obtained with a time constant of *500 femtoseconds* (figure 2a) as compared to the instantaneous response (figure 2b). A subsequent decay of the absorption at 750 nm is observed within several hundred picoseconds (figure 1). The lifetime of e_{aq}^{-} is
defined by the surfactant concentration. The kinetics of 750 nm absorption
decay may be well fitted in our time domain observation to a random walk
form $(\exp - \sqrt{t/\tau})$: $\tau \simeq 1100$ ps (figure 1a) ; $\tau \simeq 250$ ps (figure 1c).

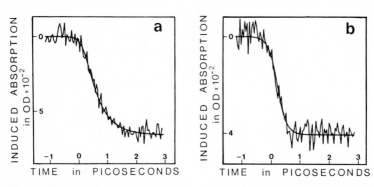

Figure 2 : Subpicosecond laser photolysis of PTH in micellar solutions.
Transient absorption at 750 nm shows the hydrated electrons
apparition in the water pool.
a) anionic micelles : PTH $(1.5 \ 10^{-4}M$, SLS (0.1M)
b) reversed micelles : AOT (0.07M), W_o = 28, PTH $(1.5 \ 10^{-4}M)$
The smooth lines indicate the calculated best fits.

Results of electron solvation in *reversed micelles*, in which water molecules are entrapped in the hydrophylic core, are shown in figure 2. When the
$[H_2O]/[AOT]$ ratio (W_O) is equal to 28, the capture of electron in water pool
and the formation of solvated electrons appear within the apparatus time
resolution (*200 fs* in this case due to group velocity dispersion between
310 nm and 750 nm in the 2 mm long sample).

324

4. Discussion

The data obtained with different micellar systems show that the photoionization of phenothiazine by *subpicosecond UV pulses* induces electron solvation process. In previous experiments, it has been demonstrated (6) that the electron solvation process in aqueous solution occurred in less than 0.3 ps. We find here that the organization of anionic micelles induces a slight delay of 0.5 ps in the electron solvation, while this delay is not found in aerosol-OT-heptane reversed micelle. More precisely, in the latter case, we give the upper limit of *100 fs* to the appearance time of the induced absorption at 750 nm. Although the kinetics of electron solvation in micellar solution is influenced by the type of the organized assemblies, the solvation process remains ultrafast and would argue for the mechanism of electron capture in water bulk similar to that observed in aqueous solutions including pre-existing deep traps (6). In agreement with the results of Pileni et al. (7), our results show that in reversed micelle with W_O higher than 20, the organization of water molecules entrapped in the hydrophylic core is comparable to that in homogeneous aqueous solutions. The lifetime of e^-_{aq} in anionic micellar aggregates is influenced by the SLS concentration and consequently by the organization of the interface formed by the polar head groups of the surfactant. The random walk form of the 750 nm absorption decay would indicate (8,9) that e^-_{aq} must diffuse through the organized assemblies before a likely quenching reaction with oxygen. In this hypothesis, the *charged micellar surface* or the *microviscosity* of the boundary layer can influence the reactivity of solvated electron with oxygen.

5. References

1. S.A. Alkaitis, G. Beck, M. Gratzel. J. Am. Chem. Soc. 97, 5723-5728 (1975).
2. M.P. Pileni. Chem. Phys. Lett. 81, 603-605 (1981).
3. A.J.W.G. Visser, J.H. Fendler. J. Phys. Chem. 86, 947-950 (1982).
4. A. Migus, J.L. Martin, R. Astier, A. Antonetti, A. Orszag, in Picosecond Phenomena III, Springer Series in Chemical Physics, 23, 6-9 (1982).
5. J.E. Aldrich, M.J. Bronskill, R.K. Wolff, J.W. Hunt. J. Chem. Phys. 55, 530 (1971).
6. J.M. Wiesenfeld, E.P. Ippen. Chem. Phys. Lett. 73, 47-50 (1980).
7. M.P. Pileni, B. Hickel, C. Ferradini, J. Pucheault. Chem. Phys. Lett. 92, 208-312 (1982).
8. R.H. Austin, K. Beeson, L. Eisenstein, H. Frauenfelder, I.C. Gunsalus, V.P. Marshall. Phys. Rev. Lett. 32, 403-405 (1974).
9. D.A. Chernoff, R.M. Hochstrasser, A.W. Steeve, Proc. Natl. Acad. Sci. USA, 77, 5606-5610 (1980).

Isomerization Intermediates in the Photochemistry of Stilbenes

F.E. Doany[1] and R.M. Hochstrasser
Department of Chemistry, University of Pennsylvania
Philadelphia, PA 19104, USA

A series of experiments were carried out to spectroscopically identify intermediates in the trans-to-cis isomerization of stilbenes. The molecule used in these investigations is a fused ring analog of stilbene (1), trans-1,1'-biindanylidene (2). The photophysical properties of this "stiff" stilbene are very similar to those of stilbene [1] with the difference accounted for by a smaller barrier (1.5 kcal/mol) to isomerization than in t-stilbene.

(1)   (2)

This low barrier results in an extremely fast (3 ps in hexane) isomerization [2], thus providing an ideal system for investigations of twisted intermediates since it would cause an increased population of short-lived intermediates compared with molecules with slower isomerization rates.

Following ultraviolet excitation using the fourth harmonic of a mode-locked Nd:phosphate glass laser (264 nm), stiff stilbene exhibits a transient absorption at 351 nm as well as a visible absorption in the 450-600 nm region of the spectrum [3, 4]. The time evolution of the visible absorption was monitored at 527 nm for stiff stilbene in hexane and in hexadecane, and the results are shown in Figure 1-(a). The data are in good agreement with calculated curves for 3 ps and 26 ps exponential decays convoluted with a 10 ps pump-probe instrument function (solid lines). These decay times correspond to the fluorescence lifetimes reported in reference 2 using quantum yields and streak camera measurements. The apparent decrease in signal amplitude with decreasing viscosity is a consequence of the sub-pulsewidth decay time.

The time evolution of the ultraviolet absorption in these two solvents is shown in Figure 1-(b). The results exhibit a rapidly evolving component during the first ∿ 50 ps and a constant absorption for delays greater than 50 ps. The latter feature was still present on a timescale of several seconds and, by means of a nanosecond absorption spectrometer [3] was attributed to the cis photoproduct.

---

[1]Present address: A. A. Noyes Laboratory of Chemical Physics, California Institute of Technology, Pasadena, CA. 91125, USA

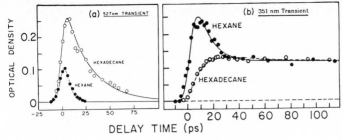

Fig. 1. Time evolution of the transient absorption at (a) 527 nm and (b) 351 nm of stiff stilbene in hexane and in hexadecane.

The early time signal amplitude increases with decreasing viscosity, which is the opposite trend as observed in the visible absorption. These results suggest that the ultraviolet and visible absorptions originate from different states. The viscosity (and temperature) dependence of the 351 nm absorption is consistent with the decay of the excited trans being the precursor of the new transient. A kinetic model consistent with these observations is presented below:

$$t \xrightarrow{\hspace{0.5cm} h\nu \hspace{0.5cm}} t^* \tag{1}$$

$$t^* \xrightarrow{\hspace{0.5cm} k_{tp}(\eta) \hspace{0.5cm}} p \tag{2}$$

$$p \xrightarrow{\hspace{0.5cm} k_p \hspace{0.5cm}} \alpha c + (1-\alpha)t \tag{3}$$

where t, c, and p are the trans, cis and twisted intermediate configurations, and * denotes electronic excitation. $k_{tp}(\eta)$ is the viscosity dependent rate constant for the formation of the intermediate and k_p is the rate constant for the decay of the intermediate. Quantitative analysis of the data using the above model suggests that after the viscosity dependent formation of the intermediate (p) it decays away with a time constant of 10 + 3 ps (see solid line fits in Fig. 1- b). The most evident assignment of this intermediate is the twisted "phantom" state of stiff stilbene.

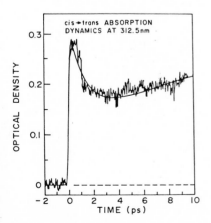

Figure 2. 0.1 ps resolved time-evolution of the cis-to-trans isomerization of stilbene monitored by absorption at 312 nm.

327

Evidence for the analogous intermediate of stilbene was also observed in recent experiments using sub-picosecond time resolution [5]. According to the above model (eqs. 1-3), the dynamics of the intermediate state can also be determined from the ground state cis or trans photoproduct formation. The appearance of trans-stilbene photoproduct absorption at 312 nm following excitation of cis-stilbene was monitored with 0.1 ps resolution. The results, shown in Figure 2, exhibit a pulsewidth limited rise with a fast (< 2 ps) decay superimposed on a slower rise. The initial decay is attributed to an excited cis absorption, which has a lifetime of about 1.3 ps [5,6]. The latter feature is the rise of the trans absorption. Kinetic analysis of the data invoking the presence of an excited cis and a phantom state absorption as well as the trans photoproduct absorption indicates that a 10 ps bottleneck is not inconsistent with the observed recovery dynamics (solid line in Fig. 2). Further studies are currently underway.

Acknowledgement: We are grateful to B. I. Greene for his invaluable collaboration resulting in Figure 2 and a forthcoming publication [5].

References

1. J. Saltiel and J. D'Agostino: J. Am. Chem. Soc. 94, 6445 (1972).
2. G. Rothenberger, D. K. Negus and R. M. Hochstrasser: J. Chem. Phys. 79, 5360 (1983).
3. F. E. Doany, E. J. Heilweil, R. Moore and R. M. Hochstrasser: J. Chem. Phys. 80, 201 (1984).
4. F. E. Doany: Ph.D. Thesis, University of Pennsylvania (1984).
5. F. E. Doany, R. M. Hochstrasser and B. I. Greene: to be published.
6. B. I. Greene and T. W. Scott, Chem. Phys. Lett. 106, 399 (1984).

Part VI

Molecular Energy Redistribution, Transfer, and Relaxation

Picosecond Laser Studies on the Effect of Structure and Environment on Intersystem Crossing in Aromatic Carbenes

E.V. Sitzmann, J.G. Langan, Z.Z. Ho, and K.B. Eisenthal

Department of Chemistry, Columbia University, New York, NY 10027, USA

Carbon is the most important atom in organic and biological chemistry. In nature and in the laboratory we find that the stable (low energy) form of carbon requires the formation of four chemical bonds, i.e., carbon is tetravalent. In this paper we will be concerned with abnormal, high energy and exceedingly reactive forms of carbon which exist only as reactive intermediates. The properties and reactions of this form of carbon must be measured on time scales ranging from a microsecond to a picosecond or less. In the reactive transients we will consider the carbon has only two bonds, i.e., carbon is in its divalent form. These divalent carbon intermediates constitute a family of transients called carbenes [1].

Of crucial importance in determining the spectroscopy, dynamics of energy relaxation and chemistry of carbenes are its two lowest electronic states [2,3]. The nature of these electronic states is dependent on groups R_1 and R_2 attached to the central carbon. The ground state for the carbenes we are considering are each found to have a spin multiplicity of three, i.e., a triplet, with a closely neighboring singlet state [4]. The occupancy of the nonbonding molecular orbitals of the carbene determines the multiplicity of the state as shown in Fig. 1 with diphenylcarbene as the example.

The intramolecular dynamics of the aromatic carbenes are dominated by intersystem crossing between the lowest singlet and ground triplet states. In the solvent acetonitrile at room temperature we found for diphenylcarbene

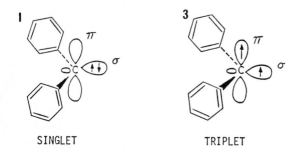

SINGLET   TRIPLET

Figure 1. Electronic configuration of lowest singlet and ground triplet states of diphenylcarbene. The σ orbital is in the plane and the π orbital is out of the plane of the carbon-phenyl bonds

330

(DPC) that the intersystem crossing time, k_{ST}^{-1}, from the singlet to the triplet was 310 ps and the reverse triplet to singlet crossing time, k_{TS}^{-1}, was 90 ns [5]. The rate constant for ^{3}DPC (ground triplet) formation (k_{ST}) was measured by means of a laser-induced fluorescence experiment following the photodissociation of the carbene precursor, diphenyldiazomethane [3b,5].

$$PH_2C=N_2 \xrightarrow[-N_2]{h\nu} {^1}PH_2C: \xrightarrow{k_{ST}} {^3}PH_2C:$$

On obtaining these kinetic results for DPC in acetonitrile, one would like to know how applicable these findings are to other solvent environments, i.e., to what extent does the solvent affect the intramolecular spin conversion dynamics of the carbene? Prior to this work it was generally assumed that the solvent had no effect on the carbene spin conversion rates.

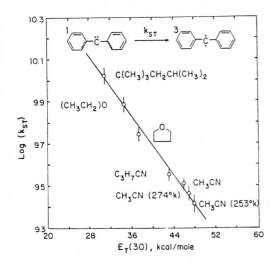

Figure 2. Diphenylcarbene intersystem crossing rate as a function of the solvent polarity parameter $E_T(30)$

To address this issue the singlet to triplet rate constant, k_{ST}, of DPC was measured in a series of solvents [5a]. The results obtained showed that the intersystem crossing rate depends strongly on the choice of solvent (Fig. 2). In isooctane, a nonpolar hydrocarbon, the time of intersystem crossing, k_{ST}^{-1}, was found to be 95 ps, a factor of three faster than in acetonitrile (310 ps). For the wide range of solvents used, it turned out that there was a clear ordering between the rate of singlet to triplet conversion and the polarity of the solvent, as measured by the solvent polarity parameter $E_T(30)$ [6], the rate being larger in the less polar solvents. No correlation was obtained with other solvent parameters which could be related to possible carbene-solvent complexes or parameters such as viscosity which could influence the dynamics of the structural change in the transition from the singlet to the triplet.

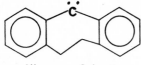

dibenzocyclohepta-
dienylidene (DCHD)

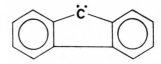

fluorenylidene (Fl)

331

To determine if the marked solvent polarity effect on singlet to triplet energy relaxation was peculiar to diphenylcarbene we studied another aromatic carbene, dibenzocycloheptadienylidene (DCHD). This molecule is similar to DPC, differing by an ethylenic bridge which connects the two phenyl rings to the central carbon atom. One effect of the restricting CH_2 groups in DCHD is to increase the angle between the central carbon and the connected phenyl rings relative to DPC. From our measurements of k_{ST} in a variety of solvents we find that DCHD also manifests an exponential dependence of k_{ST} on the solvent polarity parameter $E_T(30)$, Fig. 3. The slopes for DPC and DCHD differ but the same exponential dependence is seen. There has been a recent and most interesting report on another carbene which is in agreement with the solvent trend reported here. Fluorenylidene, in solution at room temperature, was observed to have a faster intersystem crossing rate in the nonpolar solvent cyclohexane than in the polar solvent acetonitrile [7]. From these various studies we can conclude that the solvent does indeed play a key role in singlet-triplet relaxation in aromatic carbenes, perhaps in most carbenes, due to the sharp differences in the polarities of the lowest singlet and triplet states.

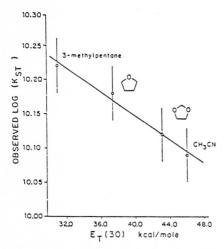

Figure 3. DCHD intersystem crossing rate as a function of the solvent polarity parameter $E_T(30)$

To understand how solvent polarity could affect spin conversion in carbenes consider the electronic nature of the singlet and triplet states. The singlet state, with both nonbonding electrons localized in the same $(\sigma)^2$ orbital of the central carbon, will be significantly more polar than the triplet state $(\sigma)^1(\pi)^1$ (Fig. 1). It is, therefore, to be expected that the polar singlet will be strongly stabilized, i.e., lowered in energy in the more polar solvents, whereas the less polar triplet will experience less stabilization. This differential stabilization between the singlet and triplet would, therefore, result in the singlet-triplet energy gap decreasing as the solvent polarity increases. This prediction that the singlet-triplet energy gap is larger in the nonpolar than in the polar solvent was confirmed experimentally in DPC. We found from measurements of DPC in nonpolar isooctane that the energy gap is 1300 cm^{-1} compared with 950 cm^{-1} for polar acetonitrile [5].

The issue now becomes why intersystem crossing, which involves the conversion of electronic energy of the singlet into vibrational energy of the triplet, increases in rate as the energy gap increases. This is a seemingly surprising result in that intersystem crossing usually decreases as the gap increases, at least for large gaps. The key factor may be that aromatic carbenes being considered here are in a small energy gap domain. The consequence of the small energy gap is that there is a sparse manifold of triplet vibronic states (ground triplet plus vibrations) with which the singlet is effectively coupled. The appropriate description is an off-resonance coupling between S_1 and a number of triplet vibronic levels [8]. The vibrations in the triplet which provide the coupling with the singlet are in turn coupled to other vibrations in the molecule and the solvent background. The level structure of these triplet vibronic levels becomes increasingly congested in the neighborhood of S_1 as the energy gap increases due to the possible contributions of higher frequency vibrations as the energy increases, and due to the decrease in the spacing between triplet vibronic energy levels as the energy gap increases. Accordingly the rate of $S_1 \rightarrow T_0$ intersystem crossing increases.

The physical picture for the inverse energy gap affect is consistent with observations of singlet to triplet intersystem crossing in three different aromatic carbenes. They are diphenylcarbene, dicycloheptadienyldene and fluorenylidene. In each case it is found that k_{ST} is larger in those solvents for which the carbene has the larger singlet-triplet energy gap. This description is not anticipated to be applicable when the energy gap increases beyond a few thousand cm^{-1} . We would expect that as the gap becomes larger the increasing density in the vibronic states of the triplet manifold which couple to S_1 will lead to a declining importance of the inverse gap effect. When the energy gap becomes large we anticipate approaching the usual energy gap effect found in the statistical limit, i.e., the rate of $S \rightarrow T$ intersystem crossing decreases with increasing energy gap due to poorer Franck-Condon factors [9].

Comparing the intersystem crossing rate k_{ST} for the three aromatic carbenes for which there is data we can consider the effects of carbene structure on the spin conversion process. The values of k_{ST} in a given solvent are in the order k_{ST}(DCHD) > k_{ST}(DPC) > k_{ST}(Fl). The same ordering is found in the magnitude of the energy gaps ΔE(DCHD, $\sim$1700 cm^{-1} [3c]) > ΔE(DPC, $\sim$950 cm^{-1} [3b]) > ΔE(Fl, $\sim$600 cm^{-1} [7]). The origin of the ordering of the k_{ST} values could be simply due to the differences in the spin-orbit matrix elements in the three molecules. However, it is tempting to speculate that the major factor is the energy gap, i.e., the larger the energy gap the faster is the intersystem crossing k_{ST}. As ΔE_{ST} increases the greater congestion of triplet vibronic states that are coupled to S_1 results in larger k_{ST} values. The physical picture is the same as we used to explain the solvent effect on the k_{ST} values for each of these carbenes.

To understand the origin of the variations of ΔE_{ST} and thus that of k_{ST} we consider how the different geometrical structures could alter ΔE_{ST}. Theoretical calculations on DPC indicate that the singlet-triplet splitting decreases as the angle defined by the two bonds to the central carbon decreases [10]. Qualitatively as the molecule goes from a linear to a bent geometry, i.e., the central angle decreases, the percent 2s character in the σ orbital increases, thereby lowering the σ orbital energy. Including steric effects it is found that the singlet $(\sigma)^2$ is more stabilized than the triplet $(\sigma)^1(\pi)^1$ leading to a decrease in energy gap as the angle decreases. The evidence that supports an increase in the central angle in the series DCHD > DPC > Fl is obtained from electron paramagnetic resonance data [11]. The value of the zero field splitting parameter $|E|$ (which is a measure of the deviation of the spin-spin interaction from cylindrical symmetry in the spin dipole-dipole Hamiltonian) is an indication of the 2s character in the orbital. Crude estimates of 2s character can also be obtained from the ratio of the $|E|/|D|$ values where $|D|$ is proportional to the separation of the unpaired

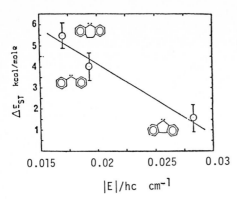

Figure 4. ΔE_{ST} as a function of the zero field splitting parameter $|E|/hc$ for aromatic carbenes

spins. Using either $|E|$ or $|E|/|D|$ we find that the central angle and thus the percent 2s character decreases as the angle increases and thus accounts for the observed trend of k_{ST} with ΔE_{ST} (fig. 4.)

Conclusions

Using picosecond laser techniques we have observed a dramatic solvent effect on the rate constant k_{ST} of intersystem crossing in the aromatic carbenes, diphenylcarbene (DPC) and dicycloheptadienylidene (DCHD). The solvent polarity is the dominant feature responsible for changes in k_{ST}. It is found that the singlet-triplet energy gap changes with polarity and thereby alters the intersystem crossing dynamics. Increasing polarity decreases the gap which leads to a slower rate constant k_{ST}. This "inverse" gap effect, i.e., the time for intersystem crossing decreases with increasing energy gap, is explained by an off-resonance intersystem crossing from the singlet to a sparse triplet vibronic manifold characteristic of a small energy gap. In comparing the intersystem crossing rate constants of the three different aromatic carbenes DPC, DCHD and fluorenylidene in a common solvent, we find that the intersystem crossing time scales inversely with the singlet-triplet energy gap. This is the same trend of k_{ST} with the solvent-induced changes in energy gap we have already noted for each of the carbenes. The different energy gaps for the carbenes are speculated to arise from the structural differences of the central divalent carbon to ligands angle which in turn alters the singlet to triplet intersystem crossing dynamics.

Acknowledgements

We wish to thank the National Science Foundation, the Air Force Office of Scientific Research, and the Joint Services Electronic Program 29-82-K-0080 for their generous support of this work.

References

1. (a) Kirmse, W.: Carbene Chemistry 2nd ed. (Academic Press, New York 1971)
   (b) Moss, R.A., Jones, M., Jr., Eds: Carbenes, Vol. 1 and 2 (Wiley Interscience, New York 1975) (c) H. Durr: Topics in Curr. Chem. 55, 87 (1975)
2. G.L. Closs, B.E. Rabinow: J. Am. Chem. Soc. 98, 8190 (1976). (b) D. Bethe E. Whittaker, J.D. Callister: J. Chem. Soc 2466 (1965) (c) D. Bethel, J.D. Stevens, P. Tickel: Chem. Comm. 792 (1970) (d) P.P. Gasper, B.L. Whitsel, M. Jones Jr., J.B. Lambert: J. Am. Chem. Soc. 102, 6180 (1980)

3. (a) K.B. Eisenthal, N.J. Turro, J.A. Aikawa, J.A. Butcher, Jr., C. Dupuy, G. Hefferon, W. Hetherington, G.M. Korenowski, M.J. McAuliffe: Am Chem. Soc. 102, 6563 (1980) (b) K.B. Eisenthal, N.J. Turro, E.V. Sitzmann, I. Gould, G. Hefferon, J. Langan, Y. Cha: Tetrahedron, accepted for publication (c) K.B. Eisenthal, N.J. Turro, J.G. Langan, I.R. Gould, E.V. Sitzmann, to be published.
4. (a) C.A. Hutchinson, B.E. Kohler: J. Chem. Phys. 51, 3327 (1969) (b) R.W. Brandon, G.L. Closs, C.E. Davoust, C.A. Hutchinson, B.E. Kohler, R. Slibey: ibid 43, 2006 (1965) (c) C.A. Hutchinson, G.A. Pearson: ibid 47, 520 (1967) (d) I. Moritani, S.I. Murahashi, M.H. Yamamoto, K. Itoh, N. Mataga: J. Am. Chem. Soc. 89, 1259 (1967)
5. (a) E.V. Sitzmann, J. Langan, K.B. Eisenthal: J. Am. Chem. Soc. 106, 1868 (1984) (b) C. Dupuy, G.M. Korenowski, M. McAuliffe, W. Hetherington, K.B. Eisenthal: Chem. Phys. Lett. 77, 272 (1981)
6. (a) C. Reichardt, in Molecular Interactions, H. Ratajcak, W.J. Orville-Thomas, Eds. (John Wiley, New York, 1982) Vol. 3, 241-282
7. P.B. Grasse, B.E. Brauer, J.J. Zupanic, K.J. Kaufman, G.B. Schuster: J. Am. Chem. Soc. 105, 6833 (1983)
8. A. Nitzan, J. Jortner: Theor. chim Acta (Berl.) 29, 97 (1973)
9. For a discussion of radiationless transitions, see: (a)C.W. Robinson, R.P. Frosch: J. Chem. Phys. 37, 1962 (1962); ibid 38, 1187 (1963) (b) J. Jortner, S.A. Rice, R.M. Hochstrasser: Adv. Photochem. 7, 149 (1969) (c) B.R. Heury, W. Siebrand, in Organic Molecular Photophysics, B.J. Birks, Ed., (Wiley, New York, 1973), Vol. 1, 153, (d) Freed, K., in Topics in Applied Physics, F.K. Fong, Ed., (Springer-Verlag, 1976) Vol. 15, 23
10. (a) J. Metcalke, E.A. Halevi: J. Chem. Soc., Perkin Trans II, 634-639 (1977) (b) R. Hoffman, G.D. Zeiss, G.W. Van Dine: J. Am. Chem. Soc. 90, 1485 (1968)
11. (a) E. Wasserman, A.M. Trozzolo, W.A. Yager: J. Chem. Phys. 40, 2408 (1964) (b) E. Wasserman, W.A. Yager, V.J. Kuck: Chem. Phys. Lett. 7, 409 (1970) (c) E. Wasserman, V.J. Kuck, R.S. Hutton, W.A. Yager: J. Am. Chem. Soc. 92, 7491 (1970) (d) J. Higuchi: J. Chem. Phys. 39, 1339 (1963)

Energy Flow from Highly Excited CH Overtones in Benzene and Alkanes

E.L. Sibert III, J.S. Hutchinson, J.T. Hynes, and W.P. Reinhardt

Department of Chemistry, University of Colorado and Joint Institute for Laboratory Astrophysics, University of Colorado and National Bureau of Standards, Boulder, CO 80309, USA

Simple classical and quantal models of the ultrarapid (50 to 250 fs) flow of energy from CH (CD) overtones in benzene (perdeuterobenzene) and in terminal alkyl methyl groups are shown to account for experimental variation in line widths, including isotopic effects.

1. Introduction

Bray and Berry [1] and Reddy, Heller, and Berry [2] have observed room temperature gas phase spectra of CH overtones in benzene, and CH and CD overtones in fully and partially deuterated benzenes. Interpretation of the observed line widths of these aryl CH overtones in terms of lifetimes (deduced by assumption of Lorentzian line shapes) of local CH excitation suggests very rapid dispersal of energy: relaxation takes place on a timescale of 50 to 100 fs. These decay rates were experimentally seen to depend neither on the overall density of states, nor monotonically on the local mode vibrational quantum number, v. Additionally, many unexpected isotope effects were observed. Perry and Zewail [3] observed comparable aryl CH line widths at 2 K in durene (tetramethylbenzene) but simultaneously observed methyl CH overtones with line widths a factor of 3 to 4 narrower, seemingly implying slower energy relaxation. Very recently, Perry and Zewail [4] have taken low temperature (1.8 K) spectra of the v=5 overtone of benzene itself, confirming that the line widths of Refs. [1,2] are at least approximately temperature independent, and presumably putting to rest the suggestion [5] that the room temperature line widths were highly contaminated by hot, or sequence bands.

Contemporaneously with the initial experimental observations, Heller and Mukamel [6] developed a phenomenological theory of aryl CH overtone broadening, and were able to account for some, but certainly not all, of the observed detail of variation of line width with excitation and isotopic substitution. In contrast, it has been the purpose of the present authors to identify specific physical mechanisms leading to the formation of such line shapes. We have found [7] that the essence of both classical and quantal pictures of relaxation of energy initially localized in an anharmonic CH stretch lies in identification of specific Fermi resonances which tune in, or tune out, as the frequency of the oscillator is varied by choice of vibrational excitation, v. As the Fourier transform of the time dependence of such energy flow is directly related to the time independent absorption spectrum, the line shapes are correspondingly explained. We discuss first the classical picture, then the quantum analog.

2. Classical Models of Aryl CH, CD Relaxation

The local mode excitation sequence of CH overtones in benzene is well modeled using a Morse oscillator with fundamental transition frequency

3043 cm^{-1}. Anharmonicity reduces the vibrational frequency to 2700 cm^{-1} at the position of the third overtone. This is nearly twice the local in-plane HCC wag frequency of 1300 cm^{-1}, leading to the possibility of 2:1 Fermi resonance. Assuming that in the bond angle coordinate system the coupling is essentially all kinetic [8], such coupling is analytically calculable from the coordinate dependence of the bend G-matrix element on the large amplitude stretch coordinate, provided that the masses, geometry, and fundamental frequencies are known. Figure 1 indicates such a pathway for energy relaxation: energy flows, via the 2:1 Fermi resonance from the stretch to the wag, and subsequently, through many near 1:1 resonances, into ring motions. As energy leaves the CH stretch, the frequency rises, the Fermi resonance is broken, a bottleneck is created for energy to flow back into the stretch: relaxation is thus classically "irreversible" on timescales of interest. The quantum dynamics is similarly "irreversible" due to the rapid increase in the density of participating ring modes available to the system as energy leaves the CH local mode. Figure 2 shows the quasi-classical probability of finding five quanta of energy in the CH stretch as a function of time. It is evident that the probability decays exponentially, with a timescale of ~60 fs.

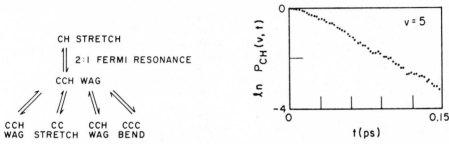

Fig. 1. Fermi resonance mechanism for the ultrarapid flow of energy from the CH overtone to the CCH in-plane wag (2:1 resonance) and subsequently into other ring motions (~1:1 resonances).

Fig. 2. Classical relaxation of the fourth CH overtone in C_6H_6. The decay is exponential with e^{-1} lifetime of ~60 fs.

A slower classical relaxation rate of the v = 5,6,8 CD overtones in perdeuterobenzene was found: the longer lifetime is in accord with the experimental observation of narrower overtone line shapes in C_6D_6 as compared to C_6H_6, and is primarily due to a smaller G-matrix coupling between the CD stretch and CCD in plane wag.

3. Classical Model for Relaxation of Alkyl CH Overtones

The detailed dynamics of CH relaxation involving a terminal methyl CH in linear alkanes is somewhat different; however, the same type of resonance analysis yields satisfactory results. In a preliminary study, Hutchinson, Hynes, and Reinhardt [9] have found that for v = 5,6,7,8 the CH stretch was in 2:1 resonance with one, or both, of two methyl terminal rocks, leading to relaxation on a 250 fs timescale, the slower relaxation being due to smaller G-matrix couplings. Relaxation from a v = 8 CH overtone into terminal CH_3 rocks is illustrated in Figure 3.

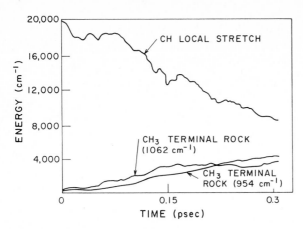

Fig. 3. Classical relaxation of energy from the v = 8 CH overtone, of a terminal alkyl CH local mode. Energy appears in the two terminal methyl rock modes, and relaxation times are independent of chain length for chains containing 3, 4 and 5 carbons.

The longer alkyl lifetime corresponds quite reasonably with the observations of Ref. [3], indicating that methyl CH line widths were 3 to 4 times narrower than aryl.

4. Quantum Dynamics of C_6H_6 and C_6D_6 and Spectra

The quantum version of the 2:1 stretch-bend Fermi resonance, followed by "irreversible" relaxation into ring modes, is simply successive transfer of quanta from the stretch to the ring normal modes, two ring quanta appear-

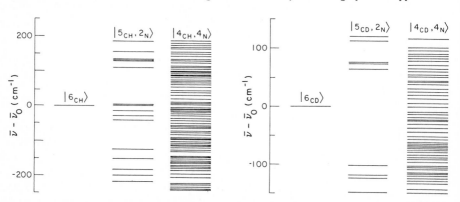

Fig. 4. Unperturbed quantum levels showing energies of the anharmonic CH overtone and the first and second "tiers" of "relaxed" states accessed by the 2:1 Fermi resonance between the CH stretch and the in plane HCC bend.

Fig. 5. Unperturbed quantum levels near the fifth CD overtone in C_6D_6. The "gap" in the level density of $|5_{CD},2_N\rangle$ states renders the possible 2:1 Fermi resonance ineffective. See, for comparison, the contrasting situation in C_6H_6, shown in Fig. 4.

ing for each stretch quantum transferred. This gives rise to the time independent picture of Fig. 4, where "N" denotes ring modes. The direct product states $|(6-i)_{CH}; (2i)_N\rangle$ are referred to as belonging to the zeroth, first, second, etc., tiers of relaxed states, for $i = 0,1,2\ldots$.

A similar level diagram for the $v = 6$ overtone of the CD stretch in C_6D_6 is shown in Fig. 5. We note at once that a "gap" in the $|5_{CD}, 2_N\rangle$ states at the energy of the $v = 6$ CD overtone immediately explains the "narrow" fifth CD overtone observed by Reddy, Heller and Berry [2].

To simulate an actual spectrum in a model with a finite number of tiers explicitly included, an empirical parameter, τ, is introduced giving the rate of flow out of the highest order tier explicitly considered. The simulated absorption spectra of Figs. 6 and 7 indicate that the CH line centers and line widths are insensitive to the value of τ, whereas the wings clearly contain information regarding such high inter-tier relaxation rates. Additionally it is to be noted that the CD overtone of Fig. 7 will likely have structure at high resolution. High resolution gas phase spectra of cold molecules are thus highly desirable.

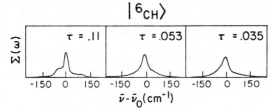

Fig. 6. Theoretical line widths of the sixth CH overtone in C_6H_6 as a function of the parameter τ, modeling decay from the third tier, τ is in psec.

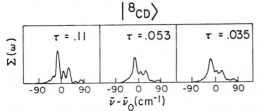

Fig. 7. As in Fig. 6, but for the eighth CD overtone in C_6D_6. In this case cold, high-resolution experiments should show structure.

Support from the National Science Foundation (grants CHE81-13240, CHE83-10122 and PHY82-00805) and the donors of The Petroleum Research Fund (administered by the American Chemical Society) is most gratefully acknowledged. JSH is a PRF Postdoctoral Research Fellow. Present addresses for three of the authors are: E.L.S., Dept. of Chemistry, Univ. of California, Berkeley, CA 94720; J.S.H., Dept. of Chemistry, Rice Univ., Houston, TX 77251; W.P.R., Dept. of Chemistry, Univ. of Pennsylvania, Philadelphia, PA 19104.

References

1. R. G. Bray and M. J. Berry, J. Chem. Phys. 71, 4909 (1979).
2. K. V. Reddy, D. F. Heller and M. J. Berry, J. Chem. Phys. 76, 2814 (1982).

3. J. W. Perry and A. Zewail, J. Phys. Chem. $\underline{86}$, 5197 (1982).
4. J. W. Perry and A. Zewail, J. Chem. Phys. $\overline{80}$, 5333 (1984).
5. G. J. Scherer, K. K. Lehmann and W. Klemperer, J. Chem. Phys. $\underline{78}$, 2817 (1983).
6. D. F. Heller and S. Mukamel, J. Chem. Phys. $\underline{70}$, 463 (1979).
7. E. L. Sibert III, W. P. Reinhardt and J. T. Hynes, Chem. Phys. Letters $\underline{92}$, 455 (1982); J. S. Hutchinson, W. P. Reinhardt and J. T. Hynes, J. Chem. Phys. $\underline{79}$, 4247 (1983); E. L. Sibert III, W. P. Reinhardt and J. T. Hynes, J. Chem. Phys. (in press).
8. P. Pulay, G. Fogarase and J. E. Boggs, J. Chem. Phys. $\underline{74}$, 401 (1981).
9. J. S. Hutchinson, J. T. Hynes and W. P. Reinhardt, J. Chem. Phys. (in preparation).

Pump-Pump Picosecond Laser Techniques and the Energy Redistribution Dynamics in Mass Spectrometry

M.A. El-Sayed, D. Gobeli, and J. Simon

Department of Chemistry and Biochemistry, University of California
Los Angeles, CA 90024, USA

In conventional mass spectrometry, studies of the dynamics of processes occurring on the time scale between excitation (10^{-15} sec) and ion analysis (10^{-7} sec) are not possible to carry out. The use of two-color picosecond lasers of variable delay times as a source in mass spectrometry enables one to probe the dynamics of rapid energy redistribution and dissociation in the parent molecule and ion. This technique will be demonstrated by applying it to 2,4-hexadiyne.

In order to understand ionic dissociation, conventional mass spectrometry has been used extensively to examine the validity of the quasi-equilibrium theory [1] (QET). This was carried out by comparing the experimental and the calculated breakdown curves. According to the QET, the electronic excited ions formed after ionization rapidly redistribute their electronic energy into a statistical distribution of vibrational energy of the ground electronic state of the parent ion prior to dissociation. So far, time resolved mass spectrometric studies have been limited to the rare cases when dissociation is comparable in time to the extraction time of the analyzer of the mass spectrometer used. In these cases, distorted (metastable) peaks appear at non-integral m/e values. The analysis [2] of these shapes yields dissociation times on the microsecond time scale. Measurements of rates of dissociation occurring on a faster time scale or of the rapid energy redistribution occurring prior to dissociation have so far been "out of reach" in conventional mass spectrometry.

The use of lasers as a source for mass spectrometry dates back to 1970 [3]. Two color lasers were also used to measure the lifetime [4,5] and ionization potentials [5] of molecules. Two color mass spectrometry on the nanosecond time scale has also been carried out [6]. One color picosecond laser as a source in mass spectrometry has also been used [7]. It is the purpose of our effort [8] to use two color picosecond lasers with variable delays to determine the rates of rapid energy redistribution in the parent ions prior to dissociation. In cases where the dissociation occurs on the picosecond time domain, this technique should enable us to determine the rates of these processes as well.

The Technique [8]:

The fourth harmonic of the YAG 400 Quantel laser (25 ps pulse width) is used as the primary pump at a few µjoules power. One photon energy of this laser is resonant with the molecular absorption. This establishes a certain population of the parent molecule and ions in different excited states separated in energy by one photon energy of the 266 nm laser. The secondary pump is a green 532 nm laser pulse which itself does not produce a mass spectrum but is capable of pumping some of the parent molecules and the ions in the different excited states to new ones. This changes the

341

level of excitation of the different excited states formed by the primary pump. Ions that were formed from the old excited states could decrease in intensity while those formed from the new excited states should increase in current intensity. If the energy redistributes between the two pulses, and if the absorption cross section for the green is different in the initial and final states involved in the energy distribution process, the effect of the green on the mass peaks should be time dependent reflecting the energy redistribution rates.

2,4-hexadiyne is selected because: a. one U.V. photon reaches S_1 of the molecule, two photons ionize with only 0.4 eV excess energy. b. three photons reach an ionic excited level whose lifetime is measured from fluorescence studies [9] and which is embedded in between the appearance potentials of many daughters [10].

Important Results:

1. The mass signal of the parent ion, produced by the absorption of one photon to S_1 and another photon to ionize, is found to increase greatly as the two pump lasers overlap in time. This is described in terms of a rapid energy redistribution process in S_1 of the neutral molecule which is over- come by the absorption of the green photon to ionize. The process occurs within our laser pulse width (25 ps). This method can thus be used to determine [8] the pulse width from the correlation of the green and U.V. pulses as shown in Fig. 1.

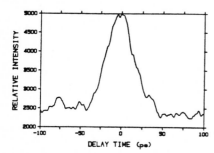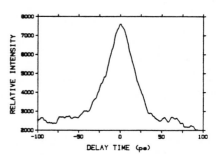

Fig. 1: left: The $C_6H_6^+$ mass signal as a function of the delay between the 266 nm and the 532 nm pulses. The observed width is comparable to the autocorrelation width of the 532 nm pulses. These results suggest the use of this molecule and method for measuring the U.V. laser pulse width.

2. The absorption of an additional photon beyond ionization is found to produce the parent ion in the first excited state at 13.6-14.0 eV above the molecular ground state. This level has a lifetime of 23 ns as measured [9] by fluorescence techniques and is above the appearance poten- tial of $C_4H_4^+$, $C_6H_5^+$ and several other ions. Monitoring the $C_4H_4^+$ ion signal with delay (see Fig. 2) gives a lifetime [8] of ~20 ns, in agreement with the fluorescence results [9].

3. Monitoring the $C_6H_5^+$ peak (Fig. 3) gave two lifetimes, one of ~20 ns and the other ~250 ps. This is described [11,12] in terms of two different chemicals (e.g., isomers) with the chemical formula of $C_6H_5^+$: one is formed from the energy redistribution of the parent ion level at the three photon level while the other is formed from the redistribution at the four U.V.

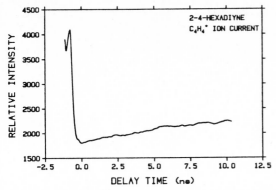

Fig. 2:   The $C_4H_4^+$ mass signal as a function of the delay between the fourth and second harmonic pulses.   The rise time is found [8] to be ~20 ns, comparable to the fluorescence lifetime [9] of the parent ion.

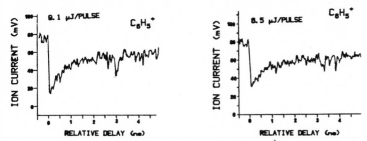

Fig. 3:   The time characteristics of the $C_6H_5^+$ mass peak   showing two components, one with 250 ps and the other with 20 ns and whose amplitudes are sensitive to the U.V. power.   This is described [11] by the presence of two different species (e.g., isomers) originating from two different excited states of the parent ion separated by one U.V. photon energy.

photon level (lifetime is 250 ps).   If this is the case, this will be the first time in mass spectrometry that different chemicals of the same m/e ratio are differentiated (through the dynamics of their formation).   According-ing to this interpretation, increasing the U.V. laser power should increase the relative contribution of the 250 ps component as shown in Fig. 4.

4.   The three U.V. photon level is located in between the appearance poten-tials [10] of a number of the daughter ions of 2,4-hexadiyne.   The appear-ance potentials of some of the daughters are below it and others are above it but can be exceeded by the absorption of one green photon.   One thus expects that as a result of the secondary pump, the daughters with AP be-low the 3-photon level decrease in signal and those above it increase in signal.   The results of Fig. 4 demonstrate this prediction [12].

It is hoped that this technique allows picosecond (and femtosecond) mass spectrometry to assist in understanding the rapid energy redistribution in neutral and parent ions prior to dissociation.   For fast dissociating ions, the technique should be able to determine the rate of these processes themselves.   It is important to point out, however, that this understanding can be accomplished for molecules which ionize and dissociate according to an ionic ladder mechanism and thus avoid complications resulting from

343

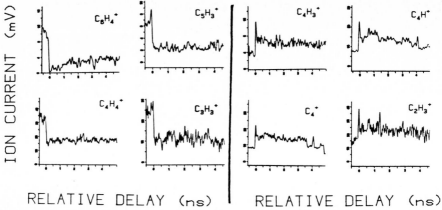

ION CURRENT (mV)

$C_6H_4^+$ $C_5H_3^+$ $C_4H_3^+$ C_4H^+

$C_4H_4^+$ $C_3H_3^+$ C_4^+ $C_2H_3^+$

RELATIVE DELAY (ns)   RELATIVE DELAY (ns)

Fig. 4: The effect of the green laser on the daughter ion signals.
Species with appearance potentials (AP) lower than the three U.V. photon
energies decrease (four graphs on left) while those with AP higher increase
(four graphs on right) in mass signal as the green pump laser is absorbed.

ladder switching mechanisms. Fortunately, this is more likely to be the
case when short pulses are used [13].

Acknowledgment: The authors wish to thank the National Science Foundation
for its support of this work.

References:
1. H. M. Rosenstock, M. B. Wallenstein, A. L. Wahrhaftig and H. Eyring:
   Proc. Natl. Acad. Sci. U.S.A. 38, 667 (1952).
2. J. A. Hipple: Phys. Rev. 71, 594 (1947).
3. N. K. Berezhetskaya, G. S. Voronov, G. A. Delone, N. B. Delone and
   G. K. Piskova: Sov. Phys. JEPT 31, 403 (1970).
4. S. V. Andreyev, V. S. Antonov, I. N. Knyazev and V. S. Letokhov:
   Chem. Phys. Lett. 45, 166 (1977).
5. D. H. Parker and M. A. El-Sayed: Chem. Phys. 42, 379 (1979).
6. V. S. Antonov, I. N. Knyazev, V. S. Letokhov, V. M. Matiuk, V. M.
   Movshev and V. K. Potapov: Opt. Lett. 3, 37 (1978).
7. P. Hering, A. G. M. Maaswinkel and K. L. Kompa: Chem. Phys. Lett. 83,
   222 (1981).
8. D. A. Gobeli, J. R. Morgan, R. J. St. Pierre and M. A. El-Sayed:
   J. Phys. Chem. 88, 178 (1984).
9. M. Allan, J. P. Maier, O. Marthaler and E. Kloster-Jensen:
   Chem. Phys. 29, 331 (1978).
10. J. Dannacher: Chem. Phys. 29, 339 (1978).
11. D. A. Gobeli, John Simon and M. A. El-Sayed: J. Phys. Chem., in press.
12. D. A. Gobeli and M. A. El-Sayed: J. Phys. Chem., in preparation.
13. D. A. Gobeli, J. Yang and M. A. El-Sayed, Advances in Multiphoton
    Processes, Vol. 1, eds. E. Schlag & S. H. Lin (Academic Press, New York)
    1984, in press.

Intramolecular Electronic and Vibrational Redistribution and Chemical Transformation in Isolated Large Molecules – S_1 Benzene

K. Yoshihara, M. Sumitani, D.V. O'Connor, Y. Takagi, and N. Nakashima

Institute for Molecular Science, Myodaiji, Okazaki 444, Japan

1. Introduction

With the advent of powerful and ultrashort optical pulses and with the progress of associated detecting techniques, many phenomena in the field of molecular dynamics become directly accessible to study. One of the important areas which came out very recently is the time-resolved measurement of the intramolecular energy redistribution and chemical transformation in the isolated molecule condition [1 - 4].

We are interested in the photochemical and photophysical properties of benzene, a prototype of aromatic hydrocarbons. Upon exciting the S_1 manifold, photochemical valence-isomerization takes place and benzvalene is confirmed. The nonradiative pathways are strongly dependent on exciting wavelengths and suddenly a new pathway opens, called channel three, by which vibrational energy in excess of about 2800 cm^{-1} is lost at an anomalously rapid rate. Very recently two-photon fluorescence [5], Doppler-free absorption [6], multiphoton ionization [7] and laser flash photolysis [8] were applied to clarify this intriguing process. However, in none of these experiments was picosecond time resolution applied. The result is that we are still ignorant of the fate of the molecule on its passage from S_1 to its arrival in S_0 some nanoseconds later [8]. It is essential to observe fluorescence directly and follow the energy decay at all times using ultrafast time-resolved spectroscopy.

2. Experimental

The experimental arrangement of picosecond tunable UV source is shown in Fig. 1 [9]. An amplified picosecond pulse was split into two lines and amplified again. The first beam was frequency quadrupled and sent to a 1 m Raman cell for generation of stimulated Raman light with conversion effeciency of ~40 %. The second beam was frequency doubled and focused onto a pair of crystals (KD*P) for optical parametric generation (OPG) and amplification. The conversion efficiency for OPG reached 40 % at the degenerate wavelength (1064 nm). This beam covers the spectral region between 850 and 1600 nm.

Upconversion was performed by mixing the optical parametric light with the fourth-harmonic light, the N_2 Raman beam and the H_2 Raman beam. In this way three different spectral ranges between 215 and 252 nm were covered. The energy exceeded 100 µJ over the entire tuning range. The spectral band width was 0.5 Å and the pulse duration was about 10 ps.

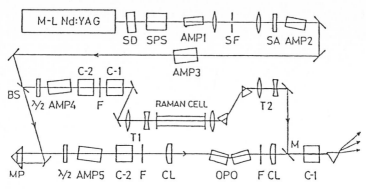

Fig. 1. Optical arrangement for generation of tunable UV light.
M-L Nd:YAG, Mode-locked Nd:YAG laser, SD; Saturable dye, SPS;
Single pulse selector, SF; Spatial filter, SA; Soft aperture,
BS; Beam splitter, $\lambda/2$; Half-wave plate, C-1,2; Nonlinear
crystal (KDP, type-1,2), F; Filter, T1; 5:2 telescope, T2; 5:3
telescope, MP; Movable prism, CL; Cylindrical lens, OPG; Optical
parametric generator, M; Half mirror for beam mixing.

Fluorescence decays were measured either with a fast channel
plate photomultiplier or with a streak camera. Fluorescence spectra were measured with a monochromator and a solid state detector
array. In order to avoid fluorescence component of long-lived
relaxed state, only early time emission was collected by gating
the detector. The vibrational temperature of benzene in the supersonic jet was estimated to be about 160 K. The rotational
temperature was assumed to be the same as the temperature of
carrier He gas, i.e., less than 10 K.

3. Results and Discussion

Fluorescence lifetimes, quantum yields [10], and spectra [11]
were measured following excitation of the strong absorption bands
above the channel three thresholds and some below it. In Table 1
some of the selected transitions are identified by the excess
vibrational energy (ΔE in cm^{-1}) of the S_1 level. We have also
indicated whether the fluorescence spectrum contained vibronic
structure characteristic of the level excited. In all levels lying
within a range of about 1000 cm^{-1} above the channel three threshold, we found double-exponential decays in both static gas and
molecular beam condition. These were analysed with double exponential decays. Above this range the decays again become exponential, and the shortest observed lifetime reach ca. 23 ps for
$6_1^2 1_0^5$ ($\Delta E=5657$ cm^{-1}). The double exponential decays are with almost the same lifetimes and relative intensities for the separate
components under bulb and beam conditions. Therefore double exponential decays do not result from hot band excitation.

In order to investigate the spectral characteristics of the
two components, we measured time-resolved spectra. In Figs. 2 and
3 are shown the spectra of 7_0^1 and $6_0^1 1_0^3$, respectively in the static
gas and cold beam condition. In the bulb experiment there is an
enhanced contribution of the broad background emission in the
early gated spectrum. This enhancement disappears completely in

Table 1. Fluorescence quantum yields, lifetimes, and spectral features

Transition	ΔE /cm^{-1}	ϕ_f	Static gas		Cold beam		Spec*
			τ_1 /ns	τ_2 /ns	τ_1 /ns	τ_2 /ns	
$6_0^1 1_0^2$	2367	0.18	47		53		s
$6_0^1 16_0^1 1_0^2$	2610	0.10	30		40		s
$6_0^1 10_0^2 1_0^1$	2614	0.10	27		38		s
$6_0^1 17_0^2 1_0^1$	2856	0.014	13	2.8	13.6	2.9	sb
7_0^1	3080	0.016	18	4.0	19	4.2	s
$6_0^1 1_0^3$	3290	0.025	2.9	0.7	3.2	0.8	sb
$6_0^1 10_0^2 1_0^2$	3537	0.0013	1.0	0.15	1.0	<0.3	b
$7_0^1 1_0^1$	4003	0.0009	0.13		<0.3		s
$6_0^1 1_0^4$	4213	0.00057	0.10				sb
$7_0^1 1_0^2$	4926	0.00047	0.080				s
$6_0^1 1_0^5$	5136	0.00028	0.038				b
$6_0^1 1_0^6$	5657	0.00025	0.023				

* Fluorescence spectra.   s: structure,   b: broad with little structure,   sb: intermediate between s and b

the beam experiment where early and late gated spectra are virtually identical. The background fluorescence we may therefore attribute to hot band excitation, especially since, at room temperature, there is a fairly large ($\sim$20 %) hot band contribution to
the absorption intensity in this wavelength region. The levels
populated via hot band absorption have a much higher vibrational
energy content and are expected to be very short-lived. That the
two lifetime components actually observed are also present under
beam conditions, where the spectrum cannot be time-resolved, suggests that they have the same origin.

    Very sharp spectra were dominantly observed at excitations of
7^1 (ΔE=3080 cm^{-1}) and $7^1 1^1$ (ΔE=4003 cm^{-1}) band in the cold beam
condition [12]. This is in great contrast to the $6^1 1^3$ (ΔE=3290
cm^{-1}) band excitation at the same condition. The latter gives a
featureless background emission superimposed with a small extent
of structured one. In general all levels above the channel three
threshold exhibit the structured fluorescence although the relative contribution of the structured to featureless fluorescence
is much smaller for levels $6^1 1^n$ than for levels $7^1 1^n$ [11]. The
featureless emission is from background vibronic levels which is
populated through intramolecular vibrational redistribution (IVR)
and the sharp one from optically prepared state. Therefore it is
evident that IVR is much less efficient in the levels involving
ν_7 vibration. This observation is consistent with the characteristic difference in absorption line widths between the transitions
$6_0^1 1_0^3$ and 7_0^1 [13].

    In spite of the characteristic difference in optical properties
and in phase decay (IVR) for $6^1 1^n$ and $7^1 1^n$ series, quantum yields
and population decay times decrease in the same manner. Figure 3
shows k_{nr} (radiationless rate constant, in s^{-1}) vs ΔE for all

347

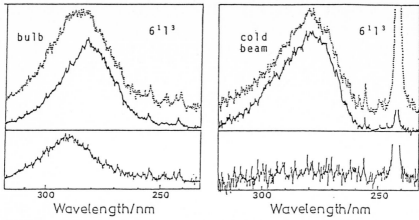

Fig. 2. Time-resolved fluorescence spectra of $6^1 1^3$ in S_1 benzene under bulb and cold beam conditions. Left; observed in the bulb condition (slit width 50 μm, spectral resolution ~1.3 nm), Upper figures: points, early gated spectrum obtained with 5 ns gate centered 1.5 ns after laser pulse; line, late gated spectrum obtained with 5 ns gate centered 6.5 ns after laser pulse. Lower figure: early-minus-late gated difference spectrum. Right; observed under the cold beam condition. Other conditions are same as in the bulb condition.

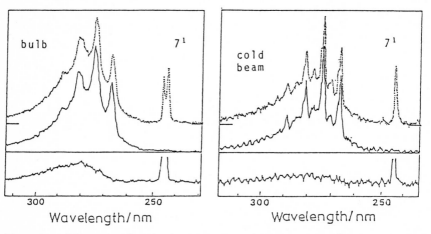

Fig. 3. Time-resolved fluorescence spectra of 7^1. Experimental conditions are same in all corresponding figures as in Fig. 2.

series of transitions observed [3], for the two-photon transitions, $14^1_0 1_0$, obtained by Wunsch et al. [5], and for one-photon transitions obtained from absorption band widths measurement by Callomon et al. [13]. The onset of channel three in C_6D_6 is nearly the same in excess energy as in C_6H_6. All of these results clearly indicate that there is an abrupt increase in k_{nr} at $\Delta E \cong$ 2800 cm^{-1} (Fig. 4).

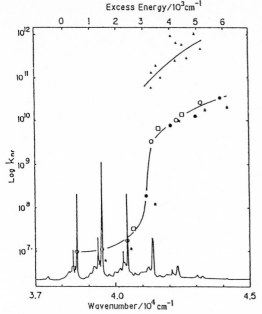

Excess Energy/10^3cm^{-1}

Fig. 4. Plots of log k_{nr} vs. ΔE for transitions $6_0^1 1_0^n$ (open circles), $6_0^1 10_0^2 1_0^n$ (squares), $7_0^1 1_0^n$ (closed circles), and $14_0^1 1_0^n$ (stars) [5]. Values of k_{nr} (triangles) obtained from line broadening in absorption are also shown (triangle) [13].

In the channel three region IVR has an important role for the decay channel, but with an added complication. We attempt to rationalise the above observations by proposing the existence of a "state" X with which all levels above $\Delta E = 2800$ cm^{-1} can couple. State X could possibly be another singlet electronic state or chemical intermediate of benzene. The levels that redistribute some of their energy rapidly can also couple with state X as illustrated in Fig. 5. In this way the rapid decrease in lifetime observed for these levels can be explained. Non-exponential decays are the result of a "communication" between state X and the precursor level or levels. For the moment this mechanism remains somewhat conjectural, but it does offer a solution to many of the apparently conflicting experimental results.

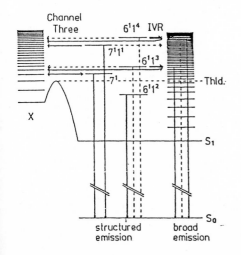

Fig. 5. Schematic diagram of a possible mechanism of energy flow in S_1 benzene.

References

1. A.H. Zewail, Disc. Faraday Soc., 75/19 (1983).
2. R. Moore, F.E. Doany, E.J. Heilweil, and R.M. Hochstrasser, ibid., 75/20.
3. K. Yoshihara, D.V. O'Connor, M. Sumitani, Y. Takagi, and N. Nakashima, "Appl. of Picosecond Spectroscopy to Chemistry", NATO Adv. Study Inst. Series, Ed., K.B. Eisenthal, Reidel Publ. Co., in press.
4. J.M. Wissenfeld and B.I. Greene, Phys. Rev. Lett., 51, 1745 (1983).
5. L. Wunsch, H.J. Neusser, and E.W. Schlag, Z. Naturforsch., 36a, 1340 (1981).
6. E. Riedle, H.J. Neusser, and E.W. Schlag, J.Phys. Chem., 86, 4847 (1982).
7. C.E. Otis, J.L. Knee, and P.M. Johnson, J. Phys. Chem., 87, 2232 (1983).
8. N. Nakashima and K. Yoshihara, J. Chem. Phys., 77, 6040 (1982), ibid., 79, 2232 (1983).
9. Y. Takagi, M. Sumitani, N. Nakashima, D.V. O'Connor, and K. Yoshihara, Appl. Phys. Lett., 42, 489 (1983).
10. M. Sumitani, D.V. O'Connor, Y. Takagi, N. Nakashima, K. Kamogawa, Y. Udagawa, and K. Yoshihara, Chem. Phys. Lett., 97, 508 (1983).
11. D.V. O'Connor, M. Sumitani, Y. Takagi, N. Nakashima, K. Kamogawa, Y. Udagawa, and K. Yoshihara, J. Phys. Chem., 87, 4848 (1983).
12. M. Sumitani, D.V. O'Connor, Y. Takagi, and K. Yoshihara, Chem. Phys. Lett. in press.
13. J.H. Callomon, J.E. Parkin, and R. Lopez-Delgado, Chem. Phys. Lett., 13, 125 (1972).

Ultrafast Intramolecular Redistribution and Energy Dissipation in Solutions. The Application of a Molecular Thermometer

P.O.J. Scherer, A. Seilmeier, F. Wondrazek, and W. Kaiser

Physik Department der Technischen Universität München
D-8000 München, Fed. Rep. of Germany

In recent years we have studied the relaxation of the vibrational energy in small molecules. Here we wish to discuss the flow of vibrational energy in large molecules of 30 to 60 atoms.

In our investigations tunable picosecond pulses in the infrared excite vibrational states around 3000 cm^{-1}. Two experimental techniques are used for measuring intra- and intermolecular energy flow. In the first technique a tunable pulse in the visible promotes vibrationally excited molecules to the fluorescing S_1 state. The fluorescence signal is a measure of the instantaneous population of the probed states. This technique gives valuable information on the intramolecular vibrational redistribution.

In the second method the instantaneous excitation is monitored via small absorption changes of the long wavelength tail of the $S_0 \rightarrow S_1$ absorption of the investigated molecules. In this way the change of the momentary temperature of the probe molecules is observed. Information is obtained on the population and depopulation of low lying modes and on the energy transfer from the solvent to the probe molecules and vice versa.

First we want to discuss fluorescence experiments performed with Coumarin 7 (see Fig. 2, insert), dissolved in C_2Cl_4 at room temperature /1/. The molecules are excited via resonant absorption of the NH vibration at 3395 cm^{-1}. In Fig. 1 the fluorescence signal, which is proportional to the excess population, is shown as a function of the delay time for $\tilde{\nu}_2$ = 19,470 cm^{-1} and 19,920 cm^{-1}. We find a rapid build-up of the fluorescence signal and a decay with a time constant of 7 ps for both frequencies $\tilde{\nu}_2$.

Information on the distribution of the excitation energy is obtained from frequency resolved experiments. Fig. 2 shows the fluorescence signal, which is due to the vibrational excess population, as a function of the monitoring frequency $\tilde{\nu}_2$. The signal was observed over four orders of magnitude (delay time t_D = 8.5 ps). The conventional absorption of the dye is plotted for comparison on the r.h.s. of the figure. Fig. 2 gives two important results: (i) We find an exponential frequency dependence of the fluorescence signal similar to the Boltzmann edge of the conventional absorption. This fact points to a considerable energy randomisation in the excited molecule. (ii) The slope of the fluorescence signal is less steep suggesting a higher temperature of T^* = 400 K.

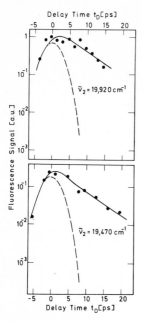

Fig. 1
Fluorescence signal as a
function of the delay time
for the frequencies $\tilde{\nu}_2$ =
19,920 cm^{-1} and 19,470 cm^{-1}.
The excess population decays
with a time constant of 7 ps
(Coumarin 7 in C_2Cl_4)

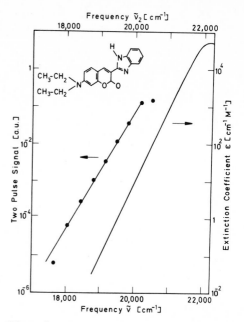

Fig. 2
Part of the absorption edge of
Coumarin 7 in C_2Cl_4. Fluorescence
signal as a function of the frequency $\tilde{\nu}_2$ for a delay time of
t_D = 8.5 ps. The different slopes
should be noted

The following picture evolves from a series of investigations.
The excitation energy of the NH vibration is rapidly redistributed (< 2 ps) within the molecule, leading to a relaxed distribution with T^* = 400 K. The relaxed distribution decays with a
time constant of 7 ps, which represents energy transfer to the
solvent.

The second technique, the study of the absorption change, is
well suited to measure intermolecular energy transfer /2/. The
absorption at long wavelengths is related to the thermal population of vibrational modes and, consequently, a measure of the instantaneous temperature of the molecules.

Our molecular thermometer is calibrated by steady-state and
picosecond measurements. Fig. 3a shows the long wavelength tail
of the absorption of Oxazine 1 (insert), dissolved in ethanol
at 293 K (solid line) and 358 K (broken line). The relative
change of the extinction coefficient per temperature interval
is plotted as a function of the frequency in Fig. 3b. In Fig. 3c
the result of a picosecond experiment is shown. Energy is supplied to the solution by an infrared pulse at $\tilde{\nu}_1$ = 3400 cm^{-1}
(resonance absorption of OH vibration of the solvent). The absorption change on account of the transient heating of the dye

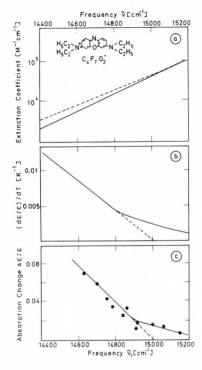

Fig. 3
(a) Extinction coefficient of Oxazine 1 in ethanol for T = 293 K (solid line) and T = 358 K (broken line)
(b) Change of the extinction coefficient per K as a function of the frequency
(c) Transient change of the extinction coefficient at t_D = 100 ps after picosecond excitation at $\tilde{\nu}_1$ = 3400 cm^{-1}

molecule is measured by a second, delayed picosecond pulse at frequency $\tilde{\nu}_2$. The frequency dependence is quite similar to Fig. 3b. Comparing the two curves we estimate a temperature increase of 10 K in agreement with an estimation using the specific heat of ethanol and the excited volume.

In a first experiment we investigated the energy transfer from low-lying modes of Oxazine 1 to the solvent. Using deuterated ethanol as solvent, vibrations of the dye molecules are excited by an infrared pulse at 2980 cm^{-1}. The time dependence of the relative absorption change is shown in Fig. 4a. The absorption increases due to the temperature rise of the dye molecules and relaxes with a time constant of 10 ps, similar to the results of the fluorescence probing experiments. The excitation energy dissipates via intermolecular processes.

In a second experiment we have addressed the question how fast energy is transferred from the solvent to the solute (probe) molecules. Energy is supplied to the solvent ethanol by infrared photons via OH modes. The ultrafast energy transfer is monitored by the change of absorption of the probing molecule Oxazine 1. The experimental result is depicted in Fig. 4b. A very rapid energy transfer is observed within the time resolution of our system (< 5 ps). The dashed line represents the cross correlation of the excite and probe pulse. After a time delay of 20 ps the absorption change remains constant for more than 200 ps. This excess temperature decreases within several milliseconds via macroscopic heat conduction. The value $\Delta\varepsilon/\varepsilon$ = 0.03 corresponds to a temperature change of 10 K (see Fig. 3).

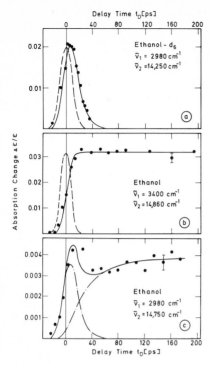

Fig. 4
Time dependence of the absorption
change of Oxazine 1.
(a) CH_3 modes of the dye are excited. The absorption change relaxes with T = 10 ps. Solvent:
ethanol-d_6
(b) OH modes of the solvent are
excited. Rapid heating of the dye
(< 5 ps) is observed. Solvent:
standard ethanol
(c) CH_3 modes of the dye and the
solvent are excited simultaneously.
The result is a superposition of
two components (dash-dotted curves).
The direct excitation of the dye
relaxes rapidly. The heating of
the solvent is delayed by the slow
depopulation of the CH_3 mode in
ethanol

In Fig. 4c the sample, standard ethanol, is excited by an infrared pulse of $\tilde{\nu}_1$ = 2980 cm^{-1}. At this frequency the dye molecules and the solvent absorb simultaneously via their CH_3 modes.
We measure a signal consisting of two processes with different
time behavior (dash-dotted lines). In the first process the dye
molecules are directly excited; they relax with a time constant
of approximately 10 ps, similar to the data of Fig. 4a. The
second component rises slowly with a time constant of 30 ps. Obviously, the long relaxation of the excited CH_3 modes slows down
the intramolecular energy redistribution in the ethanol molecule
and delays the temperature rise of the probe molecules.

In conclusion, the results show a very fast energy transfer
from the solvent to the probe molecules within 5 ps. The transfer
is delayed when the intramolecular energy redistribution in the
solvent molecules proceeds with a longer time constant.

We have collected extensive data on other molecules of different structure dissolved in various solvents. The results suggest
that the redistribution and dissipation dynamics discussed above
quite generally occur in large molecules in the liquid state.

References

1 F. Wondrazek, A. Seilmeier, W. Kaiser, Chem. Phys. Lett. 104
  (1984) 121
2 A. Seilmeier, P.O. J. Scherer, W. Kaiser, Chem. Phys. Lett.
  105 (1984) 140

Picosecond Time-Resolved Fluorescence Spectra of Liquid Crystal: Cyanooctyloxybiphenyl

N. Tamai and I. Yamazaki

Institute for Molecular Science, Myodaiji, Okazaki 444, Japan

H. Masuhara and N. Mataga

Department of Chemistry, Faculty of Engineering Science, Osaka University
Toyonaka, Osaka 560, Japan

1. Introduction

Liquid crystals are highly anisotropic molecular systems which exhibit phase transitions in several stages on going from the crystal to the isotropic liquid phases. Mesophases such as smectic A and nematic phases are defined depending on orientational and spatial ordering of molecules. Effects of molecular anisotropy of the liquid crystal on chemical reactions and photophysical processes now receive increasing attention [1-4]. It has been shown [5] that in a luminescent liquid crystal, dodecylcyanobiphenyl (12CB), two kinds of fluorescence are emitted in various phases. The dual fluorescence emission was interpreted in terms of singlet monomer and excimer formation.

We have recently examined the excited state properties of a liquid crystal, 4-cyano-4-n'-octyloxybiphenyl (8OCB), in which the phase transitions,

$$\text{crystalline} \xleftrightarrow{\text{328 K}} \text{smectic A} \xleftrightarrow{\text{340 K}} \text{nematic} \xleftrightarrow{\text{353 K}} \text{isotropic liquid,}$$
$$\text{(K)} \qquad\qquad \text{(SmA)} \qquad\qquad \text{(N)} \qquad\qquad \text{(I)}$$

have been known [6,7]. We report here the picosecond time-resolved fluorescence spectra of neat 8OCB in various phases, and discuss formation of an excimer-like state in the liquid crystal.

2. Experimental

8OCB obtained from BDH Chemicals Co. was used without further purification. The samples in quartz cells of 1 x 10 mm were dearated by repeated freeze-pump-thaw cycles. Picosecond time-resolved fluorescence spectra were obtained with a synchronously pumped, cavity-dumped dye laser and time-correlated photon-counting system. Details of this system have been described elsewhere [8]. The time resolution of the detection system is 50 ps by the use of a microchannel plate photomultiplier (Hamamatsu R1564U). Fluorescence measurements were made after the sample reached the thermal equilibrium.

3. Results and Discussion

The fluorescence spectrum of 8OCB in solution strongly depends on the solvent polarity [9]. In nonpolar solvents like n-hexane, an excimer fluorescence appears at higher concentrations, showing a broad band spectrum with the maximum at 370 nm. The fluorescence lifetime of the excimer is 8.7 ns (in 5 x 10^{-1} M), whereas that of the monomer is as short as 1.1 ns (in 10^{-5} M).

Figure 1 shows the fluorescence spectra of 8OCB in neat form at various temperatures. The fluorescence spectrum changes discontinuously at temperatures corresponding to the phase transitions. At temperatures below 311 K, the K phase spectrum appears with the maximum at 360 nm. At 311 K, the K

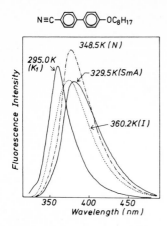

$N \equiv C - \langle \rangle - \langle \rangle - OC_8H_{17}$

Fig. 1 Fluorescence spectra of 8OCB liquid crystal in K_1, SmA, N and I phases. The K_2 spectrum closely resembles I spectrum.

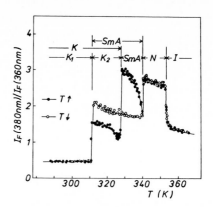

Fig. 2 Plots of ratios of the fluorescence intensity, I_F(380 nm)/I_F (360 nm), versus temperature. Full and open circles indicate the plots in increasing and decreasing temperature, respectively.

phase spectrum shifts suddenly to the red by 15 nm. At 328 K, the spectrum shifts further to the red, associated with K → SmA transition. In SmA and N phases, the spectra with the maxima at 377–380 nm are similar to each other. In I phase, the fluorescence maximum is located at 375 nm. We assign the fluorescence spectrum with the maximum at 360 nm (K_1 spectrum in Fig. 1) to the monomer-like state and those of longer wavelengths (SmA, N and I spectra in Fig. 1) to the excimer-like state.

In Fig. 2, ratios of the fluorescence intensity at 380 nm to that at 360 nm are plotted against temperature, for the cases of increasing and decreasing temperature. One can see clearly that the intensity ratio changes discontinuously at temperatures corresponding to the phase transitions, i.e., 328 K, 340 K and 353 K. In addition, we can see a discontinuous change of the ratio at 311 K, which is in parallel to the abrupt spectral shift observed in the crystalline phase at 311 k as pointed out above. This change suggests a new phase transition in the crystalline phase; hereafter the phase at temperatures lower than 311 K is referred to as K_1 and that at higher temperatures (311–328 K) as K_2. Also noteworthy is a hysteresis behavior in the $K_2 \longleftrightarrow$ SmA phase transition. A similar behavior has recently been found in the temperature dependence of heat capacity of 8OCB reported by HATTA et al. [10]. The heat capacity in SmA phase is divalent: it gives a low value in process of increasing temperature from K to N phases, but it is significantly higher in the reverse process from N to K phases. This means that there exist two states in SmA, and that the state giving higher heat capacity is more stable.

The picosecond time-resolved fluorescence spectra are shown in Figs. 3(a) and (b) for K_1 and SmA phases, respectively. The spectrum of K_1 phase shows no significant change with time, having a peak at 357 nm. This spectrum seems to be identical to that observed in the steady excitation for K_1 phase (Fig. 1). On the other hand, the spectrum in SmA phase obviously changes with time; the 0-ps spectrum, which shows a peak at 355 nm, is similar to that in K_1 phase. After 200 ps, the spectrum with a peak at 380 nm becomes dominant. Therefore, the spectrum in the initial time region can be regarded as due to

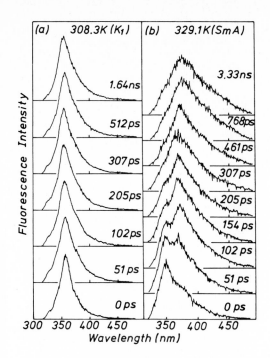

Fig. 3 Time-resolved
fluorescence spectra
of neat 8OCB in crys-
talline K_1 (a) and
smectic A (b) phases,
obtained with excita-
tion at 300 nm.

a monomer-like state, and that in the later time region to an excimer-like
state. It follows from these results that the excimer-like state is formed
from the monomer-like state within 200 ps. Similar time behaviors are observed for N and I phases. In N phase, the formation time is found to be
shorter than that in SmA phase, probably due to the lower viscosity.

Fluorescence decay curves are also affected by temperature. In SmA, N and
I phases, the decays are regarded roughly as biexponential with the lifetimes
of 100-200 ps and of 5-10 ns. The fast decaying component corresponds to the
monomer-like fluorescence, and the slow one to the excimer-like fluorescence.
The lifetime of the slow component varies depending on the phase; 10.0 ns in
SmA, 8.0 ns in N and 5.6 ns in I phase. Detailed analyses for the decay kinetics will be presented in a forthcoming paper [11].

Let us consider the nature of the fluorescence species giving the spectrum
with a peak at 375-380 nm. The intramolecular CT state can be a candidate
for elucidating the long-wavelength emission. The fluorescence spectrum of
CT state, which is observed in polar solvent like acetonitrile, shows the
maximum at 371 nm and resembles in peak position the excimer-like spectra
concerned here [11]. This result seems to suggest the electrostatic interaction mechanism as a cause for the spectral red-shift common to both acetonitrile solution and liquid crystalline state. However, the dielectric constant of 8OCB is fairly small (7-15) [12] compared to that of acetonitrile
(37.5), so that the spectral shift due to stabilization of CT state, even if
it is possible in the liquid crystal, will not be so large as that in acetonitrile solution. Alternatively, we should direct attention to the fact that
the excimer fluorescence in n-hexane solution is much the same as the long-
wavelength emission in question, in spectral shape and lifetime [11].

The structures of the liquid crystal of 8OCB and its analogues have been
examined by means of X-ray diffraction [6,7] and dielectric constant analyses

[12]. According to CLADIS et al. [6] and RATNA and SHASHIDHAR [12], the liquid crystal of 8OCB is composed of the antiparallel molecular pairs in SmA and N phases and even in the isotropic liquid phase. Note that the dipole moment of 8OCB increases five-fold on going from the ground state to the excited state [11]. When a molecule of the associated pair is photoexcited, a loosely coupled pair might be stabilized by forming an excimer-like pair in which two molecules are more closely associated. The present results show that the monomer-like state of the antiparallel dimers is relaxed within 200 ps to the excimer-like state which decays to the ground state in 10 ns. Such an association between the excited and ground state 8OCB is possible in a tight but mobile configuration of the mesophases and isotropic liquid phase. On the other hand, in the crystalline K_1 and K_2 phases, the molecular arrangement is so rigid that only monomer-like emission is observed.

The dual fluorescences of 8OCB (Fig. 1) obtained in this study are, on the whole, similar to those of 12CB reported by SUBRAMANIAN et al. [5]. Although the time-resolved experiments have not yet been made for 12CB, it is expected that the two kinds of fluorescence emissions in 12CB would show essentially the same time behavior as seen in 8OCB.

References

1. J.M. Nerbonne and R.G. Weiss: J. Amer. Chem. Soc. 101, 402 (1979).
2. V.C. Anderson, B.B. Craig and R.G. Weiss: J. Phys. Chem. 86, 4642 (1982).
3. B. Samori and L. Fiocco: J. Amer. Chem. Soc. 104, 2634 (1982).
4. H. Levanon: Chem. Phys. Lett. 90, 465 (1982).
5. R. Subramanian, L.K. Patterson and H. Levanon: Chem. Phys. Lett. 93, 578 (1982).
6. P.E. Cladis, D. Guillon, F.R. Bouchet and P.L. Finn: Phys. Rev. A23, 2594 (1981).
7. G.W. Gray: J. Phys. (Paris), 36, C1-337 (1975).
8. T. Murao, I. Yamazaki and K. Yoshihara: App. Opt. 21, 2297 (1982).
9. C. David and D. Baeyens-Volant: Mol. Cryst. Liq. Cryst. 59, 181 (1980).
10. I. Hatta, Y. Nagai, T. Nakayama and S. Imaizumi: J. Phys. Soc. Jpn. 52, Suppl. 47 (1983).
11. N. Tamai, N. Mataga and I. Yamazaki: to be published.
12. B.R. Ratna and R. Shashidhar: Pramana, 6, 278 (1976).

Excited-State Solvation Dynamics in 4-Aminophthalimide*

Sheila W. Yeh[1], L.A. Philips, S.P. Webb, L.F. Buhse[2], and J.H. Clark[3]

Laboratory of Chemical Biodynamics, Lawrence Berkeley Laboratory, and Department of Chemistry, and University of California, Berkeley, CA 94720, USA

1. Introduction

Solvation plays a major role in nearly every liquid-phase chemical reaction. Picosecond measurements of the dynamics of excited-state solvation have been exploited in the past for the study of such processes as hydrogen bond formation, "solvent-assisted" intramolecular charge transfer, and conformational isomerization[1-5]. This work explores the solvation dynamics following electronic excitation of isolated solute molecules. Molecules initially undergo a vertical transition to a Franck-Condon excited state, producing electronically excited molecules in ground state solvent environments. The equilibrium solvent configuration around an excited molecule may be significantly different than the ground state solvent configuration. Relaxation to the excited-state configuration would result in a time-dependent red shift in the emission spectrum.

2. Experimental

Time-resolved emission spectra were obtained using an ultrafast streak camera system. The excitation source was either the third (355 nm) or fourth (266 nm) harmonic of a passively mode-locked Nd:YAG laser system (Quantel, YG 400). After selecting and amplifying a single pulse from the oscillator, non-linear optical crystals were used to produce the third or fourth harmonics, which were used to excite 4-aminophthalimide (4-AP) and dimethylaminobenzonitrile (DMABN), respectively. The excitation beam was split into two parts. The main portion was used to excite the sample. Sample fluorescence was collected at a right angle to the excitation pulse, passed through a Glan-Taylor polarizing prism set at 54.7°, spectrally filtered using 10 nm (FWHM) bandpass interference filters, and combined with the minor, independently delayed part of the exciting beam. This latter pulse provided an internal time marker for signal averaging and was also used as a record of the excitation pulse profile for deconvolution purposes. These two signals were focussed together onto the photocathode of the streak camera (Hadland Photonics, IMACON 500). The output of the camera was imaged onto a linear photodiode array (Tracor-Northern, IDARSS) and transferred to an LSI 11/73 microcomputer for signal processing. Effective data collection rates were greater than 1.0 Hz. The data presented below typically consist of the average of 500 shots. Steady-state emission data were taken using a computer-controlled spectrofluorimeter (Spex, FLUOROLOG 2).

3. Results and Discussion

Previous studies of 4-AP in 1-propanol (PrOH, μ = 1.68 D) at low temperatures with nanosecond resolution by WARE [6] and coworkers show the need for higher

*Supported by the Gas Research Institute under Contract No. 5081-260-568
[1]National Science Foundation Pre-doctoral Fellow
[2]Bell Laboratories Fellow
[3]Alfred P. Sloan Research Fellow and Henry and Camille Dreyfus Teacher-Scholar

time resolution to fully resolve the excited-state dynamics. Typical temporal profiles of the emission from 4-AP in PrOH and computer-generated fits are shown in Fig. 1. From such fits, the deconvoluted fluorescence intensity as a function of time is obtained at 10 nm intervals ranging from 400 to 650 nm. The time-integrated emission intensity at each wavelength is normalized to the steady-state emission spectrum. The emission spectrum at fixed times (Fig. 2) can then be calculated from the normalized time-dependent emission data at fixed wavelength. A smooth red shift of the emission maximum with time is observed.

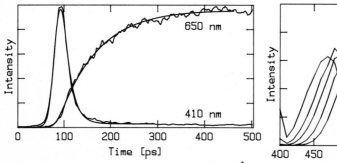

Fig. 1 Time-resolved emission from 4-AP in PrOH. Best fit falltime for blue-edge emission (410 nm) is 14 ps. Best fit risetime for red edge emission (650 nm) is 86 ps

Fig. 2 Time-resolved emission spectra from 4-AP in PrOH

Time-dependent red shifts in the emission spectra of other compounds have been previously discussed as arising from two distinct emitting states[3,4]. Blue emission is said to emanate from the initially excited electronic state. This blue-emitting state undergoes a radiationless transition to produce a second, lower energy, charge-transfer state. The population of the initially excited blue state decays away on a short time scale, and the population of the second, relaxed, red state increases with a corresponding time constant. Such a model predicts an isosbestic point in the time-dependent spectra.

DMABN is a system in which it has been argued that emission arises from two distinct emitting states[4,7]. MNDO calculations[8] show that the blue, initially excited state has the lowest energy ground-state geometry, with the plane of the dimethylamino group perpendicular to that of the benzonitrile ring. The calculations confirm earlier models, which ascribe the red emission to a twisted charge-transfer state with the dimethylamino group parallel to the ring[4,7]. Fig. 3 shows the emission spectra at fixed times for DMABN in PrOH, determined as described above for 4-AP. As with 4-AP in PrOH, blue emission from DMABN in PrOH shows a fast decay time, and the red edge of the emission spectrum has a non-zero risetime. However, time-resolved emission spectra for DMABN (Fig. 3) clearly show an isosbestic point, compelling evidence for the presence of two distinct, but chemically related, emitting states. The dramatic difference between the time-resolved emission spectra of 4-AP and DMABN in PrOH shows that the time-dependent spectral shifts in the emission of these two compounds arise from different chemical processes. The smooth shift in 4-AP emission with time suggests that, in PrOH, solvation occurs as a continuous evolution through a large number of intermediate states. MNDO calculations on 4-AP support this interpretation, in that no significant geometry change is necessary for the charge-transfer state to form upon electronic excitation.

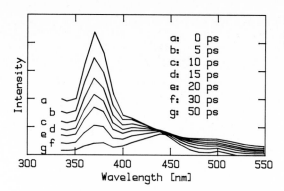

a: 0 ps
b: 5 ps
c: 10 ps
d: 15 ps
e: 20 ps
f: 30 ps
g: 50 ps

Fig. 3   Time-resolved emission spectra of DMABN in PrOH. Note isosbestic point at 440 nm

Picosecond, time-resolved emission spectroscopy has also been carried out on 4-AP in acetonitrile, a highly polar (μ = 3.92 D) but aprotic solvent. In this case the temporal history of the emission at <u>all</u> wavelengths shows an extremely short (<5-10 ps) risetime and no fast decay components. The observed differences between the excited-state dynamics of 4-AP in PrOH and acetonitrile may be understood in terms of a model in which the dominant excited-state process occurring in PrOH is the strengthening of hydrogen bonds between the carbonyl groups of 4-AP and the solvent. This explanation is consistent with the MNDO calculations, which show a higher electron density on the carbonyl groups in the excited state than in the ground state. Precisely this behavior would be predicted from the results of previous studies of excited-state proton-transfer reactions in similar systems such as the coumarins[9,10]. Experiments on 4-AP and DMABN in several different solvents as a function of temperature and pressure which are currently in progress will test the predictive ability of the proposed model for excited-state relaxation in these systems.

4.   References

1.   A. Declemy, C. Rulliere, and P. Kottis, Chem. Phys. Lett. <u>101</u>, 401 (1983).
2.   J. D. Simon and K. S. Peters, J. Phys. Chem. <u>86</u>, 4855 (1983).
3.   D. Huppert, H. Kanety, and E. M. Kosower, Faraday Discuss. Chem. Soc. <u>74</u>, 161 (1982).
4.   D. Huppert, S. D. Rand, P. M. Rentzepis, P. F. Barbara, W. S. Struve, and Z. R. Grabowski, J. Chem. Phys. <u>75</u>, 5714 (1981).
5.   S. P. Velsko and G. R. Fleming, J. Chem. Phys. <u>76</u>, 3553 (1982).
6.   W. R. Ware, S. K. Lee, G. J. Brant, and P. P. Chow, J. Chem. Phys. <u>54</u>, 4729 (1971).
7.   Y. Wang, M. McAuliffe, F. Novak, and K. B. Eisenthal, J. Phys. Chem, <u>85</u>, 3736 (1981).
8.   M. J. S. Dewar and W. Thiel, J. Am. Chem. Soc. <u>99</u>, 4899 (1977); program from Quantum Chemistry Program Exchange, Indiana University, QCPE No. 455.
9.   J. F. Ireland and P. A. H. Wyatt, Adv. Phys. Org. Chem. <u>12</u>, 131 (1976).
10.   A. J. Campillo, J. H. Clark, S. L. Shapiro, K. R. Winn, and P. K. Woodbridge, Chem. Phys. Lett. <u>67</u>, 218 (1979).

The Pyrazine Mystery: A Resolution

A. Lorincz*, F.A. Novak, D.D. Smith**, and S.A. Rice

Department of Chemistry and The James Franck Institute
The University of Chicago, Chicago, IL 60637, USA

We report new measurements of the rotational state dependence of the initial fluorescence decay of pyrazine. The apparently inconsistent data from several sources are reconciled.

I. Introduction

Pyrazine has long been considered to be a molecule with intermediate case level structure. The spontaneous emission from an excited state of such a molecule does not exhibit a simple exponential decay with time. The observed nonexponential decay is interpreted as arising from rapid evolution of the initially formed singlet wavepacket to generate mixed singlet-triplet levels, followed by slow depopulation via radiation from the singlet components of the mixed levels. The evolution of the wavepacket is, in the absence of coupling to the radiation field, a coherent process, so the mixed levels generated have well defined phase relations and can generate a beat spectrum.

II. Experimental Results

All experiments were carried out on pyrazine seeded in a supersonic free jet using He as the carrier gas. The average decay of pyrazine over the rotational contour is shown in Fig. 1. Pulse duration is 3ps, spectral width is about 7 cm^{-1}. The response function was measured by Rayleigh scattering from argon clusters.

Our results for the rotational state dependence of the initial state decay of fluorescence of pyrazine are shown in Table 1.

Table 1
Rotational state dependence of the initial decay of fluorescence of pyrazine

Branch	P						Q	R						
Level	7	6	5	4	3	2		1	2	3	4	5	6	7
Lifetime ($\bar{+}$5ps)	108	110	108	107	102	100	110	94	96	108	114	109	104	104

These data, taken by pulses of 22 ps duration and 0.7 cm^{-1} spectral width, show sensibly no rotational state dependence of the lifetime. We note that all the decay curves were found to have clean single exponential form. The

*Permanent address: Institute of Isotopes of the Hungarian Academy of Sciences, P.O. Box 77, H-1525 Hungary

**Permanent address: Department of Chemistry, Purdue University West Lafayette, Indiana 47907 U.S.A.

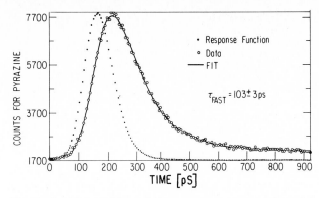

Fig. 1 Raw data, fit and instrument response function
for the fast component decay of the pyrazine

preliminary measurements of these lifetimes reported by SMITH, RICE and
STRUVE [1] were not based on a proper deconvolution of the decay curves and
were compromised by a (later discovered) defective modelocker in the
apparatus.

III. Discussion

Pyrazine has been extenisvely studied for the purpose of testing the theoretical description of the dynamical properties of intermediate case level
structure molecules. The existing data set has some apparent anomalies
which require consistent interpretation. We shall focus attention on the
following:
(i) It has been suggested that the fast component of the nonexponential decay of fluorescence is due to Rayleigh scattering and is not due to unphasing
of the initially prepared wavepacket [2]. (We use unphasing to refer to a
coherent process of evolution in an isolated molecule; dephasing will refer
to the same process in an ensemble of molecules.)
(ii) The fast component of the fluorescence decay appears to be absent when
J=0, although present for all J≠0, yet there is quantum beat structure in
the decay from J=0 [3].
(iii) As shown by the work reported in this paper, the rate of decay of the
fast component of fluorescence is independent of J.
(iv) The ratio of amplitudes of fast to slow components of the fluorescence
decay is a strong function of excitation wavelength on the scale of the width
and separation of rotational transitions [2].
(v) The presence of a relatively small magnetic field leads to alteration of
the ratio of amplitudes of fast to slow fluorescence decay components; the
effect saturates at about 100 Gauss [4,5] and suggests an increase in the
number of triplet states coupled to the singlet as the magnetic field
increases [4]. There is no effect when J=0.
   It is easiest to deal with (i) first: Our data show (see Fig. 1) that
this suggestion is not correct. The fast component in the fluorescence decay
is, as originally proposed, a coherent unphasing of the initially prepared
wavepacket.
(ii) The initial unphasing of the wavepacket state with J=0 is masked by the
beats in the decay because the recurrence time corresponding to the eigenstate distribution is comparable to the unphasing time.
(iii) The lack of dependence of the initial decay rate (unphasing rate) on J

is a consequence of sum rules [6] applicable to the fast part of the fluorescence decay of intermediate case level structure molecules [7]. The sum rules are valid if spin-orbit coupling dominates the perturbations.

(iv) Coupling of a molecule with the radiation field leads (in absorption) to both resonant and nonresonant excitation of particular states. Resonant excitation is always so much stronger than nonresonant excitation that the latter is masked except in special cases. The variation of the ratio of amplitudes of fast to slow components of the fluorescence decay is interpreted as the unmasking of the nonresonant excitation of a mixture of states by moving away from the resonant excitation of one state. The observed large ratio of fast to slow components of fluorescence for the case of off resonant excitation, and small ratio for resonant excitation of one eigenstate, is the expected effect.

(v) If a magnetic field splits the components of a level, the time required for destructive interference to develop within the wavepacket built from such split levels is of the order of $\hbar/\Delta E$, with ΔE the splitting. As the time required for development of this destructive interference decreases—with increasing magnetic field – and becomes comparable (smaller) with (than) the observation time, the ratio of fast to slow components of fluorescence decay increases (saturates). If all of the molecular eigenstates contain an equal amount of singlet component then the value of the ratio of fast to slow component amplitudes is $(2J+1)$ larger than the ratio in the absence of a magnetic field. It can be shown that this is an upper limit.

IV. Acknowledgements

This research has been supported by grants from the NSF and AFOSR. We thank Prof. D. Pratt of the University of Pittsburgh for a preprint of his recent work on the magnetic field dependence of the ratio of amplitudes of fast to slow fluorescence decay.

1. D.D. Smith, S.A. Rice, W. Struve, Faraday Dis. Chem. Soc. 75, 173(1983).
2. J. Kommandeur Workshop on Primary Photophysical Processes, Herrsching Oct. 16-20, #28 (1983).
3. J. Kommandeur Rec. Trav. Chim. PAys-Bas 102, 421 (1983).
4. P.M. Felker, W.R. Lambert, A.H. Zewail, Chem. Phys. Lett. 89,(4) 309 (1982).
5. Y. Matsumoto, L.H. Spangler, D.W. Pratt, preprint.
6. F.A. Novak, S.A. Rice, J. Chem. Phys. 73, (2) 858 (1980).
7. K.F. Freed, A. Nitzan, J. Chem. Phys. 73, (10) 4765 (1980).

Picosecond Laser Studies of Photoinduced Electron Transfer in Porphyrin-Quinone and Related Model Systems

N. Mataga, A. Karen, and T. Okada

Department of Chemistry, Faculty of Engineering Science, Osaka University
Toyonaka, Osaka 560, Japan

Y. Sakata and S. Misumi

The Institute of Scientific and Industrial Research, Osaka University
Mihogaoka, Ibaraki, Osaka 567, Japan

Mechanisms of photoinduced electron transfer (ET) in porphyrin-quinone model systems have been investigated by means of ps laser photolysis method. Transient absorption spectra were measured with a microcomputer-controlled double beam mode-locked Nd^{3+}:YAG laser photolysis system [1]. Second harmonic or third harmonic was used for excitation. The fluorescence decay times were determined by the same laser photolysis system with a streak camera. For example, the ps transient absorption spectra of EEP(ethyl-etioporphyrin)-TQ(toluquinone) system in acetone, where the fluorescence of EEP is completely quenched, are very similar to the $S_n \leftarrow S_1$ absorption spectrum of EEP itself. This result indicates that the photoinduced ET or ion-pair state of EEP-TQ system in acetone is deactivated to the original ground state very rapidly without producing dissociated ions and we are observing only the S_1 state before deactivation by collisional interaction with TQ. The decay time of this transient absorbance was obtained to be $\tau_{obs} \sim 70$ ps which was in an approximate agreement with the value ($\tau_{calc} \sim 80$ ps) estimated from the relation $\tau_{calc} = \tau_0/(1 + k_q \tau_0 [TQ])$ where τ_0 is the lifetime of S_1 state of EEP and k_q is the bimolecular rate constant of the quenching by TQ. Contrary to the above results, the ps transient absorption spectra in benzene are quite different from the $S_n \leftarrow S_1$ spectra of EEP and are similar to that of porphyrin cation. In benzene solution used for the transient absorption measurements ([TQ]=0.62 M), about 90% of EEP forms loose complex. Accordingly, the transient absorption spectra in benzene solution are due to the ET state (EEP$^+$TQ$^-$) formed by the excitation of the loose complex and its decay time was obtained to be $\tau_{obs} \sim 40$ ps. These results provide a direct connection between the porphyrin-quinone systems and usual exciplex systems as well as excited singlet state of weak EDA complexes. Although it is rather short-lived, porphyrin-quinone exciplex can be observed in relatively non-polar solvent while the photoinduced ET state undergoes solvation-induced ultrafast deactivation in polar solvents. Solvation in the ET state lowers its energy but lifts up the energy of the neutral ground state compared to that relaxed with respect to solvation, which results in a very small energy gap between two states leading to the ultrafast deactivation in the case of porphyrin-quinone system. The dissociation into solvated ion radicals can not compete with this ultrafast deactivation. However, in the case of some typical exciplex systems like pyrene-N,N-dimethylaniline and -p-dicyano-benzene as well as weak EDA complexes like tetracyanobenzene-toluene complex, the energy gap is still considerably large even in acetonitrile and acetone,leading to an efficient dissociation into separated ions [2]. A way to avoid the ultrafast deactivation of the porphyrin-quinone ET state is to construct the intramolecular exciplex type compounds where porphyrin and quinone are combined by methylene chains in order to make weaker the interaction between the chromophores and also to prevent the strong solvation by intervening chains. Some results of our ps laser photolysis studies of (octaethylporphyrin)-$(CH_2)_n$-(benzoquinone)(PnQ, n=2, 4, 6) and (etio-porphyrin)-$(CH_2)_n$-(benzoquinone)-$(CH_2)_m$-(trichlorobenzoquinone)(PnQmQ', n= 2, 4, m=4) are given here.

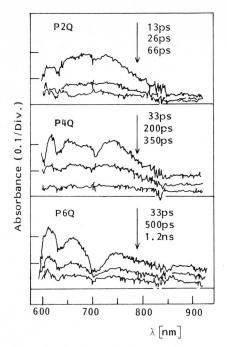

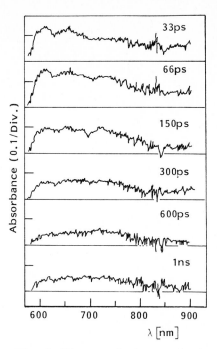

Fig. 1 Picosecond time-resolved transient absorption spectra of PnQ in benzene

Fig. 2 Picosecond time-resolved transient absorption spectra of P4Q4Q' in dioxane

In Fig. 1, P2Q shows characteristic absorption band of ET state immediately after excitation, which means that ET takes place within ca. 10 ps. The transient spectra of P6Q immediately after excitation are very similar to the $S_n \leftarrow S_1$ spectra of porphyrin itself, indicating much slower ET than P2Q. The spectra of P4Q are intermediate between those of P2Q and P6Q. The fluorescence decay times of P6Q and P4Q determined by streak camera were 590 ps and 70 ps, respectively. These values represent the time constants of photoinduced ET since the fluorescence lifetime of nonaethylporphyrin is ca. 10 ns. The transient absorption spectra of Fig. 1 can be analyzed according to the reaction scheme of (1). By using the τ values

$$\xleftarrow{1/\tau_0} S_1(P^*nQ) \xrightarrow{k} ET(P^+nQ^-) \xrightarrow{1/\tau'} \tag{1}$$

($\tau = (k+\tau_0^{-1})^{-1}$) given above, τ' values were estimated to be ca. 10 ps, 130 ps and 600 ps for P2Q, P4Q and P6Q, respectively. The rate of ET given by $1/\tau$ or $1/\tau'$, therefore, shows exponential dependence upon chain length, which is indicative of the intramolecular electron tunneling mechanism. Namely, in the well-known equation for ET probability (2), where (F.C.)

$$k_{ET} \approx (2\pi/\hbar) \cdot \beta^2 \cdot (F.C.) \tag{2}$$

is the temperature averaged Franck-Condon term, the electronic interaction term β^2 is usually assumed to be proportional to $\exp(-Const \cdot r_{PQ})$. The observed results seem to reflect this r_{PQ} dependence of β^2.

The lifetime of the photoinduced ET state of these systems increases with increase of the chain length. However,, if the chain becomes too long, the photoinduced charge separation becomes inefficient because not only τ' but also τ becomes very long. In this respect, the two step ET as indicated in (3) for the PnQmQ' system may be very important:

$$S_1(P^*nQmQ') \xrightarrow{k_1} ET_1(P^+nQ^-mQ') \xrightarrow{k_2} ET_2(P^+nQmQ'^-) \,. \tag{3}$$
$$\downarrow 1/\tau_0 \qquad\qquad \downarrow 1/\tau_1 \qquad\qquad\qquad \downarrow 1/\tau_2$$

The transient absorption spectra of P4Q4Q' in dioxane solution (Fig. 2) indicates a considerable contribution from the S_1 state at short delay times, but the spectra at several hundreds ps are characteristic of ET state. The lifetime of photoinduced ET state of this system is much longer compared with that of P4Q in dioxane ($\sim$100 ps) due to the more extensive charge separation as indicated in (3), where k_2 should be larger than $1/\tau_1$. If k_2 is much larger than k_1 and $1/\tau_1$, reaction scheme of (3) becomes practically the same as (1). The observed results of P4Q4Q' in dioxane can be reproduced satisfactorily with the scheme of (1) and τ' was obtained to be 300 ps. These results are of crucial importance for designing any bio-mimetic photosynthetic system.

References

1. H. Miyasaka, H. Masuhara and N. Mataga, Laser Chem. 1, 357 (1983).
2. (a) N. Mataga in "Molecular Interactions", eds. H. Ratajczak and W. J. Orville-Thomas, Vol. 2 (John Wiley & Sons, 1981) p. 509. (b) N. Mataga, Radiat. Phys. Chem. 21, 83 (1983). (c) Y. Hirata, Y. Kanda and N. Mataga, J. Phys. Chem. 87, 1659 (1983).

The Excited-State Proton Transfer Reactions of Flavonols in Alcoholic Solvents

K.-J. Choi

The Standard Oil Company (Ohio), Corporate Research Center
4440 Warrensville Center Road, Warrensville Heights, OH 44128, USA

B.P. Boczer and M.R. Topp

Department of Chemistry, University of Pennsylvania
Philadelphia, PA 19174, USA

The excited state proton transfer reaction of 3-hydroxyflavone has been a subject of many publications in recent years.[1-9] A recent study by McMorrow and Kasha [9] cleared early misunderstandings on the viscous barrier in non-polar solvents.

However, there have been several contradictory explanations of the dynamics of two fluorescence bands observed in hydrogen bonding solvents. The original suggestion was that there are two conformations of 3-hydroxyflavone (intramolecular hydrogen bonded species and intermolecular hydrogen bonded species).[1] The validity of this suggestion was questioned because a strong isotope effect was observed on the relative intensities of two bands.[2] Subsequent time-resolved fluorescence studies have been interpreted in two different ways: (1) One ground state conformation (intermolecular hydrogen bonded) and one channel formation of the tautomer [4], and (2) One ground state conformation and two channel formation of the tautomer from the initially excited state.[8]

We have studied the effect of hydrogen bonding solvents on the dynamics of the proton transfer reaction of flavonols. Our results suggest that there are two different conformations (intra and intermolecularly hydrogen bonded) in the ground state. The former undergoes a rapid proton transfer reaction upon excitation. However, the latter requires local solvent reorientation prior to the proton transfer process.

We have used the frequency up-conversion gating method for fluorescence time-resolution. The experimental details have been described elsewhere.[10,11] The samples were excited with the 3rd harmonic pulses from a Nd-YAG laser (354.7 nm, 10 psec) and the resulting fluorescence was gated with 1.06 micron fundamental pulses. The up-converted light was detected with a double monochromator and a photomultiplier. All experiments were performed at room temperature. The time profiles were analyzed with a nonlinear least square curve fitting program based on the Simplex method.

The decay time profiles of the blue fluorescence band (420 nm) in all selected alcohols (C1 to C10 normal alcohols) fit very well to single exponential decays. However, the time profiles of the tautomer fluorescence (520 nm) were more complex. We obtained good results by fitting the data to a two component rise. We determined the t=0 point and the instrument time resolution from the rise time of the green fluorescence of 3-hydroxyflavone in hexane.

The results of quantum yield and decay time measurements are summarized in Table I. The rise times of tautomer fluorescence indicate that the tautomer is formed through two different channels. Strandjord and Barbara attributed the fast growth to the initially excited state and the slow growth to the "relaxed" form.[8]

Recently, McMorrow and Kasha found that there is no barrier for the proton transfer process in the intramolecularly hydrogen bonded form. It is very unlikely that the relaxed and unrelaxed states have drastically different rates for a process which has no barrier.

TABLE I. Fluorescence decay time and quantum yield of 3-hydroxyflavone

Solvent	420 nm Fluorescence		520 nm Fluorescence			
					Rel. Intensities	
	Φ_F	Decay (ps)	Φ_F	Decay (ps)	Slow Rise	Fast Rise
Hexane	0.0	--	0.35	3200		1.00
Methanol	0.0031	28	0.028	268	0.66	0.28
Ethanol	0.0018	32	0.027	320	0.36	0.57
Butanol	0.0023	40	0.038	385	0.42	0.59
Decanol	0.0029	51	0.088	780	0.41	0.74

We propose a model in which there are two ground state conformations in hydrogen-bonding solvents, each having one channel for tautomer formation: one conformation with intramolecular hydrogen bonding (N_I), the other with intermolecular hydrogen bonding (N_E). The two conformations are in equilibrium before the excitation. ($N_I \leftrightarrow N_E$). The initial concentrations of the two excited state conformations should be the same as the ground state equilibrium concentrations. The intramolecularly hydrogen bonded conformation undergoes an extremely fast proton transfer reaction and is responsible for the pulse limited rise component of the green fluorescence band. The slow rising component of the tautomer fluorescence is assigned to the fluorescence from intermolecularly hydrogen bonded conformations. Unlike the intramolecular hydrogen bonded conformation, direct proton transfer is blocked for this species. Therefore, reorientation ($N_E^* \rightarrow N^* \rightarrow T^*$) must preceed the proton transfer. McMorrow and Kasha recently proposed the same mechanism to explain the results of their spectroscopic studies.[15]

We estimated the radiative lifetime of the tautomer excited state from the quantum yield and decay time of the green fluorescence. The radiative lifetime does not vary more than ±20% in all the solvents we used. The strong solvent dependence of quantum yields is mainly due to the solvent dependence of non-radiative decay rates. The estimated radiative lifetime of the tautomer excited state in hexane is 9.1 nsec. This value is very close to the limiting value of the green fluorescence decay time (8.6 nsec) at low temperature [6]. Therefore, the quantum yield of the tautomerization in hexane is close to unity. For time-resolved fluorescence measurements, the fluorescence intensity depends only on the radiative rate and the concentration of the excited state.[13] The intensities of the slow and fast components of the tautomer fluorescence normalized to the value in hexane are listed in Table I. The sums of the fast and slow components are approximately 1.0 in all solvents. Since it is reasonable to assume that the radiative lifetime of the excited tautomer has only a weak solvent dependence, the tautomer yield after the initial two channel formation should be similar in all solvents. The blue fluorescence quantum yields are only a few percent in all solvents. Therefore, we conclude that the major decay route of the initial excited state is tautomerization regardless of the ground state conformation.

Although it is not possible to deduce a detailed picture of external hydrogen bonding structure between solvent and solute, the proton transfer reaction may proceed through double proton transfer in a 5-membered ring[14] or 7-membered ring[4] which includes a solvent molecule. Neither pathway involves a net change in the number of hydrogen bonds, and both should be faster than a process involving hydrogen bond breaking followed by solvent reorientation.[12] The measured blue decay times should therefore be interpreted as the solvent reorientational time instead of the proton transfer time. Our decay time measurements show a weak dependence on solvent from methanol to decanol. The local solvation structure may determine the rate of the reorientation and the bulk properties of the solvent may have only a weak effect on the process.

Acknowledgments

We wish to thank R. L. Swofford, D. A. Chernoff and T. L. Gustafson for many helpful discussions and Professor M. Kasha for providing us with preprints.

References

1. Yu L. Frolov, et al., Izv. Akd. Nauk. USSR Ser. Khim. 10, 2364 (1974).
2. P.K. Senqupta and M. Kasha, Chem. Phys. Lett. 68, 382 (1979).
3. O.A. Salman and H.G. Drickamer, J. Chem. Phys. 75, 572 (1981).
4. G.J. Woolfe and P.J. Thistlethwaite, J. Am. Chem. Soc. 103, 6916 (1981).
5. O.A. Salman and H.G. Drickamer, J. Chem. Phys. 77, 3329 (1982).
6. M. Itoh, et al., J. Am. Chem. Soc. 104, 4146 (1982).
7. A.J.G. Strandjord, S.H. Courtney, D.M. Friedrich, and P.F. Barbara, J. Phys. Chem. 87, 1125 (1983).
8. A.J.G. Strandjord and P.F. Barbara, Chem. Phys. Lett. 98, 21 (1983).
9. D. McMorrow and M. Kasha, J. Am. Chem. Soc. 105, 5134 (1983).
10. K.-J. Choi and M.R. Topp, J. Opt. Soc. Am. 71, 520 (1981).
11. K.-J. Choi, B.P. Boczer and M.R. Topp, Chem. Phys. 57, 415 (1981).
12. S.K. Garg and C.P. Smyth, J. Phys. Chem. 69, 1294 (1965).
13. K.-J. Choi, L.A. Hallidy, and M.R. Topp in "Picosecond Phenomena II." (Hochstrasser, Kaiser and Shank eds.) Springer Series in Chemical Physics 14, 131 (1980).
14. O.S. Wolfbeis, A. Kuierzinger and R. Schipfer, J. Photochem. 21, 67 (1983).
15. D. McMorrow and M. Kasha, J. Phys. Chem. 88, 2235 (1984).

Excited-State Proton-Transfer Reactions in 1-Naphthol*

S.P. Webb, S.W. Yeh[1], L.A. Philips, M.A. Tolbert[2], and J.H. Clark[3]

Laboratory of Chemical Biodynamics, Lawrence Berkeley Laboratory, and
Department of Chemistry, University of California, Berkeley, CA 94720, USA

1. Introduction

Proton-transfer reactions provide the fundamental basis for acid-base chemistry in protic solvents. Literally hundreds of examples of excited-state proton-transfer reactions are known[1]. Picosecond, time-resolved measurements on spectrally distinct acid-base pairs yield direct kinetic information and provide insight into the effect of molecular structure on excited-state proton-transfer dynamics[2,3]. One particularly interesting and seemingly paradoxical system is 1-naphthol. From a Förster cycle calculation, 1-naphthol is predicted to have an excited state pK similar to 2-naphthol[4], for which steady-state emission from both the neutral (acidic) and anionic (basic) excited-state species is clearly observed. In contrast, the neutral form of excited 1-naphthol shows "very weak" fluorescence which is "extremely difficult" to measure[5]. Detailed studies of the mechanism and kinetics of excited-state proton transfer in 1-naphthol are needed to elucidate how the differences in molecular structure between these molecules affect excited-state proton-transfer processes.

2. Experimental

Samples of preparative-HPLC-purified 1-naphthol in degassed aqueous (H_2O or D_2O) solution were excited with 266 nm, ~20 ps pulses obtained by quadrupling the output of a passively modelocked Nd:YAG oscillator/amplifier system (Quantel, YG400). Temporal profiles of emission were monitored using an ultrafast streak camera (Hadland Photonics, IMACON 500) system described elsewhere in this volume[6]. The ~10^{-3} M solutions were prepared by dissolving 1-naphthol in a pH 12 solution of NaOH and adjusting to the desired pH with concentrated HCl. All solutions were maintained and studied under oxygen-free conditions. A similar procedure was used to prepare samples in D_2O using NaOD and DCl.

3. Results and Discussion

The ultimate goal of this work is to develop predictive models for excited-state proton-transfer reactions. This can be accomplished by isolating and quantitatively determining those rates that are significantly altered as the experimental conditions (e.g., molecular structure, solvent, temperature, and pressure) are changed. Kinetic schemes for coupled, two-state systems, Fig. 1, have been previously developed[1]. They provide an accurate description of excited-state proton-transfer reactions in 1-naphthol[4]. The clear isosbestic point at ~450 nm in the picosecond, time-resolved emission spectra, Fig. 2, demonstrates the validity of a two-state model. Kinetic expressions for the time dependence of the emission intensity from the neutral, $I_{(ROH^*)}$, and anionic,

*Supported by U.S. Dept. of Energy Contract No. DE-AC03-76SF00098
[1]National Science Foundation Pre-doctoral Fellow
[2]Present address: California Institute of Technology, Pasadena, CA 91125
[3]Alfred P. Sloan Research Fellow and Henry and Camille Dreyfus Teacher-Scholar

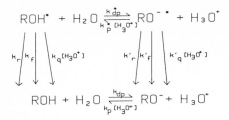

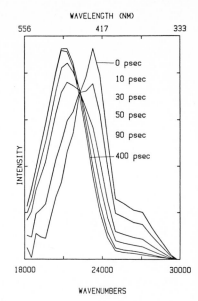

Fig. 1 Kinetic scheme for excited-state proton-transfer in 1-naphthol: k_{dp}, k^*_{dp}, k_p, and k^*_p represent the ground and excited state deprotonation and protonation rate constants, respectively; and k_r, k'_r, k_f, k'_f, k_q, and k'_q are the rate constants for radiationless processes, fluorescence, and proton-induced quenching for the neutral and anionic species, respectively

Fig. 2 Time-resolved emission spectra of 1-naphthol in pH 7 aqueous solution

$I_{(RO^{-*})}$, species, as derived from such a model, have the general form

$$I_{(ROH^*)} = A_1 e^{-\gamma_1 t} + A_2 e^{-\gamma_2 t} \tag{1}$$

$$I_{(RO^{-*})} = B(e^{-\gamma_2 t} - e^{-\gamma_1 t}). \tag{2}$$

The good agreement between the fall time of the emission from the neutral form and the risetime of the emission from the anion, Fig. 3, confirms the validity of the kinetic model. Previous steady-state results have not provided accurate rate constants for this system[5]. However, the complex relationship between the observed double exponential behavior of the emission and the various kinetic parameters also makes analysis based solely on time-dependent methods difficult. While all the needed rates can, in principle, be extracted from a measure of γ_1, γ_2, and the ratio A_1/A_2 as a function of hydrogen ion concentration, scatter in the experimental results has thus far made it impossible to obtain <u>both</u> the slopes and intercepts with high accuracy.

Combination of the results of steady-state measurements with direct, time resolved measurements yields the most accurate determination of the desired rates. For example, the rate constants τ_0 ($1/\tau_0 = k_f + k_r$) and k^*_{dp} (see Fig. 1) can be readily determined from

$$\Phi'/\Phi'_0 = k^*_{dp}/\gamma_1 = k^*_{dp}/(k^*_{dp} + 1/\tau_0), \tag{3}$$

where Φ'/Φ'_0 is the quantum yield of fluorescence of the anion at neutral pH relative to that at high pH, and γ_1 is obtained from the neutral fall time and/or the anion risetime at pH 7. A short summary of representative results from such an analysis is given in Table 1 for both H_2O and D_2O. 1-Naphthol fluorescence in D_2O shows similar overall behavior but considerably slower kinetics, by a factor of ~3.5, compared to the H_2O results. At neutral pD where quenching by D_3O^+ is negligible, there is still a substantial isotope

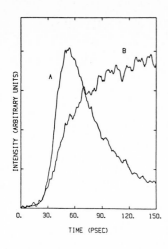

INTENSITY (ARBITRARY UNITS)

0.   30.   60.   90.   120.   150.

TIME (PSEC)

Fig. 3 Time-resolved emission from the neutral (A, 350 ± 5 nm) and anionic (B, 500 ± 5 nm) forms of 1-naphthol in pH 7 aqueous solution. Best fit fall time for curve A is 35 ± 4 ps. Best fit risetime for curve B is 30 ± 3 ps

effect. Both k^{*}_{dp} and τ_0 become markedly slower. This result strongly suggests that there is significant solvent quenching of the neutral fluorescence, and points up the importance of hydrogen bonding in these systems.

Table 1. Measured rates, lifetimes, and relative fluorescence quantum yields for 1-naphthol in H_2O and D_2O solutions. $1/\tau'_0 = k'_f + k'_r$

Solvent	Φ'/Φ'_0	γ_1^{-1} [ps]	k^{*}_{dp} [s^{-1}]	τ_0 [ps]	τ'_0 [ns]
H_2O	0.66	33	2.1×10^{10}	100	8.0
D_2O	0.69	105	6.6×10^{9}	340	20.0

4. References

1. J. F. Ireland and P. A. H. Wyatt, Adv. Phys. Org. Chem. 12, 131 (1976).
2. J. H. Clark, S. L. Shapiro, A. J. Campillo, and K. R. Winn, J. Am. Chem. Soc. 101, 746 (1979).
3. M. Gutman, D. Huppert, and E. Pines, J. Am. Chem. Soc. 103, 3709 (1981).
4. C. M. Harris and B. K. Sellinger, J. Am. Chem. Soc. 84, 891 (1980).
5. C. M. Harris and B. K. Sellinger, J. Phys. Chem. 84, 1366 (1980).
6. S. W. Yeh, L. A. Philips, S. P. Webb, L. F. Buhse, and J. H. Clark, in Ultrafast Phenomena, eds. D. H. Auston and K. B. Eisenthal, Springer, New York, 1984 (in press).

Structural and Solvent Effects on the Excited State Dynamics of 3-Hydroxyflavones

P.F. Barbara and A.J.G. Strandjord

Department of Chemistry, University of Minnesota, Minneapolis, MN 55455, USA

Much less is known about small barrier (<5 kcal/mol) proton transfer reactions than the widely studied large barrier case. Fortunately, laser studies on molecules undergoing excited state intramolecular proton transfer ESIPT offer the means to investigate the kinetics and mechanism of ultrafast small barrier proton transfer. One of the most interesting molecules which exhibits ESIPT is 3-hydroxyflavone 3HF. Several groups have been studying this molecule in recent years [1-5]. It is now clear that the stable ground-state isomer of 3HF is the normal N form. Electronically excited N molecules ($\lambda^{max} = 410$ nm) undergo a rapid proton transfer isomerization to yield a tautomer I form ($\lambda_{fl} = 543$ nm). The proton transfer kinetics (N→T) has a fast and slow component [1-2] and the kinetics are strongly dependent on solvent and temperature [1-5].

In non-hydrogen bonding solvents the proton transfer rate constant k_{PT} is extremely rapid (>10^{10} sec^{-1}) at all temperatures [3]. But, in hydrogen bonding solvents like alcohols and ethers, k_{PT} can be much slower. It seems likely that the slower kinetics in hydrogen bonding solvents can be ascribed to one or more types of 3HF/solvent complexes [3]. However, the mechanism of proton transfer of these complexes remains obscure. To help elucidate this mechanism we undertook a study of solvent and structural effects on the proton transfer kinetics of 3-hydroxyflavone, the results of which are briefly summarized in the following.

We have previously described how the rate constants for the slow component of the proton transfer kinetics of 3HF can be determined by combining information from time integrated fluorescence spectra and time resolved picosecond and nanosecond measurements [1,2]. The effect of solvent on Arrhenius plots of k_{PT} is shown in Figure 1. The activation energy E_a is considerably greater for the more strongly hydrogen bonding solvents, 2-chloroethanol and 2,2,2 trifluoroethanol. This strongly suggests that the hydrogen bond coordinates are participating in the reaction coordinate. It is important to note, furthermore, that separate experiments from our group have shown that the trends in Figure 1 are not due to viscosity effects on the proton transfer kinetics.

374

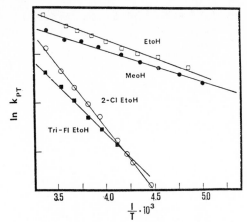

Fig. 1 Arrhenius plots for k_{PT} as determined by the methods described in Reference [2]. E_a (kcal/mole) in the various solvents are: 1.4 (MeOH), 1.7 EtOH, 5.9 (2-Cl-EtOH), and 4.6 (F_3-EtOH).

Structural effects on k_{PT} are briefly summarized in Table 1. The important trend here is that the variation of Ea is correlated with a broad range of properties of 3HF, including the absorption maximum ν_{abs}^{max}, the Pk_a of the ground state N form, the proton nmr chemical shift of the hydroxy group $\delta_{nmr}(OH)$, and the Stokes shift of each fluorescence band. This is consistent with the idea that the proton transfer dynamics are controlled by an energy barrier associated with the 3HF/solvent complex, rather than a kinetic mechanism that depends on solvent viscosity and large amplitude motion to modulate the proton transfer dynamics.

Table 1 Spectroscopic Parameters of 3HF Derivatives

Molecule	ν_{abs}^{max} cm^{-1} (MeOH)	$E_a^{(PT)}$ (MeOD) kcal/mole	PKa (MeOH)	$\delta_{nmr}^{(OH)}$ ppm(DMSO)	$\nu_{abs}^{max}-\nu_{flHE}^{max}$ cm^{-1} (MeOH)	$\nu_{abs}^{max}-\nu_{flLE}^{max}$ cm^{-1} (MeOH)
3HF	29,150	1.45	8.35	9,590	4,520	10,247
3'	29,230	1.24	8.00	9,530	4,660	10,363
4'	28,730	1.36	8.20	9,513	4,340	9,827
2'	30,580	1.79	9.80	9,100	5,828	11,623
2'4'	30,400	2.30	10.50	9,030	5,830	11,298
2'5'	30,670	2.30	10.45	9,063	5,979	11,623

One possible model for the ESIPT reaction of 3HF in alcohols (ROH) is represented by equation 1. This scheme is consistent with data presented herein and previously published kinetic and spectroscopic information [1-5]. The direct precursor to proton transfer in this model is a "loose complex" of the solvent with a N form of 3HF. The "loose" complex is in rapid equilibrium with a more strongly solvated "tight complex". Proton transfer is more rapid for the "loose complex" because it resembles an unsolvated N molecule. The observed hydrogen/deuterium isotope effect on k_{PT} is explained by an assumed mass isotope effect on the proton transfer rate of the "loose" complex.

375

"tight complex"   "loose complex"   (Eq. 1)

References

1. A.J.G. Strandjord and P. F. Barbara, Chem. Phys. Lett., 98 21 (1983).
2. A.J.G. Strandjord, S. H. Courtney, D. M. Friedrich, and P. F. Barbara, J. Phys. Chem., 87, 1125 (1983).
3. D. McMorrow and M. Kasha, J. Phys. Chem., 88, 2235 (1984).
4. G.J. Woolfe and P. J. Thistlethwaite, J. Amer. Chem. Soc., 103, 6919 (1981).
5. M. Itoh, K. Tokumura, Y. Tanimoto, Y. Okada, H. Takeuchi, K. Obi, and I. Tanaka, ibid, 104, 4146 (1982).

Electron Transfer Reactions from the First Excited Singlet State of a Polymethine Cyanine Dye

D. Doizi and J.C. Mialocq

Département de Physico-Chimie, Section Chimie Moléculaire, LRMCI (LA 331) CEN-Saclay, F-91191 Gif-sur Yvette Cedex, France

Photoactivated electron transfer reactions are of great importance for possible applications in the chemical storage of solar energy. However, whereas such reactions are very efficient "in vivo" in natural photosynthesis [1, 2] , they appear to be less efficient "in vitro" [3-7] .

The experimental results can be explained by means of a general scheme given below :

$$D* + A \longrightarrow (D*, A) \rightleftarrows (D^+, A^-)$$
$$D^+ + A^- \qquad\qquad D + A$$

D is the electron donor, the dye sensitizer,
D* is the electron donor excited in its first singlet state,
A is the electron acceptor,
(D*, A) is the encounter complex,
(D^+, A^-) is the radical ion pair created after electron transfer.

If the radical ion pair is formed in the first excited singlet state of the dye, it has the same spin multiplicity as the ground state. The so-called back reaction which gives back the reactants in their ground state is thermodynamically favoured and very efficient. It occurs in competition with charge separation which involves the diffusion of the radical ions in the bulk solution.

If the radical ion pair is created in the triplet excited state of the dye, the back reaction which is spin forbidden has a low efficiency and charge separation can occur, leading to the diffusion of the radical ions which recombine later in the bulk solution. Hence, the first problem one encounters in achieving a photosensitized electron transfer reaction is to decrease the efficiency of the recombination of the radical ions in the initial charge separated species.

In the present work, we report a study of the electron transfer reactions between a photosensitizer DODCI or 3,3'-diethyloxadicarbocyanine iodide which is excited in its first singlet state and various electron acceptors (methylviologen ($MVCl_2$), parabenzoquinone (pBQ), paradinitrobenzene (pDNB)).

Fluorescence quenching data have been calculated from the fluorescence lifetimes measured using a streak camera device and the picosecond absorption spectra of the involved intermediate species enable us to present a reaction scheme.

Although DODCI fluorescence quenching by pBQ and pDNB was very efficient, nanosecond absorption spectroscopy data gave no evidence of photoproducts [8]. The geminate recombination of the radical ions was therefore very efficient. This result agrees with previous literature findings [3-7].

Figure 1 shows the $S_n \leftarrow S_1$ absorption spectrum of DODCI in a $10^{-4}M$ methanolic solution (average of 12 shots).

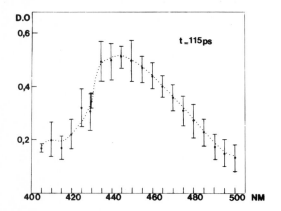

<u>Figure 1</u>   Absorption spectrum of the first excited singlet state of DODCI in a $10^{-4}M$ methanolic solution.

Figure 2 shows the differential absorption spectra obtained in a methanolic solution containing $10^{-4}M$ DODCI and $8.10^{-2}M$ $MVCl_2$ at time t = 60 ps, 390 ps and 1,29 ns.

The S_1 absorption spectrum decay (t = 60,390 ps) gave rise to a new absorption band (t = 1.2 ns) which we attribute to the oxidized $DODC^{2+}$ radical cation. The formation of the radical cation MV^+ (Fig. 3) absorbing at 395 nm [9-11] was also evidenced at t = 840 ps.

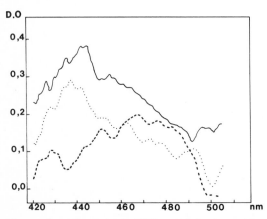

<u>Figure 2</u>   Transient differential absorption spectra in a methanol solution containing $10^{-4}M$ DODCI and $8.10^{-2}M$ $MVCl_2$.

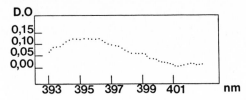

Figure 3   Transient absorption spectrum of reduced methylviologen at time
t = 840 ps in a methanol solution containing 10^{-4}M DODCI and 8.10^{-2}M $MVCl_2$

The experimental results are discussed with reference to the redox potentials of the reactants. The efficiency of the electron transfer reaction, in the case of methylviologen, is explained by the coulombic repulsion between the two radical cations created in the encounter pair [8] .

References

1   W.W. Parson, R.J. Cogdell, Biochim. Biophys. Acta, 416 (1975) 105.
2   V.A. Shuvalov, W.W. Parson, Photosynthesis III, Structure and Molecular Organisation of the Photosynthetic Apparatus, edited by G. Akoyunoglou, Balaban Int., Science Services, Philadelphia, Pa (1981) 949.
3   G.R. Seely, J. Phys. Chem. 73, (1969), 117.
4   A.K. Chibisov, Photochem. Photobiol, 10 (1969), 331.
5   D. Holten, M. Gouterman, W.W. Parson, M.W. Windsor, M.G. Rockley, Photochem. Photobiol, 23, (1976), 415.
6   D. Huppert, P.M. Rentzepis, G. Tollin, Biochim. Biophys. Acta, 440, (1976), 356.
7   D. Holten, M.W. Windsor, W.W. Parson, M. Gouterman, Photochem. Photobiol. 28 (1978), 951.
8   J.C. Mialocq, D. Doizi, M.P. Gingold, Chem. Phys. Lett., Vol. 103, N3 (1983), 225.
9   E.M. Kosower, J.L. Cotter, J.A.C.S., 86 (1964), 5524.
10 L.K. Patterson, R.D. Small Jr., J.C. Scaiano, Rad. Res., 72 (1977), 218.
11 J.A. Farrington, H. Ebert, E.J. Land, J. Chem. Soc. Farad. Trans 1, 3 (1978) 665.

Photoinduced Electron Transfer in Polymethylene Linked Donor-Acceptor Compounds: A-(CH$_2$)$_n$-D

H. Staerk, W. Kühnle, R. Mitzkus, R. Treichel, and A. Weller

Max-Planck-Institut für biophysikalische Chemie
D-3400 Göttingen, Fed. Rep. of Germany

Conformational and energetic requirements for fluorescence quenching, leading to exciplexes or radical ion pairs, and mechanisms of the back electron-transfer have been studied.

Excited species of the type A-(CH$_2$)$_n$-D (abbreviated A(n)D or A-D) exhibit intramolecular electron transfer and allow important kinetic studies when the (CH$_2$)-chain length, the solvent polarity, the viscosity, or the external magnetic field strength are varied. Our investigations are at present predominantly performed on compounds with A = Pyrene or Anthracene and D = N,N-Dimethylaniline and with n = 1 to 16.

The reaction scheme is presented in Fig. 1. ^{1}A*-D is the primary excited state which can lead to the exciplex ^{1}(A$^-$D$^+$), or due to electron transfer in solvents of higher polarity, to the radical ion pair ^{1}(^{2}A$^-$—^{2}D$^+$) with overall singlet multiplicity (SRIP) but with ^{2}A$^-$ and ^{2}D$^+$ in their doublet ground states. After picosecond excitation (third harmonic of Nd/YAG) we have measured fluorescence lifetimes τ(^{1}A*-D) with a streak camera (GEAR). Studies of compounds A(1)D and A(2)D have been described recently [1,2]. Compound A(3)D proved to be a good candidate for showing the correlation of quenching kinetics with the solvent dependent free enthalpies, ΔG_{exc} and ΔG_{rip}, of the reaction as obtained earlier for unlinked A/D systems [3]. In studies of A(3)D (A = Anthracene) we have chosen mixtures of two solvents with comparable viscosity but strongly differing dielectric constant: diethylether (DEE) $\varepsilon = 4.2$ and acetonitrile (ACN) $\varepsilon = 37$. Adjusting different mol-ratios ACN/DEE we were able to vary ΔG_{exc} and ΔG_{rip}. For pure DEE where $\Delta G_{rip} < 0$ and $\Delta G_{exc} > 0$, a fluorescence lifetime $\tau = 470$ ps was found [4]. In pure acetonitrile, where $\Delta G_{rip} >$

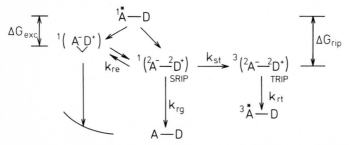

Fig. 1. Reaction scheme for linked donor acceptor compounds. ΔG_{exc} and ΔG_{rip} are the free enthalpies gained in exciplex and radical ion pair formation from the initially excited singlet state.

$\Delta G_{exc} > 0$, the measurements yielded $\tau \approx 15$ ps. In further studies of the compound A(3)D in 15 solvents of varying polarity and viscosity the quenching rate becomes already very fast when $\varepsilon \geq 7$, but the results also show that the restricted diffusion of the chain ends, influenced by the solvent viscosity, contributes to the quenching kinetics. The "harpooning" process [1] where we assumed the quenching reaction to proceed in two steps, $^1\overset{*}{A}(3)D \overset{(1)}{\longrightarrow} {}^2A^-(3){}^2D^+ \overset{(2)}{\longrightarrow} {}^1(A^-(3)D^+)$, is further evidenced by measuring the formation time of the exciplex $^1(A^-(3)D^+)$ in tetrahydrofurane. This was found to be about 40 ps while the decay time of $^1\overset{*}{A}(3)D$ was about 30 ps.

With increasing chain length, n, the average distance between A and D increases and the solvent polarity loses its importance as the rate determining parameter, so that with n = 16 the influence of the solvent polarity on the fluorescence quenching kinetics is no longer detectable.

In highly polar solvents the primary intermediate in the quenching of $^1\overset{*}{A}(n)D$ (with n = 2 - 16) is the SRIP, $^1(^2A^-(n){}^2D^+)$. The radical pair dynamics depends strongly on the chain length. With n = 3 (A = Anthracene) the electron back transfer in the SRIP takes place nonradiatively in 830 ps. With increasing chain length the direct recombination process to the ground state (rate constant k_{rg} in Fig. 1) is slowed down, whereas the decreasing spin exchange interaction in the radical ion pair opens up a fast intersystem crossing channel leading to the triplet radical ion pair TRIP, $^3(^2A^-(n){}^2D^+)$, due to nuclear hyperfine (hfi) interaction in the individual radical ions. The rate constant k_{st} connected with this nonradiative and spin-selective recombination channel is magnetic field dependent [5,2]. If the Zeemann splitting of the TRIP is tuned by an external magnetic field to coincide with the effective spin exchange interaction, $2J_{eff}$, of the system, enhanced mixing of singlet and triplet pair states takes place and leads to an increase of the molecular triplet yield. Mixing of singlet and triplet states due to hfi interaction takes place in the time range of a few hundred picoseconds to a nanosecond ($\tau_{hfi} \approx \hbar/\Delta E_{hfi}$) if not disturbed by exchange interaction [5,6]. In this time range, however, because of the very small energies (10^{-7} eV!) involved in nuclear hfi interactions uncertainty broadening also sets in [7].

In Fig. 2 are shown two illustrative examples (A = Pyrene, n = 8,9) out of a series with n = 1 - 16 [8]. The triplet extinction maximum for the system $^3\overset{*}{A}(9)D$ is found at $B_{max} = 285$ Gauss. With

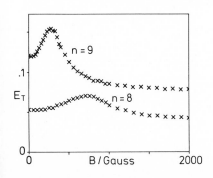

Fig.2. Extinction E_T of triplet state $^3\overset{*}{A}(n)D$ vs magnetic field strength B/Gauss. A = Pyrene, D = N,N-Dimethylaniline. $B_{max}(n=9)$: 285 Gauss; $B_{max}(n=8)$: 750 Gauss.

381

shorter chain length the effective spin exchange interaction of the system increases and consequently the triplet extinction maximum is expected at higher field strength. For $^3\overset{*}{A}(8)D$ we found $B_{max} = 750$ Gauss.

A third rather efficient decay pathway of the SRIP in A(n)D systems is the recombination to the fluorescent exciplex in polar solvents. With the application of an external magnetic field the conversion rate, k_{st}, to the TRIP is decreased as described above The SRIP concentration and the exciplex population originating from SRIP recombination then both increase. Compound A(16)D in acetonitrile, e.g., exhibits a 20 % increase of exciplex yield in a magnetic field of 300 Gauss. This is 40 times more than in the comparable unlinked system [9].

This work has been supported by the Deutsche Forschungsgemein-schaft through Sonderforschungsbereich SFB 93 "Photochemistry with Lasers".

References

[1] H. Staerk, R. Mitzkus, W. Kühnle, and A. Weller
    in "Picosecond Phenomena III" (Eisenthal, Hochstrasser, Kaiser, Laubereau eds.) Springer Series in Chemical Physics 23, 205 (1982).
[2] R. Treichel, H. Staerk, and A. Weller
    Appl.Phys. B 31, 15 (1983).
[3] A. Weller, Z.Phys.Chem. NF. 133, 93 (1982).
[4] H. Staerk, R. Mitzkus, H. Meyer, and A. Weller
    Appl.Physics B 30, 153 (1983).
[5] K. Schulten, H. Staerk, A. Weller, H.-J. Werner, and B. Nickel, Z.Phys.Chem. NF. 101, 371 (1976)
    H.J. Werner, H. Staerk, and A. Weller, J.Chem.Phys. 68, 2419 (1978)
    F. Nolting, H. Staerk, and A. Weller, Chem.Phys.Letters, 88, 523 (1982).
[6] A. Weller, F. Nolting, and H. Staerk, Chem.Phys.Letters, 96, 24 (1983).
[7] H. Staerk, R. Treichel, and A. Weller
    Chem.Phys.Letters 96, 28 (1983).
[8] A. Weller, H. Staerk, and R. Treichel,
    Faraday Discuss. 78 (1984).
[9] H. Staerk, H. Meyer, and A. Weller, in preparation.

Four Wave Mixing Studies and Molecular Dynamics Simulations

M. Golombok and G.A. Kenney-Wallace

Lash Miller Laboratories, University of Toronto, Toronto, M5S 1A1, Canada

A quantitative microscopic understanding of molecular motion in dense media at times 10^{-13} -10^{-11}s is not only the goal of experimental and theoretical studies of molecular dynamics in liquids but also a prerequisite for a comprehensive treatment of ultrafast chemical reactions, in which solute-solvent interactions, local density fluctuations and energy relaxation can all play a deterministic role. In chemical reactions driven by strong laser fields, induced dipole moments and transient orientational ordering could well influence the height and pathway over the reaction barrier. We report here new results from a nonlinear optical study of four wave mixing (4WM) via $\chi_{ijkl}^{(3)}$ in several binary organic systems, in which the solute-solvent interactions significantly modified the net nonlinear optical polarisability of the solute molecule. We then describe computer simulations of the optical field-induced anisotropy, from which we can deduce the time evolution of the subsequent relaxation and the orientational pair correlation functions ($g(\Omega)$) describing the molecular interactions, given the experimentally determined orientational component of $\chi^{(3)}$ in the 4WM data. These are novel attempts to simulate the nonlinear optical response of a dense medium to an intense but transient optical field, and, through continual refinement of the computational aspects, they will provide an important new tool for the interpretation and testing of dynamical models in the femtosecond and picosecond time domain.

Recent optical Kerr (OKE) studies of CS_2 and several aromatic molecules have identified at least two transients within the 10^{-13} to 10^{-11}s domain (1). While the slower component of picosecond duration is assumed to be Debye-like reorientational relaxation processes, the faster component has been assigned to various relaxation mechanisms, whose roots lie either in spectroscopic (electronic Kerr, two-photon or resonant enhancement of $\chi^{(3)}$) or in molecular dynamics, which include collisional interactions, small step angular diffusion, or field-driven rotation as primary mechanisms (2). Of course, these explanations are not necessarily mutually exclusive. More evidence is needed. In previous 4WM studies (3) we observed that the changing microscopic solute-solvent interactions could be revealed in the changing reorientational times with subpicosecond resolution. Thus prior to interpretation of any time-dependent molecule-field interactions, we need to establish the degree to which molecule-molecule interactions give rise to modifications in $\chi^{(3)}$, which in turn via 4WM reflect the solvation shell interactions in binary systems.

CS_2 is a triatomic $D_{\infty}h$ molecule and a liquid at 300K; intermolecular pair potential is reasonably well known and only polarizability and quadrupole terms contribute to the interaction tensor. A hard core model emphasising short-range effects reproduces those features of the potential which at least seem important to equilibrium studies (4). Thus the CS_2 nearest neighbour solvation shell interactions will predominate in our experiment, and computer simulations can be developed to predict transient optical field effects and the role of $g(\Omega)$.

The intensity of the conjugate wave $I(\omega_4)$ generated by the third-order polarization $P^{(3)}$ at ω_4 is proportional to the number density N and $\chi^{(3)}$ of the material, and to the amplitudes of the applied optical fields $E(\omega)$:

$$P^{(3)}(\omega_4 = \omega_1 + \omega_2 - \omega_3) = \chi^{(3)} E_1(\omega_1) E_2(\omega_2) E_3^*(\omega_3) \tag{1a}$$

$$I(\omega_4) = [N\chi^{(3)}]^2 \, I(\omega_1) \, I(\omega_2) \, I(\omega_3). \tag{1b}$$

The conjugate wave at ω_4 is generated via degenerate 4WM and studied as a function of density of a tagged reference molecule (a), usually CS_2, diluted in a wide range of molecular liquids (b), chosen for properties of symmetry, structure, polarizability, dipolar or quadrupolar nature, and tendency for H bonding (5).

In order to determine the *orientationally averaged* effective nonlinear susceptibility of CS_2 in these (a,b) systems, we chose to use a 25ns pulse from a 2J Q-switched ruby laser as the source of ω_1, ω_2 and ω_3. Small molecules will have sampled all orientational configurations many times at 10^{-9}s. A pump-probe geometry was used as described earlier (3) and typically power densities for $I(\omega_1) = I(\omega_2) \sim 38$ MW cm^{-2}. The intensity of $I(\omega_3)$ could be varied but was usually ~10 kW cm^{-2} and the beam interaction length in the sample was 0.7 cm. The intensity of the phase conjugate wave was monitored by a fast photodiode. In control studies, the observed reflectivity from pure CS_2 was 4.8×10^{-3}, in good agreement with a calculated value of 5×10^{-3} based on the laser operating parameters.

In the systems studied, CS_2 is the dominant source of nonlinear polarization. For constant laser intensities the variation in $I(\omega_4)$ upon dilution of CS_2 in different solvents thus reflects both the range in number density of CS_2 and any modifications of $\chi^{(3)}$ of CS_2 as its nearest-neighbour interactions change. Inspection of Fig. 1a for CS_2 illustrates the effective $\chi_{eff}^{(3)}$ calculated directly from the normalised phase conjugate reflectivity,

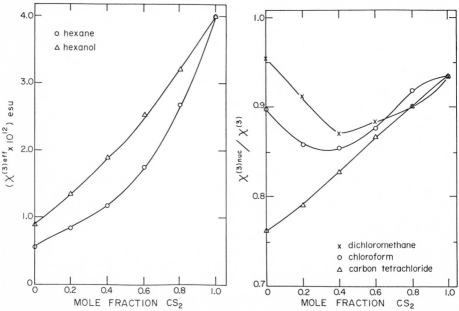

Fig. 1. Effective $\chi^{(3)}$ values calculated from 4WM data. See text for details.

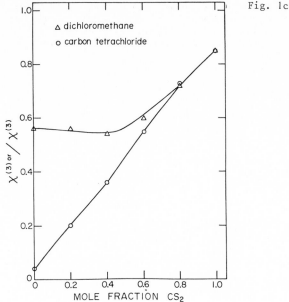

Fig. 1c

since N_a, N_b and the intensities are known. Note the influence of a OH dipole on an alkyl chain common to hexane and hexanol. From these data, if we presume that $\chi^{(3)}_{eff} = (\chi^{(3)}_{elec} + \chi^{(3)}_{nuc})$ and third harmonic generation in the same systems will generate $\chi^{(3)}_{elec}$, we may deduce $\chi^{(3)}_{nuc}$. Figure 1b shows the results for CS_2 in the series of solvents $CHCl_3$, CH_2Cl_2 and CCl_4 in which Cl successively replaces the H atom, changing both symmetry and interactions. The influence of the permanent dipole and molecular shape cause significant variation in $\chi^{(3)}_{nuc}$ in the dilute solutions, where CS_2 is fully surrounded by solvent molecules. The anisotropic part of $\chi^{(3)}_{nuc}$, which we term $\chi^{(3)or}$, is the orientationally dependent part of the total nuclear nonlinear susceptibility, and it is related to the correlations $\langle \alpha_a(o)\alpha_b(t) \rangle$ of anisotropic polarizability at the molecular level that are modified by solute-solvent interactions. Thus we can now evaluate $\chi^{(3)}_{or}$ for binary systems, as shown in Fig. 1c. The contrast in the two curves shows again the influence of local structure and interactions on $\chi^{(3)}_{or}$, a parameter of the system which we can now predict in the computer simulations.

The computer simulations using image cell boundary conditions were based on initial random configurations of molecules (hard spheroidal cylinders with experimentally based physical dimensions of CS_2) at 300 K, with an anisotropic weighting function to maintain experimentally known intermolecular separation in liquid CS_2 and avoid overlap. An applied $\vec{E}$ pump field of (typically) 200 fs, 10^3 SV/cm (simulating ~200 MW cm^{-2}, fs dye pulse) is applied to the system, and the induced-dipole moments and the correct Lorentz local field then calculated for each molecule and its nearest neighbours; the local field converged within 7 iterations in a given step of 25fs. The next step calculates the *net acting torques*, then the change $\Delta\Theta$ in the molecular orientation. When the pump field is switched "off" it is to a low probe pulse intensity of effectively kW cm^{-2}, to simulate real experiments, and we now follow the growth and relaxation of the optical anisotropy. As the pump $\vec{E}$ is off, $\Delta\Theta$ become increasingly smaller and the net acting torque gradually goes to zero. Dissipation mechanisms were modelled on collisions whose frequency was derived from a mean path approx-

385

imation. In the simulation results (5) following the laser pulse, we observe
the onset of orientational ordering in form of the growth of the static orientational pair correlation P_2 $(u_i(t)u_j(t))$, which reaches equilibrium value at
about 1.3 ps, and indicates that the molecules assume parallel alignment. The pai
correlations can be also derived from 4WM data on $\chi^{(3)}_{or}$, as noted earlier. Relaxation of the orientational time correlation was determined for $<P_1(u_i(0)u_i$
$(t))>$ and $<P_2(u_i(0) u_i(t))>$ to be of the order 1.5 ps and 280 fs respectively.
Physically, we have generated relaxation terms from initial field-imposed al-
ignments of the molecular axes, since equilibrium orientational probability
distribution functions have been significantly modified by *molecule-field*
interactions. The 280 fs P_2 relaxation time is highly reminiscent of the 240-
300 fs transients observed in OKE of CS_2 (1). We are further investigating
the origin of these responses in terms of field-induced interactions in binary
systems and comparing field-free long term relaxation to that seen in light
scattering data (6), in order to construct a full profile of transient behaviour for fs and ps experiments currently underway.

We gratefully acknowledge the National Sciences and Engineering Research
Council of Canada and the Office of Naval Research for the support of this work

1. J-M. Halbout and C.L. Tang, App. Phys. Lett. 40, 765 (1982);B.I. Greene and
R.C. Farrow, J. Chem. Phys. 77, 4779 (1982); J. Etchepare, G.A. Kenney-Wallace,
G. Grillon, A. Migus, J-P. Chambaret, IEEE JQ-18, 182 (1982), J. Etchepare,
G. Grillon, R. Muller and A. Orszag, Opt. Comm. 34 269 (1980).
2. For a summary see G.A. Kenney-Wallace in Applications of Picosecond Spectroscopy in Chemistry, ed. K.B. Eisenthal, Plenum Press (1984).
3. G.A. Kenney-Wallace and S.C. Wallace, IEEE JQ E-19, 719 (1983).
4. W.B. Street and D.J. Tildesley, Farad. Disc. 66, 27 (1978); D.J. Tildesley and P.A. Madden, Mol. Phys. 1, 129 (1983) and references therein.
5. M. Golombok and G.A. Kenney-Wallace, submitted for publication.
6. P. Madden, private communication and this volume.

Relaxation of Large Molecules Following Ultrafast Excitation

A. Lorincz*, F.A. Novak, and S.A. Rice

Department of Chemistry and the James Franck Institute,
The University of Chicago, Chicago, IL 60637, USA

Intramolecular relaxation of large molecules following ultrafast
excitation is determined by the temporal properties of the light pulse if
the density of coherently excited zero-order states within the vibronic
manifold is uniform. Consequently, longitudinal and transverse
relaxations within the excited manifold appear to be instantaneous.

I. Introduction

The time scale for relaxation of highly excited vibrational states of
large organic molecules is a subject of considerable interest [1-8].
Studies of dyes in solution, with picosecond resolution, show solvent
dependent relaxation times corresponding to a few ps. Experiments with
better time resolution give a few tens of femtoseconds as the upper limit
for the relaxation process, which now shows no solvent dependence [4,7].
We show that this apparent discrepancy may be reconciled by taking into
account the properties of coherently excited zero-order states (ZOS). We
find that the population relaxation of the superposition of excited zero-
order states to the underlying manifold is determined by the temporal
properties of the light pulse giving rise to apparently instantaneous
longitudinal and transverse relaxation.

II. Discussion and Theory

Suppose that the molecule is in the ground state. The ZOS's of the
excited surface may be divided into two sets: those which have good
Franck-Condon factors (FCF), and those which have poor FCF's with the
ground state. Anharmonic coupling connects these states so the initially
excited ZOS's with good FCF's will relax into the underlying manifold of
ZOS's with poor FCF's. Let the molecule interact with two coherent pulses
of light produced by splitting and delaying a single pulse. We assume
that the background states into which the ZOS's relax have emissions
sufficiently red-shifted that they cannot be stimulated by the laser.
     By straightforward generalization of the methods of NOVAK, FRIEDMAN
and HOCHSTRASSER [9] the amplitude of interest is (suppressing vector
indices for clarity)

$$A(t) = i \int_0^t \int_0^{t_1} dt_1 dt_2 [\varepsilon_1^*(t_1) + \varepsilon_2^*(t_1)] \langle g|\mu e^{-iH(t_1-t_2)} \mu|g\rangle [\varepsilon_1(t_2) + \varepsilon_2(t_2)], \quad (1)$$

where $\varepsilon_1(t)$ and $\varepsilon_2(t)$ are c-number functions representing pulses 1 and 2.
In (1), the laser pulses are assumed to be characterized by coherent

---

*   Permanent address: Institute of Isotopes of the Hungarian Academy
                         of Sciences, P.O. Box 77, H-1525, Hungary

states. Expression (1) is the probability amplitude for finding a molecule in its ground state due to its absorbing a photon and being stimulated to emit by another. It therefore represents pulse reshaping effects and the interaction of the two pulses. We thus focus attention on the molecule evolution and not the measurable light intensity. The views are complementary since the gain or loss in the beams is proportional to $|A(t)|^2$. Using the properties of the Laplace transform, (1) may be expressed as

$$A(t) = \frac{1}{2\pi} \int_0^t dt_1 \int_c^{t_1} dt_2 \varepsilon^*(t_1)\varepsilon(t_2) \int_c e^{-iz(t_1-t_2)} \langle g|\mu G(z)\mu|g\rangle dz , \quad (2)$$

where $G(z)$ is the Green's function and $\varepsilon(t) = \varepsilon_1(t) + \varepsilon_2(t)$. If $|y\rangle$ denotes the ZOS's accessible from the ground state, then to second order in the anharmonic coupling

$$A(t) = \frac{1}{2\pi} \int_0^t dt_2 \int_c^{t_1} dt_2 \varepsilon^*(t_1)\varepsilon(t_2) \int_c e^{-izt} \sum_y \frac{|\mu_{gy}|^2}{z - \omega_y + \frac{i}{2}\Gamma_y} dz \quad (3)$$

using the assumption that the emission from the background states is sufficiently red-shifted not to be stimulated by the laser. If we define $\rho_\mu(\omega)$ to be the Franck Condon weighted density of states and use the residue theorem we obtain, under the assumption that the Γ_y's are all roughly equal (to Γ say),

$$A(t) = \int_0^t dt_1 \int_0^{t_1} dt_2 \varepsilon^*(t_1)\varepsilon(t_2) \int_{-\infty}^\infty d\omega \rho_\mu(\omega) e^{-i\omega(t_1-t_2)} e^{-\frac{\Gamma}{2}(t_1-t_2)} . \quad (4)$$

To obtain the short time behavior of (4) we may smooth $\rho_\mu(\omega)$. If $\rho_\mu(\omega)$ is then roughly constant across the bandwidth of the pulses the ω integral in (4) becomes proportional to a δ function in time and thus, for short times,

$$A(t) \propto \int_0^t dt_1 \varepsilon^*(t_1)\varepsilon(t_1). \quad (5)$$

This short time amplitude depends only on the details of the structure of the light pulses.

The approximation used here breaks down for long enough times when recurrences become important. The transition region between time regimes is determined by the density (and/or its smoothness) of the coherently excited zero-order states. The long time behavior is governed by the poles in (3). Only this dependence is characteristic of molecular parameters and solvent properties.

The derivation presented here holds for an ensemble of molecules in liquids if the vibronic states excited have energies well above 200 cm^{-1}, the thermal energy at room temperature.

III. Conclusions

This short communication calls attention to the possibility that the discrepancies between fast and ultrafast excitation studies of intramolecular relaxation in liquids may be resolved by taking the properties of the coherently excited zero-order states into account. The use of ultrafast excitation results in apparent relaxation which follows the light pulse if a large number of uniform density zero-order states are excited, as may be seen by the following argument. If the molecule is excited with a pulse of light whose frequency width contains two zero-order states, then these will beat against each other. These beats may be interpreted as fast successive unphasings and recurrences. (Here unphasing refers to a coherent process of evolution; dephasing refers to the same process in an ensemble of molecules.) The fast unphasing depends only on the level separation, but it is independent of other molecular parameters such as the intramolecular vibrational redistribution rates of the individual levels. If the number of levels in the exciting bandwidth is increased uniformly, and the density of states is more or less constant in the exciting bandwidth, then recurrences in the beat pattern are pushed to long times so oscillations become completely suppressed. The derivation in this paper demonstrates that for short times the evolution will be strongly dependent on the nature of the exciting pulse. In fact, the characteristic molecular evolution time will be very close to the temporal width of the light pulse.

The important point for condensed phase experiments is that the rapid "population" decay described by $G(z)$ is much faster than any T_2 time for the system, and hence dominates the short time dynamics.

1. P.M. Rentzepis, M.R. Topp, R.P. Jones and J. Jortus, Phys. Rev. Lett. 25, 1742 (1970).
2. D. Ricard, J. Chem. Phys. 63, 3841 (1975).
3. A. Laubereau, A. Seilmeier and W. Kaiser, Chem. Phys. Lett. 36, 232 (1975).
4. C.V. Shank, E. Ippen and O. Teschke, Chem. Phys. Lett. 45, 291 (1977).
5. J.W. Wiesenfeld and E. Ippen, Chem. Phys. Lett. 67, 213 (1978).
6. J.J. Song, J.H. Ree and M.D. Levenson, Phys. Rev. A17, 1439 (1978).
7. A.J. Taylor, D.J. Erskine and C.L. Tang, Chem. Phys. Lett. 103, 430 (1984).
8. A.M. Weiner and E.P. Ippen Opt. Lett. 9, 53 (1984).
9. F.A. Novak, J.M. Friedman and R.M. Hochstrasser, "Resonant Scattering of Light by Molecules: Time Dependent and Coherent Effects," in Laser and Coherence Spectroscopy, J.I. Steinfeld, Ed.,Plenum Press, New York (1978).

Picosecond Time-Resolved Spectroscopy of Electronically Excited tris(2,2'-Bipyridine) Ruthenium(II) Dichloride

L.A. Philips, W.T. Brown, S.P. Webb, S.W. Yeh, and J.H. Clark

Laboratory of Chemical Biodynamics, Lawrence Berkeley Laboratory, and Department of Chemistry, University of California, Berkeley, CA 94720, USA

1. Introduction

Transition metal complexes display a rich and varied photochemistry which has found broad application in such areas as solar energy conversion, electron transfer reactions, and lasers[1-4]. Although tris(2,2'-bipyridine)ruthenium(II) dichloride (Ru(bpy)$_3^{2+}$) has been extensively studied[4-10], many details of its excited-state dynamics remain unresolved. The lowest charge-transfer excited state of Ru(bpy)$_3^{2+}$ has been the focus of some controversy. CROSBY and others [6-8] have established that initial excitation occurs via a d → π* transition. Assignment of the state(s) involved in subsequent energy transfer and emission remain in question. LYTLE and HERCULES suggested a two-state model, as shown in Fig. 1[6]. In this model energy transfer occurs via rapid radiationless relaxation (A → B), but emission occurs from both states at times short compared to τ_{AB}. At longer times emission occurs only from state B. CROSBY suggested an alternative model involving three closely spaced mixed states[9,10]. In this model, room temperature emission occurs predominately from one of these three states. Once it is formed, the emitting state has been shown to have a lifetime of 600 ns in aqueous solution at room temperature[6].

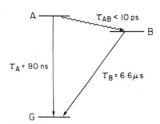

Fig. 1 Two-state kinetic model for electronically excited Ru(bpy)$_3^{2+}$, from LYTLE and HERCULES[6]

WOODRUFF et al.[5] demonstrated that the emitting state is formed in less than 7 ns using time-resolved resonance Raman spectroscopy (TR^3) of the excited state. With a time resolution limited to 7 ns, it was not possible to observe the dynamics of the energy transfer between the initially excited state and the emitting state. In the present study, picosecond TR^3 and time-resolved emission spectroscopy have been used in an attempt to answer some of these questions surrounding the dynamics of the excited state formation and decay.

2. Experimental

In both experiments the excitation source was an amplified single pulse selected from the pulse train of a passively mode-locked Nd:YAG laser (Quantel, YG400). The third harmonic at 355 nm was used to excite the Ru(bpy)$_3^{2+}$. In the time resolved emission experiments, the luminescence was imaged onto the photocathode of an ultrafast streak camera (Hadland Photonics, IMACON 500) at a right angle to the excitation beam. A fiducial pulse which accompanied the luminescence acted as a time reference and provided a monitor of the quality of the excitation pulse. Typical excitation pulses were 23 ± 7 ps in duration with an energy of ~100 µJ. The signal from the streak camera was imaged onto a diode array (Tracor-Northern, IDARSS) and read off the array into an LSI 11/2 computer. The computer evaluated each laser shot and signal averaged the accepted data.

In the Raman experiments, a single 355 nm pulse was used both to excite ground state molecules and to induce resonance Raman scattering from excited molecules. The Raman signal was imaged into a double monochromator (Spex, Model 1404) and was detected by a photomultiplier tube (RCA, 1P28). The output of the photomultiplier was sent through a gated integrator (Evans Assoc., Model 4130) and into an LSI 11/2 computer. The computer averaged the signal from 500 shots at each spectral point and scanned the monochromator. The picosecond resonance Raman spectra were taken at a resolution of 14 cm^{-1}. $Ru(bpy)_3^{2+}$ (Strem) was used without further purification. Solutions were prepared in 1.0 M aqueous Na_2SO_4 at $Ru(bpy)_3^{2+}$ concentrations of 10^{-3} to 10^{-5} M.

3. Results and Discussion

Figure 2 shows the TR^3 spectrum taken with 25 ps pulses in comparison with WOODRUFF's results using 7 ns pulses. Apparent differences in the baselines are due to instrumental effects. WOODRUFF et al. have argued that the TR^3 spectrum they observe is due to the emitting state[5]. The picosecond TR^3 spectrum suggests that the state WOODRUFF is probing at 7 ns is formed in less than 25 ps. Although there is no evidence of a separate, initially excited state in the picosecond TR^3 spectrum, the presence of another state cannot be completely

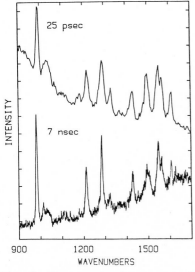

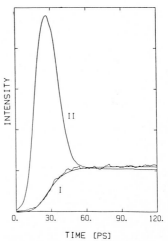

Fig. 2 Time-resolved resonance Raman spectra of electronically excited $Ru(bpy)_3^{2+}$. Upper trace taken with a pulse duration of 25 ps. Lower trace taken with a pulse duration of 7 ns[5]. The excitation wavelength was 355 nm in both cases

Fig. 3 Time-resolved emission from electronically excited $Ru(bpy)_3^{2+}$. Experimental data and two computer-generated fits are plotted. Fit I shows a single exponential rise of <5 ps. Fit II is a multiple exponential fit based on the model of [6]

391

ruled out. Multiple mixed states may have similar Raman spectra which are not distinguishable with the available sensitivity and resolution of the picosecond TR^3 apparatus.

The results of picosecond, time-resolved emission experiments provide additional information about the excited-state dynamics of the $Ru(bpy)_3^{2+}$ system. The onset of luminescence is prompt, with a risetime of <5 ps. Using the rate constants of the LYTLE and HERCULES analysis[6], one would predict that both luminescence from the initially excited state (A) as well as from the final emitting state (B) would be observed. LYTLE and HERCULES estimate the rate of transfer from A to B to be on the order of 10 ps or less. Figure 3 shows the experimental data and two calculated curves. Curve I is a single exponential fit to the data with a risetime of <5 ps. Curve II is a fit calculated using the LYTLE and HERCULES model assuming a τ_{AB} of 1 ps. The LYTLE and HERCULES model does not accurately represent the data. In contrast, a single exponential description of the data is consistent with the CROSBY model. The results of these experiments call into question the existence of any short-lived emitting states in the luminescence of $Ru(bpy)_3^{2+}$.

4. Acknowledgements

This work was supported by the Office of Energy Research, Office of Basic Energy Sciences, Chemical Sciences Division of the U. S. Department of Energy under Contract No. DE-AC03-76SF00098. Sheila W. Yeh and W. T. Brown are National Science Foundation Pre-doctoral Fellows. J. H. Clark is an Alfred P. Sloan Research Fellow and a Henry and Camille Dreyfus Teacher-Scholar. We thank Prof. Woodruff for kindly providing and allowing us to use his nanosecond TR^3 data.

5. References

1. K. Mandel, T. D. Pearson, and J. N. Demas, J. Chem. Phys. _73_, 2507 (1980).
2. M. K. DeArmond, Acct. Chem. Res. _7_, 309 (1974).
3. N. Sutin, J. Photochem. _10_, 19 (1979).
4. F. Bolletta, M. Maestri, L. Moggi, and V. Balzani, J. Phys. Chem. _78_, 1374 (1974).
5. P. G. Bradley, N. Kress, B. A. Hornberger, R. F. Dallinger, and W. H. Woodruff, J. Am. Chem. Soc. _103_, 7441 (1981).
6. F. E. Lytle and D. M. Hercules, J. Am. Chem. Soc. _91_, 253 (1969).
7. J. N. Demas and G. A. Crosby, J. Mol. Spectrosc. _26_, 72 (1968).
8. D. M. Klassen and G. A. Crosby, J. Chem. Phys. _48_, 1853 (1968).
9. R. W. Harrigan and G. A. Crosby, J. Chem. Phys. _59_, 3468 (1973).
10. K. W. Hipps and G. A. Crosby, J. Am. Chem. Soc. _97_, 7042 (1975).

Part VII

Electronics and Opto-Electronics

Modelocking at Ti:LiNbO$_3$-InGaAsP/InP Composite Cavity Laser Using a High-Speed Directional Coupler Switch

R.C. Alferness, G. Eisenstein, S.K. Korotky, R.S. Tucker, L.L. Buhl, I.P. Kaminow, and J.J. Veselka

AT & T Bell Laboratories, Holmdell, NJ 07733, USA

Short, high peak power optical pulses in the wavelength region of λ=1.3-1.6μm are potentially attractive for very-high bit-rate lightwave systems and signal processing. InGaAsP/InP lasers, the most widely used semiconductor lasers for this wavelength range, have previously been actively modelocked by current modulating the diode in a passive, extended cavity. [1] The achievement of low loss [2], high speed titanium-diffused lithium niobate waveguide optical switches has made possible a new approach to active modelocking of semiconductor lasers. We have fabricated a low-loss, broadband Ti:LiNbO$_3$ waveguide directional coupler switch that is coupled to the AR coated facet of a DC biased InGaAsP/InP diode. By driving the switch with a sinusoidal signal, we have mode-locked the composite cavity to produce a train of 22psec wide (FWHM) pulses at repetition rates as high as 7.2 GHz. This new method of mode-locking semiconductor lasers offers important advantages including the potential for combined modelocking/cavity dumping, ring laser geometry and electrically tunable modelocked pulses. In addition, the generated pulses could be multiplexed or encoded with another Ti:LiNbO3 switch/modulator integrated on the same chip.

The composite Ti:LiNbO$_3$-InGaAsP/InP device is shown schematically in Fig. 1.

MODE–LOCKED
Ti:LiNbO$_3$ –InGaAsP/InP LASER

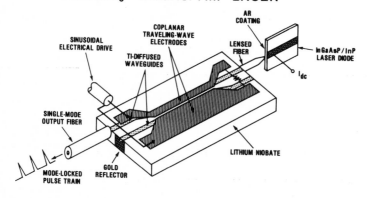

The switch was fabricated in a Z-cut, Y-propagating lithium niobate crystal. The fabrication and diffusion conditions -a ~1000Å titanium thickness diffused for nine (9) hours at 1050°C - provide low propagation and fiber coupling loss [1] at λ=1.32μm. The frequency response of the

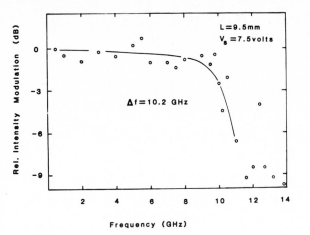

Fig. 2

9.5mm long traveling-wave directional coupler switch as measured in extra-cavity experiments is shown in Fig. 2.

The 3dB optical modulation bandwith is ^10GHz. By using thick (^3μm), highly conductive electroplated gold electrodes [3], the frequency response is limited primarily by optical-electrical velocity mismatch. The theoretical velocity mismatch limit is ^12GHz.cm.

A lensed single-mode fiber is permanently attached to the input waveguide; a gold reflector is deposited on the end of the cross-over waveguide; and single-mode output fiber is attached to the other output waveguide. The switch is coupled to the AR coated facet of a strip, index guided InGaAsP/InP laser to form a single cavity between the uncoated quaternary facet and gold reflector on the cross-over Ti:LiNbO$_3$ waveguide. The reflectivity of the AR coated facet is less than 10^{-3}.

The switch cross-over efficiency and resulting feedback from the gold reflector to the diode can be controlled continuously via the switch voltage. Intra-cavity switch control of the composite cavity Q is shown schematically in Fig. 3.

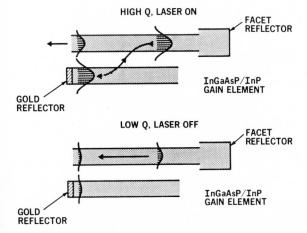

Fig. 3

With the switch voltage adjusted to provide maximum cross-over efficiency, the single pass loss through the Ti:LiNbO$_3$ switch to the reflector, including coupling from the input fiber, is only -1dB. In this case, the threshold current of the composite cavity is approximately fifty percent greater than that of the original uncoated laser. We have achieved thresholds as low as 20mA for the composite cavity. However, by increasing the switch voltage, the cross-over efficiency and resulting feedback can be reduced sufficiently to stop laser action.

Modelocking is achieved by biasing the switch for ~50 percent switching and driving it with an RF signal at the fundamental cavity round-trip frequency (1.8 GHz) or at a harmonic. The InGaAsP/InP gain medium is dc biased only. In our experiments, the output from the uncoated diode facet is detected with a very high-speed PIN InGaAs photo-diode and displayed on a sampling oscilloscope. A detector-limited optical pulse train achieved at 7.2 GHz repetition rate (fourth harmonic) is shown in Fig. 4.

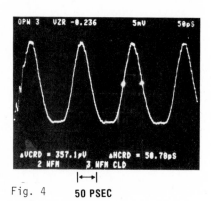

Fig. 4   **50 PSEC**

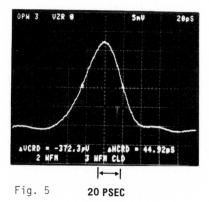

Fig. 5   **20 PSEC**

A single pulse of the train is shown in Fig. 5.

The displayed pulse width (FWHM) is ~45 psec. Deconvolving the detector plus oscilloscope response time of ~39 psec yields an actual pulse width less than 22 psec. The drive power to the switch (50 Ω input impedance) in this case was ~300 mW. Similar results were achieved at the fundamental and second harmonic as well. The measured spectrum for modelocking at 7.2 GHz is shown in Fig. 5. The spectral bandwidth (FWHM) is ~68 GHz which would corrspond to ~7 psec to 15 psec pulses depending on pulse shape, if we assumed the pulses to be transform-limited. Auto-correlation measurements to determine the pulse width more accurately are planned.

This new modelocking device structure offers several promising advantages. By reflection coating the back facet of the InGaAsP/InP diode, modelocking the high Q cavity and cavity dumping out the fiber should be possible. A ring laser configuration using a single-mode fiber loop would offer great flexibility. Furthermore, other Ti:LiNbO$_3$ guided-wave devices, such as the tunable directional coupler filter, could be used as intra-cavity elements to provide, for example, electrical tunability of the laser output. Integrating other extra-cavity devices for encoding or time division multiplexing onto the same lithium niobate crystal should also be possible.

REFERENCES

1. R. S. Tucker, G. E. Eisenstein, and I. P. Kaminow, Elect. Letts, Vol. 19, pp. 552-553, 1983; and references therein.

2. R. C. Alferness, L. L. Buhl, and M. D. Divino, Elect. Letts, Vol. 18, pp. 18, pp. 490-491, 1982.

3. R. C. Alferness, C. H. Joyner, L. L. Buhl, and S. K. Korotky, IEEE J. Quantum Electron, QE-19, pp. 1339-1342, 1983.

Picosecond Optical Measurements of Circuit Effects on Carrier Sweepout in GaAs Schottky Diodes

A. Von Lehmen and J.M. Ballantyne

Field of Applied Physics and School of Electrical Engineering
Cornell University, Ithaca, NY 14853, USA

Optical measurements of semiconductor devices can provide direct information about carrier sweepout under high field conditions [1-3]. In previous work, we exploited the nonlinearity in the luminescence in GaAs [4] to do picosecond time-resolved experiments on a GaAs photoconductor [3]. The device consisted of an n-type active layer between ohmic contacts. The speed of the device, and the characteristic time of the luminescence decay is determined by the hole transit time in the device. From the luminescence decay we were able to extract a hole velocity of 2×10^6 cm/s at a field of 20kV/cm. We have used the same technique to measure carrier sweepout in a GaAs Schottky diode incorporated into a large capacitor at room temperature. Ohmic and blocking (i.e. Schottky) contacts can be helpful in discriminating between electron and hole transport processes. At a bias voltage of -4V, we have observed pulsewidth limited luminescence decays.

In addition we have studied the effect of the intrinsic electric circuit on carrier sweepout under external fields. This was done by designing and fabricating monolithic on-chip circuits as an integral part of the measured device. We have measured luminescence decays for the diode structure with inductors included in the biasing circuit yielding decay times of 6.5 and 9.0 picoseconds for inductances of 0.6 and 1.8nH. As far as we know, this is the first direct measurement of the slowing effect on carrier sweepout due to external circuit elements.

Time resolved experiments are done using a synchronously pumped dye laser producing 3-4ps pulses at 640 nm. A standard pump-probe arrangement is used: the probe beam is chopped and includes a delay line in its path. The nonlinearity in the luminescence results in an enhancement of the probe beam luminescence which depends upon the pump-probe delay. In a bulk semiconductor, the characteristic decay time of the luminescence will be determined by material parameters, such as the doping, and surface characteristics. When a field is applied, the apparent decay time will be reduced if carrier sweepout occurs on a timescale short compared to the normal lifetime.

A cross section of the device used in these experiments is shown in Fig. 1. The active region is a .3μm thick n-type GaAs layer doped at 10^{-17} cm^{-3}. The top contact(.2μm Au) is 300μm square, and forms a low inductance circuit with an 8x8μm window of 100 Å thick gold forming a Schottky barrier at the semiconductor surface. At a bias voltage of -4V, the active layer is depleted and any carriers generated within it should be swept out in a time short compared with the normal radiative lifetime. For the high(10^5V/cm) average fields in this device, electrons travel at their saturation velocity and and sweepout should be complete in less than 3ps.

The luminescence signal as a function of pump-probe delay time is shown in Fig. 2. A field independent background signal with a time constant of about 300 ps, probably due to luminescence from the n$^+$ substrate, has been

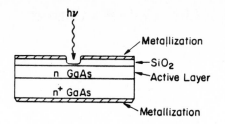

Fig. 1 Cross section of Schottky diode incorporated into a large capacitor used in measurements. The SiO_2 layer is 0.9μm thick

subtracted from the data shown here. The -4V decay is substantially faster than that at 0V. No sweepout time can be extracted without deconvolving the response from the pulse. The curve in Fig. 2 has been multiplied by a factor of 4 to normalize it to the peak height of the 0V data. Significant reduction of the peak height is additional evidence of substantially increased carrier sweepout.

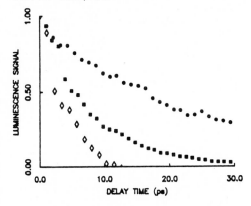

Fig. 2 Luminescence signal as a function of pump-probe delay for the device of Fig. 1 at 0V(●), and at -4V(◊). The data taken with a 1.8 nH inductor in the circuit at -4V is also shown (■). All curves have been normalized to 1.0 to facilitate comparison

The same measurements were made on devices with monolithic inductors (L=0.6 and 1.8 nH) introduced between the on-chip capacitor and the Schottky diode (cf. Fig. 3). The charge associated with the large capacitance in the device in Fig. 1 serves to hold the voltage across the device constant even at the relatively high carrier injection levels (n~5×10^{17}) resulting from photo-excitation. An inductor between the capacitor and the small Schottky diode will isolate the device from the large capacitance. The circuit simulation program SPICE was used to calculate the transient response of the circuits described above to a 3 ps pulse which injects 5×10^{17} cm^{-3} carriers. The voltage transient across the diode capacitance was computed for L=0.6 nH; the voltage drops to zero and recovers to 65% of the bias voltage after 7.5 ps. For L=1.8 nH, the recovery time is about 12 ps. No field-induced transport will occur in the active region of the device until the voltage has recovered from its initial drop to zero. Our data confirms this. As an example, the background-free data for the circuit including the 1.8 nH inductor with the bias at -4V is also shown in Fig. 2. A single time constant exponential was found to parameterize this data reasonably well, resulting in best fit decay times of 6.5 and 9.0 ps for inductances of 0.6 and 1.8 nH respectively.

A one-dimensional computer model of transport in the Schottky diode structure which includes high carrier density perturbation of the equilibrium field distribution has been analyzed. The results are summarized in Fig. 4. At a photoexcitation density of 10^{17} carriers cm^{-3}, the maximum change in the

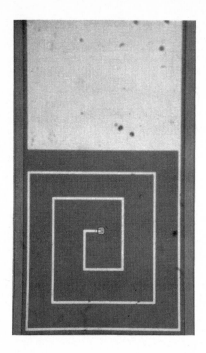

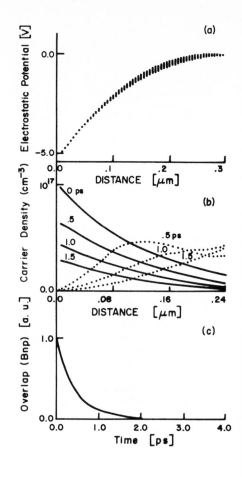

Fig. 3 Top view of the structure with an inductive circuit. The Schottky diode is at the tip of the innermost arm of the inductor. The value of this inductor was estimated to be 1.8 nH. For the device in Fig. 1, the diode was placed in the center of the large contact pad

Fig. 4 The electrostatic potential as a function of distance from the front contact is plotted in a). The magnitude of the perturbation due to photoinjected carriers is indicated by the height of the vertical bars. b) shows the carrier density for both electrons (...) and holes (—) as a function of distance. The numbers are the time in picoseconds after the photoexcitation. The luminescence resulting from the overlap of the two carrier populations in b) is shown in c)

electrostatic potential is 8%, giving rise to a maximum departure from the equilibrium field of ~10%. The perturbation is relatively small because of the small dimensions and hence short transit times across the active region of the device. The perturbation to the high fields in this device does not result in significant change in the electron velocity, and the transit time remains essentially unchanged. The predicted luminescence signal as a function of time resulting from changing electron hole population overlap is given by the product $Bn(x,t)p(x,t)$ where B is the bimolecular recombination coefficient. At -4V, the halfwidth of the luminescence signal is 0.5ps, in reasonable agreement with experimental results.

In conclusion, we report studies of carrier sweepout in very fast GaAs Schottky barrier diodes using an optical measurement technique which is appli-

cable to real device structures at room temperature and which obviates the
need for bringing fast electrical signals off the device chip. Monolithic
circuits were fabricated on-chip and used to experimentally verify circuit
effects on carrier transport measurements. These effects will be present in
any experiment involving collective macroscopic charge transport. Computer
simulations of transport inside the devices, as well as of the external circuit
are presented.

We would like to acknowledge the support of the Cornell Materials Science
Center with supplementary support by NSF Grant DMR82-17227, the Cornell Joint
Services Electronics Program, and the National Research and Resource Facility
for Submicron Structures.

References

1. B. B. Levine, C. G. Bethea, W. T. Tsang, F. Capasso K. K. Thornber, R.C.
   Fulton and D. A. Kleinman Appl. Phys. Lett. 42 104 (1983)

2. C. V. Shank, R. L. Fork, B. I. Greene, F. K. Reinhart and R. A. Logan
   Appl. Phys. Lett. 38 104 (1981)

3. A. Von Lehmen and J. M. Ballantyne Appl. Phys. Lett. 44 87 (1984)

4. D. von der Linde, J. Kuhl and E. Rosengart J. Luminescence 24/25
   675 (1981)

Color Center Formation and Recombination in KBr and LiF by Picosecond Pulsed Electrons

K. Fujii, R. Kikuchi*, S. Katagiri**, K. Tsumori, and M. Kawanishi

ISIR-SANKEN, Osaka University, Mihogaoka, Ibaraki-shi, Osaka 567, Japan

1 Introduction

As for the initial formation and recombination processes of F and H centers in ionic crystal such as KCl, KI, KBr and RbBr, etc., the excitonic mechanism proposed by HERSH[1] and POOLEY[2] has been verified by laser photolysis or from the growth of the intensity of σ emission. These processes are taken place in a very short time [3,4].

In order to investigate the very fast temporal behavior of the color centers of KBr and LiF whose band gaps are wide enough not to be excited with laser, a high energy radiation pulse whose time width is as short as possible and the electronic charge as large as possible is required. Such a pulse is able to be generated by the 38 MeV L-band electron linear accelerator which has three subharmonic prebunchers between the electron gun and the main accelerator. The total amount of electronic charge in a single 20 p-sec. pulse whose Čerenkov pattern taken by a streak camera is shown in Fig. 1 is provided from several to 50 nC with repetition frequency of 1 to 100 Hz.

2 Experimental

The system of optical density measurement, as shown in Fig.2, which is almost similar to that developed by JONAH et al. at ANL [5], has been

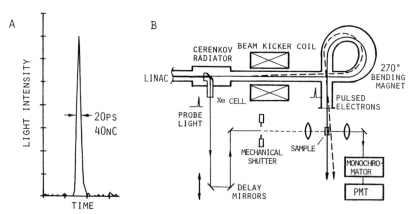

Fig.1 A:Čerenkov pattern of 20 p-sec. pulsed electrons, B:schematic diagram of the optical system for the measurement of optical density

* Present address: ULVAC, Hagizono 2500, Chigasaki-shi, Kanagawa, Japan
**Present address: JAPC, Chiyoda-ku Ootemachi 1-6-1 Tokyo, Japan

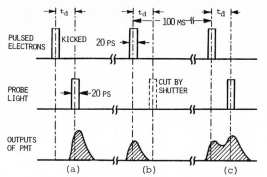

Fig.2 Three kinds of outputs from photomultiplier: (a) probe light passed through the sample (b) Čerenkov light emitted from the sample by electrons, (c) probe light transmitted through the sample and Čerenkov light emitted from it (t_d: variable)

constructed. The output of pulsed beam is divided into two ways: one part of it is led into the Čerenkov radiator cell filled with Xe gas and converted to the Čerenkov light which is used as the probe light pulse ($\sim$ 6 mmΦ), the rest beam ($\sim$ 8mmΦ) into the sample after passing through the beam kicker coil and the 270° bending magnet. The timing of probe light pulse is controlled by varying the optical path with a couple of mirrors. As both the light pulse passed through the irradiated sample [$10 \times 10 \times (2 \sim 4)$ mm^3] and the Čerenkov emission from the sample are detected at the same time by the photomultiplier (HTV R456), the following three kinds of optical detection are carried out with the beam kicker coil and the mechanical shutter of the probe light as shown in Fig. 2-(a),(b),(c). Then the optical density $OD(t_d)$ is expressed as follows:

$$OD(t_d) = \log \left[I_0(t_d) / \{ I_{e+c}(t_d) - I_e(t_d) \} \right] \qquad [1]$$

where $I_0(t_d)$, $I_e(t_d)$, $I_{e+c}(t_d)$ are the light intensities respectively and t_d is the variable time interval between the pulsed electrons entered the crystal and the probe light.

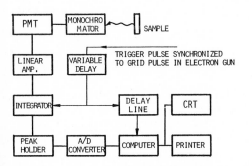

Fig.3 Block diagram of the data processing system

Figure 3 shows the block diagram of data processing system. The output of PMT differs from the real pulse shape, but it is considered that its output is proportional to the intensity of income light. Then it is amplified and integrated. The integrator circuit is operated synchronously with the grid pulse of the electron gun led through a delay line. The maximum voltage of the integrator is held and converted to digital signal with A/D converter and the optical density $OD(t_d)$ is calculated in the mini-computer after accumulating 50 data of three kinds respectvely at a fixed time t_d. To obtain the temporal variation of $OD(t_d)$, the measurements must be carried out successively by changing the timing of the probe light pulse.

3 Results and Discussion

The temporal variation of OD of color centers probably consisting of V_k and H centers (λ=380nm) produced in KBr and that of V_k (λ=365nm) in LiF^k are shown in Fig.4.

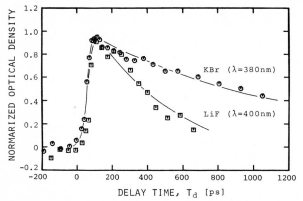

Fig.4 Temporal variation of optical density of KBr(λ=380 nm) and LiF(λ=400 nm) at room temperature : solid lines are calculated with Eq.3

From these curves, the formation and recombination time of these color centers have been estimated by assuming : (1) the thickness of the sample is so thin that the transit time of pulsed electrons and of probe light are neglected; (2) the excitation produces hole-electron pairs instantaneously when electrons come into the sample; (3) it takes, however, a finite time for them to separate each other and form V_k or H centers, some of which should recombine with electron in a finite time longer than the above; (4) both reactions have taken place independently according to the law of mono-molecular decay with the formation time τ_1, and recombination time τ_2. With these assumptions, the temporal variation of the concentration of color centers N(t) is expressed as the following equation:

$$N(t)=\{\tau_2/(\tau_2-\tau_1)\}[1-\exp\{-t(1/\tau_1-1/\tau_2)\}] .\qquad [2]$$

In order to determine τ_1 and τ_2 from the experimental OD curve, the pulse shape of both pulsed electrons and probe light which can be assumed to be Gaussians with the same width of 20 p-sec have to be considered in Eq.[2]. Then OD(t_d) is expressed as

$$OD(t_d)=\ln[\int_0^w P(t)dt/\{\int_{t_d-w}^{t_d} P(t'-t_d+w)\exp(\int_0^{\min(w,t_d)} E(t)N(t'-t)dt)dt'\}]\qquad [3]$$

where E(t) and P(t) are the pulse shape of the electrons and of light respectively and w is the practical upper limit of the integration calculated by the computer (w=20x4 P-sec.), and t_d, the time lag between the pulsed electrons and that of the probe light. Introducing the experimental data into the above equation, the formation time and the recombination one were estimated as listed in Table 1.

The experiments on the effects of temperature and of density of pulsed electrons with these centers are in progress.

Table 1 τ_1 and τ_2 estimated from OD curves in Fig.4

crystal	τ_1 [ps]	τ_2 [ps]
KBr	15 ± 10	1300 ± 20
LiF	5 ± 10	380 ± 20

References

1 H. N. Hersh : Phys. Rev., 148, 928 (1966)
2 D. Pooley : Proc. Phys. Soc., 87, 257 (1966)
3 R. T. Williams et al. : Phys. Rev. B, 18, 7038 (1978)
4 Y. Suzuki et al. : J. Phys. Soc. Jpn., 49, 207 (1980)
5 C. D. Jonah : Rev. Sci. Instrum., 46, 62 (1975)

Subpicosecond Electro-Optic Sampling Using Coplanar Strip Transmission Lines

K.E. Meyer and G.A. Mourou

Laboratory for Laser Energetics, University of Rochester, 250 East River Road
Rochester, NY 14623, USA

INTRODUCTION

A sampling technique using the electro-optic effect capable of characterizing subpicosecond electrical signals has recently been demonstrated. In this system 100 fs pulses from a colliding pulse mode-locked laser were used to generate the electrical test signal via a Cr:GaAs photoconductive switch. A second beam of pulses was used to probe the birefringence induced by the electrical pulses as they propagate down a balanced stripline fabricated on $LiTaO_3$. Signal averaging allowed submilli-volt signals to be recovered. The best temporal response of the system was achieved with the velocity matched geometry, which was obtained for a particular angle of incidence of the probe beam on the electro-optic crystal. In this configuration a risetime of 500 fs was obtained. During this work it was observed that strong dispersion effects were taking place when the wavelength of the electromagnetic signal approached the cross-sectional dimensions of the transmission line. Temporal resolution is improved and dispersion effects are reduced as the dimensions of the stripline are reduced. In order to further improve the temporal response of the sampler in this configuration, the thickness would have to be reduced to the order of 10 μm, which presents severe fabrication difficulties.

To alleviate these stringent mechanical requirements, we report on the use of coplanar striplines, which do not have the above physical limitations. Coplanar waveguide (CPW) and coplanar strip (CPS) transmission lines have been used frequently in microwave integrated circuit applications[2]. They are more easily fabricated than balanced microstrip lines because both the signal and ground lines may be made in the same process step on the same surface of the substrate. The greatest advantage, however, is that linewidths in the micron range are possible using standard photolithographic techniques. This implies that dispersion effects can be reduced even further than was previously possible.

Coplanar geometries have not been theoretically characterized as extensively as have microstrip configurations. Calculations for the

406

frequency dependence of the effective dielectric constant that were originally performed at lower frequencies by Knorr and Kuchler[3] are being extended into the terahertz range. These calculations will yield quantitative predictions for the cut-off frequency for the coplanar striplines of interest.

EXPERIMENT

The key element of the electro-optic sampler, shown in Fig. 1, is the coplanar stripline (all other details of the experimental arrangement are identical to Ref. 1). The Cr:GaAs and LiTaO$_3$ crystals were mounted side by side on a glass plate and were subsequently ground and polished together in order to present a continuous surface on which to fabricate the electrodes. The coplanar lines were then made using standard photolithographic techniques. The electrodes were made long enough (2 cm) to insure that any reflection from the terminated end of the line occurred at the sampling point long after the initial rise of the electrical signal. For relative ease of fabrication the electrode widths and separation were chosen to be 50 μm each. The crystal axis of the lithium tantalate is parallel to the direction of the electrical field between the electrodes and the probe beam is perpendicular to the electrode plane.

A typical result is shown in Fig. 2. The 50 μm gap was located 200 μm from the sampling point, the applied DC bias was + 50V, the amplitude of the switched signal is 30 mV and the probe beam was focused to a diameter of 11 μm between the two electrodes. Temporal resolution did not appear to be significantly sensitive to the angle of incidence

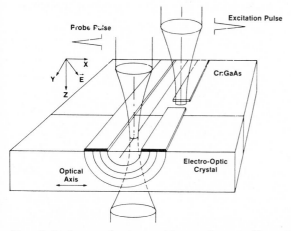

Fig. 1 Coplanar electro-optic sampler configuration.

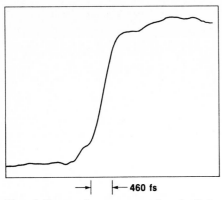

→| |←— **460 fs**

Fig. 2 Temporal response of a Cr:GaAs photoconductive switch as
resolved by the coplanar electro-optic sampler.

of the probe beam, which indicates that, as expected, the field depth in
the substrate is very small and is of the order of the electrode dimensions.
The resolved risetime (10% - 90%), neglecting the foot on the leading
edge due to dispersion, is 460 fs. The good signal/noise ratio obtained
indicates that the sensitivity of the coplanar configuration is comparable
to that demonstrated with balanced stripline.

The temporal resolution of the system is limited by τ_w, the transit
time of the electrical signal across the probe beam waist, and τ_0, the
transit time of the probe pulse across the field region. τ_w may be
reduced by reducing the focused diameter of the probe beam; the penetration
of the electric field into the dielectric scales down with the separation
of the electrodes, and to a lesser extent, with the electrode linewidths[3].
Reducing the dimensions of the stripline has the additional advantages
of improving the frequency response of the line and reducing the capacitance
of the gap. These improvements are presently being incorporated into
the experiment.

This work was partially supported by the following sponsors: The
Empire State Electric Energy Research Corp., General Electric Company,
Northeast Utilities, New York State Energy Research and Development
Authority, The Standard Oil Company (Ohio), and The University of
Rochester. Such support does not imply endorsement of the content by
any of the above parties.

1. J.A. Valdmanis, G.A. Mourou and C.W. Gabel, "Subpicosecond Electrical
    Sampling", IEEE JQE-17, pp. 664-667 (1983).
2. See, for example, K.C. Gupta, R. Garg and I.J. Bahl, "Microstrip
    Lines and Slot Lines", Chap. 7, Artech House, Inc., Dedhom, MA (1979).
3. J.B. Knorr and K-D. Kuchler, "Analysis of Coupled Slots and Coplanar
    Strips on Dielectric Substrate", IEEE MTT-23-7, pp. 541-548 (1975).

Čerenkov Radiation from Femtosecond Optical Pulses in Electro-Optic Media

K. P. Cheung, D. H. Auston, J. A. Valdmanis and D. A. Kleinman

AT&T Bell Laboratories, Murray Hill, NJ 07974, USA

It was recently suggested that under suitable conditions the propagation of femtosecond optical pulses in electro-optic materials should be accompanied by the radiation of an extremely fast electromagnetic transient [1,2]. This phenomenon, which arises from the inverse electro-optic effect [3], produces a Čerenkov cone of pulsed radiation having a duration of approximately one cycle and a frequency in the THz range.

In this report we describe the direct experimental observation of this effect and discuss its properties and potential applications for transient far-infrared spectroscopy.

Femtosecond optical pulses were obtained from a colliding pulse mode-locked dye laser [4]. Their duration, measured by second harmonic auto correlation, was 100 fs (full width at half maximum), and their center wavelength was 633 nm. A relatively low pulse energy of only 10^{-10} J at a repetition rate of 100 MHz was sufficient for the experiment.

Two optical pulses were used, both derived from a single pulse using a beam splitter. As indicated in Fig. 1, one pulse was used to generate the cone of radiation field, the other to detect it. This configuration has a novel symmetry arising from the use of the electro-optic effect for both generation and detection. In the case of the generating pulse, it is the inverse electro-optic effect that produces the nonlinear polarization responsible for the radiation field, and in the detection process, it is the direct electro-optic effect that is used to measure the small birefringence produced by the electric field of the radiation pulse [5].

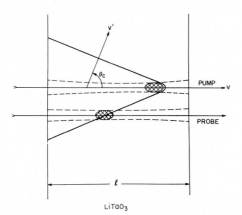

Fig. 1   Experimental arrangement

Both optical pulses were focussed on the 0.5 mm thick sample of Lithium Tantalate through a common microscope objective. They were carefully aligned to propagate parallel through the crystal with a separation that was nominally 100 micron. Their beam waists were each approximately 6 micron in diameter (1/e). The generating pulse was polarized parallel to the "c" axis of the crystal (out of the plane of Fig. 1), to produce a radiation field polarized in the same direction by the r_{33} electro-optic coefficient.

The method of detection is based on Pockel's effect suitably modified for transient measurements with femtosecond pulses as described in reference 5. The probing pulse was polarized 45 degree to the "c" axis of the crystal, and the static birefringence was compensated by a Soleil-Babinet compensator placed after the crystal. A Calcite Glan-Thomson polarizing prism was used as an analyzer to separate the two orthogonally polarized components of the transmitted pulse. Differential detection of these two signal was used to measure the rotation of the axis of polarization of the transmitted pulse arising from the birefringence produced by the radiation field.

The observed waveform for our particular experimental configuration is shown in Fig. 2. As expected it is extremely fast and approximates a single cycle of a frequency of 1 THz. Using the known electro-optic coefficient for Lithium Tantalate, the amplitude of the electric field was estimated from the measured birefringence to be 10 V/cm. The cone angle, as defined in Fig. 1, could also be measured by changing the lateral spacing between the optical pulses and observing the corresponding optical delay of the probing pulse necessary to restore synchronism with the Čerenkov wave front. This angle, θ_c, was found to be 70 ± 2 degrees. The theoretical value of this angle is given by the inverse cosine of the ratio of the group velocity of the optical pulse (.433c) to the velocity of propagation of the low frequency radiation field (.153c), and has the value of 69 degrees for Lithium Tantalate.

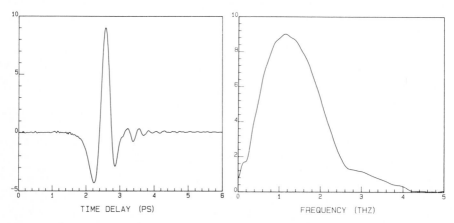

Fig. 2   EO Čerenkov waveform

Fig. 3   Frequency spectrum of the EO Čerenkov pulse

The frequency spectrum of the waveform in Fig. 2 was analyzed using Fourier transform, and is displayed in Fig. 3. It has a frequency distribution rising from zero at DC to a broad peak at 1 THz and then roll off rapidly to zero again before 4 THz.

The observed waveform has a close resemblance to the theoretically predicted shape [1,2]. Its duration, however, is broader than the optical pulse. Unlike classical Čerenkov radiation which is produced by a point charge with zero spatial extent, the radiation source in our case is a transient dipole moment produced by the focussed optical field. Being proportional to the intensity profile of the femtosecond pulse, the radiating transient dipole moment is spatially extended. Consequently, the details of the radiation field are expected to depend sensitively on both the duration and beam waist of the optical pulse. The expression for the time variation of the electric field is the following [2],

$$E(t) = -\frac{n_0^2 \, \epsilon_{33} \, r_{33} \, E_p}{c^2 \, \tau^2} \left[\frac{\sqrt{2} \cot \theta_c}{v \tau \, r_\perp} \right]^{1/2} U\left(-2, \frac{\sqrt{2}t}{\tau}\right) e^{\frac{-t^2}{2\tau^2}} \tag{1}$$

where E_p is the optical pulse energy, n_0 and v are the optical index of refraction and group velocity, $r_\perp$ is the radial distance from the beam axis, $U(-2,x)$ is the parabolic cylinder function of order -2, and the parameter τ is defined by the relationship,

$$\tau = \sqrt{\tau_p^2 + \frac{w^2 \tan^2 \theta_c}{v^2}} \tag{2}$$

where τ_p is the 1/e half-width of the duration of the optical pulse and w is the 1/e beam radius (both assumed to have Gaussian profiles). From expression 2, we see that a finite beam waist can make a substantial contribution to the duration of the electric field. Additional broadenings come from

410

absorption and dispersion of the electric pulse, and from the measurement technique. There is a spatial and temporal convolution between the electric pulse and the probing pulse. If the generating and probing pulse were identical, and no other sources of broadening were present, this would introduce an additional factor of $\sqrt{2}$ into the parameter τ. Accounting for this factor, the actual electric pulse would have a duration of 252 fs (full width at zero crossing) instead of the 356 fs shown in Fig. 2. The spectral distribution of Fig. 3 would have extended to 6 THz.

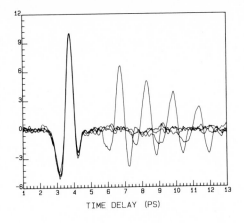

TIME DELAY (PS)

Fig. 4   Decay and broadening of EO Čerenkov pulse

Absorption and dispersion become severe as the high end of the frequency spectrum approaching the lattice vibrational frequencies. For Lithium Tantalate, the first vibrational band is at approximately 6 THz [6]. The ringing after the main pulse in the waveform of Fig. 2 may be caused by lattice vibrational resonances. The present experiment was done at room temperature where vibrational relaxation is extremely fast. The experiment will be repeated at cryogenic temperature and the vibrational relaxation will be studied.

One application of the Čerenkov radiation is to measure the absorption and dispersion of electro-optic material in the far infrared spectral range. Figure 4 is the result of one such application. Here the Čerenkov wavefront was allowed to bounce (total internal reflection) at the crystal-air interface that was parallel to the optical beams. The wavefront therefore intercepted the probe beam twice, producing two transient responses — a main pulse and a reflected pulse. The two optical beams were kept stationary in space and the crystal was translated in the plane defined by the two optical beam and perpendicular to them. Thus the distance between the reflecting crystal-air interface and the probe pulse was changed and correspondingly the reflected pulse showed up at a different delay-time. A series of these measurements were performed and the results superimposed together to generate Fig. 4. Notice the main pulse reproduces almost perfectly each time. A clear trend of decreasing amplitude and broadening is evident for the reflected pulses. Part of the amplitude decrease was due to the $r^{-1/2}$ dependence in expression 1. Taking that into account one estimates an absorption coefficient of 43 cm^{-1} at 1 THz for Lithium Tantalate, about a factor of 2 less than the results in reference 6.

Figure 5 shows another application of the electro-optic Čerenkov radiation. Once again we utilized the reflection from an interface. Instead of a crystal-air interface used above, a different material can be attached to the polished surface of the crystal and the reflection of the electric pulse from the crystal-material interface can be studied. The solid line in Fig. 5 shows the main pulse and the reflected pulse from a Lithium Tantalate - metal film (0.5 μm thick) interface. Aside from a slight broadening and a smaller amplitude, the reflected pulse has the same shape as the main pulse with a 180° phase shift. This is what one would expect from the reflectivity of a metal.

Without changing the position of either the crystal or the optical beams, the metal film was replaced with a highly doped n type Si. The experiment was repeated and the result is the dotted line in Fig. 5. The doping level of the Si sample, although not exactly known, was chosen so that it has a plasma frequency near the peak of the frequency distribution of the electric pulse. A significant amount of phase shift is clearly evidenced in Fig. 5 as a result of dispersion. Using Fourier analysis one can obtain from the waveform of Fig. 5 the real and imaginary parts of the dielectric constant of the Si sample as a

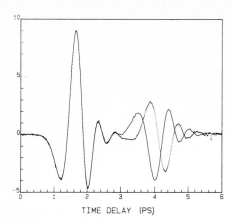

TIME DELAY (PS)

Fig. 5 Reflection of EO Čerenkov pulse

function of frequency covering the range available in the Čerenkov pulse. A more detailed account will be given in a later report.

In conclusion, we have demonstrated the experimental measurement of the electro-optic Čerenkov radiation. The resulting short electric pulse could be a very useful tool in coherent far infrared spectroscopy. Applications toward studying lattice vibrations in solids and material response to a coherent electric pulse are just two example of potential applications of this entirely new spectroscopic technique.

[1] D. H. Auston, Appl. Phys. Lett., *43* 713 (1983).

[2] D. A. Kleinman and D. H. Auston, IEEE J. Quant. Elect., Submitted.

[3] M. Bass, P. A. Franken, J. F. Ward, and G. Weinreich, Phys. Rev. Lett., *9*, 446 (1962).

[4] R. L. Fork, B. I. Greene, and C. V. Shank, Appl. Phys. Lett. *38*, 671 (1981).

[5] J. A. Valdmanis, G. A. Mourou and C. W. Gabel, IEEE J. Quant. Electr., *QE-19*, 664 (1983).

[6] A. S. Barker, Jr., A. A. Ballman, and J. A. Ditzenberger, Phys., Rev. B, *2*, 4233 (1970).

Ultraviolet Photoemission Studies of Surfaces Using Picosecond Pulses of Coherent XUV Radiation

R. Haight, J. Bokor, R.H. Storz, and J. Stark

AT & T Bell Laboratories, Holmdel, NJ 07733, USA

R.R. Freeman and P.H. Bucksbaum

AT & T Bell Laboratories, Murray Hill, NJ 07974, USA

Ultraviolet photoemission spectroscopy provides a powerful tool for studying both the bulk and surface electronic structure of solids. While previous photoemission studies of solids have concentrated on the electronic structure of filled electronic levels, i.e. the valence bands of semiconductors and the filled conduction bands of metals, the extension of this technique to study the dynamic transient processes experienced by electrons excited into otherwise unoccupied states has only begun to be exploited [1]. In this paper we describe a new experimental system which utilizes the generation of short pulsewidth coherent extreme ultraviolet light to perform time-resolved photoemission studies at solid surfaces and in the bulk.

A schematic diagram of the apparatus is shown in Fig. 1. The source of energetic photons involves the generation of harmonics of an input laser pulse in a non-linear medium, in this case a pulsed rare-gas jet. Details of these processes have been described elsewhere [2,3]. Briefly, generation of the odd multiple harmonics of short pulsewidth 248 nm light which has been amplified in a single stage, discharge pumped KrF excimer amplifier occurs at the output of a pulsed jet device which delivers a supersonic burst of He gas in a direction orthogonal to the propagating light pulse.

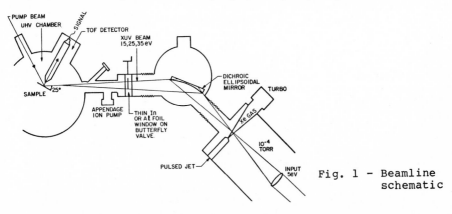

Fig. 1 - Beamline schematic

By this technique, the third, fifth and seventh harmonics have been generated, yielding photon energies of 15, 25 and 35 eV. Both the harmonic as well as the fundamental radiation propagate from the interaction region and are incident upon an ellipsoidal mirror, coated with a dichroic coating [4] designed for high reflectivity (30 - 50%) in the extreme ultraviolet and low reflectivity (< 2%) at the fundamental 248 nm wavelength. Further filtration of the fundamental is performed by a thin foil filter of indium or aluminum (depending on the harmonic of interest) which is downstream of the ellipsoidal mirror. The foil is essentially opaque in the visible and near uv (< 10^{-9} transmission) while transmitting 10-50% of the extreme uv radiation and also provides isolation of the ultra-high vacuum (UHV - 10^{-10} torr) system, where the photoemission is performed, from the moderate vacuum (10^{-4} torr) region where the harmonic generation is performed. The UHV system includes standard surface preparation and characterization instrumentations.The sample is held at the focal point of the ellipsoidal mirror by a sample manipulator which allows both heating and cooling as well as angular rotation about an axis parallel to the sample surface. Angular rotation of the sample allows study of the energy vs. wavevector dispersion of the photoemitted electrons. Photoemitted electrons are detected by a time-of-flight spectrometer. The fast response (1 nsec FWHM) of the microchannel plate electron detector at the end of the drift tube allows for high energy resolution (2% at 5 eV electron energy) as well as the recording of the entire electron energy spectrum generated with each photon pulse. The angular acceptance cone of the electron spectrometer has a full width of 6°.

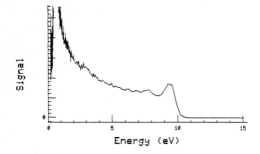

Fig. 2 - Photoemission spectrum of Si (100) 2x1 surface using 15 eV photons

Figure 2 shows the photoemission spectrum obtained from a clean silicon (100) surface exhibiting the characteristic 2x1 reconstruction. The photon energy used was 15 eV and photoelectrons were detected at 0° exit angle with respect to the surface normal. The XUV photon flux had to be kept below 10^{6} photons/pulse in order to avoid space-charge distortions in the energy spectrum. This spectrum corresponds well with previously published photoemission spectra for silicon [5]. In particular, the prominent peak at the top of the valence band is associated with an intrinsic surface state [5].

The photoemission spectrum of __photoexcited__ silicon was examined next. In this experiment, 60 psec optical pulses at 532 nm were used to create an electron-hole plasma in the sample, and the photoemission spectrum of the excited material was obtained as a function of the relative time delay between the 532 nm pump pulse and the XUV probe pulse.

The addition of electrons to the previously unpopulated
conduction band would be expected to give rise to an
additional peak in the photoemission spectrum which would
appear at an energy of 1.1 eV (the silicon band gap) above
the valence band edge. Since the pump pulse duration of 60
psec is much longer than typical intraband energy and
momentum relaxation times, a thermalized plasma is to be
expected, localized in momentum space at the conduction band
minimum. Due to the localization of the excited electrons
in k space, particular attention must be paid to the
importance of direct transitions in photoemission. In order
to have an appreciable photoemission yield, a direct
transition must be allowed between the initial band and a
final excited band lying at the XUV photon energy above the
initial band [6]. Since the present source operates at the
fixed frequencies of either 15 eV, 25 eV, or 35 eV, accurate
band structure data for the energy range of 20-40 eV above
the valence band maximum is needed. In most cases, such
data is not available. Accurate, nonlocal, semiempirical
pseudopotential band structure calculations in this energy
range are available for only a few semiconductors [7]. A
less sophisticated calculation applicable to the far
ultraviolet region for silicon is available [8] which does
indicate an allowed final band at 15 eV above the conduction
band minimum.

Figure 3 shows the results of this experiment. The pump
fluence at 532 nm was 3 mj/cm^2. At zero time delay a
distinct peak is observed at 11.7 eV. This peak corresponds
to the excited conduction electrons. The rise and fall
times of this feature as a function of the time delay
between the pump and probe beams essentially follow the 60

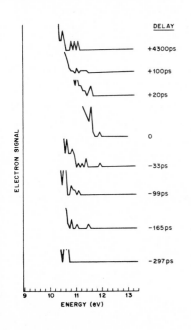

Fig. 3 — Time resolved
          photoemission from
          excited silicon.
          Vertical scale
          expanded by 1000
          compared to Fig. 2.
          Also note change
          in horizontal
          scale.

psec pump pulse. The fast decay is most probably due to a combination of Auger carrier recombination and rapid ambipolar diffusion away from the surface. A detailed analysis to assess the relative contributions of these and other decay processes has not yet been performed.

In summary, we have described a novel experimental photoemission setup for studying the transient electron dynamics at and near solid surface using harmonically generated picosecond pulses of extreme ultraviolet radiation. This technique offers the possibility of obtaining energy and wavevector data on photoexcited carriers in metals and semiconductors by making angularly resolved measurements. In addition, the surface sensitivity of ultraviolet photoemission offers the prospect of studying the dynamics of surface reconstructions, adsorbates, and surface chemical reactions. Finally, we note that this technique appears to be readily scalable into the sub-picosecond time domain.

References

1. R. T. Williams, T. R. Royt, J. C. Rife, J. P. Long, and M. N. Kabler, J. Vac. Sci. Technol. 21, 509 (1982).
2. J. Bokor, P. H. Bucksbaum, and R. R. Freeman, Opt. Lett. 8, 217 (1983).
3. P. H. Bucksbaum, J. Bokor, R. H. Storz, and J. C. White, Opt. Lett. 7, 399 (1982).
4. R. W. Falcone and J. Bokor, Opt. Lett. 8, 21 (1983).
5. F. J. Himpsel and D. E. Eastman, J. Vac. Sci. Technol. 16, 1297 (1979).
6. E. W. Plummer and W. Eberhardt, Adv. Chem. Phys. 49, 533 (1982).
7. K. C. Pandey, private communication.
8. D. Brust and E. O. Kane, Phys. Rev. 176, 894 (1968).

Synchronous Mode-Locking of a GaAs/GaAlAs Laser Diode by a Picosecond Optoelectronic Switch

J. Kuhl and E.O. Göbel

Max-Planck-Institut für Festkörperforschung, Heisenbergstraße 1
D-7000 Stuttgart 80, Fed. Rep. of Germany

Active and passive mode-locking of semiconductor lasers have been applied so far to produce optical pulses in the picosecond or even subpicosecond time regime [1,2]. We report a new method for synchronous mode-locking of a diode laser and an acousto-optically mode-locked Ar[+] ion laser by using a picosecond optoelectronic switch for synchronized excitation of the diode laser. Synchronously mode-locked pulses with 20-30 ps duration (FWHM) are generated at a repetition rate of 80.32 MHz and an average power of about 250 μW with a buried heterostructure GaAs/GaAlAs double heterostructure laser. Compared to previously published mode-locking schemes this novel technique offers the advantage of strict synchronization of the diode laser pulse train to the picosecond pulse train of a powerful gas-, dye- or solid state laser. Parallel pumping of several diode lasers or of diode and dye lasers is simply feasible because of the low driving power needed for the optoelectronic switch. The spectral range for picosecond excite and probe experiments now can be considerably extended because this short pulse electrical excitation seems to be applicable to every commercially available laser diode.

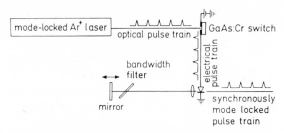

Fig.1   Scheme of the synchronously mode-locked semiconductor diode laser

The schematic diagram of the experimental set up is depicted in Fig.1. The semiconductor laser diode is excited by a current pulse train, which is generated from the mode-locked optical pulse train of the Ar[+] ion laser by a Cr doped GaAs picosecond optoelectronic switch [3,4]. The pulse duration (FWHM) of the Ar[+] laser pulse is 82 ps and the rise time 117 ps. The rise of the electrical pulses generated by the GaAs:Cr switch follows the Ar[+] laser pulse and the halfwidth amounts to about 200 ps. A commercial GaAs/GaAlAs buried heterostructure laser within an external resonator containing a bandwidth filter (narrow band interference filter or the interference filter and a 80 μm thick etalon) is used. Synchronous mode-locking is obtained by matching the external cavity length to the pulse period of the Ar[+] ion laser pulse. The spectral and temporal behaviour of

417

the output is analyzed by a 1 m grating spectrometer and a synchroscan streak camera (13 ps time resolution for the present experimental conditions), respectively.

A streak camera trace and a spectrum of the diode pulse for optimum cavity length ($\Delta L = 0$ mm) is shown in Fig.2. The halfwidth of the pulse amounts to 28 ps after deconvolution of the streak camera resolution.

Fig.2   Streak camera trace (left) and spectrum (right) of the synchronously mode-locked diode laser output at $\Delta L = 0$. The bandwidth was controlled by an interference filter. The spacing of the crystal FP-modes is 3.2 Å

The width and peak intensity of the mode-locked diode laser pulses depend strongly on external cavity length as expected for synchronous mode locking [5]. The emission spectrum clearly exhibits the longitudinal modes of the short diode laser cavity, demonstrating the strong modulation of the spectrum by the Fabry-Perot formed by the cleared diode laser crystal. This modulation could not be suppressed in spite of the antireflecting coating of the crystal surface facing the external mirror. Restriction of the emission to a single of these crystal FP modes has been achieved after insertion of an additional 80 µm Fabry-Perot etalon ($R \approx 30\%$) into the external cavity [5].

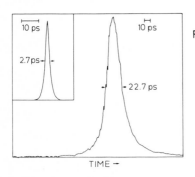

Fig.3   Profile of the diode laser pulse measured by sum frequency sampling (up conversion) with 2 ps pulses from a mode-locked dye laser in a $LiIO_3$ crystal. The temporal resolution is demonstrated by the auto-correlation trace of the dye laser pulses shown in the inset

Figure 3 displays a cross correlation measurement of the synchronously mode-locked diode laser and a synchronously mode-locked dye laser (pyridin 1 dissolved in a mixture of ethylene glycol and propylene carbonate, $\lambda = 740$ nm) pumped by the same mode-locked Ar^+ ion laser using up-conversion in a $LiIO_3$ crystal. The smooth shape and the narrow width of the cross correlation (22.7 ps FWHM) demonstrate the high synchronization of both mode-locked lasers. The cross correlation curve further reveals that only the longitudinal external resonator modes within a single Fabry-Perot mode of the diode crystal are phase-locked, because no substructure corresponding to the spacing of the crystal Fabry-Perot modes is present (weak mode-locking condition). The pulse duration thus is limited by the bandwidth of the FP modes of the crystal resonator. The time-bandwidth product amounts to 1.2 for the data shown in Fig.2 and this is more than twice the value expected for a Fourier transform limited Gaussian pulse. This broadening of the pulses results from a frequency chirp caused by the refractive index change connected with the carrier recombination [6]. This frequency chirping seems to set a limit for the obtainable pulse duration by active mode-locking of about 20-30 ps. Coherent mode-locked pulses with shorter pulse width have been generated so far only by passive mode-locking where the chirp can be compensated by the saturable absorber [7].

References

1. P.T. Ho: SPIE Proceedings, vol. 439, San Diego, 1983, p. 42.
2. J.P. von der Ziel: SPIE Proceedings, vol. 439, San Diego, 1983, p. 49.
3. E.O. Göbel, G. Veith, J. Kuhl, H.-U. Habermeier, K. Lübke, A. Perger: Appl.Phys.Lett. 42, 25 (1983).
4. C.H. Lee, C. Lunton, C. Hsin-Ming, C. Weigzhei: Chines. Laser Journ. 8, 55 (1981).
5. E.O. Göbel, J. Kuhl, G. Veith: Journ.Appl.Phys., in press.
6. J.P. van der Ziel: IEEE J. of Quant. Electr. QE 15, 1277 (1979).
7. See e.g. E.P. Ippen, D.J. Eilenberger, R.W. Dixon: Appl.Phys.Lett. 37, 267 (1980); W.A. Stallard, D.J. Bradley: Appl.Phys.Lett. 43, 626 (1983); C. Harder, J.S. Smith, K.Y. Lau, Y. Yariv: Appl.Phys.Lett. 42, 772 (1983).

Photochron Streak Camera with GaAs Photocathode

C.C. Phillips, A.E. Hughes, and W. Sibbett

Optics Section, Blackett Laboratory, Imperial College, Prince Consort Road
London SW7 2BZ, England

The electron-optical streak and framing cameras which have been used for monitoring ultrafast optical processes in the UV-NIR spectral region have incorporated conventional semitransparent, positive electron affinity photo-cathodes (usually types S1, S11, S20 or S25). Electron emission from such photocathodes consists of only those "hot electrons" that possess sufficient energy to traverse the potential barrier at the cathode/vacuum interface (Fig. 1a). This implies that any photoelectrons emitted are likely to have been excited close to the vacuum surface ($\sim$ 20nm) and the photoemission process is therefore expected to be very rapid ($\leq 10^{-13}$s). In contrast, a heavily p-type doped semiconductor can be activated to a condition of neg-ative electron affinity (NEA) such that the conduction band minimum in the bulk lies above the vacuum level. It then follows that electrons excited at comparatively deep sites within the bulk ($\sim$ 1μm) of these NEA materials, which subsequently thermalise to the conduction band minimum, can still be photo-emitted, Fig. 1b. Although this results in excellent photosensitivities (1500μA/lumen in transmission(1)), theoretical considerations imply that their response times (estimated to be $\sim$1ns (2)) will be substantially greater than their positive electron affinity counterparts but no quantitative experi-mental results on this topic have been reported. In view of the exploitable features of low dark current, high visible-to-near infrared spectral sensit-ivities and good spatial resolution of NEA cathodes, we decided to make some measurements of their temporal response characteristics in the transmission mode which is most compatible to streak cameras.

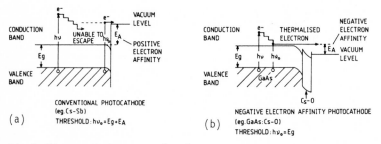

Fig 1: Photocathode energy levels

The photocathodes comprised 20mm diameter discs of Corning 7056 glass onto which a single crystal strain-matching layer of (100) orientation GaAlAs had been bonded with an epitaxially grown zinc-doped ($\sim$ 5 x 10^{18}cm^{-3}) GaAs active layer on top, Fig. 2.

Following appropriate cleaning and annealing procedures (3a) the GaAs was activated to the NEA condition by alternating cycles of exposure to caesium

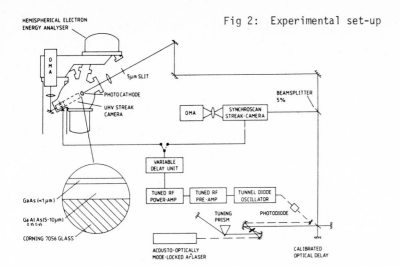

Fig 2: Experimental set-up

vapour and oxygen in a UHV electron spectrometer (3b). For the NEA photo-cathodes produced in our work the quantum efficiency spectral characteristic Fig. 3 has a typical shape with a long wavelength limit (~ 890nm) determined by the 1.4eV direct band gap absorption threshold of GaAs and a short wave-length limit ~ 560nm associated with absorption in the $Ga_{0.35}Al_{0.65}As$. Measured photoelectron energy distributions (3,4) were much narrower (~ 60meV) than those for conventional cathodes which implies a photoelectron transit time dispersion of less than 200fs in the Photochron IV streak image tube (5).

A much more serious limitation to the temporal resolution of a streak tube having an NEA photocathode is the "intrinsic emission time uncertainty" (IETU) for electrons that are photoexcited in the bulk and diffuse through the mat-erial before being emitted. To investigate the magnitude of and the influence on this time uncertainty, a demountable, miniaturised version of a Photochron IV streak tube (4,5) was incorporated into the UHV system. This streak tube

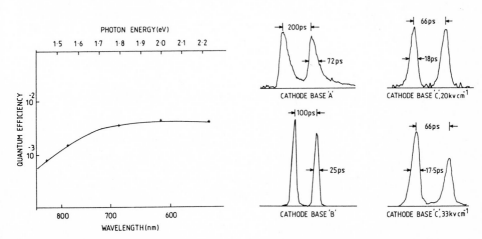

Fig 3: Quantum efficiency curve   Fig 4: Streak intensity profiles

was positioned along a radius of the spherical analysis chamber as illustrated in figure 2 in a configuration that permitted the transport of the activated NEA photocathode into a pre-arranged location relative to the extraction mesh electrode. The UHV camera and a conventional "S20" Photochron II camera monitor were operated in the repetitive (or Synchroscan) streak mode (6) in conjunction with the picosecond pulses from a synchronously mode-locked rhodamine B CW dye laser (140MHz pulse repetition rate). In both cases the intensity profiles of the streak images were recorded and analysed using optical multichannel analysers (OMA). Streak profiles obtained in this way for three GaAs cathode bases 'A','B','C' are reproduced together with optical delay calibration in Fig. 4. To deduce the intrinsic emission time uncertainty for these NEA cathodes, the streak durations measured with the monitoring S20 camera were deconvolved from the UHV camera data so that the common factors of laser pulsewidth and cumulative jitter could be eliminated. The resulting values for the IETU for cathode bases A,B,C were 71ps, 17ps and 8ps respectively.

The temporal response of cathode 'A' was approximately an order of magnitude faster than that derived from the electron diffusion calculations made by Bell (2) for cathodes of optimum thickness. Significantly, however, the photosensitivity of this cathode (reproduced in Fig. 3) was also an order of magnitude lower than the optimised case. This was also observed for all of the other cathodes.

When the electric field applied in the vicinity of a particular cathode layer was increased from 20kV cm^{-1} to 33kV cm^{-1}, there was no significant decrease in the observed streak duration, Fig. 4. The most plausible explanation of the measured fast response times and depressed sensitivities is that the GaAs films used in these studies had been etched to thicknesses that were substantially less than the optimum value $\sim 2\mu m$. In fact, when Bell's diffusion model of electron transport (2) is applied to these test cathodes, the fastest time response of 8ps would suggest that the active layer 'C' had a thickness $\sim 50nm$.

It may therefore be concluded that picosecond resolution can be obtained in streak cameras having thin NEA photocathodes and although this sacrifices photosensitivity they still have quantum efficiencies that are comparable to conventional photocathodes. For response at longer NIR wavelengths (1.5 - 1.6μm) then semiconductor compounds, such as GaInAs (7) offer scope for potential development.

References

1. J.P. Andre, P. Guittard, J. Hallais and C. Piaget: J. Cryst. Growth, 55, 235 (1981)
2. R.L. Bell: Negative Electron Affinity Devices (Clarendon Press, Oxford 1973)
3. C.C. Phillips, A.E. Hughes and W. Sibbett: J. Phys. D: Appl. Phys. (a) to be published; (b) 17, 611 (1984)
4. C.C. Phillips: PhD Thesis, London University (1983)
5. W. Sibbett, H. Niu and M.R. Baggs: Proc. 15th Inter. Cong. High Speed Photography & Photonics, SPIE 348, 271 (1982)
6. W. Sibbett: ibid. (5) 348, 15 (1982)
7. D.G. Fisher, R.E. Enstrom, J.S. Escher and B.F. Williams: J. Appl. Phys., 43, 3815 (1972)

Sequential Waveform Generation by Picosecond Optoelectronic Switching

C.S. Chang, M.C. Jeng, M.J. Rhee, and C.H. Lee

Department of Electrical Engineering, University of Maryland
College Park, MD 20742, USA

A. Rosen and H. Davis

RCA Laboratories, Princeton, NJ 08549, USA

The direct conversion of dc energy to RF pulses with high efficiency has many potential applications. They include: (1) the use of high-power electrical pulses for pulsed power devices and plasma-physics experiments; (2) various applications in high resolution radar and time domain metrology; and (3) the generation of megawatt level microwave and millimeter-wave pulses. All of these experiments require the development of an appropriate switch or an array of switches which can switch high power with extremely fast risetime and zero jitter. For sequential waveform generation several methods have been tried [1,2]. In reference [1], a series of step recovery diodes was used; however only low voltage switching has been demonstrated. The prospects of extending this technique to a high power switch are poor because the step recovery diodes encounter breakdown problems at high voltage. In reference [2], a frozen wave generator was used for sequential waveform generation. A frozen wave generator consists of many segments of transmission line charged alternately with positive and negative voltage (Fig. 1). Two adjacent segments are joined by a silicon switch which can be closed with a laser pulse. To maintain a high conversion efficiency and a good waveform, it is essential that all the switches be closed simultaneously and rapidly. Jitter in the switching process produces random frequency modulation which removes energy from the fundamental frequency. Picosecond laser activated semiconductor switches seem ideal for this application. However, in reference [2], Proud, Jr., and Norman used a nitrogen laser which produced optical pulses with a time duration of a few nanoseconds. The advantage of the photoconductive switch was not fully utilized. In this work, we report the use of pico-second laser pulses for switching. A sequential waveform of two and one half cycles has been obtained with a voltage conversion efficiency of better than 95%.

The schematic of the frozen wave generator is shown in Fig. 1. It consists of three separate silicon switches and three charge line segments. The photo-conductive switches are fabricated on an intrinsic silicon wafer with 50 Ω microstrip lines serving as electrodes. The switching region comprises the three gaps between the three electrode pairs. The microstrips are fabricated by first evaporating a 400 Å titanium layer followed by a 4 μm thick aluminum and a 4 μm layer of silver to reduce their resistance to a minimum. Semi-rigid coaxial cables are used for interconnection between electrodes. They are charged alternately with positive and negative voltage (+10V and -10V respectively). The frozen wave generator shown in Fig. 1 is a pulse forming system of one and one quarter cycles with an open end. When the switches are activated by a single picosecond optical pulse (30 ps, 30 μJ at 1.06 μm) the pulse forming network generates a two and one half cycles electrical waveform. The length of the charged line is designed in such a manner that the real time waveform can be clearly displayed on an oscilloscope (Tektronik 7834). The result is shown in Fig. 2. According to theory, the maximum amplitude of the output waveform should be equal to one half of the charging voltage if the "on-state" resistances of the switches are all equal to zero.

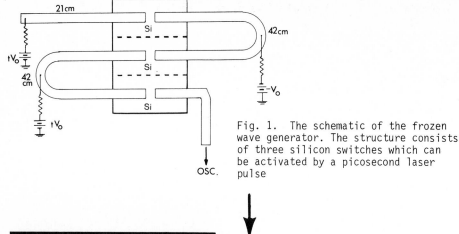

Fig. 1. The schematic of the frozen wave generator. The structure consists of three silicon switches which can be activated by a picosecond laser pulse

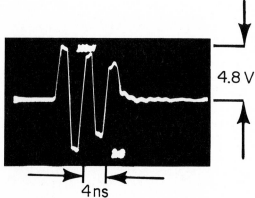

4.8 V

4 ns

Fig. 2. The oscillogram of the RF waveform. The RF waveform consists of two and one half cycles

In reality, none of these resistances become zero, resulting in a reduced amplitude. The conversion efficiency is defined as the ratio of the actual amplitude to the maximum possible amplitude in an ideal situation. From the data shown in Fig. 2, a voltage conversion efficiency greater than 95% has been achieved. Using the SPICE 2 computer code the theoretical waveform can be calculated if the "on-state" resistance of the switches is known. Good agreement between theory and experiment has been obtained. The "on state" resistances observed are attributed to the contact resistances between the electrode and the silicon wafer.

In conclusion, a direct DC to RF conversion technique has been demonstrated by optoelectronic switching. Since a bulk semiconductor is used, the voltage holding capability should scale linearly with the dimension of the device (gap length). We anticipate that the switching of a few kilovolts is possible in the immediate future as it has been demonstrated in our laboratory for the switching of a single pulse [3].

1. H. M. Cronson, "Picosecond-pulse sequential waveform generation", IEEE Trans. Microwave Theory Tech., vol. MTT-23, pp. 1048-1049, 1975.
2. J. M. Proud, Jr., and S. L. Norman, "High-frequency waveform generation using optoelectronic switching in silicon", IEEE Trans. Microwave Theory Tech., vol. MTT-26, pp. 137-140, 1978.
3. M. K. Mathur, C. S. Chang, W. L. Cao, M. J. Rhee and Chi H. Lee, "Multikolovolt Picosecond Optoelectronic switching in CdS$_{0.5}$Se$_{0.5}$", IEEE J. Quantum Electron. QE-18, pp. 205-209, 1982.

Picosecond Gain Measurements of a GaAlAs Diode Laser *

W. Lenth

Lincoln Laboratory, Massachusetts Institute of Technology
Lexington, MA 02173-0073, USA

The success of high-data-rate optical communication systems relies on
the modulation and switching capabilities of semiconductor laser oscilla-
tors and amplifiers. The fundamental physical processes which govern the
dynamic device performance depend in a complex fashion on the material and
design characteristics of diode lasers and are not yet well understood.
In this context it is important to study the transient response of the
laser gain to rapid changes of the injection current. We have used a new
direct gain measurement technique to investigate the time evolution of the
diode laser gain after the injection of an 80 ps current pulse.

The gain is probed by passing 4 ps near infrared pulses from a synchronously pumped Oxazin 750 dye laser through the active region of a laser
diode (see Fig. 1). The laser diode, a channeled-substrate-planar (CSP)
GaAlAs diode laser (Hitachi HLP1400), was dc-biased below its cw-threshold
of 63 mA and a train of 80 ps current pulses derived from a comb generator
was capacitively coupled into the device. The comb generator was driven
by the same rf oscillator which provided the mode-locking frequency for
the Kr-ion pump laser; this resulted in accurate synchronization of the
optical probe pulse and the injection current pulse with a time jitter of
less than 10 ps. Using a variable-length air line the time delay between
the two pulses could be adjusted over a range of 1200 ps, permitting investigation of the time evolution of the gain. The dye laser was tuned to
a wavelength of 835 nm approximately 10 Å away from the maximum of the
diode laser gain curve. The spectral width ($\Delta\lambda \approx 4$Å) of the dye laser
pulses was larger than the 3 Å separation between the axial modes of the
diode laser. No gain enhancement associated with cavity resonances was
observed when the dye laser was tuned or when the injection current was
varied. The polarization of the dye laser light was perpendicular to the
junction plane which is 90° from the polarization of the lasing modes.

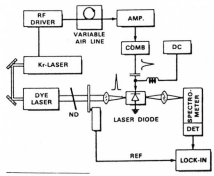

Fig. 1. Schematic
illustration of the
experimental arrangement.

*This work was supported by the Department of the Air Force.

When a dye laser pulse of low energy E_0 is coupled into the active wave-guide region of a GaAlAs laser diode it experiences a single pass unsaturated gain, $G_s = \exp(g-\alpha)\ell$, where g and α are the gain and the loss factors, respectively, and ℓ is the diode length. The total energy E_{tot} emerging from the rear facet is distributed over a train of pulses which are separated by the 8 ps round-trip time of the diode laser.

$$E_{tot} = E_0(1-\gamma)\ (1-R_\perp) \sum_{j=0}^{\infty} R_\perp^{2j}\ G_s^{2j+1} = E_0(1-\gamma)\frac{(1-R_\perp)G_s}{1-R_\perp^2 G_s^2} \quad , \quad (1)$$

where R is the facet reflectivity and γ is the net insertion loss. Due to the perpendicular polarization the dye laser light experiences a reduced facet reflectivity R compared to GaAlAs laser modes which have parallel polarization. As a result $R^2 G_s^2 < 1$ at all injection currents and the series in Eq. (1) converges.[1] Based on Eq. (1) and Ref. 1 we estimate that even at bias currents near threshold the first two pulses contribute more than 85% of the recorded signal, which is the time-averaged sum over the pulse train. Thus, due to the perpendicular polarization of the probe pulses, a high temporal resolution is maintained at all injection currents.

Figure 2 shows the effective gain $G_{eff} = E_{tot}/E_0$ for different dc-bias currents as a function of time after the injection of an 80 ps current pulse. The peak current of the pulse was approximately 25 mA. The time-response curves for the three lower dc-bias currents have a similar shape, although they differ in amplitude. Under these bias conditions the super-imposed current pulse is not strong enough to excite the GaAlAs laser above

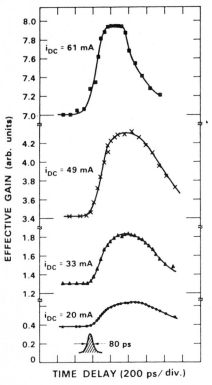

TIME DELAY (200 ps/ div.)

Fig. 2. Time evolution of the effective gain after the injection of an 80 ps current pulse. The same intensity scale is used for each bias current.

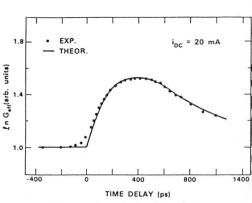

Fig. 3. ℓn G_{eff} as a function of time for i_{DC} = 20 mA.

the lasing threshold. The gain continues to increase after the termination of the current pulse reaches a
maximum after about 350 ps and then decays within several hundred picoseconds. At a dc-bias current of 61 mA the gain saturates at about 250 ps because the threshold for lasing is reached, as verified by observing the emission of a diode laser pulse in a separate experiment. The relatively flat maximum of the gain-time profile indicates that gain clamping occurs on a very short time scale.

Using Eq. (1) the temporal variation of the gain factor g(t) can be deduced from the experimental data which are shown in Fig. 2. Especially at low injection currents the analysis is very straightforward. An example is given in Fig. 3. At i_{dc} = 20 mA, R G_s << 1, and E_{tot} is proportional to the single-pass gain G_s [see Eq. (1)]. Figure 3 is a plot of $\ln G_{eff}$ vs time, i.e., g(t). Using a phenomenological model the experimental data points can be fitted with the solution of a differential equation which considers the external current pulse as a delta function and describes the time evolution of g(t) by a gain buildup time τ_b and a gain decay time τ_r. The best fit is obtained for τ_b = 325 ps and τ_r = 375 ps.

In order to understand the delayed growth of the gain factor one has to consider the junction capacitance and resistance as well as parasitic inductances due to bonding wires and connectors. Experimentally determined values for the CSP laser studied are: C_j = 40 pF, R_j = 2.75 Ω, L_p = 10 nH.[2] An estimate based upon an electrical circuit model of the diode laser suggests that these values can account for the observed gain buildup time. For the gain buildup carrier diffusion effects should be negligible compared to circuit effects in the CSP laser studied. The decay of the gain is faster than expected from the usual assumption that the spontaneous carrier recombination is of the order of 1-3 ns. Stimulated emission processes cannot cause a dramatic reduction of the gain decay time at the low bias current of i_{dc} = 20 mA. We believe that carrier diffusion in the lateral direction is the main reason for the fast decay of the gain. A comparably fast decay of the spontaneous emission of a transverse-junction-stripe laser has been explained with carrier out-diffusion.[3] The CSP laser investigated has an undoped active layer of $Ga_{0.92}Al_{0.08}As$ with no lateral carrier confinement; this could result in a relatively high ambipolar diffusion constant D_p. Experimentally determined values for undoped GaAs are of the order of $D_p \approx 125$ cm^2/s,[4] which corresponds to a diffusion distance of 2.2 μm in a time of 375 ps. Thus carrier out-diffusion can significantly contribute to the fast decay of the laser gain in the ~ 5 μm-wide active region.

In conclusion, the measurement technique described here allows direct investigation of the gain dynamics of semiconductor lasers with picosecond time resolution. The technique can be modified to employ two laser pulses so that all-optical pump-probe measurements can be performed. In the device investigated electrical circuit limitations, in part due to parasitic inductances, result in delayed gain buildup. Carrier out-diffusion in the lateral direction is probably responsible for the fast decay of the laser gain.

REFERENCES:

1.   A. P. DeFonzo, IEEE J. Quantum Electron. QE-19, 1537 (1983).
2.   S. B. Alexander and D. Welford, private communication.
3.   M. A. Dugay and T. C. Damen, Appl. Phys. Lett. 40, 667 (1982).
4.   R. J. Nelson and R. G. Sobers, Appl. Phys. Lett. 32, 761 (1978);
     B. W. Hakki, J. Appl. Phys. 44, 5021 (1973).

Transient Response Measurements with Ion-Beam-Damaged Si-on-Sapphire, GaAs, and InP Photoconductors

R.B. Hammond, N.G. Paulter, and R.S. Wagner

Electronics Division, Los Alamos National Laboratory
Los Alamos, NM 87545, USA

In recent years optoelectronic methods with the potential for making extremely high speed measurements of transient response in electronic devices and circuits have begun to be studied. In this paper we report correlation measurements using high speed photoconductor pulsers and sampling gates excited by femosecond laser pulses. We observe experimentally some of the circuit limits inherent in the photoconductors themselves, and limits imposed by the geometry of the test structures.

Test structures for the experiments reported here were fabricated on three different substrates: a) Si-on-sapphire, b) undoped semi-insulating GaAs, and c) Fe-doped semi-insulating InP. Microstrip transmission line correlation structures with 50Ω impedance were formed in the center of each wafer as described in [1]. The wafer thicknesses were 325, 340, and 400 μm, respectively.

The pulsing photoconductor gaps used the undamaged substrate material for their active regions. Each of these pulsers produced fast-rising electrical pulses a few hundred picoseconds in duration with peak amplitudes up to $\sim$1 V. The sampling photoconductor gaps were ion beam irradiated to reduce carrier lifetimes and thus produce short sampling times. Silicon samplers were damaged with 6 MeV Ne; GaAs samplers were damaged with 2 MeV deuterons; and InP samplers were damaged with 2 MeV alpha particles.

The Si-on-sapphire, GaAs, and InP samples received ion beam damage doses varying from 10^{12} to 10^{15} cm^{-2}. The characteristic response times of the damaged sampling photoconductors for these measurements were determined by the rising edge of the cross correlations. Figure 1 shows a representative series of cross correlations taken on Si-on-sapphire wafers with various ion-beam-damage doses. For each of the three materials studied, we observed a limiting value with increasing dose for the characteristic sampling time. The limiting values were $\sim$5 ps for Si-on-sapphire, $\sim$5 ps for GaAs, and $\sim$7 ps for InP. Shorter carrier lifetimes than these have been reported for both Si-on-sapphire, 3.6 ps [2], and InP, 1.2 ps [3]. Therefore, it is likely that the limiting time we observed in these experiments was due to an intrinsic circuit limit of the structures studied. Our calculations and measurements of parasitic capacitances of photoconductors in microstrip support this view.

Figure 2 shows two cross correlations taken on different InP wafers. The first wafer was 400 μm thick, while the second was 250 μm thick. The observed risetimes are approximately proportional to the wafer thicknesses. Because the RC time constants associated with photoconductor gap capacitance and transmission line impedance are also proportional to wafer thickness, it is likely that the observed risetimes are limited by circuit parasitics of the test structure and not by carrier lifetime in the sampling photoconductor.

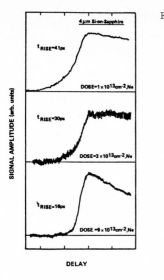

Fig.1. Cross-correlation series increasing damage

Fig.2. (a) 16 mil wafer thickness. (b) 10 mil waver

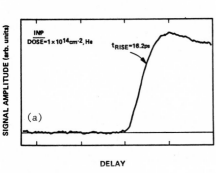

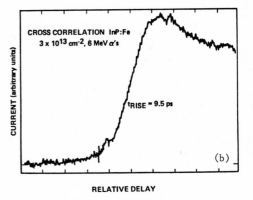

References

1. D.H. Auston, A. M. Johnson, P. R. Smith, and J. C. Bean: Appl. Phys. Lett. 37, 371 (1980).
2. D.H. Auston: IEEE J. Quantum Electron. QE-19, 639 (1983).
3. P.M. Downey and B. Schwartz: Appl. Phys. Lett. 44, 207 (1984).
4. G. Hasnain, G. Arjavalingam, A. Dienes, and J. Whinnery: Picosecond Optoelectronics Conference, San Diego, SPIE 439, 159 (1983).

Picosecond Optoelectronic Studies of Microstrip Dispersion

D.E. Cooper

The Aerospace Corporation, Chemistry and Physics Laboratory, Mail Stop: M2-253
P.O. Box 92957, Los Angeles, CA 90009, USA

1. Introduction

Picosecond optoelectronic techniques permit the impulse response of solid-state electronic devices to be measured with a temporal resolution of a few picoseconds [1]. The technique is based upon the use of optoelectronic switches, which convert picosecond optical pulses to picosecond electrical pulses and also act as picosecond-aperture sampling gates. The achievable temporal resolutions correspond to frequency bandwidths that far exceed the capabilities of conventional measurement techniques. A major application of the picosecond optoelectronic technique has been the measurement of the impulse response of fast transistors, yielding information about the transistor's frequency bandwidth and switching delay. Here we report on measurements of the propagation and distortion of short electrical pulses on microstrip lines, with an analysis of the results in terms of the microstrip dispersion and losses.

2. Experimental Procedure

Microstrip circuits incorporating optoelectronic switches were fabricated on silicon-on-sapphire wafers (Fig. 1). The substrate was 180 μm thick, with a 1.0 μm epilayer of silicon. The microstrip lines were formed with 150 nm of evaporated gold, and were 180 μm wide for approximately 50 Ω. impedance. The optoelectronic switches were formed by 25 μm gaps between the central microstrip and the microstrips on either side. The wafers were ion implanted with 400 KeV Si^+ to shorten the duration of the photoconductive response into the picosecond regime. A dose of 7 x 10^{14} cm^{-2} was adequate for optimal temporal response, with smaller doses producing greater photoconductivity with a reduction in temporal resolution.

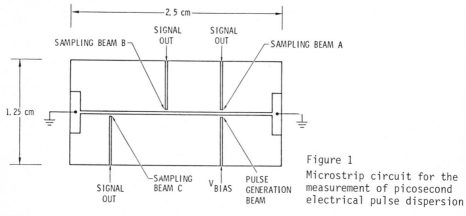

Figure 1
Microstrip circuit for the measurement of picosecond electrical pulse dispersion

A train of 3 picosecond optical pulses from a synch-pumped dye laser was focussed onto a biased optoelectronic switch to produce short electrical pulses on the central microstrip. As this voltage pulse propagated past one of the other optoelectronic switches it caused a transient bias, which could be detected by illumination of the switch with a second picosecond laser pulse. In this way the switches operated as sampling gates, with the aperture swept by delaying the second optical pulse. The current conducted by the sampling switch was amplified and used to drive the Y axis of an XY recorder. The X axis was driven by a voltage proportional to the position of the optical delay line, and the result was a record of the profile of the electrical pulse. Pulse shapes were measured at three positions labelled A, B, and C. Position A involves no propagation of the voltage pulse along the microstrip and it corresponds to an optoelectronic autocorrelation function. Position B was 0.65 cm from the switches at A, and position C involved 1.3 cm of pulse propagation.

3. Results and Discussion

Figure 2 (right) shows an optoelectronic autocorrelation function with superimposed points corresponding to a fitted Gaussian profile. To the right of the main peak is a small shoulder probably due to the electromagnetic wave radiated from the biased switch reflecting off the ground plane and modulating the bias on the sampling switch. Aside from this shoulder the autocorrelation profile is quite symmetric and fairly close to a Gaussian. The 7.0 psec width of the autocorrelation function implies that the electrical pulse width (and sampling aperture) is 5.0 psec. This temporal resolution corresponds to a frequency bandwidth of over 60 GHz.

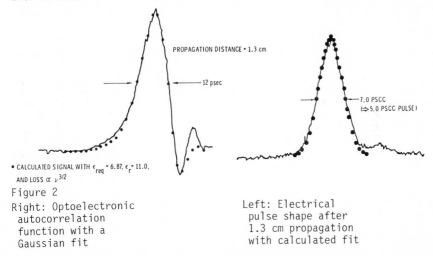

Figure 2

Right: Optoelectronic autocorrelation function with a Gaussian fit

Left: Electrical pulse shape after 1.3 cm propagation with calculated fit

Figure 2 (left) shows the electrical pulse profile after 1.3 cm propagation on the microstrip, with superimposed points calculated from a pulse dispersion model developed by WHINNERY [2]. The pulse distortion arises from the fact that low frequencies propagate on the microstrip in a quasi-TEM mode that experiences an effective substrate dielectric constant that is considerably less than the full dielectric constant ε_r that controls the propagation of very high frequencies. The crossover occurs when the wavelength on the microstrip is approximately equal to the width of the microstrip. Thus low frequency components of the electrical pulse travel relatively fast, and form the slowly rising leading edge of the pulse. The high frequency components become concentrated in the trailing edge of the pulse and cause the oscillation that is observed. These results represent the first observation of the negative-polarity region at the trailing edge of the pulse.

The pulse dispersion model generates pulse profiles based on the frequency dependence of the propagation constant (which is proportional to the square root of the effective dielectric constant). The frequency components of the initial pulse were multiplied by the propagation constant for that frequency, then a Fourier transform was performed to reconstitute the temporal pulse profile. Finally, the voltage pulse profile was convoluted with the sampling aperture to correspond to the experimental data. Frequency-dependent loss factors were added to represent ohmic, dielectric, or radiative loss. The shape of the initial pulse and the sampling aperture were approximated as a Gaussian, in accordance with the shape of the autocorrelation function (Fig. 2). The calculations were performed by numerical integration on a microcomputer.

The choice of ε_{req} and ε_r for the experimental system is complicated by the fact that sapphire is an anisotropic uniaxial crystal, and the dielectric constant varies from 9.4 to 11.6. (The effects of the silicon epilayer on the dielectric constant were neglected.) The unique axis with $\varepsilon = 11.6$ is oriented 60° from the wafer normal, and thus the dielectric constant perpendicular to the wafer $\varepsilon_{\perp} = 9.95$. The orientation of the microstrips in the plane of the wafer was unknown, so the in-plane dielectric constant ε was between 9.4 and 11.05. The in-plane component has some contribution to $\varepsilon_{\perp}$ [3], but in view of the unknown microstrip orientation we set $\varepsilon_r = \varepsilon_{\perp}$ and calculated $\varepsilon_{req} = 6.87$.[4] It proved impossible to obtain a good fit to the experimental data with these parameters, even with the inclusion of a loss factor. More dispersion was required, and a good fit was produced by raising ε_r to 11.0 and incorporating frequency-dependent loss. The increase in ε_r is justified since high frequencies propagate through non-TEM modes that involve in-plane electric fields. The 3/2-power loss factor was necessary to obtain proper pulse width and shape, although other frequency dependences also produced reasonably good fits. The 3/2 exponent would be expected from radiative losses. Other loss mechanisms, such as dielectric loss, may also be present, resulting in a complicated loss factor.

4. Summary

Picosecond optoelectronic techniques were used to study the propagation of short electrical pulses on microstrip. Optoelectronic autocorrelation functions were symmetric and indicated that the electrical pulses produced were 5.0 picoseconds long. Propagation over 1.3 cm of microstrip produced distorted pulses broadened to over 10 picoseconds. The observed pulse dispersion was greater than that expected from the out-of-plane dielectric constant of the substrate, and the non-TEM nature of the high-frequency propagation modes was used to justify a more dispersive model of pulse transport.

1. P.R. Smith, D.H. Auston, and W.M. Augustyniak: Appl. Phys. Lett. 39, 739 (1981)
2. G. Hasnain, G. Arjavalingam, A. Dienes, and J.R. Whinnery: "Dispersion of Picosecond Pulses on Microstrip Transmission Lines", in SPIE, Proceedings, Vol. 439, pp. 159-163
3. Roger P. Owens, James E. Aitken, and Terrence C. Edwards: IEEE Transactions MTT-24, 499 (1976)
4. K.C. Gupta, Ramesh Garg, and I.J. Bahl: Microstrip Lines and Slotlines (Artech, Dedham, Mass. 1979)

Measurement of the Soft X-Ray Temporal and Spectral Response of InP:Fe Photoconductors

D.R. Kania, R.J. Bartlett, and P. Walsh
Physics Division, Los Alamos National Laboratory, Los Alamos,NM 87545, USA
R.S. Wagner and R.B. Hammond
Electronics Division, Los Alamos National Laboratory
Los Alamos, NM 87545, USA
P. Pianetta
Stanford Synchrotron Radiation Laboratory, Meno Park, CA 94025, USA

X-ray pulses (300 ps FWHM) from the SPEAR storage ring at the Stanford Synchrotron Radiation Laboratory (SSRL) were used to study the energy-dependent pulse response of InP:Fe photoconductors in the soft x-ray region (0.8 to 3 keV). The detector sensitivity (2.7×10^{-3} A/W) has been measured to be independent of the photon energy in this range. The temporal response of the photoconductor to x-rays is similar to optical excitation (FWHM $\simeq$ 180 ps).

The detectors used in this study were fabricated from single-crystal, Fe-doped, semi-insulating InP. Metallic contacts were formed on opposite faces of the cube by multistep evaporation of Ge, Ni, and Au followed by a brief anneal at 450°C. The devices were cemented between the central conductor of two ultra high vacuum, high bandwidth, coaxial, feed thru connectors. The outer shell of the 50 Ω line could be split to permit mounting of the detector. A small hole was drilled in the outer shell to allow radiation to strike the detector. This system was UHV compatible, formed an impedance matched 50 Ω transmission line with a flat frequency response to 9 GHz. A pulsed semiconductor laser (830 nm, 90 ps) was used to measure the effective carrier mobility (μ = 1200 cm^2/V-s) and response time (FWHM $\simeq$ 180 ps) of the photoconductor.

Soft x-ray photons were produced by the SPEAR storage ring at SSRL and collected by a $1.5.^{\circ}$ gold coated fused silica collecting mirror. The high energy cutoff of the mirror is 3 keV. Three sets of thin foil transmission filters (beryllium, carbon, and magnesium) were used to modify the incident photon flux to yield energy resolution. The limited available flux precluded the use of a monochromator as an energy resolving element. The absolute photon intensity was measured by a photodiode with a gold photo-cathode.

The pulsed soft x-ray response was measured with a real time oscilloscope (400 MHz bandwidth) to record the photoconductor output with 29 different filter combinations modifying the incident photon flux. The photodiode measurements were used to normalize a calculation of the average incident flux. The standard deviation of this correction was 12.3%. The normalized incident flux was used to unfold the sensitivity of the photo-conductor.

We assume that a fixed amount of energy, γ, is required to create an electron-hole pair and that the electron mobility, μ, and electron-hole recombination time, τ, are unaffected by the incident photon energy. Therefore, the voltage output of a photoconductor is

$$V^i = \frac{e\mu\tau V_0 R\alpha\beta}{L^2 \gamma} \int dE\, \phi(E)\, T_i(E)\, E \tag{1}$$

where V_0 is the applied voltage, L is the detector length, R is the impedance of the oscilloscope, α is the correction for peak to average flux, β the bandwidth correction to the peak voltage, and $T_i(E)$ is the transmission of the ith filter. Using equation 1 the normalized averaged flux, $\phi(E)$, and the real time measurements of the photoconductor output for 29 different filter combination yield a γ of 8.6 eV $\pm$ 17%. This indicates a flat response. The absolute value of gamma is accurate within a factor of approximately two. A direct comparison of the x-ray sensitivity of InP:Fe to x-ray diodes (XRDs) with various photocathodes[1] is shown in Fig. 1. We note the strong effect of changing the size of the photoconductor on its sensitivity as expressed in equation 1.

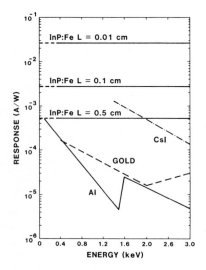

Fig. 1. A comparison of the response of InP:Fe photoconductors of different lengths, L, and three common photocathode materials used in x-ray diodes vs. photon energy.

In a separate experiment, a fast (11.5 GHz bandwidth) sampling oscilloscope and signal averager were used to measure the temporal response of the photoconductors to the 300 ps x-ray pulse. The sampling system was triggered by the radio frequency (1.28 MHz) characteristic of the storage ring. Approximately 260 ps of jitter existed between the trigger and the photon pulse which caused a broadening of the recorded pulses. Again, thin foil filters were used to modify the incident spectrum and a photon energy dependence was observed in the fall time. Filters that passed only high energy photons E $\geq$ 800 eV showed a fall time consistent with the time measured by optical excitation. Low energy transmitting filters and no filter measurements showed a fast fall with an addition long lived component. This component contained $\sim$ 10% of the peak voltage output at 1 ns after the peak. Since optical excitation of the detector showed no such tail, we conclude that very soft x-ray photons (E $\leq$ 800 eV) may cause this effect. We note that the very short penetration depth (0.1 μm) of this soft x-radiation may be a significant factor.

We have also applied these detectors to practical measurements on a z-pinch plasma device in two modes with great success. The photoconductor has proved to be superior to CsI photocathode x-ray diode for monitoring a

434

4 ns pulse of 3 - 4 keV x-ray emission from an argon plasma (as our sensitivity measurements would indicate). We have also employed the photo-conductor as a bolometer (unfiltered) monitoring the total power emitted from the plasma as a function of time assuming a constant γ in equation 1.

We have demonstrated that InP:Fe photoconductors may be used as fast (FWHM = 180 ps) and efficient soft x-ray detectors. The pulse-response is flat in the soft x-ray range with a sensitivity comparable to or better than that achievable with XRDs. We plan to study the soft x-ray temporal and spectral response in greater detail and to continue to apply photocon-ductors in practical measurements.

References

1. R. H. Day, Low Energy X-Ray Diagnostics - 1981, edited by D. T. Attwood and B. L. Henke, American Institute of Physics, New York, No. 75, p. 44, 1981.

Dynamic Response of Millimeter Waves in a Semiconductor Waveguide to Picosecond Illumination

A.M. Yurek, M.-G. Li, C.D. Striffler, and C.H. Lee

Electrical Engineering Department, University of Maryland
College Park, MD 20742, USA

1. Introduction

To construct an electronically controllable phased array system in the milli-meter-wave region, one needs switches and phase shifters which are operable at speeds of about one nanosecond with a time precision of several picoseconds [1]. Such a speed requirement is beyond the capability of current techniques. A class of devices in which the propagation parameters of millimeter-wave signals are controlled by optically induced electron-hole plasmas in semi-conductor waveguides appears to be one method of achieving such speed and precision [2-5]. A complete understanding of the physics of the interaction of millimeter-waves with an electron-hole plasma is essential for the design of such devices. An outstanding problem has been the measurement of the transient response of the millimeter-wave system after illumination by low-repetition-rate, variable energy pulses. In this paper, we report a dynamic bridge technique which has the capability to measure simultaneously the relative phase and loss response of the millimeter-wave system to illumination by frequency doubled single pulses from a mode-locked Nd:YAG laser with a potential time resolution in the picosecond range.

2. Experimental Techniques

The experiment consists of a millimeter-wave bridge with a 1.0 mm x 0.5 mm x 5 cm silicon waveguide placed in one arm and a mechanical phase shifter and a precision attenuator in the other. A schematic diagram of the experimental arrangement is shown in Fig. 1. Initially, without laser illumination, the bridge is balanced by adjusting the phase shifter and the attenuator. When a 30 ps pulse from a frequency doubled mode-locked Nd:YAG laser illuminates the broad wall of the silicon waveguide, an electron-hole plasma is generated at the surface, causing phase shift and attenuation of the millimeter-wave signal as it propagates through the plasma covered region of the waveguide. The bridge becomes unbalanced and a signal appears at the output. This signal depends strongly upon the laser intensity, which determines how much plasma is generated, and upon the laser wavelength, which determines the initial plasma depth. It also depends upon the carrier transport dynamics which determine the temporal variation of the signal. The signal persists until the excess carriers recombine. If two different initial phase offsets are

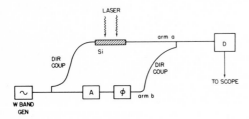

Fig. 1   Schematic diagram of the millimeter-wave bridge.

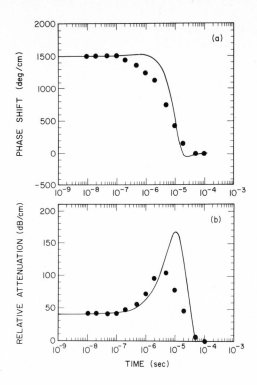

Fig. 2 Comparison of theoretical (solid line) and experimental (data points) results for the situation when the millimeter-wave bridge is illuminated by a single pulse from a mode-locked Nd:YAG laser. (a) is the phase shift with respect to time; (b) is the relative attenuation with respect to time.

chosen, and the response of the bridge is considered as the sum of two phasors, the response of the millimeter-wave system may be measured from the bandwidth limit of the measurement equipment (about 3 ns) until the end of the transient using only a few laser shots. By monitoring the energy of the laser pulses, eight shots with energies within a 3% variation were chosen and the data analyzed to give the data points of Fig. 2. Since the points were taken from photographs of the millimeter-wave response, the experimental data is actually a smooth curve connecting these points.

The solid curve is the result of a theoretical analysis using MARCATILI'S model [6] of the semiconductor waveguide as well as taking into account the diffusion and recombination properties of the optically induced plasma. The theory includes two adjustable parameters; the recombination time, and the initial plasma density. The recombination time was fitted at 20 μs, which agrees well with experimental estimates, and the initial plasma density was fitted at 8.5×10^{17} cm^{-3}, an order of magnitude different from the experimental estimate. One reason for this discrepancy may be that the theoretical model assumes an abrupt junction between the plasma layer and the rest of the semiconductor waveguide and a uniform plasma density within the plasma region. Other reasons may be that fast recombination processes such as Auger and surface recombination are not taken into account in the theory. The large peak in the theoretical attenuation is probably due to the fact that the model of the waveguide used [6] predicts a cutoff waveguide size for the lowest order semiconductor waveguide mode. A more realistic wave-guide model, such as that of GOELL [7], would be more appropriate in this case.

In conclusion, a dynamic bridge technique has been developed which can measure the response of a silicon waveguide to illumination by picosecond light pulses over a large time scale. This technique is useful in the design of millimeter-wave phase shifters, modulators, and switches because the change of phase and dynamic loss may be measured simultaneously. It is also useful in the characterization of electron-hole plasmas in semiconductors.

Acknowledgements

This work is supported in part by the Army Research Office, the National Science Foundation, and the University of Maryland Computer Science Facility.

References

1. H. Jacobs and M. M. Chrepta, IEEE Trans. MTT-22, pp. 411-417, (1974).
2. C. H. Lee, P. S. Mak, and A. P. DeFonzo, IEEE J. Quantum Electron. QE-16, pp. 277-288, (1980). ·
3. M. G. Li, W. L. Cao, V. K. Mathur, and C. H. Lee, Electron. Lett. 18, pp. 454-456, (1982).
4. C. H. Lee, A. M. Vaucher, M. G. Li, and C. D. Striffler, Presented at IEEE MTT-S International Microwave Symposium, paper C1, p. 103 (1983).
5. K. Ogusu, Electron. Lett. 19, pp. 253-254, (1983).
6. E. A. J. Marcatili, Bell Syst. Tech. J., vol. 48, pp. 2071-2102, (1969).
7. J. E. Goell, Bell Syst. Tech. J., vol 48, 2133-2160, (1969).

Part VIII

Photochemistry and Photophysics of Proteins, Chlorophyll, Visual Pigments, and Other Biological Systems

Time Resolution of Tryptophans in Myoglobin

D.K. Negus and R.M. Hochstrasser
Department of Chemistry, University of Pennsylvania
Philadelphia, PA 19104, USA

1. Introduction

Previous picosecond laser experiments aimed at detecting rapid processes in hemeproteins have involved studying changes in the heme electronic structure as a result of electronic excitation or photodeligation of CO [1-6], O_2 [1,4-6] and NO [6,7]. Recent femtosecond studies of these systems are reported elsewhere in this book [8]. A goal of the present research is to pave the way for the study of the conversion of such electronic and chemical changes into conformational changes throughout the protein. As a first step we have attempted in the present work to study the picosecond anatomy of the tryptophans in myoglobin.

Pulsed laser techniques are needed to study tryptophan fluorescence in hemeproteins as a result of very efficient energy transfer to the heme. The low fluorescence quantum yields for tryptophan in hemeproteins [9] suggest that the corresponding lifetimes are in the picosecond range. The motions of these tryptophans might also be very rapid, thus in order to obtain direct information on their behavior it was necessary to devise experiments by which the fluorescence and its polarization could be measured with picosecond accuracy.

Sperm whale myoglobin is known to contain the two tryptophan residues Trp 7 and Trp 14 [10]. Thus it is expected that the emission will be dominated by two decays. The heme is considerably further from Trp 7 than from Trp 14 (20 Å compared with 15 Å) so that the effects of energy transfer on these excited states are expected to be different. Tuna myoglobin has an X-ray structure very similar to sperm whale [11] but has only one tryptophan, Trp 14, and should be a simpler system with which to understand the fluorescence properties.

The elucidation of a full dynamical structure for a protein such as myoglobin requires measurements of the relative motion of its constituent parts. Molecular dynamics simulations suggest that side chain amino acids can undergo rather rapid motions in certain proteins [12, 13], but direct studies of actual systems were previously limited to time resolution in the range 0.5 - 1 ns [14, 15]. The time evolution of the fluorescence anistropy can yield important information about the molecular motion, but the analysis of tryptophan fluorescence requires consideration of the complexities of its photophysics [16-19]. Two electronic states having differently oriented transition dipoles are involved in the absorption process [20-23], and frequently there were observed two fluorescent states [16,18,24,25].

440

2. Fluorescence Decay and Anisotropy Measurements

The picosecond experiments were carried out using single pulses
from a passively modelocked $Nd:^{+3}$ glass laser in conjunction
with a streak camera which was described previously [25-28]. The
combination of the spatial resolution of the camera and associated optics allowed decays of 5 ps. to be discriminated. The
fluorescence was imaged through a quartz window onto the streak
camera photocathode (S-25 type). Distortion of the measured
decay due to velocity dispersion within the sample or collection
optics was determined to be negligible.

The output of the streak camera was coupled to an optical
multichannel analyzer. The data collection technique involved
the simultaneous recording on separate tracks of the time profile of the fluorescence decay and a second harmonic laser
pulse after passage through an etalon.

A calcite crystal and Wollaston prism placed in the fluorescence path enabled the acquisition of both the decay profile
of fluorescent light polarized parallel ($F_{\parallel}(t)$) and perpendicular ($F_{\perp}(t)$) to the excitation polarization for each laser
shot since these two perpendicularly polarized beams were
physically separated when focused onto the photocathode (see
Fig. 1). The experimental anisotropy r(t) was computed on
each shot as $r(t) = [F_{\parallel}(t)-F_{\perp}(t)]/[F_{\parallel}(t)+2F_{\perp}(t)]$. With a
magic angle polarizer in the fluorescence path a function pro-
portional to $[F_{\parallel}(t) + 2 F_{\perp}(t)]$ also was measured. The instrument sensitivity was determined to be independent of channel
number and wavelength over the region of interest (320 → 450 nm).
The excitation consisted of a single 1-10 μJ, 8 ps pulse at
265 nm. The ns gated spectra and lifetimes were studied using
a time correlated photon counting apparatus and 272-277 nm
excitation. The fluorescence was dispersed through a monochromator with 6 nm bandpass.

3. Results

The dominant contribution to the observed signals is from tryptophan. At 265 nm, the exciting light is absorbed by the heme
($\sim$ 60%), tyrosines ($\sim$ 8%), tryptophans ($\sim$ 30%) and phenylalanines
($\sim$ 2%) in sperm whale Mb. The hemes undergo rapid radiationless
deactivation to states that can emit significantly only in the
visible or infrared.

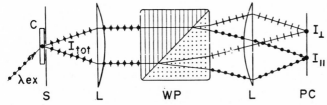

Fig. 1. Schematic representation of the double imaging via
double refraction principle which allows simultaneous acquisition of both fluorescence components. C = cell, S = slit.
λ_{ex} = excitation beam, L = lens, WP = Wollaston prism, PC =
photocathode, $I_{\parallel}$, $I_{\perp}$ = perpendicular and parallel intensity.

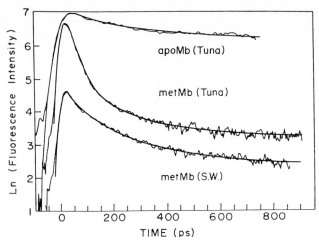

Fig. 2. Ultraviolet fluorescence intensity of apoMb (tuna),
metMb(tuna) and metMb(SW). The metMb data are a concatina-
tion of fastest streak speed data (for t < 200 ps) with slowest
streak speed data (for 200 < t < 900 ps). The fitting procedure
and parameters are described in detail in reference [29].

The fluorescence decays of Mb, MbCO, metMb and apoMb from
sperm whale, metMb and apoMb from tuna are all nonexponential.
Figure 2 shows some typical results. For the sperm whale samples
the data were consistent with two decays having about equal
weight, with one a few tens of ps and the other about 100 ps,
and a third decay representing ca. 5% of the emission having a
nanosecond life time [29]. For metMb (tuna) the fluorescence
decay was approximately analyzable into 90% of a 30 ps component, 8% of a 130 ps component and a couple percent of a few ns

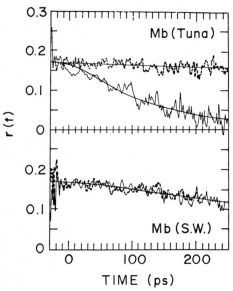

Fig. 3. Streak camera
measured early time fluo-
rescence anisotropies for
the apo(---) and met (——)
derivatives of Mb(tuna)
and Mb(SW). The smooth
curves drawn through the
experimental data are convolutions described in detail in reference [29].
Note the fast (110 ps) decay of the metMb(tuna) correlation exposed by energy
transfer processes.

component. The heme-free (apo) proteins show ns. fluorescence decays which appear to be double exponentials. The fluorescence anisotropy decays were ca. 1 ns or longer for all systems except metMb (tuna) for which the anisotropy decayed in 110 ps. The anisotropy averaged over the pulsewidth at time zero, $r_e(0)$, was 0.15 ± 0.02. Typical results are given in Fig. 3.

4. Discussion

The Trp's are clearly transferring their excitation to the heme on a timescale that is short compared with their normal fluorescence lifetime but comparable with some of their motions as seen in Fig. 3. The fluorescence decay characteristic is therefore expected to be nonexponential with the number of excited state molecules at time t being given by:

$$n(t) = \langle n(o) e^{-\int_o^t (k(t')+k_f) \, dt'} \rangle \qquad (1)$$

where $\langle \cdots \rangle$ indicates an average over the initial distribution of heme-Trp configurations. The energy transfer rate constant $k(t)$ will depend on time because the Trp is undergoing motion which can alter the magnitude of the dipole-dipole interaction. Clearly studies of the fluorescence decay provide a direct probe of protein dynamics in this case.

In a first analysis the distances and angles needed to evaluate the Förster energy transfer rate were obtained from the X-ray coordinates [30-31] assuming each Trp to be fixed at its location in crystalline sperm whale Mb. The radiative lifetime for tryptophan was estimated to be 20 ns from the fluorescence lifetime and fluorescence yield. The direction of the tryptophan emission dipole was used as reported by Yamamoto and Tanaka [22], and the heme was considered to be a circular oscillator [32]. The spectral overlap integral was calculated for MbCO and Mb to be 2.14 x 10^{14} quanta cm^6 $mole^{-1}$, and 2.60 x 10^{14} quanta cm^6 $mole^{-1}$ from fluorescence and absorption spectra corrected for overlapping components. Based on these values the reciprocal energy transfer rate constants for Trp 7 and Trp 14 respectively are: 125 ps and 31 ps for Mb; 121 ps and 41 ps for MbCO. These results describe qualitatively what is seen. For metMb (tuna) a single energy transfer limited decay of ca. 30-40 ps would be expected as a result of the absence of Trp 7. Again this is approximately what is observed.

This comparison of our results with predictions based on long-range dipolar energy theory provides a qualitative picture of the fluorescence properties of these hemeproteins. It is apparent that the subnanosecond fluorescence lifetimes are determined by energy transfer to the hemes from excited states of tryptophan associated with hemeproteins. With this interpretation, the lifetimes in the range 15 - 30 ps would then be associated with Trp 14, and the lifetimes of 106-136 ps, with Trp 7.

It was possible with the MbCO system to prove directly that Trp to heme energy transfer is occurring to the extent indicated by the fluorescence experiments. After 265 nm excitation the transient absorption in the Soret region could be measured using the picosecond continuum methods presented in earlier work [2, 4-7]. When 353 nm or 530 nm radiation excites HbCO the Hb-like spectrum appears instantaneously when 6 ps pulses are used. On the other hand, with 265 nm excitation 30% of the light is absorbed by tryptophan. We observe that 25-30% of the MbCO dissociates on a timescale which matches the fluorescence decay of the Trp's, thus proving that energy transfer is occurring.

The anisotropy decays because of the loss of the correlation between transition dipoles, introduced by the light pulse, such as would result from motion of the tryptophans. The motion could involve hindered rotation about the single bonds that hold the tryptophyl residue to the polypeptide backbone and/or motions of the backbone itself. On the other hand the energy transfer rate constant is a function of the angular relationship and distance between the tryptophan and the heme. We therefore have an interesting situation in which the transfer of excitation, being faster for certain configurations than for others, can cause the fluorescence signal at longer times to reflect a population of excited states with configurations that are unfavorable for energy transfer. Thus it is quite possible for the anisotropy to decay much faster in the presence than in the absence of energy transfer. This, we suggest, is the explanation of the anisotropy decay of 110 ps observed for Mb (tuna), whereas apo-tuna gave a 3.8 ns decay. In this case the rapidly decaying anisotropy is caused by only ca. 10% of the molecules, the other 90% being essentially motionless on this timescale but nevertheless more favorably positioned for energy transfer. A more detailed interpretation of these decay curves will require some knowledge of the potential function of the protein. Mataga and coworkers [33] have considered theoretically a model situation involving the processes considered here. The model potential implicit in the foregoing discussion would have to confine the Trp 7 in mainly one position corresponding approximately to that found in the crystal, for times less than ca. 1 ns. Trp 14 is essentially confined also at near crystal configuration but the equilibrium distribution should have ca. 10% of the molecules in positions having considerably more freedom of movement.

Tryptophans are useful optical probes of protein structure because of their relatively long wavelength absorption compared with polypeptides and other common amino acid residues. A first step in realizing the full potential of such probes is the development of techniques to distinguish the different tryptophans in a protein by means of their optical responses. Our picosecond laser experiments on myoglobin were therefore aimed at characterizing the fluorescent behavior of each of the tryptophans.

The technique introduced in this paper of exploiting energy transfer gating to highlight the contribution of certain spatial configurations to the fluorescence anisotropy decay may be useful in obtaining more detailed information of the equilibrium distributions of molecular transition dipoles and in testing numerical simulations of molecular motion. In addition, the present

444

work has concerned motions and properties of excited electronic states of tryptophans whereas in future studies it will be important to evaluate ground state properties by other techniques such as polarization spectroscopy.

It is apparent that femtosecond methods might yield extremely interesting information about protein dynamics by using methods such as demonstrated here using 6ps pulses. It should be possible to observe the evolution of the true $r(o)$ into $r_e(o)$ and learn about parts of the equilibrium distribution that could provide critical tests of theoretical calculations of protein dynamics.

5. Acknowledgement

We are indebted to Professor W. D. Brown and Mr. Mark Levy (U. C. Davis) for the generous gift of Tuna myoglobin; to Professor N. Kallenbach for help in purifying the samples; to Guido Rothenburger for his contribution to the experiments; and to W. A. Eaton for giving us access to important data on the crystal structures.

6. References

1. C. V. Shank, E. P. Ippen, R. Bersohn, Science 193 50-51 (1976).
2. B. I. Greene, R. M. Hochstrasser, R. B. Weisman, W. A. Eaton, Proc. Natl. Acad. Sci. USA 75, 5255-5259 (1978).
3. W. G. Eisert, E. O. Degenkolb, L. J. Noe, P. M. Rentzepis, Biophys. J. 25, 455-464 (1979).
4. D. A. Chernoff, R. M. Hochstrasser, A. W. Steele, Proc. Natl. Acad. Sci. USA 77, 5606-5610 (1980).
5. P. A. Cornelius, A. W. Steele, D. A. Chernoff, R. M. Hochstrasser, Proc. Natl. Acad. Sci. USA 78, 7526-7529 (1981).
6. P. A. Cornelius, R. M. Hochstrasser in Picosecond Phenomena III, eds. K. B. Eisenthal, R. M. Hochstrasser, W. Kaiser, A. Laubereau, (Springer-Verlag, New York 1982) pp. 288-293.
7. P. A. Cornelius, A. W. Steele, R. M. Hochstrasser, J. Mol. Biol. 163, 119-128 (1983).
8. J. L. Martin, A. Migus, C. Poyart, Y. Lecarpentier, A. Astier, A. Antonetti, in Ultrafast Phenomena IV, eds. D. H. Auston, K. B. Eisenthal (Springer-Verlag, New York 1984).
9. G. Weber, F. J. W. Teale, Disc. Farad. Soc. 27, 134-141 (1959).
10. A. B. Edmundson, Nature 205, 883-887 (1965).
11. F. E. Lattman, C. E. Nockolds, R. H. Kretsinger, W. E. Love, J. Mol. Biol. 60, 271-277 (1971).
12. M. Karplus, J. A. McCammon, Crit. Rev. Biochem. 9, 293-349 (1981).
13. T. Ichiye, M. Karplus, Biochem. 22, 2884-2893 (1983).
14. I. Munro, I. Pecht, L. Stryer, Proc. Nat. Acad. Sci. USA 76, 56-60 (1979).
15. J. R. Lakowitz, G. Weber, Biophys. J. 32, 591-601 (1980).
16. R. J. Robbins, G. R. Fleming, G. S. Beddard, G. W. Robinson P. J. Thislethwaite, G. J. Woolfe, J. Am. Chem. Soc. 102, 6271-6279 (1980).
17. E. P. Kirby, R. F. Steiner, J. Phys. Chem. 74, 4480-4490 (1970).
18. A. G. Szabo, D. M. Rayner, J. Am. Chem. Soc. 102, 554-563, (1980).

19. S. R. Meech, D. Phillips, A. G. Lee, Chem. Phys. 80, 317-328 (1983).
20. S. V. Konev, Fluorescence and Phosphorescence of Proteins and Nucleic Acids, ed. S. Undenfriend, (Plenum, New York 1967) p. 34.
21. L. J. Andrews, L. S. Forster, Biochem. 11, 1875-1879 (1972).
22. Y. Yamamoto, J. Tanaka, Bull. Chem. Soc. Jap. 45,1362-1366 (1972).
23. B. Valeur, G. Weber, Photochem. Photobiol. 25, 441-444 (1977).
24. A. Balcavage, Mol. Photochem. 7, 309-323 (1976).
25. G. R. Fleming, J. M. Morris, R. J. Robbins, G. J. Woolfe, P.J. Thistlethwaite, G. W. Robinson, Proc. Natl. Acad. Sci. USA 75, 4652-4656 (1978).
26. Y. Liang, D. K. Negus, R. M. Hochstrasser, M. Gunner, P. L. Dutton, Chem. Phys. Letts. 84, 236-240
27. F. E. Doany, B. I. Greene, Y. Liang, D. K. Negus, R. M. Hochstrasser, in Picosecond Phenomena II, eds. R. M. Hochstrasser, W. Kaiser, C. V. Shank (Springer-Verlag, New York 1980) pp.259-265.
28. G. Rothenberger, D. K. Negus, R. M. Hochstrasser, J. Chem. Phys. 79, 5360-5367 (1983).
29. R. M. Hochstrasser and D. K. Negus, Proc. Nat. Acad. Sci. USA (1984); in press.
30. J. C. Kendrew, R. E. Dickerson, B. E. Strandberg, R. G. Hart, D. R. Davies, D. C. Phillips, V. C. Shore, Nature 185, 422-427 (1960).
31. H. C. Watson, in Progress in Stereochemistry, eds. B. J. Aylett, M. M. Harris, (London Butterworths, 1969) 299-333.
32. W. A. Eaton, J. Hofrichter, in Methods in Enzymology, eds. E. Antonini, L. Rossi-Bernardi, E. Chiancone, (Academic Press, New York 1981) Vol. 76, pp. 175-261.
33. F. Tanaka, N. Mataga, Biophys. J. 39, 129-140 (1982).

Resolution of the Femtosecond Lifetime Species Involved in the Photodissociation Process of Hemeproteins and Protoheme

J.L. Martin[#] , A. Migus, C. Poyart[+], Y. Lecarpentier, A. Astier and A. Antonetti.

Laboratoire d'Optique Appliquée. Ecole Polytechnique - ENSTA
Batterie de l'Yvette. 91128 Palaiseau Cedex. France.

Femtosecond absorption spectroscopy of photodissociated hemoproteins and protoheme indicates that a transient species with a 350 femtosecond lifetime is involved in the photodissociation process whatever the ligand is. The spectral properties of the deoxy-like species are analyzed.

1. Introduction

A major change is occurring in our view of globular proteins. The classic picture of a static protein structure is being supplanted by a dynamic view. Until recent years, the explanations of enzyme functions have been based on the analysis of the average structure obtained from X-ray crystallography. The high specificity of the enzyme-substrate reaction was interpreted in terms of complementarity of their static conformations. Ligand induced conformational changes, known from X-ray data, led to the concept of allosteric transitions (e.g. R and T states in hemoglobin)[1] where the protein is treated as a bistable system.

From experiments[2, 3, 4] and theory[5] several clues suggest that dynamic properties are crucial to the biological function of proteins. The dynamics of internal motion in proteins range over a large spectrum from hundreds of femtoseconds for the rattling motion of groups in their cages, to milliseconds for the bending of whole domains about molecular hinges. Most of these estimations come from molecular dynamics simulation where the equations of motion are solved for all the atoms. There are no actual experimental observations of the motion of a given part of a protein. However, several experimental data are, at different degrees, correlated to atomic displacements. The high correlation between rms fluctuation of atoms from molecular dynamics simulation and the atomic temperature factors from X-ray diffraction,[6] is an example of dynamic infomation obtained from a static technique. Time resolved data come from NMR relaxation, fluorescence depolarization and absorption spectroscopy experiments. The general trend of these techniques is that, the higher the degree of correlation between data and structure, the lower the time resolution. Nevertheless, spectroscopic studies which provide the best time resolution could give information, in favorable cases, on local structural rearrangement.[7] A favorable case is when a part of the protein has a spectrum sensitive to its ligation state and/or to its conformational structure. Such a behavior is found in hemeproteins and explains in part the extensive study of this model in time resolved spectroscopy. Hemeproteins, like myoglobin or hemoglobin, represent a case where side chain fluctuations are vital in the ligand binding process. The way in which oxygen or carbon monoxide molecules migrate through the protein suggests transient packing defects generated by thermal motion of the atoms, since the heme region is closely packed[8].

By extending the absorption spectroscopy of hemeproteins into the femtosecond time scale where few events are able to occur, we are trying to provide data on the rates of local motions inside the heme pocket, following ligand detachment.

+I.N.S.E.R.M. U.27 - 92150 Suresnes. France.

Present address: UCSD . Department of Chemistry B-014, La Jolla, California 92093

2. Picosecond absorption spectroscopy

Picosecond absorption spectroscopy has been extensively used in recent years to probe both the transient species and the ligand rebinding process after photoexcitation of the liganded heme proteins. From most of these experiments, it appears that the photodissociation process occurs in less than 6 ps[9, 10, 11] or even in a subpicosecond time scale.[12] This conclusion is supported by the appearance of a deoxy-like difference spectrum with kinetics limited by the pulse duration (6 - 8 ps). As it has been pointed out by others[13] , most of the transient spectra reported differ significantly from the equilibrium spectra. Broadened, red shifted Δ A spectra seem to be the most common characteristic of these picosecond studies. Multiphoton excited species[9] and possible systematic errors in the picosecond spectroscopic methods [13] could explain the poor agreement among the various published studies concerning the precise shape and the time evolution of the transient spectra.[14, 13]

In an attempt to clarify the dynamics of the heme structural rearrangement after ligand detachment, we have recently extended the range of the transient absorption spectroscopy into the femtosecond time scale.[15, 16, 17] We have established that the deliganded species appears with a 300 fs time constant for both oxygen (O_2) and carbonmonoxide (CO) as ligand.

We now report recent results in photodissociated CO, O_2 and NO liganded hemoglobin and myoglobin using an improved spectrophotometric technique. This experimental apparatus has made it possible, for the first time, to resolve the 300 fs lifetime of the intermediate species involved in the photodissociation process. The precise shape and time evolution of the transient spectra have revealed the influence of the protein on the transient heme conformation.

3. Experimental

Femtosecond kinetics and Δ A spectra were recorded after photoexcitation of liganded complexes by pulses of 100 fs duration. Generation and amplification of these pulses have been described elsewhere.[18, 19, 17] The 1 mJ output pulses of 100 fs duration at 620nm are split in two parts. Each of these beams generates a broadband stable continuum . One continuum beam is used as the spectroscopic source in a two-beam arrangment providing a reference source and a probe source. From the other femtosecond continuum, a selected spectral part is further amplified to GW peak power and used as the photodissociation beam. The transverse energy profile of the pump and probe beam are adjusted to be identical. This system is tunable all over the visible spectrum. In the present experiments, the excitation wavelength is centered at 580 nm with a 6 nm bandwidth, that is above the lowest presumed dissociation channel in liganded hemoproteins.[20]

Both the kinetics and the spectra were recorded with a detection system using a two photodiode arrangement at the output slit of a spectrometer. Kinetics are recorded at a given wavelength with 20 fs and 100 fs delay increments between the pump and the probe beams for the first 2 and 10 ps respectively. The Soret absorption band is scanned between 390 nm and 500 nm by increments of 1 nm near the isosbestic point and in the region of the maximum positive and negative ΔA . The value of the increment is increase to 5 nm in the flatter region (440 nm - 500 nm). At a given delay and wavelength, about 20 laser pulses of discriminated energy are integrated. For each pulse the probe pulse is normalized to the reference pulse. This represents 2×10^5 registered laser pulses for each sample scan. The group velocity dispersion in the continuum beam is automatically compensated using a shift function which has been measured in a separate experiment. At each wavelength the delay line is adjusted to its zero delay. With this procedure, we obtain a 50 fs time resolution associated with a 0.5 nm wavelength accuracy and a precision of $\pm$ 2. $\times 10^{-3}$ O.D in the amplitude of the ΔA spectra.

Purified Hb solutions were prepared from fresh human blood [16] and the protoheme was prepared from hemin (Fe^{3+}. Protoporph. Cl^-).[16, 17] The concentration is adjusted to obtain absorbances of 2.2 - 2.4 in the Soret band with a cuvette of 0.05 cm light path. The cuvette is moved so that each pulse at 10 Hz excited a new region of the sample.

4. Photodissociation process

We have examined the changes in absorption of the Soret band after photoexcitation of CO, O_2 and NO liganded myoglobin and hemoglobin solutions. In all the cases the bleaching of the Soret liganded band is instantaneous (i.e. limited by our 100 fs pulse duration) while the induced absorption in the liganded region appears with a time constant of 350 fs (fig. 1a,b). This time constant has been attributed to the appearance of a deoxy like product (denoted $Hb^\dagger$).

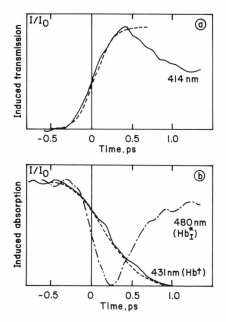

Fig. 1.
Femtosecond kinetics of the normalized absorption changes following photoexcitation of HbO_2 solution. **Upper** : *(fig. 1a) solid line : induced transmission at 414 nm. The rise time is limited by the pulse duration as indicated by the dotted line which represents the integral of the cross correlation as determined in a separate experiment (pulse duration = 100 fs).* **Lower** : *(fig. 1b): induced absorption at the maximum positive ΔA spectrum (431 nm)(-----). The best fit gives a time constant of 350 fs (-----). The induced absorption at 480 nm (- - -) reveals the presence of a short-lived state (denoted Hb_I^* in the text). This species appears in less than 50 fs and has a lifetime of about 350 fs .*

In contrast the kinetics at 480 nm revealed an instantaneous induced absorption which relaxes with a time constant of $\sim$ 350 fs. (fig. 1b). Transient ΔA spectra in the femtosecond time scale indicate the presence of a broadband extending from 450 nm to 500 nm. The amplitude of this absorption band is larger when the ligand is O_2 or NO but still present with CO (fig. 2).

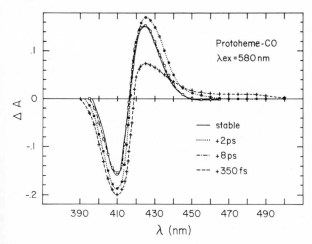

Fig. 2. *Transient difference spectra of protoheme-CO in the Soret region. Time delay in femtosecond (fs) or picosecond (ps) is indicated on each curve. The equilibrium ΔA spectrum is also drawn and closely matches the +8 ps transient spectrum.*

449

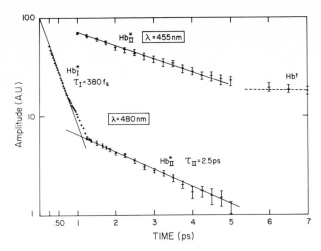

Fig. 3. *Time evolution of the induced absorption at 480 nm and 450 nm after photoexcitation of HbO₂ at 580 nm by pulses of 100 fs duration. The two-component decay is associated with the presence of two short lived excited states. The shortest one (Hb$_I^*$) is an intermediate step in the process of dissociation. The Hb$_{II}^*$ state is a competitive channel which relaxes to the ground liganded state .*

We have obtained evidence that two forms of Hb^* coexist in this spectral region. One of them (Hb_I^*) absorbs in the 470-500 nm region and is characterized by a 350 fs lifetime. The other one (Hb_{II}^*) has a lifetime of ∼ 2.5ps and absorbs in the 445 - 475 nm region (fig. 3). The first one is a good candidate for the intermediate species between the liganded product (HbO_2 for example) and $Hb^†$. The Hb_{II}^* state is much larger when the ligand is O_2 or NO , as compared to CO . The Hb_{II}^* state appears to be a competitive channel to the dissociative process as it does not produce species in the $Hb^†$ state. Along this line, the Hb_{II}^* state is in part responsible for the lower quantum yield of dissociation for O_2 and NO .

5. The deoxy-like species ($Hb^†$ or $Mb^†$)

The $Hb^†$ is characterized by a Δ A spectrum whose main difference as compared to the equilibrium spectrum is the presence of a red shift. The magnitude of this red shift, measured at the isosbestic point, is in the 1 to 3 nm range, depending upon the ligand (table 1). The remarkable result is that no red shift of the Δ A spectrum is detectable after dissociation of CO liganded protoheme ,as indicated by the +8 ps Δ A spectrum identical to the equilibrium one (fig. 2).

An increase in the Fe-N_ϵ His(F8) bonding interaction either by Fe spin transition or steric factors is known to lead to significant red shift in the Soret band.[21, 22] The $Hb^†$ species could then be attributed to a high spin state where the porphyrin ring is already in its domed conformation. This assignment is also supported by the observation of a very weak shift when the ligand is NO , which can be related to a weaker Fe-N_ϵ His(F8) bond due to the presence of the non paired electron of NO in the antibonding d_{z^2} orbital.

Table 1. *Wavelengths of the isosbestic point at equilibrium and 8 ps after photoexcitation ($Hb^†-HbX$) for various ligands (X).*

	HbCO	HbO₂	HbNO	MbNO	Protoheme-CO
λ_{isos} stable (nm)	425.5	422.0	423.5	426.0	416.0
λ_{isos} +8 ps (nm)	428.5	425.5	424.5	428.5	416.0
Δλ (nm)	+ 3.0	+ 3.0	+ 1.0	+ 2.5	0.0

Acknowledgement:

This research was supported by grants from INSERM and ENSTA .

References

1. J. Monod, J. Wyman, and J.P. Changeux, *J. Mol. Biol.* **12**, 88-118 (1965).
2. H. Frauenfelder, G.A. Petsko, and D. Tsernoglou, *Nature (London)* **280**, 558 (1979).
3. A. Kasprzak and G. Weber, *Biochemistry* **21**, 5924 (1982).
4. J.R. Lakowicz, B.P. Maliwal, H. Cherek, and A. Balter, *Biochemistry* **22**, 1741 (1983).
5. M. Karplus and J.A. McCammon, *Ann. Rev. Biochem.* **53**, 263 (1983).
6. S.H. Northrup, M.R. Pear, J.D. Morgan, J.A. McCammon, and M. Karplus, *J. Mol. Biol.* **153**, 1087 (1981).
7. F.R.N. Gurd and T.M. Rothgeb, *Adv. Protein Chem.* **33**, 73 (1979).
8. S.E.V. Phillips, *J. Mol. Biol.* **142**, 531 (1980).
9. B.I. Green, R.M. Hochstrasser, R.B. Weisman, and W.A. Eaton, *Proc. Natl. Acad. Sci. USA* **75**, 573-577 (1978).
10. P.A. Cornelius, A.W. Steele, D.A. Chernoff, and R.M. Hochstrasser, *Proc. Natl. Acad. Sci. USA* **78**, 7526-7529 (1981).
11. P.A. Cornelius, R.M. Hochstrasser, and A.W. Steele, *J. Mol. Biol.* **163**, 119-128 (1983).
12. C.V. Shank, E.P. Ippen, and R. Bershon, *Science* **193**, 90 (1976).
13. P.A. Cornelius and R.M. Hochstrasser, in *Picosecond Phenomena III*, edited by K.B. Eisenthal, R.M. Hochstrasser, W. Kaiser and A. Laubereau (Springer-Verlag, Berlin, 1982) p. 288-293.
14. A.H. Reynolds and P.M. Rentzepis, *Biophys. J.* **38**, 15-18 (1982).
15. J.L. Martin, A. Migus, C. Poyart, Y. Lecarpentier, A. Antonetti, and A. Orszag, *Biochem. Biophys. Res. Comm.* **107** (3), 803-809 (1982).
16. J.L. Martin, A. Migus, C. Poyart, Y. Lecarpentier, R. Astier, and A. Antonetti, *Proc. Natl. Acad. Sci. USA* **80**, 173-177 (1983).
17. J.L. Martin, A. Migus, C. Poyart, Y. Lecarpentier, R. Astier, and A. Antonetti, *The EMBO Journal* **2** (10), 1815-1819 (1983).
18. A. Migus, J.L. Martin, R. Astier, A. Antonetti, and A. Orszag, in *Picosecond Phenomena III*, edited by K.B. Eisenthal, R.M. Hochstrasser, W. Kaiser and A. Laubereau (Springer-Verlag, Berlin, 1982) p. 6-9.
19. J. L. Martin, C. Poyart, A. Migus, Y. Lecarpentier, R. Astier, and J. P. Chambaret, in *Picosecond Phenomena III*, edited by K. B. Eisenthal, R. M. Hochstrasser, W. Kaiser and A. Laubereau (Springer-Verlag, Berlin, 1982) p. 294.
20. A. Waleh and G.H. Loew, *J. Am. Chem. Soc.* **104**, 2346-2356 (1982).
21. M.F. Perutz, *Annu. Rev. Biochem.* **48**, 327-386 (1979).
22. W.R. Scheidt and C.A. Reed, *J. Am. Chem. Soc.* **104**, 2352-2356 (1981).

Picosecond Vibrational Dynamics of Peptides and Proteins

T.J. Kosic, E.L. Chronister, R.E. Cline, Jr., J.R. Hill, and D.D. Dlott

School of Chemical Sciences, University of Illinois at Urbana-Champaign
505 S. Mathews Avenue, Urbana, IL 61801, USA

Fast structural fluctuations are known to be essential to the function of proteins. [1] For example,the motion of small ligands such as CO or O_2 through myoglobin is greatly facilitated by structural fluctuations [1b]. Temperature dependent x-ray studies of proteins [2] show that they consist of fluid-like regions and crystal-like regions. The latter are rigid hydrogen bonded structures such as α-helices or β-pleated sheets [2]. The atomic packing, bond lengths and bond angles in these rigid structures are very similar to amino acid and peptide crystals [3]. Therefore, it should be possible to gain insight about the behavior of these rigid sections of proteins by ps Raman studies of such crystals [4].

We have performed detailed ps time delayed coherent anti-Stokes Raman scattering (ps CARS) experiments on a variety of amino acid and peptide crystals [4,5]. This technique measures the time dependence of vibrational dephasing of modes coherently excited by two simultaneous pulses from a dual dye laser system. In low temperature crystals the dephasing is due to vibrational relaxation or defect scattering [5,6]. The former is recognizable by exponential decay while the latter results in non-exponential ps CARS decays [6b]. In the great majority of cases, exponential decays are observed which give directly T_1, the vibrational lifetime [5]. In histidine-$HCl \cdot H_2O$ we observe two modes which give decays with Gaussian components. These are probably localized histidine-chloride stretches [5]. In a case such as this we can set a lower limit to T_1.

Figure 1 shows a "ps CARS spectrum" of a single ℓ-alanine crystal at 10K [4]. The vertical lines represent intense CARS resonances, and each is labelled by its lifetime T_1. The lowest frequency modes ($\Omega = 40$ and 50 cm^{-1}) are long-lived, and T_1 is observed to decrease rapidly with frequency. All of the Raman active modes between 130 and

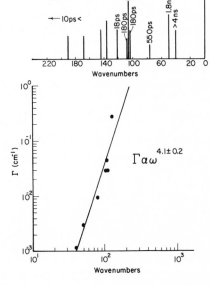

Figure 1. A picosecond CARS spectrum of single crystal ℓ-alanine at 10K. The long-lived modes are librons. The lifetimes decrease as ω^4. The modes between 130 and 1600 cm^{-1} decay faster than 10 ps.

1600 cm^{-1} were investigated by this technique, and all decayed faster than instrumental resolution, 10 ps [4]. The long-lived modes in Fig. 1 are best described as torsional oscillations, or librons (L) [7]. At low temperature, librons will decay to two lower frequency phonons. There are three possible pathways-decay to two acoustic (A) phonons, one acoustic phonon and one libron, or two librons [5]. Using the density of states for isotropic solids and the theory of anharmonic interactions [8] we obtain for the three processes at low temperature [5],

$$(T_1)^{-1}_{L \to 2A} = \frac{36}{\hbar^2} \left| B^{(3)} \right|^2_1 \left[\frac{3\sqrt{6}(\Omega/2)^2}{\omega_D^3} \right]^2 \propto \Omega^4 \tag{1}$$

$$(T_1)^{-1}_{L \to L'+A} = \frac{36}{\hbar^2} \left| B^{(3)} \right|^2_2 (\frac{27\Omega^2}{\omega_D^3})\rho_L \propto \Omega^2 \tag{2}$$

$$(T_1)^{-1}_{L \to L'+L''} = \frac{36}{\hbar^2} \left| B^{(3)} \right|^2_3 \rho_L^2 \propto \Omega^0, \tag{3}$$

where $\left| B^{(3)} \right|$ is a cubic anharmonic matrix element, ω_D is the Debye frequency, and ρ_L is the average libron density of states.

Figure. 1 shows that for ℓ-alanine $T_1 \propto \Omega^{-4}$, so the emission of two acoustic phonons is dominant. This process is illustrated in Fig. 2. The crystal is a rigid hydrogen bonded network similar to a parallel β-sheet [4]. The modes in Fig. 1 at ca. 105 cm^{-1} are librations along the axis of heavy arrows in Fig. 2 [7]. The emitted acoustic phonons, represented by light arrows, are translatory motions of the unit cells. We have determined the value of the B-coefficient for this process and set upper limits on the coefficients for the other two processes, and we find for ℓ-alanine $|B^{(3)}|_1 = 3$ cm^{-1}, the matrix element in Eq. 2 is less than 20% of this value, and the matrix element in Eq. 3 is less than 5% of this value [5]. The coupling of librons to acoustic phonons illustrated in Fig. 2 is the most efficient decay process. The other two processes involve emission of at least one libron which oscillates along an axis perpendicular to the decaying mode. The orthogonal librons couple poorly, which

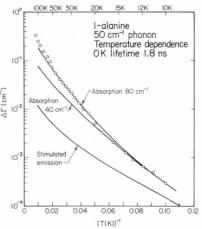

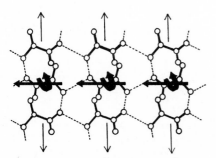

Figure 2. The alanine crystal consists of interlocking hydrogen bonded chains of amino acids. The ca. 105 cm^{-1} librons, represented by heavy arrows, decay into two acoustic phonons which are translatory modes of the chains (light arrows).

Figure 3. The temperature dependence of one ℓ-alanine libron shows the lifetime decreases rapidly at high temperature. At 100K, T_1 is 10^{-2} of the 10K value, 1.8 ns. The mechanism is probably scattering by thermally excited acoustic phonons.

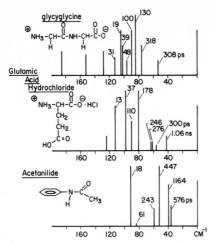

Figure 4. Picosecond CARS spectra of glycylglycine, glutamic acid, and acetanilide, which forms hydrogen bonded helices. Like ℓ-alanine, the lifetimes decrease as Ω^4 [5].

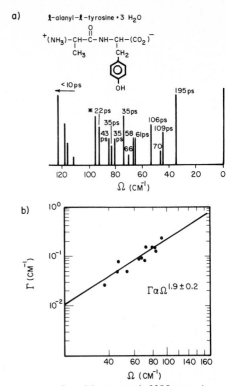

Figure 5. Picosecond CARS spectrum of the complex peptide ℓ-alanyl-ℓ-tyrosine·$3H_2O$. The lifetimes decrease as Ω^2 which is predicted by Eq. 2 for crystals with very large, complex unit cells.

accounts for the decreased efficiency of these processes [5]. As shown in Figure 3, the temperature dependence of the lifetime is quite steep. By 100K, T_1 is less than 20 ps. The decreased lifetime is most likely caused by absorption of thermally populated acoustic phonons [5].

Figure 4 shows 10K ps CARS spectra of a simple peptide, glycylglycine, glutamic acid·HCl, and acetanilide, which is not an amino acid but which crystallizes in a hydrogen bonded helix. For all the substances $T_1 \propto \Omega^{-4}$, just as in ℓ·alanine [5].

The much more complex peptide, ℓ-alanyl-ℓ-tyrosine trihydrate, shows a different behavior than the simpler crystals. As seen in Figure 5, the spectrum is quite dense. In this material we find $T_1 \propto \Omega^{-2}$, which is characteristic of decay by emission of one acoustic phonon and one libron, as seen in Eq. 2 [5]. Despite the smaller B coefficient for this process, it can dominate emission of two acoustic phonons when ρ_L is large and the acoustic density of states is small. Those conditions occur in materials with very large complex unit cells [5]. A protein molecule would correspond to a very large unit cell, and we would expect the processes analogous to those in Eqs. 2 and 3 to dominate. Thus we do not predict the dramatic dependence of T_1 on Ω observed in amino acid crystals. Since proteins have lower symmetry than crystals, we would expect torsional oscillations to have more available relaxation channels, resulting in faster decay rates than in crystals.

454

Using ps CARS we have determined the efficiency of possible decay pathways of librons in amino acids and peptides. X-ray data on proteins at ambient temperature show that the atomic mean square displacements are dominated by torsional oscillations which are similar to librons [2,3]. These oscillations may be functionally important. In myoglobin, the distal histidine (E7) is the gateway to the ligand binding site, and it is attached to a rigid α-helix [1b,2]. Small displacements of the histidine can greatly decrease the energy of the barrier to ligand binding [1b]. Below 200K, flash photolysis experiments on myoglobin-CO rebinding kinetics show there is a distribution of barrier heights [9]. The protein molecules are frozen in a number of conformational substates which have different barrier heights [9]. This distribution may correspond to various positions of the α-helices [10]. At 298K, the barrier is still present but is no longer distributed [11]. This could be caused by rapid motion of the α-helices which narrows the distribution of barrier heights [10,11].

This research was supported by the National Science Foundation. DDD acknowledges support from an Alfred P. Sloan fellowship, and thanks Professors P.G. Wolynes and Hans Frauenfelder for many helpful conversations.

References

1. a) J. A. McCammon and M. Karplus: Ann. Rev. Phys. Chem. 29(1980); b) M. Karplus and J. A. McCammon: Ann. Rev. Biochem. 263 (1983).
2. H. Hartmann, F. Parak, W. Steigemann, G. A. Petsko, D. Ringe Ponzi, and H. Frauenfelder: Proc. Nat. Acad. Sci. USA, 79, 4967(1982); G. A. Petsko and D. Ringe: Ann. Rev. Biophys. (1984).
3. P. G. Debrunner and H. Frauenfelder: Ann. Rev. Phys. Chem. 243(1982).
4. T. J. Kosic, R. E. Cline, Jr., and D. D. Dlott: Chem. Phys. Lett. 103, 109(1983).
5. T. J. Kosic, R. E. Cline, Jr., and D. D. Dlott: Submitted to J. Chem. Phys.
6. a) P. L. Decola, R. M. Hochstrasser and H. P. Trommsdorff: Chem. Phys. Lett. 72, 1(1980); b) I. Abram and R. M. Hochstrasser: J. Chem. Phys. 72, 3617 (1980).
7. E. Loh: J. Chem. Phys. 63, 3192(1975).
8. S. Califano, V. Schettino and N. Neto: Lattice Dynamics of Molecular Crystals (Springer, Berlin, 1981).
9. R. H. Austin, K. W. Beeson, L. Eisenstein, H. Frauenfelder, and I. C. Gunsalus: Biochemistry 14, 5355(1975).
10. H. Frauenfelder in Structure and Dynamics of Nucleic Acids, Proteins and Membranes (New York: Adenine Press, 1984).
11. D. D. Dlott, H. Frauenfelder, P. Langer, H. Roder and E. E. DiIorio: Proc. Nat. Acad. Sci. USA 80, 6239(1983).

New Investigations of the Primary Processes of Bacteriorhodopsin and of Halorhodopsin

H.-J. Polland, W. Zinth, and W. Kaiser

Physik Department der Technischen Universität München
D-8000 München, Fed. Rep. of Germany

Halobacterium halobium contains several retinal protein units which show light induced ion pumping. Best known is the light driven proton pump bacteriorhodopsin (bR) /1/ which has been extensively studied during the past decade. There are several picosecond investigations on bR, unfortunately with contradictory results. A detailed study of the primary photochemical processes appears to be necessary. Another ion pumping retinal protein, halorhodopsin (hR) /2/, has been prepared only recently. Up to now, there is no investigation of its primary photochemical processes.

In this letter we present new extensive data of the first processes in bacteriorhodopsin and also the first results on the picosecond photochemical events of halorhodopsin. Our measurements were made with an advanced Nd:glass laser system: (i) The time resolution was better than 1 ps; (ii) excitation and probing pulses were available over a wide frequency range, and (iii) a highly sensitive detection system provided accurate data with very low excitation of the specimen. The excitation was kept much smaller than one photon per molecule and pulse.

A series of different measurements was made: Fluorescence spectra, quantum efficiencies, and lifetimes were studied in the visible and near infrared. Transient absorption changes and optical amplification were investigated over a large wavelength range.

1. Bacteriorhodopsin

Data from time resolved measurements of light induced absorption changes are shown in Fig. 1. Excitation wavelength was 540 nm; the sample was kept at room temperature (25°C). Different curves are found depending on the wavelength of the probing pulse. At 490 nm a rapid absorption increase (excited-state absorption) is followed by a rapid absorption decrease remaining constant for $t_D > 10$ ps. At 555 nm a rapid and strong absorption decrease is followed by a relaxation process ($\tau = 5$ ps) ending in a weaker absorption decrease. At 675 nm the absorption rises rapidly to a peak value and decays afterwards to a smaller absorption increase (with a time constant of 5 ps).

Combining these results with a large number of absorption and fluorescence measurements on bR, deuterated bR, and on bR containing a modified, i.e. a sterically fixed, retinal chromophore, we obtain the following picture for the primary photochemical processes in bR:

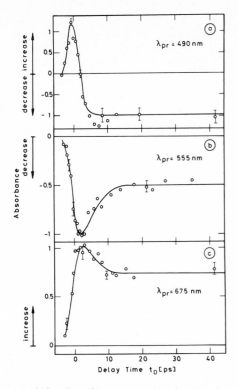

Fig. 1
Transient absorbance changes
of bacteriorhodopsin with
excitation wavelength $\lambda = 540$ nm.
Probing wavelengths:
(a) 490 nm, (b) 555 nm,
(c) 675 nm.

(i) The first excited singlet state has a lifetime of 0.5 to
1 ps. (ii) A first intermediate K' appears rapidly decaying
with a time constant of 5 ± 1 ps to the second intermediate K.
(iii) The absorption of K' and K peaks at 615 nm and 600 nm, respectively. (iv) The intermediate K is unchanged for at least
1 ns. (v) Replacing the hydrogen at the Schiff base by deuterium
gives exactly the same experimental effects on the picosecond
time scale as in bR.

Summary: The data support that the formation of K' and K is
accompanied by isomerization of the retinal from the all-trans
to the 13-cis configuration.

2. Halorhodopsin

Halorhodopsin is an interesting protein unit pumping chlorine
ions through the bacterial membrane. Until now, the photoreaction
cycle in halorhodopsin has been studied on the time scale of
microseconds /3/. We present the first investigations on halorhodopsin on a picosecond time scale.

Purified halorhodopsin /4/ (λ_{max} = 578 nm) in high salt concentration (1 M) was used. Data on transient absorbance changes
following a picosecond excitation are shown in Fig. 2. At early
times a bleaching of the ground state of hR causes the absorption
decrease. A long-lived red-shifted intermediate state generates
the later absorption changes. The absorption data and additional
fluorescence measurements give the following picture.

457

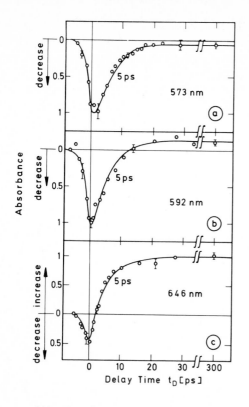

Fig. 2
Transient absorbance change
of halorhodopsin with excitation wavelength λ = 540 nm.
Probing wavelengths:
(a) 573 nm, (b) 592 nm,
(c) 646 nm.
The rapid absorbance decrease
is due to the population of
the excited singlet state.
A red-shifted photoproduct is
formed with a time constant
of 5 ps.

(i) The first excited singlet state of retinal in halorhodopsin
has a lifetime of 5 ps. (ii) A new photoproduct is generated
within the same time constant of 5 ps. (iii) The photoproduct is
red-shifted and lives for at least 1 ns. (iv) The absorption of
the first intermediate peaks at 600 nm.

Summary: In spite of differences in the kinetics there appear
to be important similarities between the K intermediate of
bacteriorhodopsin and the first photoproduct of halorhodopsin.

Acknowledgement:
The authors acknowledge valuable contributions of Professor D.
Oesterhelt and coworkers.

References

1 D. Oesterhelt, W. Stoeckenius, Proc. Nat. Acad. Sci. USA
   70 (1973) 2853
2 A. Matsuno-Yagi, Y. Mukohata, Biochem. Biophys. Res. Commun.
   78 (1977) 237
3 H.J. Weber, R.A. Bogomolni, Photochem. Photobiol. 33 (1981) 601
4 M. Steiner, D. Oesterhelt, The Embo J. 2 (1983) 1379

Primary Events in Vision Probed by Ultrafast Laser Spectroscopy

A.G. Doukas and R.R. Alfano

The Institute for Ultrafast Spectroscopy and Lasers, Department of Physics
The City College of New York, New York, NY 10031, USA

I. INTRODUCTION

A decade of picosecond absorption spectroscopy on visual pigments has unraveled a number of fast dynamical properties about the primary photochemistry of the visual process. The visual pigment consists of a single chromophore, 11-cis retinal covalently bound to a protein in a form of a protonated Schiff base. The visual process is initiated by the absorption of a photon and the subsequent formation of bathorhodopsin. It was established very early that the primary event of vision takes place in less than 6 psec [1]. In fact, this first step still remains unresolved. The observation of weak fluorescence from rhodopsin [2,3] shows that the fluorescence lifetime can be below a picosecond. In this work we will summarize some recent work in our laboratory on picosecond fluorescence spectroscopy in order to understand the underlying molecular changes occurring during vision.

II. EXPERIMENTAL

The fluorescence kinetic measurements were taken in the picosecond fluorescence shown in fig. 1. A single pulse (1054 nm) from the output of a mode-locked Nd:glass laser was selected and amplified. The second harmonic at 527 nm was used to excite the sample. The samples were frontally excited at average density of 10^{16} photons/cm^2. The fluorescence was detected by a streak camera (Hamamatsu C-979) coupled to a GBC video camera. The time resolution (FWHM) of the system was 15 psec.

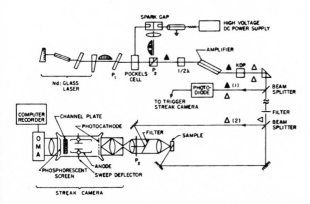

Figure 1: Schematic of the picosecond fluorescence apparatus

Samples of rhodopsin, squid or bovine,either solubilized or in suspension were prepared as described in the literature [4,5]. The fluorescence of rhodopsin was compared to the fluorescence of erythrosin dissodium salt in water taken under identical conditions of scattering geometry, diameter of exciting beam, slit width, microchannel plate gain and sweep speed. Low temperature measurements were taken in liquid helium dewar, and in the range of 5-40 K. The ratio of the time integrated areas is proportional to the ratio of quantum yields after correcting for the streak camera frequency response.

III. RESULTS and DISCUSSION

The typical fluorescence kinetic profiles at different temperatures are shown in fig. 2. A fluorescence kinetic profile of rhodopsin at room temperature is shown in fig. 2c. The emission of the fast component peaks at 600 nm with a bandwidth of about 50 nm (FWHM). The fluorescence is fitted to a double exponential: A resolution limited, less than 15 psec, "fast" component and a "slow" component of the order of 200-300 psec, depending on the sample preparation. Upon bleaching of the sample in the presence of hydroxylamine the fast component disappears and only the slow component remains. We have attributed the slow component to impurities of the sample since the intensity depends on the history of the sample. The quantum yields of both squid and bovine rhodopsin were measured. Using a quantum yield for erythrosin 0.02 [6] we found that quantum yields of squid and bovine rhodopsins are $(1.3 \pm 0.6) \times 10^{-5}$ and $(1.3 \pm 0.4) \times 10^{-5}$, respectively. Since the estimated radiative lifetime of rhodopsin is 5 nsec, the fluorescence lifetime must be of the order of 0.1 psec.

A temperature dependence study of the rhodopsin fluorescence [7] showed that the fluorescence remained unresolved even at 4K (fig. 2a). The value of the quantum yield also remained within the experimental error. What is remarkable is that substitution of H_2O with D_2O had no effect on either the fluorescence kinetics or the quantum yield (fig.

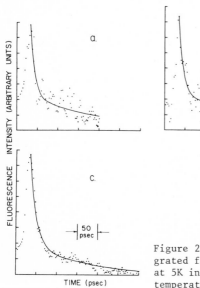
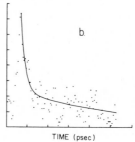

Figure 2: Typical fluorescence kinetics integrated from 550-800 nm of bovine rhodopsin: (a) at 5K in H_2O, (b) at 5K in D_2O, and (c) at room temperature in H_2O

2b). PETERS et al.[8], using picosecond absorption spectroscopy, demonstrated that the formation of bathorhodopsin is temperature dependent. The formation of bathorhodopsin increases to 36 psec at 4K. In addition, rhodopsin samples in D_2O show an increase in the formation of bathorhodopsin to 256 psec at 4K. They suggested that a proton translocation is the rate limiting step in the formation of bathorhodopsin. It is clear from our measurements that proton translocation is not the process that competes with the fluorescence. In fact, at low temperatures the relaxation of the excited state is orders of magnitude faster than the formation of bathorhodopsin. There is a subpicosecond kinetic process competing with the relaxation that does not involve the translocation of a proton. Given the large quantum yield for the chromophore this process is likely to involve the rotation of the retinal chromophore around the 11-12 double bond. In a recent paper [3] we have argued that it is the cis-trans isomerization that competes with the fluorescence of rhodopsin. In fact, our temperature dependent fluorescence measurements [7] show the existence of a barrierless excited state as it has been proposed by Honig et al. [9]. Finally, our measurements are corroborated by the picosecond kinetic studies of bovine rhodopsin with a fixed 11-ene [10], a chromophore where the C-10 and the C-13 of the retinal polyene backbone have been connected by a three carbon alkyl group. This chromophore cannot isomerize but otherwise has similar optical properties as bovine rhodopsin. We found that the quantum yield of this pigment is at least an order of magnitude larger than the quantum yield of rhodopsin.

The fluorescence of rhodopsin also contrasts the relaxation kinetics of retinal [11] where the lifetime is found to increase from resolution limited at room temperature to about 400 psec at 77K with an activation energy of a 1kcal/mole. In the case of rhodopsin the protein must modify the excited state providing a more efficient relaxation process.

IV. ACKNOWLEDGEMENTS

Supported by NIH EYO2515

V. REFERENCES

1. G. Busch, M. Applebury, A. Lamola and P. Rentzepis, Proc..Nat. Acad. Sci. (USA) 69, 2802 (1972).
2. A. G. Doukas, P. Y. Lu and R. R. Alfano, Biophys. J. 35, 547 (1981).
3. A. G. Doukas, M. R. Junnarkar, R. R. Alfano, R. H. Callender, T. Kakitani and B. Honig, Proc. Nat. Acad. Sci. (USA), in press (1984).
4. T. Suzuki, K. Oji and V. Kito, Biochim. Biophys. Acta 428, 321 (1976).
5. D. S. Papermaster and W. J. Dryer, Biochem. 13, 2438 (1973).
6. P. Bowers and G. Porter, Proc. R. Soc. (London) A299, 348 (1967).
7. A. G. Doukas, M. R. Junnarkar, R. R. Alfano, R. H. Callender, V. Balogh-Nair and K. Nakanishi (submitted 1984).
8. K. Peters, M. Applebury and P. M. Rentzepis, Proc. Nat. Acad. Sci. (USA) 74. 3119 (1977).
9. B. Honig, T. G. Ebrey, R. H. Callender, U. Dinur and M. Ottolenghi, Proc. Nat. Acad. Sci. (USA) 76, 2503 (1979).
10. J. Buchert, V. Stefancic, A. G. Doukas, R. R. Alfano, R. H. Callender, J. Pande, H. Akita, V. Balogh-Nair and K. Nakanishi, Biophys. J. 43, 279 (1983).
11. A. G. Doukas, M. R. Junnarkar, D. Chandra, R. R. Alfano and R. H. Callender, Chem. Phys. Letts. 100, 420 (1983).

Picosecond Time-Resolved Polarized Emission Spectroscopy of Biliproteins (Influence of Temperature and Aggregation)

S. Schneider, P. Hefferle, P. Geiselhart, T. Mindl, and F. Dörr

Institut für Physikalische und Theoretische Chemie, Technische Universität
Lichtenbergstraße 4, D-8046 Garching, Fed Rep. of Germany

W. John and H. Scheer

Botanisches Institut der Universität, Menzingerstraße 67
D-8000 München, Fed. Rep. of Germany

1. Introduction

Phycobiliproteins are photosynthetic light-harvesting pigments in blue-green and red algae. They consist of 2-3 polypeptide subunits, each bearing up to 4 covalently bound linear tetrapyrrolic chromophores. In vivo the biliproteins are organized into complex structures, the phycobilisomes, which are attached to the outer thylakoid surface. Within the phycobilisome the excitation energy is transferred from the "outer" biliproteins with higher excitation energy to "inner" lying ones with lower excitation energy. In the intact alga the last step in the energy transfer chain leads to chlorophylls within the membrane, i.e. the reaction center. It is generally assumed that the energy transfer is based upon dipole - dipole interaction (Förster mechanism), but details are yet insufficiently understood. In an approach complementary to the study of energy transfer in functionally intact phycobilisomes or large fractions thereof (1), we are currently investigating by time-resolved fluorescence spectroscopy C-Phycocyanin (PC) isolated from <u>Mastigocladus laminosus</u> and its subunits, which are prepared according to procedures described earlier (2,3). All samples are dissolved in potassium phosphate buffer (80 mM, pH 6.0).

2. Measurements and Data Analysis

The fluorescence decay curves were measured using a synchronously pumped mode-locked ring dye laser (rhodamine 6G, 80 MHz repetition rate, pulse width ≤ 1 ps) in conjunction with a repetetively working streak camera (for details see, e.g., (4)). The apparent time resolution of this system is approximately 25 ps without deconvolution procedure; it allows measurements with low excitation intensities (10^{13} photons per pulse and cm^2). Fluorescence decay curves measured with the analyzer parallel, ($I_p(t)$), and orthogonal, ($I_s(t)$), to the polarization of the exciting beam are transferred to a minicomputer where, after proper correction for the systems response, the expressions $I(t) = I_p(t) + 2 I_s(t)$ and $D(t) = I_p(t) - I_s(t)$ are calculated. $I(t)$ measures the decay of the excited state population (electronic lifetime) and $D(t)$ the product of the former with the correlation function of absorption and emission dipoles (2,5). In contrast to the anisotropy function $R(t)$ the difference function $D(t)$ is additive and can be evaluated if more than one emitting species is present. Lacking better information, we approximate the correlation function by an exponential. The best fits for both functions (I and D) are determined under the assumption of a biexponential response function (two emitting species) by means of a Marquardt algorithm.

Depending on the S/N ratio of the recorded fluorescence decay curves and their relative magnitude, the fit parameters derived may be subject to considerable error. We will, therefore, discuss their trends rather than their absolute magnitude.

3. Results and Discussion

It is found that in all cases the decay curves can be fitted sufficiently well as convolutions of biexponentials. The fit parameters, e.g., the decay times (T_1, T_2 in psec) and the relative amplitudes (A_1, A_2 in %) of the short- and long-lived component, resp., are given in the inserts in Fig. 1. The measurements were performed at three different temperatures, namely at 18° (A), at 36° (B), the temperature at which the algae are grown, and at 52° (C), where irreversible thermal denaturation starts to become effective. Partial denaturation takes place already at lower temperatures. Static measurements show a drastic loss in fluorescence yield (up to four orders of magnitude) which is much larger than the decrease in absorption connected with a conformational change of the chromophore (6). The time-integrated fluorescence intensities expressed as $A_1 * T_1 + A_2 * T_2$ also confirm the reduction at higher temperature. It is found as a general rule that the decrease is more pronounced in the alpha than in the beta subunit and larger for the monomer than for the trimer. The normalized fluorescence decay curves show also small but distinct variations with temperature. For this reason the results presented in Fig. 1 must be taken as evidence for an intermediate state being present during the process of thermal denaturation.

The alpha subunit of PC contains only one chromophore. If it is stabilized by noncovalent interaction with the protein to adopt only one conformation, a single exponential decay is expected with a lifetime of 1.5 to 2.5 ns (lifetime of the chromophore in a native environment). Instead, an additional short-lived component is found, whose lifetime varies with temperature between 690 and 1060 psec. A similar behaviour was verified for the alpha subunit of Spirulina platensis (2) and Anabaena varia-bilis (7). Since aggregation of the subunits is unlikely, one must assume at least two different sets of emitting species, i.e. chromophore-protein-arrangements. The long-lived species must be close to that in native environment, whilst the short-lived form should be closer to the denatured, less interacting species. The faster decay in the difference function D(t), furthermore, signals that the faster component is subject to a depolarization mechanism with $T \simeq 1500$ psec. Since no acceptor molecules are present, the depolarization should be due to orientational relaxation of the less rigidly bound chromophores.

The beta subunit contains two chromophores in different protein environment. The respective absorption maxima are separated by about 20 nm. The stationary fluorescence spectra of both subunits are essentially equal, a fact which indicates an efficient energy transfer from the "sensitizing" to the "fluorescing" chromophore. The energy transfer is also manifested in the fluorescence decay curves. The short-lived component ($T_1 \simeq 300$ps) is interpreted as "leakage" fluorescence from the s chromophore, whose lifetime is shortened due to energy transfer to the f chromophore in the same subunit. The depolarization time of the fast component is much shorter than that of the alpha subunit and de-

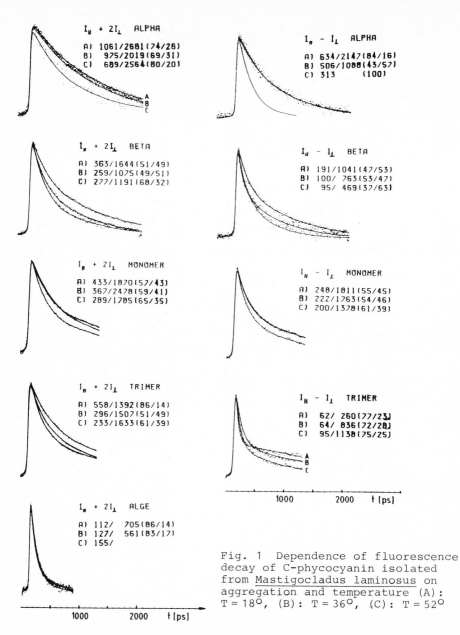

$I_\parallel + 2I_\perp$ ALPHA
A) 1061/2681(74/26)
B) 975/2019(69/31)
C) 689/2564(80/20)

$I_\parallel - I_\perp$ ALPHA
A) 634/2147(84/16)
B) 506/1088(43/57)
C) 313   (100)

$t_\parallel + 2I_\perp$ BETA
A) 363/1644(51/49)
B) 259/1075(49/51)
C) 277/1191(68/32)

$I_\parallel - I_\perp$ BETA
A) 191/1041(47/53)
B) 100/ 763(53/47)
C) 95/ 469(37/63)

$I_\parallel + 2I_\perp$ MONOMER
A) 433/1870(57/43)
B) 367/2478(59/41)
C) 289/1785(65/35)

$I_\parallel - I_\perp$ MONOMER
A) 248/1811(55/45)
B) 222/1763(54/46)
C) 200/1378(61/39)

$I_\parallel + 2I_\perp$ TRIMER
A) 558/1392(86/14)
B) 296/1507(51/49)
C) 233/1633(61/39)

$I_\parallel - I_\perp$ TRIMER
A) 62/ 260(77/23)
B) 64/ 836(72/28)
C) 95/1138(75/25)

$I_\parallel + 2I_\perp$ ALGE
A) 112/ 705(86/14)
B) 127/ 561(83/17)
C) 155/

Fig. 1 Dependence of fluorescence decay of C-phycocyanin isolated from <u>Mastigocladus laminosus</u> on aggregation and temperature (A): $T = 18^\circ$, (B): $T = 36^\circ$, (C): $T = 52^\circ$

creases with increasing temperature from 400 to 150 psec. The longer lifetime is close to the shorter one in the alpha subunit; a lifetime of 2 ns, which would be expected for the f chromophore in native environment is not detected, possibly for experimental reasons. An unambiguous interpretation is presently not possible, because different subsets of chromophore-protein arrangements can not be excluded in view of the preparation procedure which involves a denaturation - renaturation sequence.

The isotropic decay curves of monomers and trimers are similar
to each other. The short lifetime is in the range of 200-600 psec
and represents most likely the lifetime of the s chromophores,
which are quenched by energy transfer.The longer one,which varies
between 1600 and 2500 psec, characterizes the terminal acceptor,
i.e. the f chromophore in the native environment. Chromophores
excited via energy transfer rather than directly by photon absorption should emit a less polarized fluorescence. Only the
short- lived leakage fluorescence is partly polarized, but the
depolarization times are moderately short. In contrast to <u>Spiru-
lina platensis</u> (2) we observe for PC from the thermophilic algae
no significant increase of the depolarization time with temperature. In the intact alga, the energy is efficiently transferred
to the nonfluorescing reaction center. The observed emission is
only leakage fluorescence from PC and Allophycocyanin (APC).
Since the fraction of emission from directly excited chromophores
is small, the emission is essentially unpolarized.

References

1 For a recent review see, e.g., "Biological events probed by
   ultrafast laser spectroscopy", ed. by R.R. Alfano (Academic
   Press, New York, 1982)
2 P. Hefferle, W. John, H. Scheer and S. Schneider,
   Photochem. Photobiol. <u>39</u>,221 (1984)
3 W. Kufer, H. Scheer, Hoppe-Seyler´s Z. Physiol. Chem.
   <u>360</u>,935 (1979)
4 P. Hefferle, M. Nies, W. Wehrmeyer, S. Schneider
   Photobiochem. Photobiophys. <u>5</u>,41 (1983),id.<u>5</u>,325 (1983)
5 G.R. Fleming, J.M. Morris, G.W. Robinson, J.Chem.Phys.
   <u>17</u>,91 (1976)
6 H. Scheer, H. Formanek, S. Schneider, Photochem.Photobiol.
   <u>36</u>,259 (1982)
7 S.C. Switalski and K.Sauer, Photochem. Photobiol. in press

Dynamics of Energy Transfer in Chloroplasts and the Internal Dynamics of an Enzyme

R.J. Gulotty, L.Mets*, R.S. Alberte*, A.J. Cross, and G.R. Fleming

Department of Chemistry and James Franck Institute and * Department of Biology, The University of Chicago, Chicago, IL 60637, USA

Origin of Fluorescence Decay Components in Chloroplasts.

The form of the fluorescence decay function from the light harvesting system of photosynthetic organisms has been the object of intensive study over the last ten years [1]. The advent of low intensity picosecond lasers coupled with time correlated single photon counting detection has caused optimism that a detailed understanding of both the structural organization and energy transfer pathways in the photosynthetic unit will be possible. Several groups have found it necessary to use a sum of three exponential components to fit the chloroplast fluorescence decay curves [2-5]. An essential prerequisite for a mechanistic description of the fluorescence decay in terms of the excitation transfer and trapping processes is an assignment of the various decay components in terms of the functional constituents of the photosynthetic unit. Closely related is the question of whether the true fluorescence decay is really a triple exponential or is different - for example, a more complex function that, due to the limitations of real data (i.e.,finite time resolution, dynamic range,etc.), can be statistically well fit as a triple exponential decay.

We present here measurements of the fluorescence decay kinetics of photosynthetic mutants of C. reinhardii. We believe that the fluorescence decay measurements of the PSI and PSII mutants are the first subnanosecond kinetic measurements of excitation transfer in isolated PSI and PSII where genetic mutation and selection have been used instead of detergent or mechanical extraction to prepare membranes containing only one type of higher plant photosystem. We analyze our data in terms of the excitation dynamics and use the decay characteristics of the various mutants to synthesize the behavior of the whole photosynthetic unit. Applying the approach of Pearlstein [6] to our data we calculate the single step transfer time and the average number of visits an excitation makes to the reaction center before it is finally trapped. Our simulations of the wild type decays give excellent agreement with the experimental parameters when the relative absorption cross sections of PSI and PSII are assumed approximately equal. The simulations also reveal that single photon counting fluorescence decay data of wild type chloroplasts of higher plants and algae do not contain sufficient information for detailed a priori analysis in terms of the isolated parts of the photosynthetic unit.

Results

Fluorescence decay measurements were carried out using a mode locked cavity dumped dye laser and Hamamatsu R1645 microchannel plate detector. The instrument function has a FWHM of 130 ps and a full width at tenth maximum of 250 ps. Figure 1 shows typical fluorescence decays for C. reinhardii wild type strain 2137 (curve a), PSII mutant strain 8-36C (curve d), PSI mutant strain 12-7 (curve b), and a standard dye solution of oxazine in water (curve c). The figure illustrates the nonexponential kinetics of the C. reinhardii strains compared with the single exponential decay of the dye solution. Both the PSII and PSI mutant strains have a higher quantum yield than the wild type strain, and the PSI mutant (lacking PSI) is missing the major short component present in both the wild type and PSII mutant strains. The PSII mutant (lacking PSII) has a higher proportion of long lifetime components. Full details of the decay kinetics may be found in ref. [7].

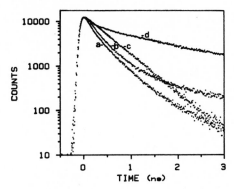

Fig.1 Fluorescence decay curves of C. reinhardii strains a) wild type 2137 b) PSI mutant 12-7 d) PSII mutant 8-36C and c) oxazine 725 in water

Energy Transfer in the PSII Mutant

The PSII mutant 8-36C decay curve observed at 680nm fits well to the exponential components with lifetimes 53 ps, 424 ps and 2197 ps, and weights 0.503, 0.191 and 0.306 respectively. No variation in the decay was observed as a result of pre-illuminating the sample, confirming the view of Butler et al. [8] that variable fluorescence is associated with PSII.

The weight of the short lifetime component is higher at 730nm than at 680nm, suggesting that this component is associated with longer wavelength chlorophyll close to the PSI reaction center. The large magnitude of this component, however, suggests that excitations originating in the chl a/b protein, which comprises 71% of the total chlorophyll in this mutant, must also contribute to this component.

An estimate of the time scale of energy transfer between communicating chlorophyll molecules in this mutant can be obtained by combining our biochemical and fluorescence data. The expressions of Pearlstein [6] enable an estimate of both the transfer time between adjacent antenna molecules and of the number of visits an excitation makes to the reaction center before trapping. The important parameters [6,7] are

the excitation lifetime, the number of chlorophyll molecules
involved, the rate of photochemical reaction at the trap, the
ratio of trapping and detrapping rates and the ratio of
trapping and antenna-antenna hopping rates. We use $(3 \text{ ps})^{-1}$ for
photochemistry and follow Shipman [9] in assuming Boltzmann
weighted reverse energy transfer between different spectral
forms. An important feature of Pearlstein's expression is
that it accounts for multiple visits to the reaction center
without requiring any assumption about the actual number of
visits. Combining the parameters with a lattice size of 106
and a lifetime of 53 ps leads to a single step transfer time,
[(the coordination number)x(the Forster rate constant)]$^{-1}$,of
about 0.1 ps. This is significantly shorter than the earlier
estimates of Campillo and Shapiro [10].

The assumption of N = 106, used above, is based on the
weight of the 53 ps component for 680 nm. In the presence of
significant detrapping the interpretation of the weights of
the various components may be complex. Bearing this in mind
an alternative assumption should be considered: that the 53 ps
component originates only in closely coupled antenna
molecules. This leads to N = 60,giving a single step transfer
time of 0.4 ps and an average of about two visits to the
reaction center. These values are consistent with a similar
analysis carried out on a second PSII mutant (A-4d) containing
significantly less chlorophyll per PSI reaction center than
the 8-36C mutant (86 vs 220).

Simulation of Wild Type Fluorescence Decays

The fluorescence decay kinetics of the entire chloroplast
should reflect the contributions of the isolated parts of the
photosynthetic unit. In the absence of significant interphotosystem couplings the fluorescence decay properties should
add to the whole when weighted by their relative absorption
cross sections. Figure 2 summarizes the results of a simulation in which we sum the decay properties of our photosystem I and photosystem II mutants and use the lifetime of
the PSI-PSII mutant to represent the lifetime of decoupled
light harvesting chl a/b protein. The three decays contain
five exponential components in all. These components are
convoluted with a real instrument function, Gaussian noise is
added and the resulting curve is fitted,in the same way as our

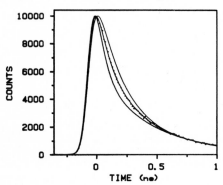

Fig.2 Simulations of wild
type decay curves using PSI,
PSII, and PSI-PSII mutants:
lower curve, PSII mutant +
decoupled a/b protein; upper
curve, PSI mutant + decoupled
a/b proten; middle curve, PSI
+ PSII mutants + decoupled
a/b protein. Dots are decay
with experimentally observed
parameters.

experiments, to three exponential components. Simulations
with 10^4 counts in the peak channel give χ^2 = 1.0 -1.2
for all ratios of contributions from the three curves.
Evidently the information content is inadequate to reveal the
presence of additional components.

Figure 2 shows simulated data (smooth lines-noise is not
shown for clarity) and real data (dots). Shown in Fig. 2 are
simulations for PSI-PSII mutants only, PSII and PSI-PSII
mutants only and a simulation weighting PSII, PSI and PSI-PSII
mutants 0.6, 0.39, 0.0075 respectively. This latter simulation
fits our measured curve well and in general we find that the
wild type data is best simulated from the isolated parts when
the ratio of excitations distributed between PSI and PSII is
approximately equal.

Conclusions

Our study of the C. reinhardii mutants combined with a
simulation of the wild-type fluorescence decays leads us to the
following conclusions: (1) Fluorescence associated with the
presence of PSI reaction centers must be included in analyses
of wild-type chloroplast decays, (2) The true wild-type decay
is considerably more complex than a sum of three exponential
components, (3) In simulations which neglect interphotosystem
couplings the best correspondence with experimental data is
found when the ratio of excitations distributed between PSI and
PSII is approximately equal.

Using the formalism of Pearlstein and data for the PSII
mutant we estimate the single step transfer time in the array
to be between 100 and 400 fs. We also find that the excitation
makes between 2 and 4 visits to the reaction center before the
photochemical event occurs.

Internal Motions of Lysozyme

Concerted motions of residues in proteins have been suggested
as an important kind of internal motion which may play an
essential role in biological activity [11,12]. We have made
time resolved fluorescence anisotropy measurements of the
internal motions of lysozyme using an extrinsic probe, eosin,
which binds in the hydrophobic box region of the enzyme [13].
Our measurements provide evidence that the residues in this
region undergo significant motions on the time scale of 100 ps.
The extent of the motion as measured by the model independent
order parameter S^2 shows a different temperature dependence when the inhibitor $(GlcNAc)_3$ is bound to the active
site. In both cases the order parameter has a stronger
temperature dependence than can be explained by activation in a
harmonic potential. However, the observed temperature dependence and changes in S^2 upon binding are reproduced well by
a nonharmonic model of the effective potential which is
consistent with the picture of concerted motions in the
protein. The values of the parameters of the potential which
reproduce the data with and without the bound inhibitor imply
that $(GlcNAc)_3$ binding causes an increase in the rigidity of
the protein.

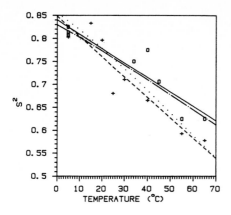

Fig.3 Order parameter vs temperature for LE (O) and LEN_3 (+). Least squares fits for LE (-) and LEN_3 (--). Curves generated from potential in Fig.4: LE (_.) and LEN_3 (...).

The order parameter which describes the angular restriction in the rapid motion [14,15] is given by $S^2 = r(0^+)/r(0)$ where $r(0^+)$ is the value of $r(t)$ extrapolated back to time zero from the long time (overall tumbling) behavior and $r(0)$ is the true critical value. The experimental values of S^2 for lysozyme-eosin (LE) and lysozyme-eosin-$(GlcNAc)_3$ (LEN_3) are shown in Fig. 3. Our data is well accounted for by the potential shown in Fig. 4. This potential has three parameters V_0, the height of the initial step; θ_0, the angular range over which free motion is possible; and V_1, the value of the potential when $\theta = \pi$. Full expressions may be found in [16]. From the fits in Fig. 4 we find V_0 = 2.8±0.3 kcal/mole (LE), 3.6±0.2 kcal/mole (LEN_3); and θ_0 = 8°±3° (LE), 6°±2° (LEN_3).

We interpret the potential we used (Fig. 4) as follows. The probe molecule can move unhindered over the range O to θ_0. Further motion requires accumulation of energy in collective modes of the protein corresponding to a concerted motion. After accumulation of V_0 in the collective mode the reaction coordinate can again become θ and probe motion becomes relatively unrestricted with an approximately harmonic restoring force. The projection of this multidimensional reaction coordinate onto one dimension, i.e. θ, gives the analytic form of the potential [16]. The decrease of θ_0 and increase of V_0 give a quantitative feel for the tightening of the hydrophobic box region of the enzyme on inhibitor binding.

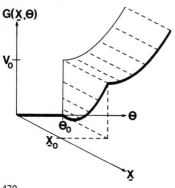

Fig.4 Schematic illustration of the nonharmonic potential used to interpret the data. The vertical axis is the free energy, G. The angle of the transition dipole with respect to a protein-fixed axis is given by θ. X represents additional modes of the protein into which energy must be channeled before movement to angles greater than θ_0 can occur.

Acknowledgements

This work was supported by grants from the USDA and from SOHIO.

References

1. J. Breton and N. Geacintov, Biochim. Biophys. Acta, 594, 1
    (1980).
2. R.J. Gulotty, G.R. Fleming, and R.S. Alberte, Biochim.
    Biophys. Acta, 682, 420 (1982).
3. W. Haehnel, A.R. Holzworth, and J. Wendler, Photochem.
    Photobiol. 37, 435 (1983).
4. J.A. Nairn, W. Haehnel, P. Reisberg, and K. Sauer, Biochim.
    Biophys. Acta, 682, 420 (1982).
5. S. Berens, Ph.D. Dissertation, University of California at
    San Diego (1984).
6. R.M. Pearlstein, Photochem. Photobiol. 35, 835 (1982).
7. R.J. Gulotty, L. Mets, R.S. Alberte, and G.R. Fleming,
    submitted to Photochem. Photobiol. (1984).
8. W.L. Butler and M. Kitajima, Biochim. Biophys. Acta, 396, 72
    (1975).
9. L.L. Shipman, Photochem. Photobiol. 31, 157 (1980).
10. A.J. Campillo and S.L. Shapiro, in Topics in Applied
    Physics, vol 18 (Ultrashort Light Pulses) 318 (1978).
11. M. Karplus and A. McCammom, CRC Crit. Rev. Biochem. 9, 293
    (1981).
12. P.J. Artimiuk et al., Nature 280, 563 (1979).
13. J.F. Baugher, L.I. Grossweiner, and C. Lewis, J. Chem. Soc.
    Faraday Trans II 70, 1389 (1974).
14. G. Lipari and A. Szabo, J. Amer. Chem. Soc. 104, 4559
    (1982).
15. M.C. Chang, A.J. Cross, and G.R. Fleming, J. Biomolec.
    Struct. Dynam. 1, 299 (1983).
16. A.J. Cross and G.R. Fleming, in preparation.

Picosecond Single Photon Fluorescence Spectroscopy of Nucleic Acids

R. Rigler, F. Claesens, and G. Lomakka

Department of Medical Biophysics, Karolinska Institutet, Box 60400
S-104 01 Stockholm, Sweden

1. Introduction

Fluorescence spectroscopy of nucleic acids at room temperature has been limited by the short lifetimes of the excited states of purines and pyrimidines (1) and scarce information is available from measurements of the stationary emission. Recently attempts have been made by using the time structure of synchrotron radiation in order to obtain time resolved emission spectra (2,3). The large pulse width (1.7 ns) of the synchrotron pulse used has however prevented the measurement of response times in the ps domain. Inherently better time resolution is provided by mode locked and synchronously pumped dye laser pulses for the excitation of nucleoside fluorescence. Their detection however has been limited by the time response of photodetectors usually in the range of a few hundred ps. Here we demonstrate the advantage of using multichannel plate detectors with short transient time jitter for single photon detection (4) together with deconvolution procedures for measurements of decay times of a few ps as well as for time resolved fluorescence spectroscopy of nucleic acids.

2. Experimental

For generation of ps laser pulses a large frame Kr-ion laser (Coherent 3000K) with an acousto-optic mode locker (Coherent 467) was used. Rhodamine

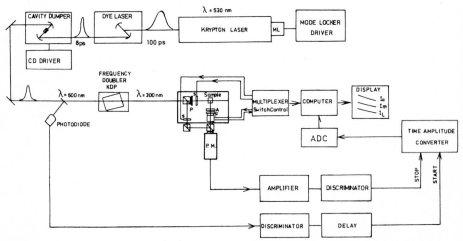

Fig. 1 Setup for time correlated single photon spectroscopy. Mode locked synchronously pumped and cavity dumped dye laser and beam splitting device for measurement of excitation pulse (I_L) and polarized emitted components of probe fluorescence (I_m, $I_{||}$). For spectrum analysis the monochromator is placed before the detector.

6G was pumped synchronously at 530.9 nm in a cavity dumped dye laser (Coherent 599). For frequency doubling an angle tuned KDP crystal was used.

A computer controlled split beam arrangement was developed allowing simultaneous detection of the (attenuated) excitation pulse together with the polarized components of the fluorescence (Fig. 1) in order to eliminate variations in intensity and timing of the exciting laser pulse as well as of the convoluted response signal. The time dependent emission was detected by a Hamamatsu R1564-U multichannelplaté detector (MCP) using time correlated single photon counting. Timing of the time amplitude converter (TAC, Ortec 457) was performed by low time jitter constant fraction discriminators (Tennelec TC453) using the amplified MCP pulse as start and the laser pulse as stop signal. The probability distribution of single photon events was accumulated in a ND66 multichannel analyzer (MCA).

For time resolved fluorescence spectroscopy the MCA was used in the multichannel scaling mode and gated by a window discriminated output of the TAC. For analysis of the fluorescence decay non-linear least square fit procedures involving deconvolution routines (5) were used.

The FWHM of the detector response could be reduced to 43 ps (Fig. 2A) after optimal adjustment of CFD walktimes and discrimination levels. For performance tests we used cresyl-violet for which rotational relaxation time and lifetimes were measured by optical autocorrelation and sum frequency generation techniques (6). Rotational relaxation times for cresyl-violet (lifetime 2.38 ns) in H_2O (Fig. 1B) and in other solvents (Table 1) were determined in agreement with published values. The unconvoluted anisotropy decay function (Fig. 2B) is close to the deconvoluted value due to the fast detector response.

3. Results

The emission spectra of constituent purine nucleotides as well as of corresponding synthetic DNA molecules are shown (Fig. 3A-D). All spectra were gated with a time window of about 2 ns and no background was detectable. For ATP and GdP an emission maximum above 400 nm with a pronounced shoulder at 330 nm for GdP was observed. Analogous spectra were observed for

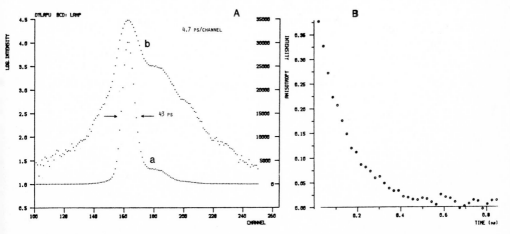

Fig. 2   A) Detector response function to a 6 ps dye laser pulse at 300 nm. Hamamatsu R1564-U multichannelplate detector in single photon counting mode, time scale 4.7 ps/channel. Linear scale (a), log scale (b). B) Anisotropy decay function of emission from 10^{-7} M cresyl-violet in H_2O, excited at 590 nm, OG 630 cutoff filter. Uncorrected data without deconvolution.

473

Table 1 Rotational relaxation times of cresyl-violet in various solvents as measured by time correlated single photon counting (SPC) and sum frequency generation (SFG) at 20°C.

Solvents	Viscosity (η) (cP)	Rotational relaxation time τ_R (ps)	
		SPC	SFG*
Acetone	0.32	63±2	78±4
Methanol	0.55	176±10	134±4
Water	1.03	124±4	130±5
Ethanol	1.2	350±7	350±14
Propanol	2.2	669±12	696±15

*Data of Beddard,G.S.,Doust,T.&Porter,G. (1981), Chemical Physics 61, 17-23

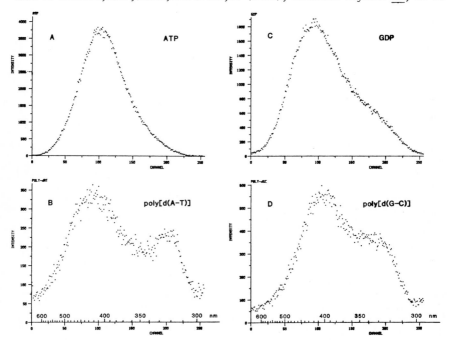

Fig. 3 Fluorescence spectra of nucleic acids in H_2O. Zeiss MQIII monochromator, bandwidth 15 nm. Excitation at 290 nm. Concentrations given in OD units at 290 nm for 4.4 mm sample thickness. A) ATP, conc. 0.035; B) poly[d(A-T)], conc. 0.19; C) GdP, conc. 0.20; poly[d(G-C)], conc. 0.18.

Table 2 Lifetime of excited states of nucleosides and synthetic DNAs excited at 290 nm at room temperature.

Sample	Lifetimes (ps)		Relative amplitudes *		χ^2
	τ_1	τ_2	A_1	A_2	
ATP	4170±3	289±5	0.64	0.36	1.1
poly d(A-T)	872±15	94±1	0.03	0.97	2.1
GdP	1409±41	211±1	0.01	0.99	1.4
poly d(G-C)	2926±30	255±27	0.24	0.76	1.4

*S.E. <5% of mean.

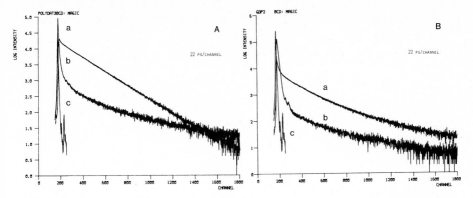

Fig. 4 Fluorescence decay of nucleic acids. Same conditions as in Fig. 3, time scale 22 ps/channel. WG 305 nm cutoff filter. A) ATP (a), poly[d(A-T)] (b), laser pulse (c). B) GdP (a), poly[d(G-C)] (b), laser pulse (c). Background subtracted after correction for probe absorption according to (7).

poly[d(A-T)] as well as for poly[d(G-C)], with the short wave emission band accentuated.

The decay profiles of the integrated emission spectrum of the corresponding samples are shown in Fig. 4 and the lifetimes of emission in Table 2.

4. Discussion

Essential for successful analysis of short lifetime components is a fast detector response mainly limited by the transient time spread (TTS). With electron trajectories equivalent for each channel, MCPs are expected to have the smallest TTS. Our experimental values (Fig. 1A) which include a 10 ps time jitter for each of the CFDs are in agreement with theoretical predictions by OBA and ITO (4).

The time resolution provided by single photon counting MCPs for the measurement of rotational relaxation of small aromatic molecules as shown by our data is equivalent to the sum frequency generation technique which is limited only by the width of the laser pulse (8). For the analysis of rotational motions of residues and segments of biopolymers, time correlated photon counting, which is orders of magnitudes more sensitive, provides a decisive advantage. Analysis of rotational motions of individual aromatic residues in nucleic acids as well as in proteins and peptides is performed at present (Claesens and Rigler, in preparation) and can be compared with the results from molecular dynamic simulations (9).

Comparison of purine nucleotides and synthetic DNAs shows two main emission maxima around 400 and 330 nm with the short wave maximum prevailing in the latter. Similar spectra were found for ATP and GdP (10) and for polyadenylic acid (2) and ApA (11) at room temperature as well as for AMP, GMP and GpC at 77 K (12). The existence of monomer and excimer fluorescence and phosphorescence bands in the actual wavelength range has been demonstrated previously (12,13,14,15,2), their weighting being dependent on solvent, temperature and pH conditions. Further investigations are requested to decide on the contributions of various singlet and triplet emission bands to the spectra observed and their relation to different conformations (stacking and hydrogen bonding) of purine and pyrimidine bases. For this purpose time gated spectroscopy is particularly suitable, permitting analysis of the time evolution of individual spectral bands as well as an effective suppression of disturbing instantaneous scattering processes.

The lifetime analysis shows two main components (Table 2) with values similar to those reported from synchrotron experiments (3). A 4 ns component has been found for poly-rA and has been ascribed to triplet transitions (2). From quantum yields and absorption cross-sections lifetimes of a few ps are predicted (16). In the present analysis the lowest ps range has not been penetrated in sufficient detail; the values of the weighted residuals (χ^2) indicate additional components.

The aim of this study is to provide relevant information on the dynamics of excited states in nucleic acids and natural as well as modified constituents under biological conditions. Their knowledge is also essential for the analysis of molecular motion from polarized emission (17).

5. Acknowledgement

The cooperation of Dr. K.Oba, Hamamatsu Photonics K.K., is gratefully acknowledged. This work was supported by grants from K. and A. Wallenberg foundation and the Swedish Natural Science Research Council.

6. References

1. Daniels,M.: in "Physico-chemical properties of nucleic acids" (I.Duchesne, ed.), Academic Press, London-New York (1973), p. 99.
2. Ballini,J.P.,Daniels,M.,Vigny,P.: J. Luminescence 27, 389 (1982).
3. Ballini,J.P.,Vigny,P.,Daniels,M.: Biophys. Chem. 19, 61 (1983).
4. Oba,K.,Ito,M.: Computer analysis of timing properties of micro channel plate photomultiplier tube. Proceed. 8th Symposium on Photoelectronic image devices, Imperial College of Science and Technology, London, Sept. 1983.
5. Rigler,R.,Ehrenberg,M.: Quart. Rev. Biophys. 9, 1 (1976).
6. Beddard,G.S.,Doust,T.,Porter,G.: Chem. Phys. 61, 17 (1981).
7. Ehrenberg,M.,Cronvall,E.,Rigler,R.: Febs Letters 18, 199 (1971).
8. Mahr,H.,Hirsch,M.D.: Opt. Commun. 13, 96 (1975).
9. Karplus,M.,McCammon,J.A.: Ann. Rev. Biochem. 53, 263 (1983).
10. Börresen,H.C.: Acta Chem. Scand. 17, 921 (1963).
11. Morgan,J.P.,Daniels,M.: Photochem. Photobiol. 31, 101 (1979).
12. Hélène,C.,Michelson,A.M.: BBA 142, 12 (1966).
13. Eisinger,J.,Guèron,M.,Shulman,R.G.,Yamane,T.: Proc. Natl. Acad. Sci. 55, 1015 (1966).
14. Eisinger,J.,Shulman,R.G.: J. Mol. Biol. 28, 445 (1967).
15. Hélène,C.: in "Physico-chemical properties of nucleic acids (I.Duchesne, ed.), Academic Press, London-New York (1973), p. 119.
16. Shapiro,S.L.: in "Biological Events Probed by Ultrafast Laser Spectroscopy" (R.R.Alfano, ed.), Academic Press, New York-London (1982), p. 161.
17. Ehrenberg,M.,Rigler,R.: Chem. Phys. Letters 14, 539 (1972).

Excited-State Dynamics of NADH and 1-N-Propyl-1,4-Dihydronicotinamide

D.W. Boldridge, T.H. Morton, and G.W. Scott
Department of Chemistry, University of California, Riverside, CA 92521, USA
J.H. Clark, L.A. Philips, S.P. Webb, and S.M. Yeh
Laboratory for Chemical Biodynamics, Lawrence Berkeley Laboratory, and
Department of Chemistry, University of California, Berkeley, CA 94720, USA
P. van Eikeren
Department of Chemistry, Harvey Mudd College, Claremont, CA 91711, USA

The reduced form of nicotinamide-adenine dinucleotide (NADH) is an important enzyme cofactor containing two chromophores - a dihydronicotinamide (lowest absorption band at 340 nm) and an adenine (absorption at 265 nm). Fluorescence of NADH has a λ_{max} at 460 nm and has been used for in vivo assays of NADH. The photophysics of NADH has been extensively studied [1-9]. In room temperature aqueous solution, the fluorescence quantum yield is 0.02 and the lifetime ∼ 0.40 ns [3-5]. Recently, biphotonic induced electron ejection by NADH has been reported [9]. NADH reportedly exists in aqueous solution in two conformations--extended and folded [3,5,10].

The present study reports time-resolved emission and transient absorption studies on the disodium salt of NADH and on a simple analog, 1-N-propyl-1,4-dihydronicotinamide (NPNH), (shown below) in several solvents.

NADH   NPNH

Purity of NADH (Boehringer-Mannheim) and of NPNH (prepared according to ANDERSON and BERKELHAMMER [11]) was tested using a reverse phase HPLC analysis, which showed for each a single peak with a homogeneous uv absorption spectrum. The photophysics of directly excited dihydronicotinamide was observed following excitation by a 355-nm pulse from a modelocked neodymium laser. Fluorescence decay kinetics were obtained using Nd:YAG laser excitation and 2-ps resolution streak camera detection [12]. Transient absorption spectra and kinetics were obtained using Nd:glass laser excitation and a picosecond continuum probe with polychromator/vidicon detection [13]. Energy transfer from adenine to dihydronicotinamide in NADH was studied by fluorescence kinetics after excitation at 266 nm.

Following excitation of NADH, the growth of a broad, unstructured absorption spectrum was observed (see Fig. 1). The similarity of these spectra to that of the solvated electron, e_{aq}^{-} [14], suggests this identification, but does not exclude a solvated electron-ion pair. A deconvolution analysis indicates a buildup time of 40 ± 10 ps for the transient

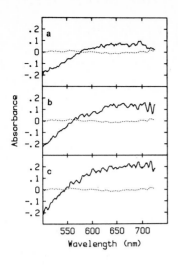

Figure 1. Transient absorption spectra of NADH in aqueous solution at room temperature taken at (a) 7 ps, (b) 20 ps, and (c) 376 ps after excitation at 355 nm. Apparent negative absorbance values at short wavelengths are due to sample fluorescence. Analysis shows that this spectrum grows linearly with intensity and not quadratically [9].

absorption spectrum. No perceptible decay occurs within 2 ns following excitation. Identical observations were made on NPNH. The spectral buildup rate is slower than electron solvation in water [14-16] and may be limited by electron ejection [17,18]. Since the electron ejection time does not match the known NADH fluorescence lifetime [3-5], multiple species and decay routes must be involved. Using the known extinction coefficient for the solvated electron [19], the transient absorbance observed upon excitation of NADH with an actinometrically calibrated pulse gave a quantum yield for photon-induced electron ejection from NADH of ≈0.5. This yield is significantly higher than NADH photodecomposition [20], implying a high yield of ion recombination.

NADH fluorescence kinetics data with detection at 380 nm and 460 nm show a previously unreported, fast decaying emission component on the blue edge of the spectrum (see Fig. 2). Fluorescence kinetics detected at 420 to 700 nm yielded only longer fluorescence lifetimes and no risetime (<10 ps). The decay time of the fast component (≈30 ps) is similar to the absorption buildup. Fluorescence lifetimes of NADH and NPNH are summarized in the Table.

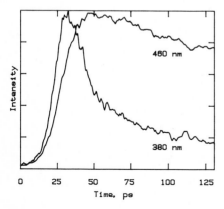

Figure 2. Fluorescence kinetics of NADH in aqueous solution at room temperature using excitation at 355 nm. Detection was at the indicated wavelengths through 10 nm wide bandpass filters. Shortlived fluorescence at 380 nm was also observed for NPNH in water and for both NADH and NPNH in methanol.

Table. Lifetimes of the longer-lived fluorescence of NADH and NPNH at room temperature broad-band detected.

Solute[1]	Solvent[2]	λ_{exc}	λ_{detect}	τ (ps)
NADH	H_2O	266	> 345 nm	470 ± 50
NADH	D_2O	266	> 345 nm	720 ± 70
NADH	H_2O/Buffer	266	> 345 nm	520 ± 50
NADH	H_2O/Buffer	355	> 375 nm	440 ± 40
NADH	D_2O/Buffer	355	> 375 nm	620 ± 60
NADH	H_2O	355	> 375 nm	450 ± 40
NADH	EtOH/Buffer	355	> 375 nm	850 ± 80
NPNH	H_2O/Buffer	355	> 375 nm	320 ± 30
NPNH	D_2O/Buffer	355	> 375 nm	460 ± 50
NPNH	EtOH/Buffer	355	> 375 nm	940 ± 90

[1]NADH = Reduced form of nicotinamide-adenine dinucleotide, NPNH = 1-N-propyl-1,4-dihydronicotinamide, [2]Buffer = 10^{-2} M Tris (pH 8.6 in water)

The fluorescence kinetics show these features: (1) The kinetics are relatively independent of excitation wavelength or the presence of buffer. (2) In D_2O and ethanol lifetimes are longer than in H_2O. Since the fluorescence decay kinetics following excitation at 266 nm and at 355 nm are similar, this fluorescence is ascribed to a "folded" conformation of NADH, which allows energy transfer from the adenine [3] . Therefore, folded forms of NADH exist in H_2O and D_2O and contribute to the longer-lived fluorescence. The results on NPNH suggest that open forms of NADH also contribute to this longer-lived fluorescence.

Possible explanations of the observed photophysics of NADH and NPNH include (1) two different excited-state forms of reduced nicotinamide arising from two ground state conformers or (2) a branched excited state decay mechanism involving two fluorescent forms. For case (1), the two planar rotamers of reduced nicotinamide with different amide orientations were investigated by INDO/CI M.O. calculations, which indicated a small energy difference between these two forms ($\sim$ 3kJ/mole). For case (2), an excited state mechanism could include the following kinetic scheme:

$$S_0 \xrightarrow{h\nu} \{S_1\} \xrightarrow{k_1} [D^{\cdot +} \cdots e^-] \xrightarrow{k_2} D^{\cdot +} + e^-_{aq}$$

with k_3, k_4, K connecting to S_1.

S_1 is the relaxed, fluorescent, excited singlet state of reduced nicotinamide, $[D^{\cdot +} \cdots e^-]$ is a solvated, nonfluorescent ion pair, and $D^{\cdot +} + e^-_{aq}$ are a separated radical cation and a solvated electron. Initially fluorescence comes from "vertically" excited molecules, $\{S_1\}$, with, for example, ground-state solvent organization or hydrogen bonding. $\{S_1\}$ could exhibit a slightly blue-shifted fluorescence, relative to S_1, with a lifetime determined by the sum of k_1 and k_3. The ion pair and the solvated electron may well have similar absorption spectra. Whether the equilibrium with constant K would need to be established during the lifetimes of S_1 and $[D^{\cdot +} \cdots e^-]$ is unclear. However, this could explain the complex multiexponential decay reported for the longer lived fluorescence [4, 5]. If the lifetime of S_1 were determined by k_2, then the rate of ion pair separation would have to depend on solvent deuteration (see table). A high probability of recombination of $D^{\cdot +}$ with e^-_{aq} is consistent with its low reactivity observed in the gas phase. NPNH has an ionization potential < 8 eV, and molecular ion dominates its mass spectrum at low ionization energies. In a Fourier Transform Mass Spectrometer, $NPNH^{\bullet +}$ does not react with neutral NPNH or with 10^{-6} torr of NH_3 on the 0.1 s timescale (during which $NPNH^{\bullet +}$ experiences $\sim$ 50 collisions with NH_3). Further work to refine the kinetic model is in progress.

This work was supported by the Committee on Research, University of California, US Department of Energy Contract DE-AC03-76SF0098, and NIH grants NS 14992 and BRSG RR 07010-15. J. H. Clark is an Alfred P. Sloan Foundation Fellow and a Camille and Henry Dreyfus Foundation Teacher-Scholar. We thank Ms. Marian Hawkes for preparing the manuscript.

1. Dehydrogenases Requiring Nicotinamide Coenzymes, J. Jeffery ed., Birkhauser Verlag, 1980.
2. G. Blankenhorn in Pyridine Nucleotide Dependent Dehydrogenases, H. Sund, ed., De Gruyter, 1977, pp 185-205.
3. T.G. Scott, R.D. Spencer, N.J. Leonard, and G. Weber: J. Am. Chem. Soc. 92, 687 (1970).
4. A. Gafni and L. Brand: Biochem. 15, 3165 (1976).
5. A.J.W.G. Visser and A. van Hoek: Photochem. Photobiol. 33, 35 (1981).
6. M.F. Powell, W.H. Wong, and T.C. Bruice: Proc. Natl. Acad. Sci., USA 79, 4604 (1982).
7. F.M. Martens, J.W. Verhoeven, C.A.G.O. Varma, and P. Bergwerf: J. Photochem. 22, 99 (1983).
8. A. Ohno and N. Kito: Chem. Lett., 369 (1972).
9. B. Czochralska and L. Lindqvist: Chem. Phys. Lett. 101, 297 (1983).
10. N. J. Oppenheimer, L. J. Arnold, and N. O. Kaplan: Biochem. 17, 2613 (1978).
11. A. G. Anderson and G. Berkelhammer: J. Am. Chem. Soc. 80, 992 (1958).
12. S.P. Webb, S.W. Yeh, L.A. Philips. M.A. Tolbert, and J.H. Clark: "Excited-State Proton-Transfer Reactions in 1-Naphthol," in THIS VOLUME.
13. D.W. Boldridge and G.W. Scott: J. Chem. Phys. 79, 3639 (1983).
14. M.J. Bronskill, R.K. Wolff, and J.W. Hunt: J. Chem. Phys. 53, 4201 (1970).
15. G. A. Kenney-Wallace and D.C. Walker: J. Chem. Phys. 55, 447 (1971).
16. W.J. Chase and J.W. Hunt, J. Phys. Chem. 79, 2835 (1975).
17. Y. Wang, J.K. Crawford, M.J. McAuliffe, and K.B. Eisenthal: Chem. Phys. Lett. 74, 160 (1980).
18. G.A. Kenney-Wallace and C.D. Jonah: J. Phys. Chem. 86, 2572 (1982).
19. G.E. Hall and G.A. Kenney-Wallace: Chem. Phys. 32, 313 (1978).
20. V.V. Nikandrow, G.P. Brin, and A. Krasnovskii: Biokhimiya 43, 507 (1978).

Primary Process in the Photocycles of the Low pH Bacteriorhodopsin

T. Kobayashi, H. Ohtani, and J. Iwai

Department of Physics, Faculty of Science, University of Tokyo
Hongo 7-3-1, Bunkyo, Tokyo 113, Japan

A. Ikegami

Institute of Physical and Chemical Research, Wako, Saitama 351, Japan

1. Introduction

The photocycle of the light-adapted purple membrane (bR_{568}) has extensively been studied on the aspects of proton pumping and analogy to visual pigments [1]. Four intermediates, K(batho), L, M, and O, have been established [2-3]. The comparison between the photocycle of bacteriorhodopsin in low pH suspension and that in neutral pH suspension will be helpful for understanding the mechanism of proton pumping. The absorption maximum of bR_{568} shifts to 605 nm in low pH suspension [1]. The dynamical behavior of acidified purple membrane (bR_{605}) in the time region between 80 μs and 10 ms has been reported [4]. Recently we studied the dynamical behaviors of acidified purple membrane (bR_{605}) in the time region between 21 ps and 50 μs and reported that formation rates of K and L were identical with those of bR_{568} and that the formation of M, Schiff-base of which is unprotonated [5], was blocked in low pH suspension [6]. In the present work we discuss the effect of pH on the photocycle of bacteriorhodopsin in picosecond regime.

2. Experimental

Acidified bacteriorhodopsin (bR_{605}) was prepared with ion-exchange regin (Dowex 50) from the purple membrane [6]. The 630 nm pulse (the stimulated Raman scattering of 532 nm by acetone) was used for the excitation of bR_{605} without possible

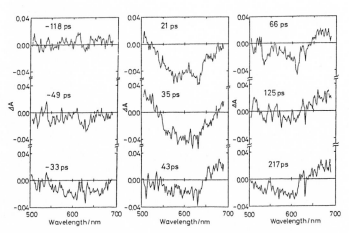

Fig.1 Picosecond difference absorption spectra of bR_{605} following 630 nm excitation at room temperature.

excitation of contaminated bR_{568}. All measurements were performed at 20-22 °C.

3. Results and Discussion

Figure 1 shows transient difference absorption spectra at room temperature. A bleaching was observed in 550-680 nm wavelength region within 21 ps. An increase in absorbance in the longer wavelength region was observed 43-217 ps after excitation. The measured difference absorption spectra were nearly identical with the spectrum 150 ns after excitation which was measured with a nanosecond spectroscopy apparatus [6]. The difference absorption spectra observed 43-217 ps and 150 ns after excitation are due to the appearance of K(batho) intermediate in low pH suspension (K_{acid}). Figure 2 shows the kinetics of transient difference absorption. The time constant of the formation of K_{acid} is estimated to be between 10-20 ps. Figures 1 and 2 clearly show the existence of a precursor of K_{acid}.

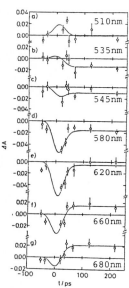

Fig.2 Kinetics of the difference absorption of bR_{605} following 630 nm excitation at room temperature.

Fluorescence lifetime of bR_{568} at room temperature has been reported to be 2-3 ps [7,8]. Ippen et al. found a rapid rise of a transient absorption with a time constant of 1.0 ps monitoring at 615 nm [9]. The transient may be assigned to be K. Kaufmann et al. and Applebury et al. found a transient S which has a lifetime of 11 ps and assigned to a precursor of K [10]. But Gillbro et al. reported that the transient S was not a physiologically meaningful intermediate in the photocycle but was an excited state of a photoproduct [11]. The time constant of the conversion from the lowest excited state of bR_{568} to K is 1-3 ps. The formation time constant of K_{acid} (10-20 ps) is one order larger than that of K. The result is consistent with the result that the fluorescence quantum yield of bR_{605} is one order larger than that of bR_{568} [8]. The precursor of K_{acid} is assigned to the lowest excited state of bR_{605}.

In the present study of bacterio-rhodopsin in low pH suspension the excitation energy (<90 µJ) is much less than that usually used (3 mJ). The effects of the photoproduct which are formed within excitation pulse width can be neglected. Furthermore there is a latent time before the appearance of the 680 nm band. The formation yield of K_{acid} was measured to be 0.11 ± 0.01 with the aid of nanosecond spectroscopy. This value is smaller than that of K (0.3) [12]. The branching ratio of trans-cis isomerization depends on pH. Reaction schemes are shown as follows,

$$bR_{568} \xrightarrow{h\nu} bR_{568}(FC) \longrightarrow bR_{568}(S_1) \longrightarrow P \xrightarrow{1\ ps} K$$

$$bR_{605} \xrightarrow{h\nu} bR_{605}(FC) \rightarrow bR_{605}(S_1) \rightarrow P_{acid} \xrightarrow{10ps-20ps} K_{acid}$$

where P and P_{acid} are branching states of trans-cis isomerization.

Our newly proposed reaction mechanism is as follows. The first relaxation process of Franck-Condon state to S_1 is the change of structures of chromophore and near residues. The $S_1 \rightarrow P$ conversion process is a pH dependent rate-controlling step which determines the quantum yield of fluorescence and that of the formation of P. The next step is a fast branching process which is also dependent on pH.

References

1. D. Oesterhelt and W. Stoeckenius: Nature New Biol. 233, 149 (1971).
2. R.H. Lozier, R.A. Bogomolni, and W. Stoeckenius: Biophys. J. 15, 955 (1975).
3. T. Iwasa, F. Tokunaga and T. Yoshizawa: Biophys. Struct. Mech. 6, 253 (1980).
4. P.C. Mowery, R.H. Rozier, Q. Chae, Y-W. Tseng, M. Taylor and W. Stoeckenius: Biochem. 18, 4100 (1979).
5. A. Lewis, J. Spoonhower, R.A. Bogomolni, R.H. Lozier, and W. Stoeckenius: Proc. Nat. Acad. Sci. USA 71, 4462 (1974).
6. T. Kobayashi, H. Ohtani, J. Iwai, A. Ikegami, and H. Uchiki: FEBS Lett. 162, 197 (1983).
7. R.R. Alfano, W. Yu, R. Govindjee, B. Becher, and T.G. Ebrey: Biophys. J. 16, 541 (1976).
8. T. Kouyama, K. Kinoshita, Jr., and A. Ikegami: to be published.
9. E.P. Ippen, C.V. Shank, A. Lewis, and M.A. Marcus: Science 200, 1279 (1978).
10. K.J. Kaufmann, P.M. Rentzepis, W. Stoeckenius, and A. Lewis: Biochem. Biophys. Res. Commun. 68, 1109 (1976); M. L. Applebury, K.S. Peters, and P.M. Rentzepis: Biophys. J. 23, 375 (1978).
11. T. Gillbro and V. Sundström: Photochem. Photobiol. 37, 445 (1983).
12. C.R. Goldschmidt, O. Kalisky, T. Rosenfeld, and M. Ottolenghi: Biophys. J. 17, 179 (1977); B. Becher and T.G. Ebrey: ibid. 185.

Picosecond Spectroscopy on the Primary Process in the Photoconversion of Protochlorophyllide to Chlorophyllide *a*

T. Kobayashi and J. Iwai

Department of Physics, Faculty of Science, University of Tokyo
Hongo 7-3-1, Bunkyo, Tokyo 113, Japan

M. Ikeuchi and Y. Inoue

Institute of Physical and Chemical Research, Wako, Saitama 351, Japan

1. Introduction

A key step in the development of chloroplasts in higher plants is the reduction of the precursor protochlorophyllide (PChlide) into chlorophyllide (Chlide), which is then esterified by phytol. The reduction process is activated by light and it involves the attachment of two hydrogen atoms. Spectroscopic studies of etiolated leaves or isolated etioplasts at low temperatures have been extensively performed for the investigations on the primary states of the chlorophyllide formation. However little is established about the detailed mechanism of this process at the present stage of research.

Besides the well-known absorption shift: P-630$\longrightarrow$P-650$\xrightarrow{h\nu}$C-678$\longrightarrow$ C-684 $\longrightarrow$ C-672 a nonfluorescent intermediate between P-650 and C-678 with absorption maximum near 690 nm has been detected [1,2]. The intermediate (X-690) was also found by transient spectroscopy with the use of nanosecond Nd:YAG laser with the resolution of 0.5 μs [3] and of 0.2 μs [4]. Low temperature experiment suggests that very complicated reactions are taking place in the P-650 $\xrightarrow{h\nu}$ X-690 $\longrightarrow$ C-678. At this stage of research where X-690 is found to be the precursor of C-678, a key problem in the reduction process of protochlorophyllide is whether or not the first photoproduct of protochlorophyllide is X-690. In order to elucidate the problem we have performed picosecond transient spectroscopy and found that there is at least one more intermediate between the excited P-650 and the intermediate X-690 [5].

2. Experimental

A mode-locked Nd:YAG laser (Quantel, Model 472) was used for the generation of both excitation and probe light sources. Excitation pulse at 630 nm is generated in the process of stimulated Raman scattering by focussing the second harmonic in a cell containing acetone. The pulse width (FWHM) and energy of the excitation pulse is about 25-30 ps and 70-90 J, respectively. Probe light is generated by focussing the fundamental in a cell containing water or heavy water.

3. Results and Discussion

Picosecond time-resolved difference spectra at the delay time of 100 ps (a), 700 ps (b), 1.5 ns (c), and 4.4 ns (d) are shown in Fig.1. The sharp spikes at 630 nm in the difference spectra are due to the excitation light scattered by the sample. The dotted curves in the figure were obtained by removing the effect of scattered excitation light. The fluorescence from the sample was much weaker than the monitoring light and it was

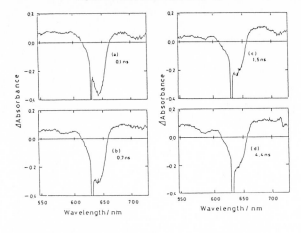

Fig.1: Time-resolved difference spectra
of PChlide at -the delay time of
100 ps (a), 700 ps (b), 1.5 ns (c), and
4.4 ns (d) after the excitation by the
stimulated Raman light (630 nm) of the
second harmonic (532nm) of Nd:YAG laser
with 30 ps width. The measurement was
performed at room temperature.

Fig.2: Time dependence
of absorbance change
due to 30 ps pulse (630
nm) excitation of PChlide
observed at 570 nm (a),
600 nm (b), 640 nm (c),
and 675 nm (d). The
measurement was performed
at room temperature.

found not to affect the time-resolved absorption spectrum by
blank-test experiments.

The typical features of the difference spectra at the delay
time ranges between 100 ps and 4.4 ns are as follows: (1) No
characteristic absorption peak was found in 690 nm region, where
X-690,which was found in the previous paper,has an absorption
maximum. (2) The negative absorbance change between 620 and
665 nm showed a maximum at 644 nm at 100 ps after excitation.
This wavelength is slightly longer than that of the absorption
maximum (639-642 nm) of PChlide. (3) The negative maximum
shifts toward shorter wavelengths with increasing delay time
with a time constant of about 1-2 ns. (4) The two wavelengths
for zero absorbance change initially observed at 620 and 665 nm
shift toward shorter wavelengths with increasing delay time
with a time constant of about 1-2 ns. (5) The amplitude of the
negative absorbance change at around 640 nm decreased with
increasing delay time with a time constant of about 1-2 ns.
(6) The ratio (R) of absorbance change at 690 nm (A_{690}) to that
at 640 nm (A_{640}) increases with delay time after excitation,
the time constant of the growth in R is about 1-2 ns.

The following view can be obtained from the above-mentioned
results: there are at least three transient species which
precede X-690. They may be either an excited state of PChlide
or chemically different intermediate species. We denote the
first one as X_{-1}, which has a growth time of <50 ps and the
second as X_0 which has a lifetime of about 1-2 ns and an

485

absorption maximum at <640 nm, and the third as X_1, which has a much longer lifetime than a few nanoseconds. The process can be expressed in the following way:

$$PChlide(S_0) \xrightarrow{h\nu} X_{-1} \xrightarrow{<50ps} X_0 \xrightarrow{1-2\ ns} X_1 \longrightarrow X\text{-}690 \longrightarrow Chlide(C\text{-}678).$$

The formation time constant of X_0 is shorter than 50 ps. The relaxation time (<50 ps) in the present system is probably due to the relaxation associated with the change in the configuration of protein moiety in PChlide. Therefore X_{-1} and X_0 can be assigned to $S_1(FC)$ and $S_1(EQ)$ in PChlide, respectively. The fluorescence lifetime of X_0 was measured to be 1-2 ns.

As a conclusion, the primary process of chlorophyll biosynthesis is given as follows:

$$PChlide(S_0) \xrightarrow{h\nu} PChlide(S_1(FC)) \xrightarrow{<50ps} PChlide(S_1(EQ)) \xrightarrow{1-2ns}$$
$$X_1 \xrightarrow{>>4ns} X\text{-}690 \longrightarrow Chlide(C678) \qquad \downarrow 1\text{-}2\ ns$$
$$PChl\ (S_0) + h\nu_F$$

4. References

1. F.F. Litvin, N.V. Ignatov and O.B. Belyaeva: Photochem. Photobiol. 2, 233 (1981).
2. O.B. Belyaeva and F. F. Litvin: Photosynthetica 15, 210 (1981).
3. F. Frank and P. Mathis: Photochem. Photobiol. 32, 799 (1980).
4. Y. Inoue, T. Kobayashi, T. Ogawa and K. Shibata: Plant & Cell Physiol. 22, 197 (1981).
5. J. Iwai, M. Ikeuchi, Y. Inoue and T. Kobayashi: Protochlorophyllide Reduction and Greening, eds. C. Sironval and M. Brouers, Martinus Nijhoff/Dr W. Junk Publishers, pp.99-112 (1984, The Hague).

Energy Transfer in Photosynthesis: The Heterogeneous Bipartite Model

S.J. Berens, J. Scheele, W.L. Butler, and D. Magde
University of California at San Diego, La Jolla, CA 92093, USA

Time-resolved fluorescence measurements let us measure energy transfer from the absorbing pigments called the antenna to the reaction center in chloroplasts, the structures responsible for photosynthesis in green plants. It is accepted that reaction centers capture excitation energy from antennae and quench fluorescence efficiently only when they are in a resting or "open" state. This results in a minimal fluorescence yield which is termed Fo. Once they have captured an exciton, reaction centers cannot readily capture another and are said to be "closed." Fluorescence is a maximum under this Fm condition. Steady state fluorescence measurements at Fo, Fm, and intermediate conditions, with and without a variety of perturbing agents present, have led to specific models for the photosynthetic apparatus. These have implicit predictions for time-resolved fluorescence decays as well.

We worked with spinach chloroplasts. Broken chloroplasts were isolated by the method of BUTLER and YAMASHITA [1]. The temperature was about 15 °C for the measurements. For excitation, we used a synch-pumped dye laser which generated 8 ps pulses at 635 nm. Solutions flowed continuously. For measurements at Fm, the solutions contained 10 μM DCMU and 20 mM hydroxylamine (pH 7.8) and were preirradiated for 3 minutes. The single photon time-correlation electronics were standard. Instrument response was 440 ps. Deconvolution was by iterative reconvolution. A sum of exponentials was assumed.

Even under extreme Fo and Fm conditions, we require three exponentials to fit our data. The yield of fluorescence associated with each component is calculated as $\phi_i(t) = a_i T_i$, where T_i is a characteristic lifetime and a_i is the pre-exponential amplitude. The yields assigned are normalized to unity for Fo total emission. Typical data are shown in Table 1.

Table 1. Fluorescence under Fo and Fm conditions

	T_1	T_2	T_3	a_1	a_2	a_3	ϕ_1	ϕ_2	ϕ_3
Fo	.08	.42	3.2	.74	.26	.002	.35	.61	.04
Fm	.09	1.2	2.8	.60	.18	.22	.30	1.1	3.2

The long component at Fo contributes very little to the amplitude, only 3 parts per thousand. It may involve a little free chlorophyll, a few closed reaction centers, and a few antennae deficient in reaction centers. This component probably has nothing to do with authentic Fo conditions.

Part of the fast component at both Fo and Fm has a simple interpretation: Chloroplasts contain two types of photosynthetic units, which have different reaction centers, carry out different photochemistry, and are linked to

487

different sorts of antennae. One of these, PSI, is thought to have a very short decay time. Part of the fast component we attribute to PSI. The remainder we assign to a fast initial transient associated with PSII.

At Fo, the middle component must be assigned to PSII. At Fm both the longer two components must surely be associated with emission from PSII units, but then what distinguishes the middle and long components?

To interpret our results, we propose [2,3] a model based upon the "bipartite" model [4]. The governing kinetic equation is

$$\text{Antenna} \underset{k_t}{\overset{k_T}{\rightleftarrows}} \text{Reaction Center} \xrightarrow{k_p}$$
$$\downarrow k_F + k_D \qquad\qquad\qquad \downarrow k_d$$

where k_F is the radiative rate constant, k_D and k_d are nonradiative decay rates, and k_p is an effective rate for an irreversible step leading ultimately to photochemistry. At Fm, $k_p = 0$; at Fo it is finite and large. Only PSII is considered; PSI is assumed to have a fast decay which may be added to give the overall fluorescence.

The first important point to emphasize is that even when one monitors only the emission from the antennae, this model predicts two exponential decays, in general. Furthermore, the decay times cannot be associated with any single rate constant of the model. We have developed the mathematics in detail and tested a broad range of parameter values for the rate constants [5]. We require that the parameters predict not only the kinetic data but also the absolute fluorescence and photochemical quantum yields. The model can readily explain the observation that a portion of the fastest component at both Fo and Fm is due to a fast initial transient from PSII. However, there is no way to account for the existence of two long components at Fm as long as all the PSII units are identical.

Having in mind that certain conventional experiments in recent years have suggested that there may be two types of PSII units, we propose [2,3] that the two longer components in the Fm data may be attributed to decay from two different sorts of PSII systems. We conclude that there are two exponentials mixed together, unresolved, in the middle component of the Fo data. In the spirit of the original model, we demand that the only change allowed between Fo and Fm conditions is the drop in k_p to zero at Fm.

Table 2. Rate parameters of the bipartite heterogeneous model (ns^{-1})

	k_T	$k_F + k_D$	k_t	k_d	k_p
PSII α	3.0	.20	3.3	.53	13.5
PSII β	7.0	.33	5.6	1.2	21

We interpret the smaller value of k_T in PSII α as reflecting a larger antenna size resulting in slower average transfer to the reaction center. The larger values for k_D and k_d in PSII β may reflect coupling of PSIIβ to PSI units, resulting in an extra decay channel quenching beta units.

Other groups have made related measurements [6-8]. The decays measured are similar to ours. However, an alternative explanation has been proposed, based upon the tripartite model [3], which postulates that the antennae of

PSII units have an inner core tightly coupled to the reaction center and a more loosely coupled auxiliary antenna called the light harvester. One may account for the fast decay component as coming from PSII cores (along with any PSI contribution), the middle component as coming from the light harvesters as they transfer energy to the cores, and the long component as coming from excitation which has visited a closed reaction center and then returned to the antenna pool. We have not succeeded in fitting our data to such a model, although we can come close by allowing all parameters to change between Fo and Fm and introducing additional processes.

In order to generate more evidence, we carried out experiments in which we modified the system to influence selectively the reaction center or the antenna. We irradiated with ultraviolet light, which is supposed to affect only the reaction centers. We treated the chloroplasts with DBMIB, which increases k_D, and we "phosphorylated" the chloroplasts, which is supposed to convert $PSII_\alpha$ into $PSII_\beta$. We are driven to the conclusion that the middle component at both Fo and Fm involves excitation which has already visited the reaction center. Space does not permit reproducing the figures and tables displayed on our poster. Details will be published elsewhere.

We conclude that the heterogeneous bipartite model is the simplest model able to account for currently known data. We believe that it incorporates a significant feature. However, there are already reasons to believe that the complete picture is yet more complicated. It may be that the assignment of three exponentials is simply a parameterization of the curves. Still they are a convenient way to quantify the effect of perturbations and to describe the curve that must be reproduced by quantitative models.

Acknowledgement: This work was supported in part by the NSF.

References.

1.   T. Yamashita and W. L. Butler: Plant Phys. 43, 1978 (1968).

2.   D. Magde, S. J. Berens, and W. L. Butler: S.P.I.E. 322, 80 (1982).

3.   W. L. Butler, D. Magde, and S. J. Berens: PNAS USA, 80, 7510 (1983).

4.   W. L. Butler: in Chlorophyll Organization and Structure, Ciba Foundation Symposium (Excerpta Medica, Elsevier/North-Holland, 1979) p. 237.

5.   S. J. Berens: Ph. D. Dissertation, Univ. of Calif., San Diego (1984).

6.   W. Haehnel, J. A. Nairn, P. Reisberg, and K. Sauer: Biochem. et Biophys. Acta 545, 496 (1982).

7.   W. Lotshaw, R. Alberte, and G. R. Fleming: Biochem. et Biophys. Acta 682, 75 (1982).

8.   W. Haehnel, A. R. Holzwarth, and J. Wendler: Photochem. and Photobiol. 37, 435 (1983).

Excitation Energy Transfer in Phycobilin-Chlorophyll. A System of Algal Intact Cells

I. Yamazaki, N. Tamai, and T. Yamazaki

Institute for Molecular Science, Myodaiji, Okazaki 444, Japan

M. Mimuro and Y. Fujita

National Institute for Basic Biology, Myodaiji, Okazaki 444, Japan

1. Introduction

A light-harvesting antenna system of red and blue-green algae consists of phycobilisomes and thylakoid membrane: the former is a supramolecular unit involving several kinds of phycobiliproteins, as shown in Fig. 1 [1]. In the course of the energy transfer from the initially photoexcited phycobiliprotein to the reaction centers (RC) I and II, fluorescence is emitted from almost every type of pigment and can be used as a probe to examine the mechanism of energy transfer within the pigment system [2-4].

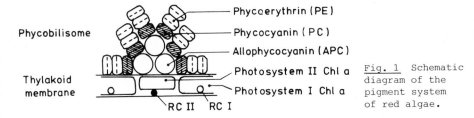

Phycobilisome — Phycoerythrin (PE)
— Phycocyanin (PC)
— Allophycocyanin (APC)

Thylakoid membrane — Photosystem II Chl a
— Photosystem I Chl a

RC II   RC I

Fig. 1 Schematic diagram of the pigment system of red algae.

We report here the picosecond time-resolved fluorescence spectra of some algae and the energy-transfer kinetics in a phycobilin-chlorophyll $\underline{a}$ (Chl $\underline{a}$) system $\underline{in}$ $\underline{vivo}$. A comparison of the rate constants of the energy transfer processes is made between the systems chromatically adapted, i.e., red-grown and green-grown systems.

2. Experimental

$\underline{Flemyella}$ $\underline{diplosiphon}$ (M-100) was grown under red light (2.5 W/cm^2) for establishing the phycoerythrin (PE)-less system and under green light (1.0 W/cm^2) for the PE-rich system. Picosecond time-resolved fluorescence spectra were obtained with a synchronously pumped, cavity-dumped dye laser and time-correlated single-photon counting system with 50-ps time resolution [5,6]. Fluorescence decay curves were measured successively at different monitoring wavelengths, usually with a 0.625 nm interval. Time-resolved spectra with the minimum time difference of 12.8 ps were obtained from a set of fluorescence decay curves with the aid of computer.

3. Results and Discussion

Figure 2 shows time-resolved fluorescence spectra of two kinds of $\underline{Flemyella}$ $\underline{diplosiphon}$ intact cells, i.e., red-grown cells (PE-less) and green-grown cells (PE-rich). The spectrum of PE-less system (Fig. 2(a)) changes with time as follows: (1) Phycocyanin (PC) spectrum appears at 0-100 ps, with its peak being shifted gradually to the red, (2) at 100-400 ps, allophycocyanin (APC)

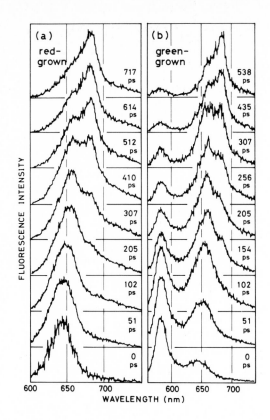

(a) red-grown
717 ps
614 ps
512 ps
410 ps
307 ps
205 ps
102 ps
51 ps
0 ps

(b) green-grown
538 ps
435 ps
307 ps
256 ps
205 ps
154 ps
102 ps
51 ps
0 ps

FLUORESCENCE INTENSITY

WAVELENGTH (nm)

600   650   700   600   650   700

Fig. 2 Time-resolved fluorescence spectra of Flemyella
diplosiphon: (a) red-grown
and (b) green-grown intact
cells. The fluorescence
bands appear with peaks at
578 (PE), 645 (PC), 660
(APC) and 683 nm (Chl a).

spectrum becomes dominant and then (3) at 500 ps, Chl a spectrum appears
clearly and no longer changes after 700 ps. The PE-rich system (Fig. 2(b))
exhibits PE spectrum in the initial time region (0-200 ps) in addition to
sequential appearance of the PC, APC and Chl a spectra.

The time-resolved spectra were analyzed into components to obtain the rise
and decay curve of the pigment component itself. The results are summarized
as follows: (1) the fluorescences from various pigments appear sequentially
in the order of PE - PC - APC - Chl a, i.e., from the outer surface to the
inner core of the pigment system (Fig. 1). (2) In every case the rise time
is significantly short compared with the decay time. These sequential time
behaviors can be simulated with the decay kinetics of $\exp(-2kt^{1/2})$ type as
proposed by PORTER et al. [2] for Porphyridium cruentum. The rate constants
thus obtained are presented in Table 1. The energy transfer at PC level is
faster in PE-rich system, whereas those at APC and at Chl a levels are almost
identical. The decay kinetics of $\exp(-2kt^{1/2})$ type can be derived from the
Förster kinetics under the condition that the energy transfer occurs with an
extremely high efficiency between the donor and acceptor chromophores. This
means that the chromophores are closely packed in the phycobilisomes.

It is seen from Table 1 that there is regularity among various species:(1)
In the case of PE-less systems, the energy transfer of PC $\longrightarrow$ APC is slower
than that of APC $\longrightarrow$ Chl a, and the former takes almost the same rate constant
except for P. cruentum. (2) In the case of PE-rich systems, the process PC $\longrightarrow$
APC is faster than that in PE-less system and is even faster than the process

491

Table 1. Rate constants of the energy transfer in phycobilin-Chl a system of the algal intact cells

| Pigment system | Algae | Rate constants (k, ps$^{-1/2}$) | | | |
		PE	PC	APC	Chl a
PC-APC-Chl a	Flemyella diplosiphon	–	0.071	0.13	0.042
	Anabaena cylindrica	–	0.065	0.33	0.059
(PE-less)	Anabaena variabilis	–	0.077	0.14	0.057
	Anacystis nidulans	–	0.065	0.13	0.067
	Porphyridium aerugineum	–	0.12	0.13	0.091
PE-PC-APC-Chl a	Flemyella diplosiphon	0.11	0.40	0.13	0.056
(PE-rich)	Nostoc sp.	0.17	0.29	0.13	0.063
	Porphyridium cruentum	0.14	0.24	0.27	0.10

Rate constants (k) were calculated with the decay kinetics of $\exp(-2kt^{1/2})$.

APC $\longrightarrow$ Chl a in blue-green systems. It is to be noted that the transfer rate is not necessarily higher at later steps within phycobilisomes. (3) The transfer from APC takes an almost identical rate constant except for two cases. (4) Decays of Chl a fluorescence are similar among blue-green systems, but it is slightly faster in red algal systems. It will depend on the primary photoreaction at the reaction center and the spill-over to photosystem I. Some difference in the phycobilin structure may be suggested between the two algal groups. In conclusion, deviation in kinetic feature is generally small, or negligible, in algal systems of the same type of pigment composition.

The energy transfer in PE-rich systems appears to be characterized by a rapid step at PC level when compared with PE-less systems (Table 1). We should note here that difference in the global structure of phycobilisome little affects the energy transfer kinetics; phycobilisome of P. cruentum is hemispherical, whereas that of Nostoc sp. is hemidiscoidal [7]. Therefore, it is most likely that the presence of PE in the outer surface of phycobilisome is an important factor in the energy transfer process PC $\longrightarrow$ APC. This is directly proved from the present experiments with the two systems of Flemyella diplosiphon chromatically adapted.

References

1. E. Gantt: Int. Rev. Cytol., 66, 45 (1980).
2. G. Porter, C.J. Tredwell, G.F.W. Searle and J. Barber: Biochim. Biophys. Acta, 501, 232 (1978).
3. W. Haehnel, J.A. Nairn, P. Reisberg and K. Sauer: Biochim. Biophys. Acta, 680, 161 (1982).
4. W. Haehnel, A.R. Holzwarth and J. Wendler: Photochem. Photobiol. 37, 435 (1983).
5. T. Murao, I. Yamazaki and K. Yoshihara: App. Opt. 21, 2297 (1982).
6. I. Yamazaki, M. Mimuro, T. Murao, T. Yamazaki, K. Yoshihara and Y. Fujita: Photochem. Photobiol. 39, 233 (1984).
7. E. Gantt and S.F. Conti: J. Cell. Biol. 29, 423 (1966).

Analysis of Fluorescence Kinetics and Energy Transfer in Isolated α Subunits of Phycoerythrin from *Nostoc* Sp

A.J. Dagen and R.R. Alfano
Institute for Ultrafast Spectroscopy and Lasers, Department of Phyiscs
The City College of New York, New York, NY 10031, USA
B.A. Zilinskas
Department of Biochemistry and Microbiology, Cook College, Rutgers University
New Brunswick, NJ 08903, USA
C.E. Swenberg
Radiation Sciences, Department, Armed Forces Radiobiology Research Institute
Bethesda, MD 20814, USA

I. INTRODUCTION

Photosynthetic organisms have evolved a number of light harvesting antenna systems for the primary purpose of absorbing sunlight and transferring the absorbed energy to reaction centers. In red and blue-green algae, the light harvesting system consists of an aggregation of phycobiliproteins, namely phycoerythrin, phycocyanin, and allophycocyanin, which collectively form the phycobilisomes. Each phycobiliprotein is composed of two dissimilar polypeptide chains, the α and β subunits, to which chromophores are covalently bonded. The number and chemical nature of the chromophores depend on the origin and spectroscopic class of the phycobiliproteins. For *Nostoc* Sp, the α and β subunits of phycoerythrin have two and four chromophores, respectively. The basic unit is the trimer form $(\alpha\beta)_3$ which has dimensions of approximately a right circular disk of radius 60 Å and height 30 Å [1].

In this report, picosecond laser spectroscopy has been used to characterize the intramolecular energy transfer and fluorescence properties of the α subunit of phycoerythrin isolated from *Nostoc* sp. The measurements indicate that both 's' and 'f' chromophores within the α subunit absorb and fluorescence with lifetimes of approximately 84 psec and 1.1 ns, respectively. The s→f transfer rate of the α unit is on the order of 16 psec. In contrast to experiments on intact phycobilisomes and isolated phycobiliproteins [2,3,4] where singlet-singlet exciton annihilation occurs at high excitation intensities, no indications of exciton fusion within the small α subunit was observed at the highest pump excitations.

II. METHODS

A single 6 ps, 530 nm excitation pulse from a Nd:glass laser system was used to excite the sample. The sample was frontally excited and the fluorescence was focused onto the entrance slit of the streak camera. The output from the streak camera was digitized by an OMA and stored in a computer. The intensity of the excitation beam at the sample site and the fluorescence signal were measured by PMT's connected to an integrator. A photodiode located behind the sample, also connected to the integrator, measured the transmitted intensity of the beam.

III. RESULTS AND DISCUSSION

The absorption spectrum of α unit is displayed in fig. 1. The spectrum can be deconvoluted into bands associated with the s and f chromophores. The deconvoluted curve was obtained by assuming mirror symmetry between the absorption and emission spectra for the 'f' chromophore. The deconvolution was checked by comparing the experimental values of the polari-

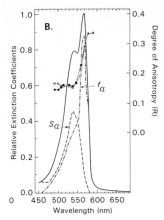

Figure 1: Absorption spectrum of α subunit and the theoretical deconvolution of the spectrum into its s and f components. Theoretical predicted polarization anisotropy R (closed circles) using deconvoluted spectrum and experimental polarization anisotropy values (open circles).

zation anisotropy with the calculated weighted sum of the anisotropy using the absorption bands and assuming values of anisotropy for direct excitation of the s and f chromophores. For wavelengths greater than 580 nm direct excitation of the 's' chromophore is negligible.

The fluorescence decay profile for laser excitation of 8.3×10^{13} photons/cm^2 is shown in fig. 2a with a 3-67 filter in the fluorescence. The shape of the fluorescence profiles was found to be intensity-independent over the range of 4×10^{13} to 4×10^{15} photons cm^{-2} per pulse. The decay profiles are doubly exponential with an e^{-1} time of approximately 1.1 ns. Fig. 2b shows the effect of adding a 2-59 filter in front of the streak camera on the time-dependent emission. The 2-59 filter transmits light of wavelengths beyond 590 nm only. In this case, the decay exhibits single exponential behavior with an e^{-1} time of 1.1 ns. In contrast

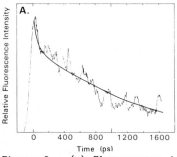
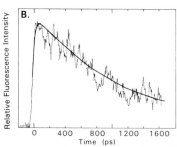

Figure 2: (a) Fluorescence kinetic profile, excitation wavelength 530 nm, 3-67 filter, 200 psec per division. Solid line denotes fit to $0.79 \exp(-K_1 t) + 0.19 \exp(-K_2 t)$ with $K_1 = 8.76 \times 10^8 \text{s}^{-1}$ and $K_2 = 3.68 \times 10^{10} \text{s}^{-1}$. Peak of fluorescence intensity is defined as t=0; (b): excitation wavelength 530 nm, 2-59 and 3-67 filters, 200 psec per division. Solid line denotes $0.99 \exp(-K_2 t) - 0.56 \exp(K_3 t)$ fit to data with $K_2 = 8.76 \times 10^8 \text{ sec}^{-1}$, $K_3 = 3.68 \times 10^{10} \text{ sec}^{-1}$.

494

to the decay profiles in the absence of the 2-59 filter, the kinetics display a slight rounding in the emission profiles for t<200 psec when a 2-59 filter is present, indicative of energy transfer from the s to f chromophore.

The dependence of the relative quantum yield and the ratio of transmitted light at high intensity to that at low intensity are shown in fig. 3. The two curves are approximately mirror images of each other. Usually a decrease in yield at high intensity would suggest annihilation of excitons.

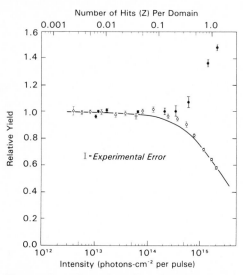

Figure 3: Relative fluorescence quantum yield and transmission as a function of laser single pulse intensity (photon cm^{-2}), O measured relative quantum yield, ● measured relative transmission. Solid line is fit of relative fluorescence quantum yield to the Paillotin-Swenberg annihilation theory.

The intensity-independent fluorescence kinetics and mirror image of quantum yield and transmission data indicate the absence of exciton annihilation. We, therefore, have analyzed the data using rate equations in terms of fluorescing 's' and 'f' chromophores with energy transfer from 's' to 'f'. Back transfer was neglected because the activation energy is on the order of 1000 cm^{-1} between the s and f units. The parameters which best fit both kinetic curves correspond to radiative lifetimes of 333 ps and 2.16 ns for the 's' and 'f' chromophores, respectively. These are solid curves fitting the data. When the time response of the system is taken into account, the 's' to 'f' transfer rate is calculated to be 16 ps. We attribute the increase in relative transmission to ground state depletion and upper excited state absorption.

Similar kinetic studies of the β and (αβ) forms of the same pigment from the same alga also indicate the absence of annihilation in those small units. We have been able to describe the fluorescence kinetics in all three forms of the pigment by linear transfer processes and found that the (αβ) kinetics can be explained by assuming the same transfer rates as in the isolated α and β units with an appropriate coupling constant.

The lack of exciton annihilation in the small forms of the pigment, while existing in the larger forms, and the ability to explain the kinetics in terms of linear energy transfer schemes is of fundamental interest and opens future avenues of research in small biological units.

IV. ACKNOWLEDGEMENT

This research is supported by a grant from AFOSR and a PSC/BHE of CCNY award.

V. REFERENCES

1. Bryant, D. A., G. Guglielmi, N. Tandeau deMarsac, A. M. Castets, and G. Cohen-Bazire (1979) Arch. Microbiol. 123, 113-127.
2. Porter, G., C. J. Tredwell, C. F. W. Searle and J. Barber (1978) Biochem. Biophys. Acta. 501, 232-245.
3. Doukas, A. G., V. Stefancic, J. Buchert, R. R. Alfano, B. A. Zilinskas (1981), Photochem. Photobio. 34, 505-510
4. Pellegrino, F., D. Wong, R. R. Alfano, B. A. Zilinskas (1981), Photochem. Photobio. 34, 691-696.

Picosecond Time-Resolved Fluorescence Spectra of Hematoporphyrin Derivative and Its Related Porphyrins

M. Yamashita, M. Nomura, S. Kobayashi, and T. Sato

Electrotechnical Laboratory, 1-1-4 Umezono, Sakura-mura, Niihri-gun
Ibaraki 305, Japan

K. Aizawa
Department of Physiology, Tokyo Medical College, 6-1-1 Shinjuku
Shinjuku-ku, Tokyo 160, Japan

As it is recognized that photoradiation therapy using hematoporphyrin derivative (HpD) as a photosensitizer is effective in treatment of human cancers, its photophysical, photochemical and biological behaviors are being widely investigated. The mechanisms enabling the selective accumulation of HpD in cancerous cells and of the subsequent energy relaxations and photochemical reactions are not well understood. For the purpose of elucidation of the mechanisms, the fluorescence decay kinetics from the first excited singlet state has been investigated in the time region of nanoseconds [1] and picoseconds [2-3]. Consequently, it has been found that a fluorescence decaying curve of HpD in a phosphate buffer saline aqueous (PBS) solution shows two components of fast and slow decays. The fluorescence decay is lengthened with decreasing HpD concentration. It is generally thought that the fast and slow decay components are due to the aggregates (including dimers) and monomers, respectively [1,3]. On the other hand, from biological studies on the relation between HpD components and treatment efficacy [4,5], it has been found that the aggregates are essential to treatment. The selective accumulation of HpD is thought to involve the aggregational properties of HpD components. However, the direct spectroscopic information on the energy relaxation from the excited state through the interaction between the aggregates and the monomers and the accumulation properties is not yet obtained. In this paper, we report the first investigation of the picosecond time-dependent fluorescence spectra $I(\lambda, t)$ from HpD and photofrin II (the effective aggregate fraction separated by gel filtration from the HpD solution: HpD II) in the PBS solutions by using a two-dimensional synchroscan streak camera method.

The HpD and HpD II solutions were offered from Dougherty's group. It is said that the HpD II solution mainly contains ether bonded dimers of hydroxyethylvinyl-deutroporphyrin IX and hematoporphyrin IX [4]. The experimental apparatus for measuring picosecond time-dependent fluorescence spectra is improved in comparison with the previous one [2]. A polychrometer is attached between a sample optics and a synchroscan streak camera. For data recording and processing, an OMA II (two dimension) is used. A sychronously mode-locked CW dye (R6G) laser generates continuous trains of pulses at 82 MHz with the average power of 30 mW and the duration of less than 5 ps at 570 nm. The overall time resolution of the system was examined by measuring the duration of pulses from the dye laser. The recorded pulse duration at the present operational conditions was 19 ps at the accumulation of $\approx 10^9$ pulses. The reabsorption effect of the fluorescence at the high concentration was carefully avoided by measuring the front fluorescence near the surface of the sample cell pumped at the same angle in respect to its surface.

The picosecond time-dependent fluorescence spectra $I(\lambda, \Delta t_n)$ were measured for HpD and HpD II solutions at different concentrations. Correspondingly the one-dimensional picosecond decays $I(t)$ of the total

497

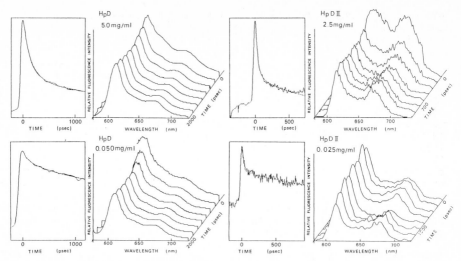

Fig. 1. Picosecond time-dependent fluorescence spectra and the corresponding total fluorescence decaying curve from the HpD & HpD II solutions at the different concentrations.

fluorescence spectra were also measured. In Fig.1, the typical time-dependent fluorescence spectra and the corresponding total fluorescence decays at the highest (5 mg/ml) and low (1/100) concentrations are shown. It is found that there is a new emission band (Y band) around 635 nm between the usual two bands around 615 and 675 nm for both the HpD and HpD II solutions at the highest concentration. The Y band decays faster than 200 ps, while the other bands decay slower than 3 ns. With decreasing the concentration the Y band disappears rapidly. These behaviors of the Y band are more remarkable for the HpD II solution than for the HpD solution. The one-dimensional fluorescence decaying curves also show to be compris of fast and slow components. From the comparison between the time-dependent spectra and the one-dimensional decaying curves, it is found that the time and concentration variations of the fast component correspond to the behavior of the Y band. Therefore, it can be concluded that the fast decay appearing in the total fluorescence decay I(t), which is thought to be due to the aggregates [1,3], is assigned to the Y band. This was also confirmed by measuring the wavelength-resolved fluorescence decays $I(\Delta\lambda, t)$ for the peak of each band.

According to Kasha's theoretical analysis on the energy relaxation from the excited states of the molecular aggregates [6], the head-to-tail dimer shows the red shift relative to the monomer excited-state level. This suggests that the Y band originates in the emission of the head-to-tail dimer. We suppose for the head-to-tail dimer that two types of porphyrin dimers exist, the one is a covalenty ether-bonded dimer [4] and the other is a van der Waals dimer (non-convalent dimer kinetically in equilibrium with monomer), since the Y band depends on the total concentration. It should be noted that the formation of the ether-bonded dimer necessitates the pre-formed non-convalent dimer properly aligned [4].

Since the static fluorescence spectrum (which corresponds to the fluorescence spectrum of the slow component) from the monomer overlaps partially the absorption spectrum at the highest concentration due to the dimers, the fluorescence of the slow component can be dynamically quenched through the Förster type resonance energy transfer from the monomer to the dimers. Therefore, the total fluorescnece decay $I(t)$ is expressed by the equation $I(t) = A1\exp[-t/\tau_1] + A2\exp[-t/\tau_2 - \sqrt{\pi}\gamma\sqrt{t/\tau_2}]$. The first and second terms describe the fast fluorescence decay of the Y band from the dimers and the slow and dynamically quenched fluorescence decay of the usual bands from the monomer, respectively. A result of numerical fits (the dotted lines in Fig.1) to the experimental decaying curves for the HpD solution gives τ_1, τ_2, A1, A2 , and γ as the following values; $\tau_1 = 120$ ps, $\tau_2 = 3.6$ ns, A1=0.48, 0.07(1/100), A2 =0.52, 0.93(1/100), γ =0.38, 0.03(1/100). The numerical fitting curves agree well with the experimental ones. The combination of the total dimer concentration Cd (calculated by A1, A2, τ_1, τ_2 and the additional parameters of the ratios of the molar extinction coefficient $\varepsilon_1/\varepsilon_2$ and the fluorescence quantum yield ϕ_1/ϕ_2) with the value of γ enables us to determine the apparent critical distance Ro for the resonance energy transfer. It was confirmed by using the Ro that the true critical distance Rot estimated by extrapolating the Ro towards the higher concentration is almost equal to the value of 24 $\overset{\circ}{A}$ evaluated on the basis of the spectral overlapping integral.

The static fluorescence spectra from the gastric cancers, the erosion of the cancers, and the surrounding normal cells after 48 hours injection of the HpD solution (5 mg/kg weight) were simultaneously measured in vivo using a fiber endoscope with a polychrometer and an OMA I. The result showed that only the static fluorescence spectra from the gastric cancers have a third emission band which corresponds to the Y band. Therefore, it is deduced that the face-to-head dimers play a role important for the accumulation of HpD in cancerous cells and the fluorescence decaying time of the dimers is remarkably lengthened by binding to some biomolecule in vivo.

Reference

1. A. Andreoni, R. Cubeddu, S. De Silvestri, P. Laprota, G. Jori & E. Reddi, Chem. Phys. Lett. 88 (1982) 33.
2. M. Yamashita, T. Sato, K. Aizawa & H. Kato, 'Picosecond Phenomena III' Ed. by K. B. Eisenthal, R. M. Hochstrasser, W. Kaiser & A. Laubereau, (Springer-Verlag, Berlin, 1982) p. 298, and IEEE Trans. Instru. & Meas. IM-32 (1983) 124.
3. C. Hanzlik, W. Knox, T. Nordlund, R. Hilf & S. Gibson, Biological Physics Group Technical Report 37 (Dep. of Physics & Astronomy,Univ. of Rochester) May 2 (1983).
4. T. J. Dougherty, D. G. Boyle, K. R. Weishaupt, B. A. Henderson, W. R. Potter, D. A. Bellinier & K. E. Wityk, 'Porphyrins Photosensitization' Ed. by D. Kessel & T. J. Dougherty, (Plenum Press. N.Y. 1983) p.3, Cancer Res. 42 (1982) 1188, and Photochem. & Photobiol. 38 (1983) 377.
5. J. Moan & S. Sommer, Photobiochemisry & Photobiophysics 3 (1981) 93 and Cancer Lett. 15 (1982) 161.
6. M. Kasha, Radiation Research 20 (1963) 55.

Femtosecond Spectroscopy of Bacteriorhodopsin Excited State Dynamics

M. C. Downer, M. Islam, and C. V. Shank
AT&T Bell Laboratories
Holmdel, New Jersey 07733

A. Harootunian and A. Lewis
School of Applied and Engineering Physics
Cornell University, Ithaca, New York 14853

The dynamics of the photochemistry of bacteriorhodopsin have been investigated extensively.[1] Bacteriorhodopsin is a pigment related to visual rhodopsins which is extracted from the cell membrane of the bacterium **Halobacterium halobium**. This pigment acts as a light-driven proton pump converting light energy into electrochemical energy in the form of a proton gradient across the cell membrane. In this paper we report time-resolved ($\Delta t \sim 100$ fsec.) spectroscopic measurements of the initial step in the photochemical cycle of light-adapted bacteriorhodopsin (bR_{570}), namely the transformation of bR_{570} into bathobacteriorhodopsin (K_{610}). Our results confirm the hypothesis[2] that the ground state of K_{610} forms directly from the excited state of bR_{570}. In addition, we present the first measurements of the influence of deuteration on the primary photochemistry at physiological temperatures.

Early investigations with microsecond time resolution[3] could not resolve the formation time of the first intermediate (K_{610}). Later measurements using 6 psec. optical pulses[4] were still instrumentally limited. Ippen et al.[5] finally reported measuring a K_{610} formation time of 1.0 ± 0.5 psec. using 0.5 psec. pulses at 615 nm. Since only a single laser frequency was used, however, the evolution of the induced absorption spectrum was not determined.

In our experiment we employed 130 fsec. white light continuum pulses[6] as ultrafast spectral probes. Photoexcitation was provided by amplified[7] 90 fsec. pulses from a colliding-pulse mode-locked laser.[8] Continuous sample flow was maintained to remove photochemical products from the interaction region between pulses. The optically induced normalized transmittance spectra are plotted in Fig. 1 for bR_{570} in D_2O for 0.2 psec. and 5.0 psec. pump-probe time delays. Strong bleaching is observed near 570 nm and increased absorbence is observed near 620 nm which is clearly delayed in time. These spectra support the model in which K_{610} forms directly from the initially excited bR_{570}. To determine the rate of batho formation we plotted the transmittance change as a function of time for two frequencies, 570 nm and 620 nm. The results are shown in Fig. 2 where the points are experimental and the curves represent least squares fits of a single exponential to each set of data points. Both fits yield the decay constant $\tau = 1.0 \pm 0.1$ psec.

The data shown in Figs. 1 and 2 can be interpreted as the instantaneous formation of the excited state bR^*_{570} of bacteriorhodopsin followed by the slower formation of the ground state of bathobacteriorhodopsin from this excited state, as shown schematically in Fig. 3. Some recovery of the original ground state may also occur. We can symbolically denote the sequence of events by the chemical equation below:

$$bR_{570} \underset{h\nu}{\overset{\tau_1}{\underset{\longrightarrow}{\longleftarrow}}} bR^*_{570} \overset{\tau_2\,(H_2O)}{\underset{\tau_2\,(D_2O)}{\longrightarrow}} K_{610} . \qquad (1)$$

The sharp initial rise in transmittance around 570 nm corresponds to the instantaneous bleaching of the absorption maximum. The observed magnitude of the signal indicates that approximately 20 percent of the ground state molecules are excited. Recovery of the 570 nm bleaching from both the formation of bathobacteriorhodopsin, which has a significant absorption coefficient at 570 nm, and from relaxation back to the ground state of bR_{570} causes 570 nm transmittance to decay exponentially with an effective time constant $\tau^{-1} = \tau_1^{-1} + \tau_2^{-1}$, which is the time constant measured by our experiment. Furthermore, the induced 620 nm absorption exactly follows the decay of the 570 nm bleaching in time, as expected from this simple model.

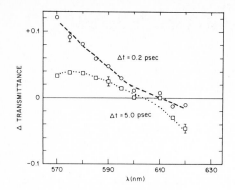

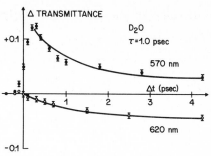

Fig. 1 - Changes in transmittance (LOG T(Δt)/To) through bacteriorhodopsin in D_2O at all probe wavelengths between 570 nm and 620 nm at time delays of 0.2 psec. (circles) and 5.0 psec. (squares) following photo-excitation.

Fig. 2 - Changes in transmittance as a function of time at probe wavelengths 570 nm (closed circles) and 620 nm (open circles) through bacteriorhodopsin in D_2O following photo-excitation by a 90 fsec. optical pulse. Solid curves are theoretical fits (see text) which yields a time constant $\tau = 1.0 \pm 0.1$ psec.

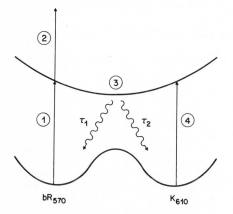

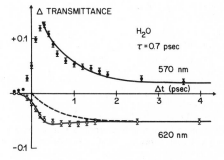

Fig. 3 - Energy surfaces of ground and excited states of bR_{570} and K_{610}. Vertical scale denotes energy, horizontal scale denotes a combined retinal and protein reaction coordinate. Arrows depict path of ultrafast photochemical dynamics: 1) Excited state of bR_{570} is populated. 2) Excited state absorption. 3) Excited state relaxation. 4) Ground state of K_{610} forms.

Fig. 4 - Changes in transmittance as in Fig. 2, but for bacteriorhodopsin in H_2O and pump intensity twice as large. Solid curves are theoretical fits (see text) which yield a time constant $\tau = 0.7 \pm 0.1$ psec.

The experiment was repeated with H_2O as the solvent with the result that $\tau = 0.7 \pm 0.1$ psec. This result suggests that the formation rate of the batho complex may depend on deuteration at room temperature.

At higher excitation intensities, evidence of nonlinear excited state absorption is seen, as shown by the data in Fig. 4, where the pump intensity is twice as high as for the data in Fig. 2. The 570 nm induced absorption still decays exponentially with $\tau = 0.7$ psec. The induced 620 nm absorption, however, deviates significantly from the 0.7 psec. rise time shown by the dashed curve. We can explain this apparent discrepancy by assuming nonlinear excited state absorption from bR_{570} at 620 nm (arrow 2 in Fig. 3).

Using an excited state lifetime of 0.7 psec., the fit shown by the lower solid curve in Fig. 4 is obtained. At still higher pump intensities clear evidence of similar nonlinear absorption is seen also at 570 nm.

We have shown that the ground state of the first photochemical intermediate of bacteriorhodopsin forms within a picosecond at physiological temperatures. We have also been able to follow the development and decay of the pigment excited state while simultaneously measuring the generation of the photochemical product. The data clearly indicate the power of femtosecond spectroscopy in the elucidation of bacteriorhodopsin photochemistry.

M. Islam is a Fannie and John Hertz fellow.

REFERENCES

[1] For a review see W. Stoeckenius, R. H. Lozier, and R. A. Bogomolni, Biochim. Biophys. Acta. *505*, 215 (1979).

[2] A. Lewis, Proc. Nat'l. Acad. Sci. USA 75, 549 (1978).

[3] W. Stoeckenius and R. H. Lozier, J. Supramol. Struct. 2, 759 (1974); R. H. Lozier, R. A. Bogomolni, W. Stoeckenius, Biophys. J. *15*, 955 (1975).

[4] K. J. Kaufmann, P. M. Rentzepis, W. Stoeckenius, and A. Lewis, Biochem. Biophys. Res. Commun. *68*, 1009 (1976).

[5] E. P. Ippen, C. V. Shank, A. Lewis, and M. A. Marcus, Science *200*, 1279 (1978).

[6] R. L. Fork, C. V. Shank, C. Hirlimann, R. Yen, and W. J. Tomlinson, Opt. Lett *8*, 1 (1983).

[7] R. L. Fork, B. I. Greene, and C. V. Shank, Appl. Phys. Lett. *38*, 671 (1981).

[8] R. L. Fork, C. V. Shank, and R. T. Yen, Appl. Phys. Lett. *41*, 223 (1982).

Time-Resolved Picosecond Fluorescence Spectra of the Antenna Chlorophylls in the Green Alga Chlorella Vulgaris

J. Wendler, W. Haehnel*, and A.R. Holzwarth

Max-Planck-Institut für Strahlenchemie
D-4330 Mülheim/Ruhr, Fed. Rep. of Germany

*Biochemie der Pflanzen, Ruhr-Universität Bochum
D-4630 Bochum 1, Fed. Rep. of Germany

I. Introduction

In green plants and algae light energy is absorbed by a large array of 13 light-harvesting chlorophyll (Chl) a/b proteins and by Chl a proteins associated with the reaction centers of either PS I or II. Still a great deal of uncertainty exists on the detailed organization and function of the photosynthetic apparatus and the processes relevant for the first steps in photosynthesis after absorption of a photon. During the last few years agreement has been reached between different laboratories as to the general features of the decay kinetics in chloroplasts and green algae /1-3/. At least three decay components have been required whose amplitudes and also lifetimes respond differently to the redox-state of PS II. A fast decay component of 80 - 150 ps lifetime has been ascribed to open PS II centers /1-3/. An intermediate component of several hundred ps was attributed mainly to LHC II /1,3,4/ or alternatively to PS II in a different redox state /2/. A long-lived decay component of 1 - 2 ns was recognized as being related to the amount of closed PS II centers. Its amplitude is zero at fully open centers /3/ (F_o-state) and maximal when all PS II centers are closed either by light /1,4/ or inhibitors of PS II /3/. It was concluded that this component was almost exclusively responsible for the variable fluorescence F_{var} observed upon closing PS II centers /1,3,4/. These basic features are now fairly well established. Many questions remain to be solved, however. Measuring the time-resolved emission and excitation spectra of the various decay components should provide a powerful method to elucidate the characteristics of the emitting pigments and the excitation or absorption spectra of the connected antennae.

II. Materials and Methods

Chlorella vulgaris, strain 221-11b, has been grown as described /3/. The algae were harvested during the logarithmic growth phase. For the measurements the algae were diluted to give a concentration of 10 μg Chl/ml. For closing PS II reaction centers (F_{max}) 20 uM DCMU and 10 mM hydroxylamine hydrochloride were added to the algae and the samples were preilluminated with white light of saturating intensity. During all measurements the samples were pumped through a 3*3 mm flow cuvette at variable flow rates up to 300 ml/min. All measurements were carried out at room temperature. Time-resolved fluorescence measurements were performed using a synchronously-pumped dye laser system and a single-photon timing apparatus with picosecond resolution for detection as described recently /5/. Fluorescence was selected by a double-monochromator with slits set to give a 4 nm bandwidth. A width of <130 ps (FWHM) was measured by single-photon timing for the excitation function. Excitation intensity was less than 10^{10} photons/cm^2 at a repetition rate of 800 KHz.

III. Results and Discussion

The time-resolved fluorescence and excitation spectra of flowing cells of the green alga Chlorella vulgaris have been measured with picosecond resolution. All decays have been fitted to a sum of three exponential components. As indicated by the residual plots a fit with a sum of three exponentials is not always sufficient, however, to describe the measured decay functions well. The amplitudes of the time-resolved components as a function of emission or excitation wavelength have been plotted to give time-resolved spectra. All data have been corrected for changes in excitation intensity and wavelength dependent detection efficiency. Considerable variations in the lifetime values are observed across the emission or excitation bands in the time-resolved spectra for all three decay components. However, the relative variations are strongest for the fast decay component. Excitation spectra recorded at 685 and 706 nm, respectively, with PS II reaction centers closed by incubation of the algae with DCMU and hydroxylamine and preillumination show significant differences in the preexponential factors of the different decay components. At λ_{em} -685 nm long-lived species with lifetimes of 2.1-2.4 and 1.2-1.3 ns, respectively, are the main components. A short-lived component with lifetimes of 0.1-0.16 ns of relatively small but nonvanishing amplitude is also found. When the emission is detected at 706 nm a short-lived component with a lifetime of less than 0.1 ns predominates. Time-resolved emission spectra using λ_{exc}=630 and λ_{exc}=652 nm had a spectral peak of the two longer-lived components at about 682 nm whereas the fast component was red-shifted with a maximum near 692 nm. At almost open reaction centers the lifetime of the fast component decreased from 150-160 ps at 682 nm to less than 100 ps at 720 nm emission wavelength. The emission spectrum observed by excitation at 696 nm with closed PS II reaction centers showed a large increase in the amplitude of the fast component with a lifetime of 80 to 100 ps as compared to that at 630 nm excitation. All these findings indicate that the origin of the fast component is inhomogeneous. We have therefore resolved this decay into two contributions with constant lifetimes of 80ps (τ_f') and 180px (τ_f''). The main reason for choosing these values comes from the fact that they approx. represent the shortest and longest values obtained directly by three exponential deconvolution for the fast τ_f decay under the various conditions. The spectra of the two components thus obtained from the resolution of the fast decay component show interesting features. At open PS II centers (F_o) the 180ps component (τ_f'') shows an emission maximum at 682-685nm. Its amplitude is larger when excited at 652 as compared to excitation at 630nm. This latter observation indicates that the resolved 180ps (τ_f'') component is connected to proteins with a high Chl b/Chl a ratio, i.e. the LHC II complex. We conclude that at least two pigment pools contribute to the fast component. One originates from PS II connected pigments with a lifetime of approx. 180 ps and a spectral peak around 682-605nm. We now turn to a discussion of the 80ps (τ_f') component. Its spectrum is strongly red shifted with a maximum slightly below 700nm under all conditions. The peak intensity does not depend, within the error limits, on the state of the PS II centers. Its amplitude at 682nm is low. These features all indicate that this component should be attributed to antenna pigments of PS I. Its main emission maximum situated slightly below 700nm ideally fits the energy of the P700 reaction center. Our analysis thus allows the contributions of PS I to the room temperature fluorescence of algae to be accounted for in an overall consistent manner using four decay components. Contributions of PS I to the fast decay component have not been resolved before although it was assumed that they should contribute to a significant extent /1-3,6/. The middle component τ_m is not related to LHC II but arises mainly from Chl a pigments and is tentatively assigned to PS II β centers. This is concluded from the finding that the excitation spectrum of this component indicates a low CHl b content of the connected antenna, not consistent with the earlier interpretation as LHC II fluorescence. More work is necessary, however, to finally confirm this new assignment. The amplitudes of the fast (180 ps,

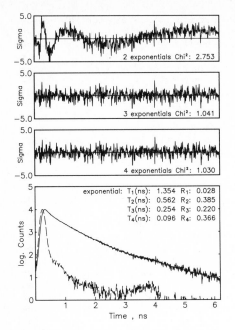

Figure 1:

Semilogarithmic plot of the fluorescence decay of Chlorella vulgaris detected at λ_{em}=686 nm and excitation at 630 nm. No inhibitors were added and the algae were pumped at a high rate. The frames on top show the residuals for model functions with different numbers of components.

PS II) component and the long-lived decay show an opposite dependence on the state of the PS II centers and confirm our earlier contention that the fast PS II contribution disappears at the F_{max} state /3/. Our data also show that the fast (180 ps) and slow components are connected to the same pigment pool and that they correspond to each other.

IV. References

(1)   Haehnel, W., Nairn, J.A., Reisberg, P., and Sauer, K.
       Biochim. Biophys. Acta 680, 161-173 (1982).

(2)   Gulotty, R.J., Fleming, G.F., and Alberte, R.D.
       Biochim. Biophys. Acta 682, 322-331 (1982).

(3)   Haehnel, W., Holzwarth, A.R., and Wendler, J.
       Photochem. Photobiol 37, 435-443 (1983).

(4)   Nairn, J.A., Haehnel, W., Reisberg, P., and Sauer, K.
       Biochim. Biophys. Acta 682, 420-429 (1982).

(5)   Wendler, J., Holzwarth, A.R., and Wehrmeyer, W.
       Biochim. Biophys. Acta 765, 58-67 (1984).

(6)   Holzwarth, A.R., Haehnel, W., Wendler, Jr., Suter, G.W.
       and Ratajczak, R. (1984) Adv. Photosynthesis Research
       1, 73-76, Eds: C. Sybesma.

Index of Contributors

Abul-Haj, A. 292
Aizawa, K. 497
Akhmanov, S.A. 278
Alberte, R.S. 466
Alfano, R.R. 459,493
Alferness, R.C. 394
Anfinrud, P. 193
Antonetti, A. 126,166,
  323,447
Arjavalingam, G. 211
Asaka, S. 226
Astier, A. 447
Attwood, D.T. 63
Auston, D.H. 409

Bado, P. 296
Baer, T.M. 19,96
Balant, A.C. 220
Ballantyne, J.M. 398
Band, Y.B. 102
Barbara, P.F. 374
Bartlett, R.J. 433
Becker, M.F. 187
Berens, S.J. 487
Berg, M. 52,300
Bergsma, J.P. 296
Bloembergen, N. 111,118,
  122
Bloom, D.M. 19
Boczer, B.P. 368
Boggess, T.F. 184,202
Bokor, J. 413
Boldridge, D.W. 477
Bor, Zs. 60
Boyd, I.W. 184,202
Brown, J.K. 52,300
Brown, W.T. 390
Bucksbaum, P.H. 413
Buhl, L.L. 394
Buhse, L.F. 359
Burdge, G.L. 81
Busch, G.E. 170
Butler, W.L. 487

Campillo, A.J. 289

Catherall, J.M. 75
Chambaret, J.P. 196
Chang, C.S. 423
Chase, L.L. 126
Chemla, D.S. 162
Chernoff, D.A. 266
Cheung, K.P. 409
Cho, Y. 93
Choi, K.-J. 368
Chronister, E.L. 452
Claesens, F. 472
Clark, J.H. 359,371,390,
  477
Cline, R.E.,Jr. 452
Cooper, D.E. 430
Corkum, P.B. 38
Cotter, D. 78
Crackel, R.L. 193
Craig, B.B. 173
Cross, A.J. 466

Dagen, A.J. 493
Davis, H. 423
De Silvestri, S. 23,230
Diels, J.-C. 30,233
Dietel, W. 30
Dlott, D.D. 452
Doany, F.E. 326
Doizi, D. 377
Doukas, A.G. 459
Downer, M.C. 27,106,162,
  500
Döpel, E. 30
Dörr, F. 462
Duppen, K. 224
Dupuy, C.G. 296

Eesley, G.L. 143
Egger, H. 42
Eisenstein, G. 394
Eisenthal, K.B. 216,330
El-Sayed, M.A. 341
Elsaesser, T. 49
Erskine, D.J. 137
Etchepare, J. 126,166,196

Evenor, M. 181

Fabricius, N. 258
Fauchet, P.M. 129
Faust, W.L. 173
Feldberg, S. 181
Fleming, G.R. 466
Flytzanis, C. 270
Fontaine, J. 30
Fork, R.L. 7,27,106,
  162
Fox, A.M. 199
Freeman, R.R. 413
Fujii, K. 402
Fujimoto, J.G. 11,111
Fujimoto, S. 84
Fujita, Y. 490
Fujiwara, M. 226

Gale, G.M. 270,274
Gauduel, Y. 323
Geirnaert, M.L. 270
Geiselhart, P. 462
Genack, A.Z. 72
Glownia, J.H. 211
Gobeli, D. 341
Goldberg, L.S. 87,289
Golombok, M. 383
Gordon, J.P. 7
Gossard, A.C. 162
Gottesfeld, S. 181
Göbel, E.O. 150,417
Greene, B.I. 308
Grillon, G. 126,166,196
Grischkowsky, D. 220
Guha, S. 205
Gulotty, R.J. 466
Guo, F.-C. 93
Gustafson, T.L. 266
Guyot-Sionnest, P. 270,
  274

Haehnel, W. 503
Haight, R. 413
Hammond, R.B. 428,433

Haner, M. 99
Harootunian, A. 500
Harris, A.L. 52,300
Harris, C.B. 52,300
Harter, D.J. 102
Hartmann, H.-J. 252
Harzion, Z. 181
Hasselbeck, M. 208
Hauser, J.J. 133
Hefetz, Y. 176
Hefferle, P. 462
Heinz, T.F. 216
Hill, J.H. 452
Ho, P.-T. 81
Ho, Z.Z. 330
Hochstrasser, R.M. 326, 440
Holzwarth, A.R. 503
Höger, R. 150
Hughes, A.E. 420
Hulin, D. 126(166
Huppert, D. 181
Hutchinson, J.S. 336
Hynes, J.T. 336

Ikegami, A. 481
Ikeuchi, M. 484
Inoue, Y. 484
Ippen, E.P. 11,111,230
Ishida, Y. 69,239
Islam, M. 500
Iwai, J. 481,484

Jalenak, W.A. 320
Jeng, M.C. 423
Jhee, Y.K. 187
John, W. 462
Johnson, A.M. 16
Johnson, B.C. 35
Johnson, R.R. 170

Kafka, J.D. 19
Kaiser, W. 49,263,351, 456
Kaminow, I.P. 394
Kania, D.R. 433
Karen, A. 365
Kash, K. 147
Katagiri, S. 402
Kawanishi, M. 402
Keller, J.-C. 236
Kelley, D.F. 292
Kenney-Wallace, G.A. 383
Kikuchi, R. 402
Kleinman, D.A. 409
Knox, W.H. 162
Kobayashi, S. 497

Kobayashi, T. 84,93,481, 484
Kolner, B.H. 19
Koroteev, N.I. 278
Korotky, S.K. 394
Kosic, T.J. 452
Kuhl, J. 150,417
Kurz, H. 118,122
Kühnle, W. 380
Kwok, H.S. 208

Langan, I.G. 330
Langan, J.G. 330
Laporta, P. 23
Laubereau, A. 252
Le Gouet, J.-L. 236
Lecarpentier, Y. 447
Lee, C.H. 81,423,436
Lee,J. 313
Lemaitre, J.M. 323
Lenth, W. 425
Leupacher, W. 190
Lewis, A. 500
Li, J.C.M. 114
Li, M.G. 436
Lin, L.H. 69
Liu, J.M. 111,122
Loaiza-Lemos, F. 99
Lomakka, G. 472
Lompré, L.A. 122
Lorincz, A. 362,387
Luk, T.S. 42

Madden, P.A. 244
Magde, D. 487
Magnitskii, S.A. 278
Malvezzi, A.M. 118
Manning, R.J. 199
Manring, L. 304
Maris, H.J. 133
Marsh, J.H. 199
Martin, J.L. 323,447
Martinez, O.E. 7
Masuhara, H. 317,355
Masumoto, Y. 156
Mataga, N. 317,355,365
Matsuoka, M. 226
McDonald, D.B. 66
McMichael, I.C. 30
Mets, L. 466
Meyer,K.E. 406
Mialocq, J.C. 377
Migus, A. 126,166,323, 447
Miller, A. 199
Miller, D.A.B. 162
Mimuro, M. 490
Mindl, T. 462

Misumi, S. 365
Mitzkus, R. 380
Miyasaka, H. 317
Mollenauer, L.F. 2
Mooradian, A. 35
Moore, R.A. 313
Morimoto, A. 84,93
Morita, N. 239
Morozov, V.B. 278
Morton, T.H. 477
Moss, S.C. 184,202
Moulton, P.F. 35
Mourou, G. 114,406
Mukherjee, P. 208
Müller, A. 60
Müller, W. 42
Mysyrowicz, A. 126

Naganuma, K. 69
Nakashima, N. 345
Nakatsuka, H. 226
Nattermann, K. 258
Negus, D.K. 440
New, G.H.C. 75
Nikolaus, B. 60
Nomura, M. 497
Novak, F.A. 362,387
Nurmikko, A.V. 176
Nuss, M.C. 263

O'Connor, D.V. 345
Ohtani, H. 481
Okada, T. 365
Orszag, A. 196
Oudar, J.L. 166

Palfrey, S.L. 216
Palmer, J.F. 266
Paulter, N.G. 428
Penzkofer, A. 190
Peters, K. 304
Philips, L.A. 359,371, 390,477
Phillips, C.C. 420
Pianetta, P. 433
Pohlmann, J.L.W. 205
Polland, H.-J. 49,456
Poyart, C. 447
Pummer, H. 42
Putnam, R.S. 90

Rácz, B. 60
Reinhard, W.P. 336
Rhee, M.J. 423
Rhodes, C.K. 42
Rice, S.A. 362,387
Rigler, R. 472
Roberts, D.M. 266

Robinson, G.W. 313, 320
Rosen, A. 423
Rosenbluh, M. 35
Rothenberg, J.E. 220
Rudolph, W. 30

Sabersky, A.P. 63
Sah, R.C. 63
Sakata, Y. 365
Sato, T. 497
Schäfer, F.P. 56
Scheele, J. 487
Scheer, H. 462
Scherer, P.O.J. 351
Schneider, S. 462
Schoen, P.E. 87,289
Scott, G.W. 477
Seilmeier, A. 49,351
Shah, J. 147
Shank, C.V. 46,106,162, 500
Sharp, E.J. 205
Sheikbahae, M. 208
Shepard, C.L. 170
Shumay, I.L. 278
Sibbett, W. 420
Sibert, E.L. III 336
Siegman, A.E. 129
Simon, J. 341
Simoni, F. 30
Simpson, W.M. 16
Sitzmann, E.V. 330
Smirl, A.L. 184,202
Smith, D.D. 96,362
Soileau, M.J. 205
Sorokin, P.P. 211
Spano, F. 99
Staerk, H. 380
Stark, J. 413

Statman, D. 320
Stolen, R.H. 2,16,46
Storz, R.H. 413
Strait. J. 133
Strandjord, A.J.G. 374
Striffler, C.D. 436
Strobel, S.A. 81
Struve, W.S. 193
Sueta, T. 84,93
Sumitani, M. 345
Svelto, O. 23
Swenberg, C.E. 493
Szabó, G. 60
Szatmári, S. 56

Takagi, Y. 345
Tamai, N. 355,490
Tang, C.L. 137
Tarassevich, A.P. 278
Tauc, J. 133
Taylor, A.J. 137
Thomazeau, I. 196
Thomsen, C. 133
Tolbert, M.A. 371
Tomlinson, T.W. 46
Topp, M.R. 368
Torti, R. 30
Treichel, R. 380
Tsumori, K. 402
Tucker, R.S. 394
Tunkin, V.G. 278

Valdmanis, J.A. 409
van Eikeren, P. 477
Van Stryland, E.W. 205
Vanherzeele, H. 30,205, 233
Vardeny, Z. 133
Veselka, J.J. 394

von der Linde, D. 258
Von Lehmen, A. 398

Wagner, R.S. 428,433
Walser, R.M. 187
Walsh, P. 433
Warren, W.S. 99
Webb, S.P. 359,371, 390,477
Weiner, A.M. 11,230
Weitekamp, D.P. 224
Weller, A. 380
Wendler, J. 503
Wiegmann, W. 162
Wiersma, D.A. 224
Wilhelmi, B. 30
Williams, R.T. 173
Williams, W.E. 205
Williamson, S. 114
Wilson, K.R. 296
Wondrazek, F. 351
Wood, G.L. 205

Yajima, T. 69,239
Yamashita, M. 497
Yamazaki, I. 355, 490
Yamazaki, T. 490
Yeh, S.M. 477
Yeh, S.W. 359,371, 390
Yoshihara, K. 345
Yurek, A.M. 436

Zewail, A.H. 284
Zhang, X.-C. 176
Zheng, W.Q. 274
Zilinskas, B.A. 493
Zinth, W. 263,456

Picosecond Phenomena

Proceedings of the First International Conference on
Picosecond Phenomena, Hilton Head, South Carolina,
USA, May 24–26, 1978

Editors: **C. V. Shank, E. P. Ippen, S. L. Shapiro**

1978. (Springer Series in Chemical Physics, Volume 4)
Out of print

Picosecond Phenomena II

Proceedings of the Second International Conference
on Picosecond Phenomena, Cape Cod, Massachusetts,
USA, June 18–20, 1980

Editors: **R. Hochstrasser, W. Kaiser, C. V. Shank**

1980. 252 figures, 17 tables. XII, 382 pages.
(Springer Series in Chemical Physics, Volume 14)
ISBN 3-540-10403-8

Contents: Advances in the Generation of Picosecond
Pulses. – Advances in Optoelectronics. – Picosecond
Studies of Molecular Motion. – Picosecond Relaxation
Phenomena. – Picosecond Chemical Processes. –
Applications in Solid State Physics. – Ultrashort
Processes/Biology. – Spectroscopic Techniques. –
Index of Contributors.

Picosecond Phenomena III

Proceedings of the Third International Conference on
Picosecond Phenomena Garmisch-Partenkirchen,
Federal Republic of Germany, June 16–18, 1982

Editors: **K. B. Eisenthal, R. M. Hochstrasser, W. Kaiser,
A. Laubereau**

1982. 288 figures. XIII, 401 pages.
(Springer Series in Chemical Physics, Volume 23)
ISBN 3-540-11912-4

Contents: Advances in the Generation of Ultrashort
Light Pulses. – Ultrashort Measuring Techniques. –
Advances in Optoelectronics. – Relaxation Phenom-
ena in Molecular Physics. – Picosecond Chemical
Processes. – Ultrashort Processes in Biology. – Appli-
cations in Solid-State Physics. – Index of Contributors.

Springer-Verlag
Berlin
Heidelberg
New York
Tokyo

Laser Spectroscopy III

Proceedings of the Third International Conference, Jackson Lake Lodge, Wyoming, USA, July 4–8, 1977
Editors: **J.L. Hall, J.L. Carlsten**
1977. 296 figures. XI, 468 pages.
(Springer Series in Optical Sciences, Volume 7)
ISBN 3-540-08543-2

Laser Spectroscopy IV

Proceedings of the Fourth International Conference, Rottach-Egern, Federal Republic of Germany, June 11–15, 1979
Editors: **H. Walther, K. W. Rothe**
1979. 411 figures, 19 tables. XIII, 652 pages.
(Springer Series in Optical Sciences, Volume 21)
ISBN 3-540-09766-X

Laser Spectroscopy V

Proceedings of the Fifth International Conference, Jasper Park Lodge, Alberta, Canada, June 29 – July 3, 1981
Editors: **A.R.W. McKellar, T. Oka, B.P. Stoicheff**
1981. 319 figures. XI, 495 pages.
(Springer Series in Optical Sciences, Volume 30)
ISBN 3-540-10914-5

"...Most of the papers have excellent figures and extensive references cited for background material. It is a welcome addition to the literature on laser spectroscopy.
...this book and the background material should be of interest to all spectroscopists. The editors made a most appropriate gesture by dedicating this volume to their colleagues and friends Nicolaas Bloembergen of Harvard University and Arthur L. Schawlow of Stanford University. The announcement that they would share half of the 1981 Nobel Prize in Physics for their contributions to laser spectroscopy came just three months after the end of this conference. Contributions were made by each of these Nobel Laureates to this conference and are contained in this book...."
Applied Optics

Laser Spectroscopy VI

Proceedings of the Sixth International Conference, Interlaken, Switzerland, June 27–July 1, 1983
Editors: **H.P. Weber, W. Lüthy**
1983. 258 figures. XVII, 442 pages.
(Springer Series in Optical Sciences, Volume 40)
ISBN 3-540-12957-X

Contents: Photons in Spectroscopy. – Spectroscopy of Elementary Systems. – Coherent Processes. – Novel Spectroscopy. – High Selectivity Spectroscopy. – High Resolution Spectroscopy. – Cooling and Trapping. – Collisions and Thermal Effects on Spectroscopy. – Atomic Spectroscopy. – Rydberg-State Spectroscopy. – Molecular Spectroscopy. – Transient Spectroscopy. – Surface Spectroscopy. – NL-Sprectroscopy. – Raman and CARS. – Double Resonance and Multiphoton Processes. – XUV – VUV Generation. – New Laser Sources and Detectors. – Index of Contributors.

Springer-Verlag
Berlin
Heidelberg
New York
Tokyo